Lecture Notes in Computer Science 8177

Commenced Publication in 1973
Founding and Former Series Editors:
Gerhard Goos, Juris Hartmanis, and Jan van Leeuwen

Michael Floater   Tom Lyche
Marie-Laurence Mazure   Knut Mørken
Larry L. Schumaker (Eds.)

Mathematical Methods for Curves and Surfaces

8th International Conference, MMCS 2012
Oslo, Norway, June 28 – July 3, 2012
Revised Selected Papers

 Springer

Volume Editors

Michael Floater
University of Oslo, Norway
E-mail: michaelf@ifi.uio.no

Tom Lyche
University of Oslo, Norway
E-mail: tom@ifi.uio.no

Marie-Laurence Mazure
Université Joseph Fourier, Grenoble, France
E-mail: mazure@imag.fr

Knut Mørken
University of Oslo, Norway
E-mail: knutm@ifi.uio.no

Larry L. Schumaker
Vanderbilt University, Nashville, TN, USA
E-mail: larry.schumaker@vanderbilt.edu

ISSN 0302-9743   e-ISSN 1611-3349
ISBN 978-3-642-54381-4   e-ISBN 978-3-642-54382-1
DOI 10.1007/978-3-642-54382-1
Springer Heidelberg New York Dordrecht London

Library of Congress Control Number: 2014931390

Typesetting: Camera-ready by author, data conversion by Scientific Publishing Services, Chennai, India

Printed on acid-free paper

Springer is part of Springer Science+Business Media (www.springer.com)

Preface

The 8th International Conference on Mathematical Methods for Curves and Surfaces took place June 28 – July 3, 2012, in Oslo, Norway. The earlier conferences in the series took place in Oslo (1988), Biri (1991), Ulvik (1994), Lillehammer (1997), Oslo (2000), Tromsø (2004), and Tønsberg (2008). The conference gathered 170 participants from 31 countries who presented a total of 135 talks. This includes nine invited talks and six mini-symposia. This book contains 28 original articles based on talks presented at the conference. The topics range from mathematical analysis of various methods to practical implementation on modern graphics processing units. The papers reflect the newest developments in these fields and also point to the latest literature. The papers have been subject to the usual peer-review process, and we thank both the authors and the reviewers for their hard work and helpful collaboration. We wish to thank those who have supported and helped organize the conference. First and foremost it is a pleasure to acknowledge the generous financial support from the Department of Informatics and the Centre of Mathematics for Applications (CMA) at the University of Oslo, and the Research Council of Norway. We would also like to thank Georg Muntingh for his help with with technical matters, Katja Elisabeth Andersson for help with the registration, and our students Helene Norheim Semmerud and James Trotter for help with various practical matters.

October 2013

Michael Floater
Tom Lyche
Marie-Laurence Mazure
Knut Mørken
Larry Schumaker

Organization

Organizing Committee

Morten Dælen	University of Oslo, Norway
Michael Floater	University of Oslo, Norway
Tom Lyche	University of Oslo, Norway
Marie-Laurence Mazure	UJF Grenoble, France
Knut Mørken	University of Oslo, Norway
Larry L. Schumaker	Vanderbilt University, Nashville, USA

Invited Speakers

Albert Cohen, France	Vin de Silva, USA
Oleg Davydov, UK	Gabriele Steidl, Germany
Yaron Lipman, Israel	Michael Unser, Switzerland
Charles Loop, USA	Johannes Wallner, Austria
Carla Manni, Italy	

Mini-Symposia Organizers

Rick Beatson, New Zealand	Jens Gravesen, Denmark
Peter Binev, USA	Tomas Sauer, Germany
Constanza Conti, Italy	Tatyana Sorokina, USA

Sponsoring Institutions

Department of Informatics, University of Oslo
Centre of Mathematics for Applications, University of Oslo
Research Council of Norway

Table of Contents

Vibrational Error Extraction Method Based on Wavelet Technique

Loay Alkafafi[1], Carsten Hamm[1], and Tomas Sauer[2]

[1] Siemens AG, Erlangen 91056, Germany
{loay.alkafafi,carsten.hamm}@siemens.com
[2] University of Passau, Passau 94032, Germany
tomas.sauer@math.uni-giessen.de

Abstract. A key factor in developing and assessing any vibration attenuation technique for elastic systems is the measure that quantifies the occurring vibrations. In this paper, we propose a general and instantaneous
vibration measure which allows for more subtle methods of localized vibration attenuation techniques. This measure is based on extracting the
vibrational part from the conventional tracking error signal using wavelet
technique. The paper also provides a method for constructing a wavelet
function based on the system impulse response. This wavelet outperforms
the existing ones in representing the system behavior while guaranteeing
admissibility and providing sufficient smoothness and rate of decay in
both time and frequency domains.

Keywords: elastic system, vibration attenuation, vibrational error, impulse response, wavelet transform, admissibility.

1 Introduction

Suppressing the vibrations in numerically controlled mechanical systems is a
challenge in many industrial applications. Vibrations typically occur when an
elastic system is driven by positioning commands that contain frequencies close
to the system's critical frequency. There already exists a variety of techniques to
treat such a problem, the most popular of which is the input shaping technique
[10]. In general, all existing techniques share the same idea of filtering out the
system's critical frequency from the driving commands.

A primary requirement for any such vibration suppression technique is to derive a measure that quantifies the occurring vibrations. In literature, discussions
are often limited to performance measures for the techniques in use rather than
for the existing vibrations. A very good survey on the available key performance
measures is found in [3]. One of the limitations of the existing performance measures is that they "represent" the system vibrations merely after the command
completion and only for a specific type of input commands. On the other hand,
an instantaneous measure of vibration allows for more subtle methods for local
corrections of the input command. Providing such a measure will indeed be useful in many practical applications, e.g., the attenuation of vibrations in CNC

M. Floater et al. (Eds.): MMCS 2012, LNCS 8177, pp. 1–12, 2014.

machine tools via local modification of their reference motion commands [5], where a localized algorithm is needed to allocate the critical oscillation regions, and the assessment of the influence of motion commands on the vibrations of industrial CNC machine tools [1], where an instantaneous measure for the oscillations of the machine is needed.

A general vibration measure that characterizes the system vibrations at any point in time is presently not available. An exception to this can be found in Barre [1]. In this paper, we propose a general vibration measure based on extracting the vibrational part from the conventional tracking error signal using system adapted wavelet technique.

2   Elastic Systems and Vibrational Error

Elastic systems tend to vibrate whenever they are excited by fast motion. Their oscillatory behavior is described by the *vibrational modes* which in turn are characterized by an oscillation frequency ω_0 and a damping ratio ζ that defines how fast the vibration will decrease in amplitude. The behavior of each vibrational mode can be generally described by a second order underdamped system which we will use, without loss of generality, as a representative model for elastic systems. As a mechanical system, a single-degree-of-freedom mass-spring-damper system as shown in Fig. 1 will be used. The system is driven by a time varying displacement input function $x(t)$. Using Newtons second law, the ordinary differential equation (ODE) of the system is given by

$$m\ddot{y}(t) + d\dot{y}(t) + Ky(t) = Kx(t). \tag{1}$$

The dynamical behavior of such a system is described by the quotient between its output signal $y(t)$ and input signal $x(t)$ in the Laplace domain which is known as the *transfer function* of the system. For the above system, the transfer function is given by

$$G(s) = \frac{Y(s)}{X(s)} = \frac{\omega_0^2}{s^2 + 2\zeta\omega_0 s + \omega_0^2}, \tag{2}$$

where $\omega_0 = \sqrt{\frac{K}{m}}$ and $\zeta = \frac{d}{2\sqrt{mK}}$. Whenever an elastic system is driven with an input that contains frequencies close to its natural frequency, oscillating behavior is seen in its response. A conventional measure for the performance of such a

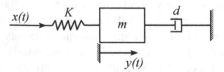

Fig. 1. Single-degree-of-freedom mass-spring-damper system

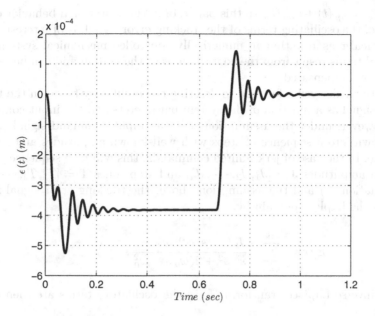

Fig. 2. Tracking error signal for second order underdamped system in response to jerk-limited step command

system is the *tracking error signal* $\epsilon(t)$ which measures the deviation of the system response from the input command:

$$\epsilon(t) := y(t) - x(t). \tag{3}$$

By (2), the tracking error signal (3) can also be described in the Laplace domain as the *error model transfer function*

$$\frac{E(s)}{X(s)} = -\frac{s(s + 2\zeta\omega_0)}{s^2 + 2\zeta\omega_0\,s + \omega_0^2}. \tag{4}$$

Fig. 2 shows an example of a tracking error signal for a second order underdamped system in response to a jerk-limited step command. In technical applications, *jerk* is used for the third derivative with respect to time, i.e., the variation of acceleration. In addition to the vibratory behavior of the system, the tracking error signal also shows various static deviations from the input command. In general, two types of error can be distinguished in the tracking error signal [1]:

1. Aperiodical terms representing errors related to the tracking characteristics of the system, denoted as $\epsilon_{ap}(t)$,
2. Oscillatory periodical terms $\epsilon_{vib}(t)$ related to the vibrational behavior of the system,

where $\epsilon(t) = \epsilon_{ap}(t) + \epsilon_{vib}(t)$. In this paper, only the vibrational behavior of the system, i.e. the oscillating terms of the tracking error signal, is of interest to us. The practical reason is that in numerically controlled mechanical systems the aperiodical terms result from the control loops of the system and can be neither avoided nor compensated.

In [1], Barre derived a formula to describe the oscillatory terms in the tracking error signal as a function of the system parameters and the input command signal, however, under the assumption that the input command signal can be broken down into a sequence of steps with well-known amplitudes and time of occurrence. In the case of jerk-limited commands, this will be a sequence of jerk steps with amplitudes $\mathbf{J} = [J_1, J_2, \cdots, J_n]$ and step times $\mathbf{T} = [T_1, T_2, \cdots, T_n]$. Using equation (4) and the assumption above, the tracking error signal is described in the Laplace domain by:

$$E(s) = \sum_{k=1}^{n} -\frac{J_k}{s^3} \frac{s + 2\zeta\omega_0}{s^2 + 2\zeta\omega_0 s + \omega_0^2} e^{-\sum_{j=1}^{k} T_j \cdot s}. \tag{5}$$

From an inverse Laplace transform of (5) the oscillatory terms are then easily extracted.

Barre's approach has two main pitfalls. First, the existence of steps with well-known amplitude and time locations can only be satisfied in theoretical cases and for very simple input commands. In practical reality, the jerk signal is often noisy and hard to describe. Thus, describing the jerk signal as needed by this approach is practically impossible. Secondly, the approach assumes an *exact* knowledge of the system transfer function which is also quite impossible for real applications.

3   The Continuous Wavelet Transform

The continuous wavelet transform (CWT) is the correlation of a time signal with a dictionary of translated and dilated versions of the analyzing (mother) wavelet ψ. It decomposes a signal into a time-scale representation that elucidates the transient characteristics of that signal. To fix notation, we recall that for a finite energy signal $f(t) \in L^2(\mathbb{R})$ the continuous wavelet transform is defined by

$$Wf(b,a) = \langle f, \psi_{b,a} \rangle = \int_{-\infty}^{\infty} f(t) \frac{1}{\sqrt{a}} \overline{\psi\left(\frac{t-b}{a}\right)} dt, \tag{6}$$

cf. [6] where a and b are the scaling and translation parameters, respectively.

Since the continuous wavelet transform is complete and maintains an energy conservation, an inverse wavelet transform exists and given by

$$f(t) = \frac{1}{C_\psi} \int_{0}^{\infty} \int_{-\infty}^{\infty} Wf(b,a) \frac{1}{\sqrt{a}} \psi\left(\frac{t-b}{a}\right) db \frac{da}{a^2}. \tag{7}$$

This inverse formula exists as long as ψ satisfies the *admissibility condition*

$$C_\psi = \int\limits_0^\infty \frac{\left|\hat{\psi}\left(\omega\right)\right|^2}{\omega}\, d\omega < \infty, \tag{8}$$

which in particular implies that $\hat{\psi}\left(0\right) = 0$, i.e., $\int \psi = 0$.

4 Balanced Impulse Response Wavelet

One feature that makes wavelets attractive for practical use is the possibility to select the wavelet function ψ such that it fits the application at hand. Indeed, in the literature a number of well-developed wavelet functions can be found which cover a wide range of applications. For the extraction of vibrational error signals we tried several wavelet candidates, e.g. Morlet wavelet, Mexican hat wavelet and impulse response wavelet. All tested wavelets suffer from two main pitfalls: first, an additional optimization method is always required to optimize the wavelet shape parameters in order to achieve satisfactory results. Second, most of the available wavelets are (relatively) symmetric and two-sided wavelets. Thus, whenever such wavelets are used for reconstructing purposes, additional spurious oscillating parts will show up in the shape of reconstruction error. Therefore, we intend to design a new wavelet that overcomes such pitfalls.

Since the oscillatory behavior of an elastic system is characterized by its impulse response, we construct the wavelet function ψ from the system impulse response as a template whose scaled and dilated occurrence we wish to detect in the given signal. We call such a specific machine adapted wavelet *balanced impulse response wavelet*. The impulse response wavelet itself is not new; different forms of such a wavelet are available in the literature. The starting point for building such a wavelet is the impulse response of an underdamped second order system

$$h\left(t\right) = \frac{\omega_0}{\sqrt{1-\zeta^2}}\, e^{-\zeta\omega_0 t}\sin\left(\omega_d t\right), \tag{9}$$

where $\omega_d = \omega_0\sqrt{1-\zeta^2}$ is the damped natural frequency of the system. Since the system impulse response usually does not satisfy the admissibility condition, modifications have to be applied. Junsheng [2], for example, modified the impulse response via direct mirroring to achieve the admissible wavelet

$$\psi\left(t\right) = \begin{cases} e^{-\frac{\beta\omega_c t}{\sqrt{1-\beta^2}}}\sin\left(\omega_c t\right), & t \geq 0, \\ e^{\frac{\beta\omega_c t}{\sqrt{1-\beta^2}}}\sin\left(\omega_c t\right), & t < 0, \end{cases} \tag{10}$$

where ω_c is the wavelet center frequency and β is a damping or control parameter. In practice these values are directly related to the damped natural frequency of the system ω_d and the system's damping ratio ζ which are normally estimated from an experimental analysis of the dynamic system.

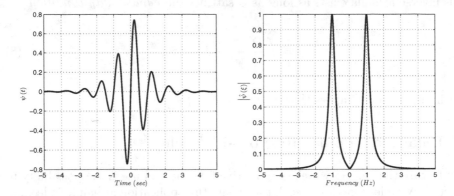

Fig. 3. Impulse response wavelet in time (left) and frequency (right) domain: with $\omega_c = 1\,Hz$, $\beta = 0.2$

The Fourier transform of the impulse response wavelet is given by

$$\hat{\psi}\left(\xi\right) = \frac{\beta\omega_c}{i\sqrt{1-\beta^2}} \left[\frac{1}{\frac{\beta^2\omega_c^2}{1-\beta^2} + i\left(2\pi\xi - \omega_c\right)^2} - \frac{1}{\frac{\beta^2\omega_c^2}{1-\beta^2} + i\left(2\pi\xi + \omega_c\right)^2} \right]. \tag{11}$$

The resulting impulse response wavelet in time and frequency domains is shown in Fig. 3. By construction, ψ is an odd function and thus has zero mean which is the essential part of the admissibility condition. On the other hand, the reflection clearly results in a two-sided wavelet that will not reproduce the original system response which, for example, had no symmetry in the beginning.

Our construction will complete a function with a damped oscillation behavior in $\mathbb{R}_+$ by adding a function with controllable support in $\mathbb{R}_-$ such that the resulting function satisfies not only the admissibility condition but also provides a certain amount of smoothness, so that ψ and $\hat{\psi}$ both decay sufficiently fast. Consequently, with a generalized form of the system impulse response $h(t)$ where the damping ratio ζ is replaced with a general control parameter β as $g(t) = \exp\left(-\beta\omega_c\left(1-\beta^2\right)^{-1/2}t\right)\sin\left(\omega_c t\right)$, we define the wavelet function as

$$\psi\left(t\right) = \begin{cases} g(t), & t \geq 0, \\ f(t), & \tau \leq t \leq 0, \\ 0, & t < \tau, \end{cases} \tag{12}$$

where $\tau < 0$ is a freely chosen parameter that defines the support extension to the negative axis and controls the time localization properties of the resulting wavelet. Moreover, f is a function from a finite dimensional space that has to satisfy the balancing condition

$$\int_\tau^0 f(t)\,\mathrm{d}t = -\int_0^\infty g(t)\mathrm{d}t = \frac{-1}{\omega_c\left(\frac{\beta^2}{1-\beta^2}+1\right)}, \tag{13}$$

as well as for $k = 0, \ldots, n$ the smoothness conditions

$$f^{(k)}(\tau) = 0, \tag{14}$$

and

$$f^{(k)}(0) = g^{(k)}(0) = \omega_c^k \sum_{j \le (k-1)/2} (-1)^{k-j-1} \binom{k}{2j+1} \left(\frac{\beta}{\sqrt{1-\beta^2}}\right)^{k-2j-1}. \tag{15}$$

The easiest way to build f is to use polynomial completions. The $2n+2$ Hermite conditions in (14) and (15) always have a unique solution in Π_{2n+1}, thus the complete problem defined by (13), (14) and (15) can be solved in Π_{2n+1} if the solution p of (14) and (15) happens to satisfy the balancing condition of (13). Otherwise, a solution in Π_{2n+2} is given by

$$f = p - \frac{q}{\int_\tau^0 q(t)dt} \left(\int_\tau^0 p(t)dt + \int_0^\infty g(t)dt\right), \qquad q(t) = t^{n+1}(t-\tau)^{n+1},$$

where $q > 0$ on $(\tau, 0)$, hence, $\int_\tau^0 q(t)dt > 0$.

For example, for the case $n = 1$ the coefficients $a_0, \ldots, a_4$ of the polynomial completion are the solutions of the system

$$\begin{bmatrix} 1 & 0 & 0 & 0 & 0 \\ 0 & 1 & 0 & 0 & 0 \\ 1 & \tau & \tau^2 & \tau^3 & \tau^4 \\ 0 & 1 & 2\tau & 3\tau^2 & 4\tau^3 \\ -\tau & \frac{-\tau^2}{2} & \frac{-\tau^3}{3} & \frac{-\tau^4}{4} & \frac{-\tau^5}{5} \end{bmatrix} \begin{bmatrix} a_0 \\ a_1 \\ a_2 \\ a_3 \\ a_4 \end{bmatrix} = \begin{bmatrix} 0 \\ \omega_c \\ 0 \\ 0 \\ \frac{-1}{\omega_c\left(\frac{\beta^2}{1-\beta^2}+1\right)} \end{bmatrix}, \tag{16}$$

and the Fourier transform of the resulting wavelet is

$$\widehat{\psi}(\xi) = \frac{1}{2i} \left(\frac{1}{\frac{\beta\omega_c}{\sqrt{1-\beta^2}} + i(2\pi\xi - \omega_c)} - \frac{1}{\frac{\beta\omega_c}{\sqrt{1-\beta^2}} + i(2\pi\xi + \omega_c)}\right)$$

$$+ a_0 \frac{i\left(1 - e^{-i2\pi\xi\tau}\right)}{2\pi\xi} + a_1 \frac{1 - e^{-i2\pi\xi\tau}\left(1 - 2i\pi\xi\tau\right)}{(2\pi\xi)^2}$$

$$- a_2 \frac{2i + e^{-i2\pi\xi\tau}\left(4\pi\tau\xi + i\left(4\tau^2\pi^2\xi^2 - 2\right)\right)}{(2\pi\xi)^3}$$

$$- a_3 \frac{6 + e^{-i2\pi\xi\tau}\left(12\tau^2\pi^2\xi^2 - 6 + i\left(8\tau^3\pi^3\xi^3 - 12\tau\pi\xi\right)\right)}{(2\pi\xi)^4}$$

$$+ a_4 \frac{24i - e^{i2\pi\xi\tau}\left(32\tau^3\pi^3\xi^3 - 48\tau\pi\xi + i\left(16\tau^4\pi^4\xi^4 - 48\tau^2\pi^2\xi^2 + 24\right)\right)}{(2\pi\xi)^5},$$

which has a (removable) singularity at $\xi = 0$, since, due to the balancing property (13) we have $\widehat{\psi}(0) = \int \psi = 0$. Since the practical computation of the wavelet transform requires a sampling of $\widehat{\psi}$, we recall that the Fourier transform in this approach can be explicitly computed as

$$
\widehat{\psi}(\xi) = \left(\chi_{[\tau,0]} \sum_{j=0}^{2n+2} a_j(\cdot)^j \right)^{\wedge}(\xi) + \left(\chi_{[0,\infty]} e^{-\beta\omega_c\left(1-\beta^2\right)^{-1/2}} \cdot \sin\left(\omega_c\cdot\right) \right)^{\wedge}(\xi)
$$

$$
= \sum_{j=0}^{2n+2} a_j \left(\frac{-j!}{(i2\pi\xi)^{j+1}} + \frac{e^{-i2\pi\xi\tau}}{(i2\pi\xi)^{j+1}} \sum_{l=0}^{j} \frac{j!\,(i2\pi\xi\tau)^l}{l!} \right) + \tag{17}
$$

$$
\frac{1}{2i} \left(\frac{1}{\frac{\beta\omega_c}{\sqrt{1-\beta^2}} + i\left(2\pi\xi - \omega_c\right)} - \frac{1}{\frac{\beta\omega_c}{\sqrt{1-\beta^2}} + i\left(2\pi\xi + \omega_c\right)} \right).
$$

The first part of this expression is singular at $\xi = 0$ and therefore hard to sample in the neighborhood of the origin. The singularity is only removable due to the choice of the coefficients a_j which guarantees that $\widehat{\psi}$ is uniformly continuous. This dependency of the coefficients which requires that the numerator is *precisely* zero in order to apply the l'Hôpital rule cannot be maintained in floating point computations, hence this formula is numerically very unstable in the neighborhood of the origin. Fortunately, there is a series expansion of the truncated polynomial which can be used close to the origin.

Lemma 1. *For the truncated polynomial function*

$$
f = \chi_{[\tau,0]} \sum_{j=0}^{2n+2} a_j\,(\cdot)^j,
$$

we define the convergent series representation as

$$
\widehat{f}(\xi) = -\sum_{k=0}^{\infty} \frac{(i\tau\xi)^k}{k!} \sum_{j=0}^{2n+2} a_j \frac{\tau^{j+1}}{j+k+1}. \tag{18}
$$

Proof. We first note that

$$
\widehat{f}^{(k)}(0) = \int_{\mathbb{R}} (it)^k\,f(t)\,\mathrm{d}t = i^k \int_{\tau}^{0} \sum_{j=0}^{2n+2} a_j\,t^{j+k}\,\mathrm{d}t = -(i\tau)^k \sum_{j=0}^{2n+2} a_j \frac{\tau^{j+1}}{j+k+1}.
$$

Substituting this into the Taylor series

$$
\widehat{f}(\xi) = \sum_{k=0}^{\infty} \frac{\widehat{f}^{(k)}(0)}{k!}\,\xi^k,
$$

which exists since f is compactly supported, hence $\widehat{f} \in C^\infty(\mathbb{R})$, gives (18). The sum

$$
\sum_{j=0}^{2n+2} |a_j| \frac{|\tau|^{j+1}}{j+k+1},
$$

is bounded independently of k and the remainder of the series is the series expansion of $e^{i\xi\tau}$, hence the series converges absolutely.   □

For small values of $|\tau\xi|$, the series in (18) converges very fast and so (18) is suitable and a very stable way for sampling $\widehat{\psi}$ close to the origin, while for large values of $|\tau\xi|$, (17) is the more appropriate expression to evaluate. This observation suggests the use of small values of $|\tau|$ which is in accordance with our application of completing a single-sided wavelet without adding too much support on the negative side. Of course, smaller values of $|\tau|$ will lead to larger coefficients $|a_j|$, and these numbers will diverge for $\tau \to 0$.

In addition, Lemma 1 can be used to derive a convenient formula for the Fourier transform of a spline function, which is easily obtained by shifting each polynomial piece of the spline to a support interval of the form $[\tau, 0]$.

Corollary 1. *Let $t_0 < \cdots < t_m$ be a knot sequence and f a piecewise polynomial of the form*

$$f = \sum_{\ell=1}^{m} \chi_{[t_{\ell-1}, t_\ell)} \sum_{j=0}^{n} a_{\ell j}(\cdot)^k.$$

Then the Fourier transform of f is

$$\widehat{f}(\xi) = \sum_{\ell=1}^{m} e^{t_\ell \xi} \sum_{k=0}^{\infty} \frac{(i\tau_\ell\xi)^k}{k!} \sum_{j=0}^{2n+2} a_{\ell j} \frac{\tau_\ell^{j+1}}{j+k+1}, \qquad \tau_\ell := t_{\ell-1} - t_\ell. \qquad (19)$$

Again, (19) is an alternative to the formula in [7], in particular for frequencies such that $\tau_\ell\xi$ is small.

Returning to our application, we first show the resulting balanced impulse response wavelet for $n = 1$ in time and frequency domain in Fig. 4. The wavelet is indeed a real single-sided one that satisfies the admissibility condition.

To compare this to the behavior of the conventional impulse response wavelet given by (10), we use a test signal containing two impulse responses of a second

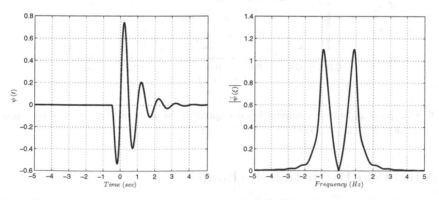

Fig. 4. Balanced impulse response wavelet in time (left) and frequency (right) domain with $\omega_c = 1\,Hz$, $\beta = 0.2$, $\tau = -0.5$

order underdamped system with damped natural frequency $f_d = 25\,Hz$ and damping ratio $\zeta = 0.1$, defined as

$$g\left(t\right) = \mathcal{H}\left(t - 0.2\right) e^{-\frac{\zeta 2\pi f_d}{\sqrt{1-\zeta^2}}(t-0.2)} \sin\left(2\pi f_d\left(t - 0.2\right)\right) \tag{20}$$
$$+ \mathcal{H}\left(t - 1\right) e^{-\frac{\zeta 2\pi f_d}{\sqrt{1-\zeta^2}}(t-1)} \sin\left(2\pi f_d\left(t - 1\right)\right),$$

where $\mathcal{H}\left(\cdot\right)$ is the Heaviside function. For this signal, we consider the two wavelet transforms with an analyzing frequency of $25\,Hz$ and identical center frequency and damping parameter, $\omega_c = 25\,Hz$ and $\beta = 0.1$. For comparing the behavior of the two wavelets in terms of their localization capabilities, it is sufficient to consider the wavelet transforms with a single analyzing frequency and a noiseless test signal so that we can highlight only the effects of the choice of wavelet. A normalized version of the test signal and the modulus of the wavelet coefficients is shown in Fig. 5. As the results demonstrate, the wavelet from our above construction outperforms the conventional one in catching the impulse amplitude envelope and their time locations. Thus, it provides a much better alternative for applications where accurate detection of impulses amplitude and time location are needed.

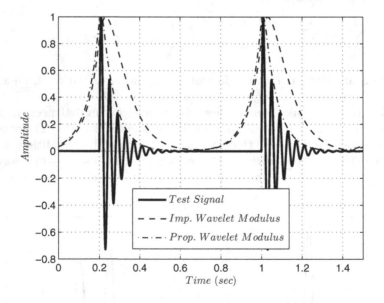

Fig. 5. Comparison between conventional impulse response wavelet and the new proposed wavelet: with $\omega_c = 25\,Hz$, $\beta = 0.1$, $\tau = -0.5$

5   Vibration Extraction Technique

To extract and remove the vibrational part of the tracking error signal we finally use a method based on forward and inverse wavelet transforms, which makes use of the redundant representation of the wavelet transform and its ability to localize the signal information on the time-scale grid. The use of forward and backward transforms requires a bit of care, cf. [8,9], but can be performed in an efficient and stable way. The method extracts the relevant error information by performing forward and inverse continuous wavelet transforms on the tracking error signal at a small number of selected scales only and consists of the following steps:

1. Define the wavelet shape parameters (ω_c and β) as the system damped natural frequency ω_d and damping ratio ζ, respectively.
2. Perform a forward wavelet transform on the tracking error signal $\epsilon(t)$. The analyzing scales should cover a small band around the wavelet center frequency, i.e. $\omega \in [\omega_c \pm \upsilon]$ where ω is the analyzing frequencies and υ is a small percentage from the wavelet center frequency, typically in the range of 5 %.
3. If necessary, a simple soft thresholding can be applied to the resulting wavelet coefficients for reducing the noise and highlighting interesting error features.
4. Perform inverse wavelet transform on the thresholded wavelet coefficients to reconstruct the vibrational part of the error signal.

Fig. 6 shows the extracted vibrations from the tracking error signal shown in Fig. 2 by means of our method. The results shows the capability of this method to precisely extract the vibrational behavior from the tracking error signal. Using the balanced impulse wavelet eliminates the need to optimize any obscure wavelet shape parameters as the parameters are now adapted *exactly* to the

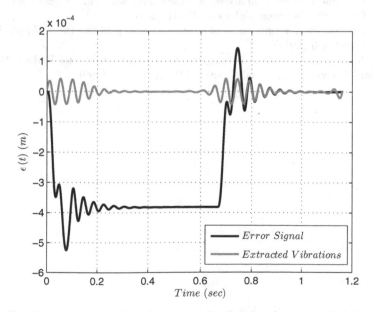

Fig. 6. Tracking error signal and extracted vibrational error using the proposed method: with $\omega_c = 25\,Hz$, $\beta = 0.2$, $\tau = -0.5$, $\upsilon = 5\,Hz$

problem. Furthermore, since the constructed wavelet is essentially single-sided, the effect of extra side oscillations introduced by the reconstruction in the inverse wavelet transform is significantly reduced. This can be nicely seen from the second oscillation in Fig. 6 where the extracted vibration only has one oscillation prior to the peak of the error signal.

References

1. Barre, P.-J., Bearee, R., Borne, P., Dumetz, E.: Influence of a Jerk Controlled Movement Law on the Vibratory Behaviour of High-Dynamics Systems. J. of Intell. Robot. Systems 42, 275–293 (2005)
2. Junsheng, C., Dejie, Y., Yu, Y.: Application of an Impulse Response Wavelet to Fault Diagnosis of Rolling Bearings. Mechanical Systems and Signal Processing 21, 920–929 (2007)
3. Kozak, K., Singhose, W., Ebert-Uphoff, I.: Performance Measures For Input Shaping and Command Generation. J. Dyn. Syst., Meas., Control 128, 731–736 (2006)
4. Krantz, S.G.: 4.1.5 The Riemann Removable Singularity Theorem. In: Handbook of Complex Variables, pp. 42–43. Birkhäuser, Boston (1999)
5. Alkafafi, L., Hamm, C., Sauer, T.: A Strategy for Suppressing Residual Vibrations in Motion Control. In: Proc. PCIM, Nuremberg, pp. 76–82 (2012)
6. Mallat, S.: A Wavelet Tour of Signal Processing The Sparse Way, pp. 102–112. Academic Press, San Diego (2009)
7. Polge, R.J., Bhagavan, B.K.: Fourier Transform and Ambiguity Function of Piecewise Polynomial Functions. IEEE Trans. Aerospace Electron. Systems 13, 403–408 (1977)
8. Sauer, T.: Time–frequency analysis, wavelets and why things (can) go wrong. Human Congnitive Neurophysiology 4, 38–64 (2011)
9. Sauer, T.: Transformations, implementations, and pitfalls. Proc. Appl. Math. Mech. 11, 863–866 (2011)
10. Singer, N.C., Seering, W.P.: Preshaping Command Inputs to Reduce System Vibration. J. Dyn. Syst., Meas., Control 112(1), 76–82 (1990)

A Mathematical Model for Extremely Low Dose Adaptive Computed Tomography Acquisition

Oren Barkan[1], Amir Averbuch[1], Shai Dekel[2,3], and Yaniv Tenzer[3]

[1] School of Computer Science, Tel-Aviv University, Tel-Aviv 69978, Israel
[2] School of mathematical sciences, Tel-Aviv University, Tel-Aviv 69978, Israel
[3] GE Healthcare, Israel

Abstract. One of the main challenges in Computed Tomography is to balance the amount of radiation exposure to the patient at the time of the scan with high image quality. We propose a mathematical model for adaptive Computed Tomography acquisition whose goal is to reduce dosage levels while maintaining high image quality at the same time. The adaptive algorithm iterates between selective limited acquisition and improved reconstruction, with the goal of applying only the dose level needed for sufficient image quality. The theoretical foundation of the algorithm is nonlinear Ridgelet approximation and a discrete form of Ridgelet analysis is used to compute the selective acquisition steps that best capture the image edges. We show experimental results where the adaptive model produces significantly higher image quality, when compared with known non-adaptive acquisition algorithms, for the same number of projection lines.

Keywords: Adaptive compressed sensing, Ridgelets.

1 Introduction

In the last decade, several studies have shown that radiation exposure during Computed Tomography (CT) scanning is a significant factor in raising the total public risk of cancer deaths [3], [29], [34]. To balance image quality with these concerns, radiologists use the protocol As Low as Reasonably Achievable (ALARA). It meant to ensure that "... CT dose factors are kept to a point where risk is minimized for maximum diagnostic benefit..", where the dose can be determined by the product of the CT tube current and the time the patient has been exposed to the radiation (see [26] for an overview). Currently, there are several state-of-the-art technologies that attempt to achieve dose reduction. There are the Iterative Reconstruction (IR) methods which are successful in reducing artifacts, improving resolution and lowering the noise in the reconstructed images ([10], [35]). More recently, Model Based Iterative Reconstruction (MBIR) [2], [36] was introduced. It improves upon the IR methods by incorporating accurate system physics models coupled with statistical noise models and prior models.

However, dosage levels during CT exams are still at the focus of attention and any new method that can reduce them is considered highly valuable. This paper

M. Floater et al. (Eds.): MMCS 2012, LNCS 8177, pp. 13–33, 2014.
© Springer-Verlag Berlin Heidelberg 2014

describes an adaptive acquisition model that theoretically is superior to existing non-adaptive acquisition methods and allows minimal and optimal dosage levels. The method can be considered a significant generalization of existing two step adaptive acquisition methods [20], [27] and can potentially use the same hardware configurations that are capable of changing their geometric configuration and acquisition protocols on-the-fly (see also [33]). For example, in [20], the authors describe an imaging C-arm system where a low-dose overview (OV) scan is used to dynamically identify an arbitrary Volume Of Interest (VOI). The OV and VOI scans are then registered and reconstructed together. In [27], the authors develop a flexible x-ray micro-CT system, named FaCT, capable of changing its geometric configuration and acquisition protocol in order to best suit an object being imaged for a particular diagnostic task. In their system, a fast, sparse-projection pre-scan is performed, the data are reconstructed, and the region of interest is identified. Next, a diagnostic-quality scan is performed where, given the region of interest, the control computer calculates an illumination window for on-line control of an x-ray source masking aperture to transmit radiation only through the region of interest throughout the scan trajectory. In going further, for our adaptive CT approach to work, it is required that the system will be able to configure on the fly an iterated projection scheme, beyond the existing predefined uniform parallel or fan beam acquisition patterns.

Observe that adaptive acquisition should not be confused with adaptive reconstruction. In the latter, the acquisition model is a non-adaptive uniform sampling scheme, where over a discrete set of pre-determined angles, projections lines are computed at equal intervals. In this setup, the adaptive elements, if exist, are part of the post-acquisition reconstruction step.

The outline of the algorithm is as follows: First, the system projects the object with an extreme low dose according to a uniform predetermined pattern and reconstructs an initial low quality image. Next, the system predicts from the reconstructed low quality image where the significant edges of the true objects are and projects along them. Then, the system iterates by incorporating the newly added line projections in order to obtain a refined approximation of the true image. The algorithm continues to iterate between estimation of locations of finer significant features, adaptive acquisition and reconstruction until a convergence criterion is met. The goal is to quickly converge until a high quality reconstruction is achieved with minimal dose. Moreover, by using the mathematical model of Ridgelets [4], the algorithm has a natural multiresolution capability, where the significance of edges is analyzed at different scales. We show, in the experimental results section that this approach yields significantly higher image reconstruction quality, when compared with known non-adaptive acquisition algorithms, for the same number of projection lines.

It is important to clarify the following fundamental assumption we make on the acquired images. To illustrate, let I be a bi-level image, i.e. with pixels that are either '0' or '1', where the '1' values are sparse. Even on this simple image, our approach would be rendered useless if the '1' values are scattered in random locations against the background of zeros. In such a case, as clearly explained in

[1], adaptive acquisition has absolutely no advantage over non-adaptive methods (e.g [7], [17]). However, if I is what is called a 'cartoon' image, where the '1' values are grouped into 'nicely' connected subdomains with piecewise smooth boundaries, then the situation changes dramatically. Our method relies on the mathematical theory of [4] which quantifies in the setup of Computed Tomography the geometric 'structure' of the image and how fast a Ridgelet approximation converges to the image. Our algorithm, whose goal is to acquire an unknown image, regards the adaptive Ridgelet approximation of this image as the 'optimal' benchmark and is designed to match its performance. This approach has strong ties with the waveform analysis presented in [31], that allowed the authors to classify singularities and quantify the 'stability' of limited angle tomography. Indeed, although in our work we limit the number of line projections, but do not limit the angles, the fundamental understanding of the relationship between a function's edge singularities and its Radon representation as explained in [31] is at the core of our algorithm (see Fig. 3 and the accompanying explanation).

The paper is organized as follows: Section 2 overviews necessary mathematical background. Section 3 describes in detail our adaptive acquisition algorithm. Experimental results and comparisons with non-adaptive methods are given in Section 4. In the last section we drew conclusions and discuss future work.

2   Preliminaries

2.1   Fast Algorithms for Total Variation Functionals with 'Sparse' Constraints

For a given image $I \in \mathbb{R}^{m \times m}$, with pixels values $\{I_{i,j}\}$, we define the gradient of I by $(\nabla I)_{i,j} = (I_{i,j} - I_{i-1,j}, I_{i,j} - I_{i,j-1})$. The Total Variation (TV) norm of the image is given by

$$|I|_{TV} := \sum_{i,j=2}^{m} (|I_{i,j} - I_{i-1,j}| + |I_{i,j} - I_{i,j-1}|).$$

Denote $N = m^2$, and let $x \in \mathbb{R}^N$, be a one-dimensional representation of I by concatenating the rows of I into a single column vector

$$x = (I_{1,1},\ I_{1,2}, \ldots, I_{1,m}, \ldots, I_{m,1}, \ldots I_{m,m})^T.$$

Given an $n \times N$, $n \ll N$, sampling matrix $A \in \mathbb{R}^{n \times N}$ and corresponding observations vector $y \in \mathbb{R}^n$, generated by $Ax = y$, the so-called TV-minimization is concerned with solving one of the following optimization problems

$$\min_{U} |U|_{TV} \quad s.t.\ Au = y, \quad \min_{U} |U|_{TV} + \mu \|Au - y\|_2^2, \tag{1}$$

where $u \in \mathbb{R}^N$ is the one-dimensional representation of $U \in \mathbb{R}^{m \times m}$ and μ is a given weight parameter. The right hand side minimization problem is applied in the presence of noise in the sampling process and the weight μ depends, in part, on the expected noise level. This model is difficult to solve directly due to non-differentiability and non-linearity of the TV term. During the last few years there has been an explosion of new numeric iterative methods (see the papers in the "Compressive Sensing Recovery Algorithms" section of [8]).

Although conceptually our method may use such solvers as black boxes, its unique features allow us to apply critical modifications that not only accelerate the iterative methods, but also make them feasible in large datasets problems when N is large. In this work, we implemented a modified version of the TVAL3 solver [23], [37]. Our modified version utilizes the fact that in our special case the matrix A is highly sparse. This is in complete contrast to the usual setup in compressed sensing, where the theory typically promotes a dense matrix (usually of pseudo-random nature). As we shall see in Section 3, in our case the sparsity is due to the fact that each row of A is associated with an integration over a digital line in the image and therefore a vector of '0's and '1's. The values '1' are located in entries associated with the pixels of the digital line and thus each row in matrix A has $\leq \sqrt{2m} = \sqrt{2N}$ non-zero entries. We note that even if we use a more accurate model based interpolation, where the line is given some width and then the result is a weighted sum of pixels, the matrix A would remain sparse. This structure allows us to store, to adaptively update a sparse data structure for A and to implement fast linear algebra operations. This idea is not new to the CT community. Moreover, for practical clinical data sizes in 3D helical uniform acquisition, the matrix A can be too large to hold in memory and must be computed on the fly. Also, its form is carefully determined from the geometry of the focal spots and detectors [11]. In this work we focus on the 2D model and in future work we plan to investigate whether in the 3D case our smaller adaptive sampling set can be stored in memory or computed on the fly.

We now explain, for the sake of completeness, our modification of the TVAL3 algorithm. For the constrained optimization problem such as (1), there are a number of methods that approach the original constrained problem by a sequence of unconstrained subproblems. One of them is the Quadratic Penalty Method [9]. This method puts a quadratic penalty term instead of the constraint in the objective function where each penalty term is a square of the constraint violation with multiplier. However, this method requires to increase the multipliers to infinity so as to guarantee the convergence, which may cause the ill-conditioning problem, numerically. Another method concerning the constrained optimization problem is the Augmented Lagrangian method [15] (an augmented Lagrange method has been already used in CT reconstruction [32]). According to this method, the corresponding Augmented Lagrangian of the left-hand side minimization in (1), is given by

$$L_A(w, u, v, \lambda, \mu, \beta) := \sum_{s=1}^{N} \left(\|w_s\|_1 - \langle v_s, (Du)_s - w_s \rangle + \frac{\beta_s}{2} \|(Du)_s - w_s\|_2^2 \right)$$
$$- \langle \lambda^t, Au - y \rangle + \frac{\mu}{2} \|Au - y\|_2^2,$$

(2)

where $w_s, v_s \in \mathbb{R}^2$, $\|w_s\|_1 := |w_s(1)| + |w_s(2)|$, $(Du)_s := \nabla U_{i(s), j(s)}$, $1 \leq s \leq N$, and the two vectors λ, v, are the Lagrangian multipliers. To solve (2), the following Alternating Direction scheme is used: Denote the approximate minimizers of (2) at the kth inner iteration by $w^{(k)}$ and $u^{(k)}$. Then $w^{(k+1)}$ and $u^{(k+1)}$ can be attained by solving two separated subproblems. The first is the 'w-subproblem':

$$w^{(k+1)} = \arg\min_w L_A(w, u^{(k)}) =$$
$$\sum_{s=1}^N \left(\|w_s\|_1 - \langle v_s, (Du^{(k)})_s - w_s \rangle + \frac{\beta_s}{2} \left\| (Du^{(k)})_s - w_s \right\|_2^2 \right). \tag{3}$$

Note that the 'w-subproblem' is separable with respect to each w_s, $1 \le s \le N$, and has a closed form solution [23]. The second subproblem, also known as the 'u-subproblem' is:

$$u^{(k+1)} = \arg\min_u L_A(w^{(k+1)}, u) =$$
$$\sum_{s=1}^N \left(\left\| w_s^{(k+1)} \right\|_1 - \left\langle v_s, (Du)_s - w_s^{(k+1)} \right\rangle + \frac{\beta_s}{2} \left\| (Du)_s - w_s^{(k+1)} \right\|_2^2 \right) \tag{4}$$
$$- \langle \lambda^t, Au - y \rangle + \frac{\mu}{2} \|Au - y\|_2^2.$$

The 'u-subproblem' can be solved using a steepest decent method, but since this might be too costly for large scale problem, an aggressive 'one-step' of the steepest decent can be computed as an iteration (see the details in [23]). After attaining $w^{(k+1)}$ and $u^{(k+1)}$, the multiplier updating is performed based on the analysis of [18], [30]

$$v_s^{(k+1)} = v_s^{(k)} - \beta_s \left((Du^{(k+1)})_s - w_s^{(k+1)} \right), \quad 1 \le s \le N,$$
$$\lambda^{(k+1)} = \lambda^{(k)} - \mu(Au^{(k+1)} - y).$$

This second update step is exactly an example of where our modification accelerates significantly the TV minimization, by either storing and applying the matrix A in a sparse form or by computing and applying the sparse rows of A on the fly. Finally, choose new penalty parameters $\beta_s^{(k+1)} \ge \beta_s^{(k)}$ and $\mu^{(k+1)} \ge \mu^{(k)}$. The stopping criteria are one of the following:

(i) The quantities

$$\left| \nabla L_A(w^{(k)}, u^{(k)}, v^{(k)}, \lambda^{(k)}, \mu^{(k)}, \beta^{(k)}) \right|, \quad \sum_{s=1}^N \left\| (Du^{(k)})_s - w_s^{(k)} \right\|_2, \quad \left\| Au^{(k)} - y \right\|_2,$$

are sufficiently small.

(ii) The relative change $\|u^{(k+1)} - u^{(k)}\|_2$, is sufficiently small.

Inside the main loop of the Alternating Direction scheme, the number of rows in A is increased by a predetermined fixed constant M at each iteration (see Section 3), where the rows are projection lines determined from Ridgelet analysis of the approximant of the image.

2.2 Fundamentals of Ridgelet Theory

Let $\psi \in L_2(\mathbb{R})$ be a wavelet [24]. For the purpose of this paper it is sufficient that the wavelet function has two properties: compact support and vanishing moments. The latter implies that for some $r \geq 1$,

$$\int_{\mathbb{R}} \psi(x) x^l dx = 0, \quad l = 0, \ldots, r - 1.$$

The classical example for a wavelet function is the Haar wavelet with one vanishing moment

$$\psi(x) := \begin{cases} 1, & 0 \leq x \leq 1/2, \\ -1, & 1/2 < x \leq 1, \\ 0, & else. \end{cases} \tag{5}$$

A bivariate Ridgelet function [4], [14], is defined by

$$\psi_{a,b,\theta}(x_1, x_2) := a^{-1/2}\psi((x_1 \cos\theta + x_2 \sin\theta - b)/a),$$

where a, b and θ are the parameters determining the scale, transition and rotation of the Ridgelet function, respectively (see Fig. 1).

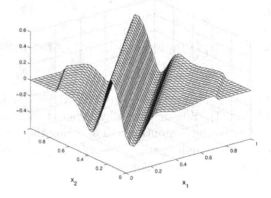

Fig. 1. A Ridgelet function $\psi_{a,b,\theta}(x_1, x_2)$

Given $f \in L_1(\mathbb{R}^2)$, its Continuous Ridgelet Transform (CRT) is defined by

$$CRT_f(a, b, \theta) := \int_{\mathbb{R}^2} \psi_{a,b,\theta}(x)f(x)dx. \tag{6}$$

The continuous Radon transform [19], [28] of a bivariate function f at direction θ is defined as

$$R_f(\theta, t) := \int_{\mathbb{R}^2} f(x_1, x_2)\delta(x_1 \cos\theta + x_2 \sin\theta - t) \, dx_1 dx_2, \tag{7}$$

where δ is the Dirac function. The Radon and the Ridgelet transforms are related by

$$CRT_f(a, b, \theta) = \int_{\mathbb{R}} \psi_{a,b}(t) R_f(\theta, t) \, dt, \tag{8}$$

where $\psi_{a,b}(x) := a^{-1/2}\psi((x-b)/a)$. In applications, this means that the Ridgelet transform can be computed by the application of the Radon transform at a given angle, followed by 1D fast wavelet transform.

It is interesting to point out that Ridgelets [4] did not previously find too many applications in image processing. Their 'descendants' Curvelets [5], [6] and Shear-lets [21], [22], which capture directional information as well, were found to be more useful due to their better time-frequency localization. In the context of CT reconstruction, Curvelets have been used as a regularization tool [16]. However, we find that Ridgelets are the right mathematical tool in the setup of Computed Tomography, because the acquisition device is not able to capture through its sampling process, well localized functionals such as Curvelet coefficients.

From approximation theoretical perspective, the mathematical foundation of our adaptive algorithm follows the framework of characterizing the images by the appropriate function smoothness spaces and then providing an estimate for the order of convergence.

Definition 1. *[4] For $\alpha > 0$, and $p, q > 0$, we say that $f \in \dot{R}_{p,q}^{\alpha}(\mathbb{R}^2)$, if $f \in L_1(\mathbb{R}^2)$ and*

$$\|f\|_{\dot{R}_{p,q}^{\alpha}} := \left(\sum_{j=-\infty}^{\infty} 2^{j(\alpha+1/2)q} \left(\frac{1}{\pi} \int_0^{\pi} \|CRT_f(2^j, \cdot, \theta)\|_p^p \, d\theta \right)^{q/p} \right)^{1/q} < \infty.$$

We note that this definition requires certain conditions on the wavelet ψ. It is sufficient to assume ψ is compactly supported, is in C^r and has r vanishing moments, with $r > \max(2/p, \alpha + 5/2)$. These conditions ensure that membership in the smoothness space $\dot{R}_{p,q}^{\alpha}$ does not depend on the particular wavelet used in (6). A typical non trivial example for a function in $\dot{R}_{p,q}^{\alpha}$ is a function with a singularity along a line such as

$$f(x_1, x_2) = 1_{\{x_1 > 0\}}(x_1, x_2)(2\pi)^{-1/2} e^{-(x_1^2 + x_2^2)/2}.$$

This function is in the Besov class [12] $B_{1,1}^{\alpha}$ only for $\alpha < 1$, which means that it almost has a first derivative in the classical sense. In contrast, this function is contained in $\dot{R}_{1,1}^{\alpha}$, for any $\alpha < 3/2$ [4], which implies that it is smoother in the scale of Ridgelet spaces than in the scale of Classical Besov spaces. This is a direct consequence of the fact that its singularity has simple lower dimensional structure.

In this work we assume that the functions we analyze are compactly supported in a 'standard' compact domain such as $[-1, 1]^2$ and attain the value zero on its boundary. Indeed, CT images satisfy this requirement (see the examples below). Therefore, by a simple zero extension argument, a function $f \in L_2([-1, 1]^2)$

of this nature can also be regarded as a function in $L_1\left(\mathrm{I\!R}^2\right) \cap L_2\left(\mathrm{I\!R}^2\right)$. By sampling the CRT, one may obtain a discrete Ridgelet Frame system $\{\psi_\gamma\}$ with a dual system $\left\{\tilde{\psi}_\gamma\right\}$, for a countable index $\{\gamma = (a, b, \theta)\}$, such that for $f \in L_2\left([-1, 1]^2\right)$,

$$f = \sum_\gamma \left\langle f, \tilde{\psi}_\gamma \right\rangle \psi_\gamma = \sum_\gamma \langle f, \psi_\gamma \rangle \tilde{\psi}_\gamma.$$

Recall, that the frame property guarantees 'stability' of the representation, in the sense that there exist constants $0 < A \le B < \infty$, such that

$$A \|f\|_2^2 \le \sum_\gamma |\langle f, \psi_\gamma \rangle|^2 \le B \|f\|_2^2, \quad \forall f \in L_2\left([-1, 1]^2\right).$$

Let us rearrange the Ridgelet coefficients based on the size of their absolute values

$$|\langle f, \psi_{\gamma_1} \rangle| \ge |\langle f, \psi_{\gamma_2} \rangle| \ge \cdots,$$

and denote the n-term adaptive approximation to f by

$$f_n := \sum_{i=1}^n \langle f, \psi_{\gamma_i} \rangle \tilde{\psi}_{\gamma_i}.$$

Then, we have the Jackson-type estimate [4] for $\alpha > 1/2$ and $1/\tau = \alpha - 1/2$,

$$\|f - f_n\|_{L_2([-1,1]^2)} \le cn^{-\alpha/2} \|f\|_{\dot{R}_{\tau,\tau}^\alpha}.$$

Thus, under certain assumptions on the input function, not only is the convergence of the adaptive approximation is ensured, but its rate is also estimated. The outcome of the theory is that the approximation rate of an adaptive Ridgelet approximation depends on the smoothness of the function in a given Ridgelet smoothness space, much in the same manner that adaptive wavelet approximation is characterized by Besov space smoothness [12].

As we shall see in Section 3, our adaptive acquisition method relies on adaptive Ridgelet approximation to predict, at each iteration, the next significant acquisition set.

3   Adaptive Tomography Acquisition

Before presenting the details of the algorithm, we first provide an instructive and useful example: Assume we had access to an optimal 'oracle'. We then ask, how many line projections are needed as rows in the matrix A, such that the 'Square' image of Fig. 2 can be reconstructed with high precision, using the TV functional (1)?

In fact, equipped with an 'oracle', this image can be reconstructed with extremely high quality, where the matrix A in (1) contains only 8 rows associated with 8 line projections. Thus, the numbers of samples, satisfies $n = 0.000122N$,

Fig. 2. : 'Square' image of size 256×256

which is a tiny fraction of the size of the image $N = 256 \times 256$. This is achieved by selecting the unique four pairs of line projections that are the immediate neighbors of each of the four lines associated with the edges of the white square. Fig. 3 shows the locations of the line projections and the reconstructed image.

The moral of this example, which correlates well with the theory reviewed in Section 2.2, is that during the acquisition process, we should try to adaptively sample the line projections that are aligned and centered on the edges of the image. Obviously, the image to be acquired is unknown and we do not have access to an 'oracle'. As we shall see in the next subsection, this is exactly where the multiresolution nature of the Ridgelet model is useful.

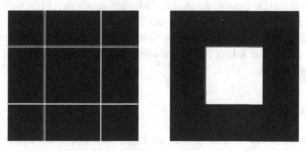

Fig. 3. 'Square' image: On the left, the acquired 8 line projections (using an 'oracle') and on the right, a reconstruction computed from the 8 projections using TV minimization, PSNR=61.85dB

We now present in detail the steps of the algorithm. After initialization, at the kth iteration, we have an adaptive sampling matrix $A^{(k)}$ whose last rows are the new samples obtained at the previous step. We use $A^{(k)}$ to solve a TV minimization problem and obtain the iterative approximation $U^{(k)}$.

3.1 Initialization

First, we create an initial sampling matrix $A^{(0)}$ by using a relatively very small uniform set of line projections. The number of line projections is relative to the image size. For example, in our experimental results, for images of size 256×256, we measured 8 uniformly spaced line integrals at eight uniformly spaced angles, which gives a total of 64 initial measurements that are about 0.1% of the image size. In Fig. 5(a), we see an illustration of this non-adaptive sampling pattern for images of size 256×256. We also initialize the current approximation to the unknown image I with $U^{(-1)} = -\infty$ (e.g. each entry equal to -10^8).

3.2   TV Minimization Step

At the beginning of the kth iteration of our algorithm, we are equipped with the sampling matrix $A^{(k)}$ whose last rows contain a subset of newly acquired samples and the previous approximation $U^{(k-1)}$ as the initial guess. Therefore, we use this updated matrix and proceed with a TV minimization step (1) to compute $U^{(k)}$. Recall that in our setup, the sparse nature of $A^{(k)}$ allows us to achieve this computation on large images by using a sparse representation of the matrix $A^{(k)}$. Our modified TVAL algorithm (see Section 2.1) stores $A^{(k)}$ as a MATLAB sparse matrix of Boolean values, which reduces significantly the memory access overhead.

Here, we have an option to select a tradeoff between reconstruction quality and performance. We do not necessarily need to completely solve the TV minimization problem by iterating an algorithm such as in Section 2.1 until it converges. Instead, we may apply only a fixed and limited number of iterations of the TV solver, or terminate the iterations using a less demanding stop criterion and then proceed to the next step of the Ridgelet analysis. This will speed up the algorithm, but in some cases, its effect on the next analysis step will imply that we will need to acquire more line projections for the same reconstruction image quality. In any case, our adaptive acquisition process terminates at an iteration of this step if we obtain $\left\| U^{(k)} - U^{(k-1)} \right\|_2 \leq \varepsilon$, for some prescribed threshold ε.

3.3   Ridgelet Analysis Step

Now that we have, at the kth iteration, an improved approximation $U^{(k)}$ to I, we compute a discrete set of its Ridgelet coefficients. Recall, that a Ridgelet transform can be computed by the application of the Radon transform (7) followed by a wavelet transform, as shown in (8). Since in our application, we only require Ridgelets for analysis, we do not need to use an invertible transform as in [14], which simplifies the implementation. In practice, we found out that if we chose the number of angles to be a quarter of the image length, then our sampling scheme is sufficiently dense for the purpose of our algorithm, but not too dense so as to lead to subsequent unnecessary acquisition, as will become clear. Thus, for an images of size 256×256, the allowable set of projection lines corresponds to only 64 angles $\{0, \pi/64, ..., 63\pi/64\}$, with 256 line projections per direction. For our experimental results, we computed Ridgelet coefficients using the univariate discrete Haar wavelet (5). The discretization of angles is related to the scale parameter of the Ridgelets, so as to avoid subsequent unnecessary acquisition. Specifically, we calculate the Ridgelet coefficients $\alpha_{a,b,\theta}^{(k)} := \langle U^{(k)}, \psi_{a,b,\theta} \rangle$, using the Haar wavelet function $\psi_{a,b}(x)$, with $a = 2^j$, $j = 0, ..., J_\theta$, where J_θ depends on the angle θ. The discrete sampling of the Ridgelet coefficients is controlled in the following way:

Angle, $0 \leq l < 8$	J_θ
$8l\pi/64$	3
$(8l + 1)\pi/64$	0
$(8l + 2)\pi/64$	1
$(8l + 3)\pi/64$	0
$(8l + 4)\pi/64$	2
$(8l + 5)\pi/64$	0
$(8l + 6)\pi/64$	1
$(8l + 7)\pi/64$	0

3.4   Adaptive Sampling of New Line Projections

Based upon the analysis of the Ridgelet coefficients $\{\alpha_{a,b,\theta}^{(k)}\}$, computed at the previous step, we make our decision on which new line projections are added to $A^{(k)}$ as new rows to create the matrix $A^{(k+1)}$. Specifically, we chose these line projections to be associated with the M coefficients with largest absolute values that have not yet been marked as sampled by the algorithm. In our experiments, we select $M = 0.1n$, which is a tenth of the image row size.

The goal of the line projections is to roughly approximate (6) where ψ is the Haar wavelet. In Fig. 4 we see an illustration of a support of a Haar Ridgelet function (outer dotted lines) and the associated two line projections (inner lines) within its support that we compute on the unknown image .

Fig. 4. Line integrals acquired per a significant Ridgelet coefficient: The support of the Ridgelet lies within the area bounded by the external dotted lines. The inner lines are the sampled line projections.

Let us look closer at the implication of using only two line projections to approximate the value of a Haar Ridgelet. Assume that the Ridgelet coefficient $\alpha_{a,b,\theta}^{(k)} = \langle U^{(k)}, \psi_{a,b,\theta} \rangle$ has not been marked as sampled yet, but is significant enough to be sampled at the current iteration. Let $R_I(\theta, \cdot)$ be the Radon transform of the unknown image I at the fixed angle θ. In such a case, the two values of the line projections that we acquire are $R_I\left(\theta, b + a/4\right)$ and $R_I\left(\theta, b + 3a/4\right)$. These values should be considered as the approximation

$$a^{-1/2}\left(R_I\left(\theta, b + a/4\right) - R_I\left(\theta, b + 3a/4\right)\right) \approx CRT_I\left(a, b, \theta\right).$$

We summarize the adaptive acquisition algorithm by 'pseudo-code':

1. *Initialize the approximated image $U^{(-1)}$ with entries corresponding to $-\infty$ (e.g. -10^8). Obtain initial small number of line projection samples determined by a 'uniform' sampling matrix $A^{(0)}$.*
2. *For $k = 0, 1, \ldots$ Iterate:*
   (a) *Compute $U^{(k)}$ using $A^{(k)}$ by solving (1).*
   (b) *If $\left\|U^{(k)} - U^{(k-1)}\right\|_2 \leq \varepsilon$, go to step 3.*
   (c) *Apply Ridgelet analysis on $U^{(k)}$ to obtain next set of M new candidate projection lines.*
   (d) *Sample the (unknown) image I at the new M projection lines.*
   (e) *Add the new projection lines to the matrix $A^{(k)}$ to create an updated sampling matrix $A^{(k+1)}$.*
3. *Output the most updated approximated image $U^{(k)}$.*

3.5   Analysis and Examples

In Fig. 5, we see a few iterations of the adaptive acquisition algorithm on the Ellipse image. We see in (a) the small number of uniform, non-adaptive line projection measurements that are used for the initialization step. In (b) we see the reconstructed approximation $U^{(0)}$. In (c), we show the new set of line projections that were determined by the Ridgelet analysis on $U^{(0)}$, to be the most significant. The next subfigures show further iterations of newly acquired line projections associated with the next unsampled M largest Ridgelet coefficients and then the approximations $U^{(k)}$ produced by solving the TV functional after adding these new samples as last rows of $A^{(k)}$. Note that the algorithm quickly identifies the edges of the ellipse and only takes line measurements that are aligned with them, where more samples are taken along the longer axis first. Moreover, initially, when the approximation $U^{(k)}$ is still blurry, the algorithm finds through the Ridgelet analysis that it should first acquire line projections associated with low resolution Ridgelet coefficients. Only after the approximation contains sufficiently sharp edges, higher scale Ridgelet coefficients become significant and the line projections associated with them are acquired. In summary, the algorithm attempts to acquire only line projections around and aligned with edge singularities and ordered by scale.

Next, we demonstrate the effectiveness of the estimate for the significant Ridgelet coefficients of the unknown image. The test is conducted on the well-known 256×256 CT Zubal Head test image [38]. To this end, we use the standard Peak Signal to Noise Ratio (PSNR), measured in dB, to quantify an approximation $\tilde{I}$ to the image I where the images pixels take values in $[0, 1]$,

$$PSNR\left(I, \tilde{I}\right) := 10\log_{10}\frac{1}{\frac{1}{N}\sum_{i,j}\left|I_{i,j} - \tilde{I}_{i,j}\right|^2}. \tag{9}$$

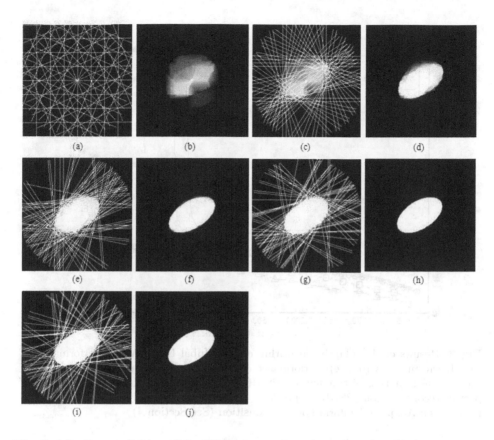

Fig. 5. Adaptive acquisition of the Ellipse image: Iterations of newly added projection lines and approximations $U^{(k)}$

In Fig. 6, we show a graph with number of line projections used by the adaptive algorithm and the PSNR obtained. We compare the performance of our adaptive algorithm with the performance of a benchmark method that uses Ridgelet analysis of the actual real image instead of using the Ridgelet analysis performed on the iterated image. We see that despite of not having the true image available at the time of acquisition, our algorithm manages to perform almost as well as an algorithm equipped with an 'oracle' that uses the Ridgelet analysis of the true image. In general, this property of the algorithm depends on the size M of the set newly sampled projection lines at each iteration. That is, the algorithm manages to trace and collect more accurately the significant line projections associated with largest Ridgelet coefficients of the real image, if it runs in more iterations, adding each time a small set of new line projections.

We also see (Fig. 7) that the algorithm obtained perfect reconstruction using 3834 line projections. In standard CT acquisition models, 256 line projections are acquired at 256 orientations, a total of 65,536 line projections. In comparison, our algorithm achieves perfect reconstruction using about 6% of that total.

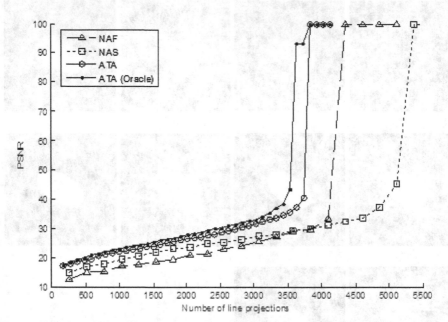

Fig. 6. Results of the adaptive algorithm on the Zubal Head. ATA - performance of the algorithm. ATA (oracle)- performance of the algorithm equipped with an 'oracle' (i.e. Ridgelet analysis of true image). PSNR=100 is actually a graphical cut-off line of perfect reconstruction (PSNR=∞). NAS - Non Adaptive Equally Spaced acquisition. NAF - Non Adaptive Uniform Fourier acqusition (See Section 4)

Under the assumption that the algorithm manages to almost accurately identify the most significant Ridgelet coefficients using only the approximate iterated image, we may analyze the rate of convergence using the theory discussed in Section 2.2. Observe that a function belongs to the Radon 'smoothness' space of Definition 1, with a high value of 'smoothness' index α, if it has 'sparse' directional information, which decreases sufficiently fast as $a = 2^j$ is smaller. As an example, consider 'cartoon' functions which are piecewise constants over polygonal domains. At a fixed angle θ, the Radon transform $R_f(\theta, t)$ is a simple piecewise linear function of the variable t, with compact support and a bounded number of discontinuities, depending of the number of segments in the polygonal boundary. Assume the wavelet ψ has at least two vanishing moments, then for small values of a in (8), the Ridgelet transform $CRT_f(a, b, \theta)$ will be non-zero only in segments of total length $\leq ca$, where c is an absolute constant depending on the choice of ψ and the geometry of the polygonal boundary. Therefore, in the p-norm, for $a = 2^j$, $j < 0$, we get an estimate $\left\| CRT_f\left(2^j, \cdot, \theta\right) \right\|_p^p \leq c(f, \psi, p)\, 2^{j/2}$. This implies that $f \in \dot{R}_{p,q}^{\alpha}\left(\mathbb{R}^2\right)$, for any 'smoothness' $\alpha > 0$. Therefore, from the Jackson estimate we may conclude that the adaptive algorithm will converge for these simple prototype functions at the rate $n^{-\alpha/2}$ for any $\alpha > 0$, which matches the perfect reconstruction results we obtain for these functions

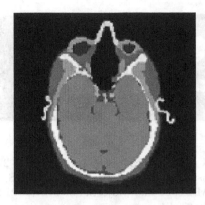

Fig. 7. Perfect reconstruction of the Zubal Head from 3834 adaptive line projections

in this work. For more complex images, the convergence analysis is similar, but more involved. One needs to estimate the Radon smoothness of the function in order to understand the rate of convergence. This type of analysis of adaptive methods has been carried out for wavelet image compression by characterizing images as functions in Besov spaces [13].

4 Experimental Results

In this section we compare our adaptive approach with standard non-adaptive methods. We show that for a given number of projection lines measured on the unknown image I, our adaptive method provides a significantly better approximation to I. To this end, given an $m \times m$ image, we prescribe a target of n samples. Denote $d = n/m$ (assuming $n \bmod m = 0$). We compare four acquisition and reconstruction methods:

1. Filtered Back Projection (FBP): For the FBP method we sample $60 \times m$ line projections (regardless of the target limit), which are m equally spaced line integrals over the angles $0, \pi/60, ..., 59\pi/60$. We then used the MATLAB implementation ('iradon') to obtain an approximate image.
2. Non Adaptive Equally Spaced (NAS): We use equally spaced rotations and a fixed number of line integrals at each angle such that the total number of line integrals matched the prescribed budget. We then applied TV minimization to this sampled data. Specifically, $m/2$ (equally spaced) line projections are acquired over the angles $0, \pi/2d, 2\pi/2d, ..., (2d-1)\pi/2d$.
3. Non Adaptive Uniform Fourier (NAF): This method is used in [7]. It is mathematically equivalent to NAS, but produces slightly different results in digital implementation. In this mode, we uniformly select lines in the Fourier domain of the image and use Fourier coefficients on these lines as the entries of the sampling matrix A. Specifically, m Fourier coefficients were taken on the lines associated with the angles $0, \pi/d, 2\pi/d, ..., (d-1)\pi/d$.

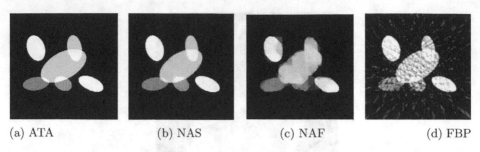

(a) ATA   (b) NAS   (c) NAF   (d) FBP

Fig. 8. Non-adaptive and adaptive acquisition on the 256 × 256 6-Ellipse image. (a) ATA. 971 line projections. Perfect reconstruction. (b) NAS. 1024 line projections. PSNR=29.73 dB. (c) NAF. 1024 line projections. PSNR=21.93 dB. (d) FBP. 5120 line projections. PSNR=19.64 dB.

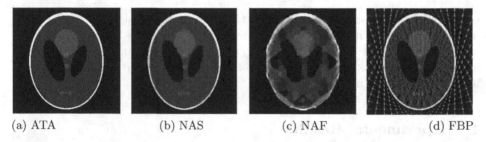

(a) ATA   (b) NAS   (c) NAF   (d) FBP

Fig. 9. Non-adaptive and adaptive acquisition on the 256 × 256 Shepp-Logan image. (a) ATA. 1630 line projections. Perfect reconstruction. (b) NAS. 1792 line projections. PSNR=26.44 dB. (c) NAF. 1792 line projections. PSNR=19.53 dB. (d) FBP. 5120 line projections. PSNR=18.04 dB.

4. Adaptive Tomography Acquisition (ATA): Our proposed adaptive method. Line projections were acquired adaptively as described in Section 3.

For the first set of noise-free phantom test images, we used in the iterations the left-hand side constraint in (1), $Au = y$, so that our solutions satisfy the sampling equations exactly. We see below results on well-known tests image. In Fig. 8 we see that for an equivalent number of line projections, our adaptive algorithm achieves prefect reconstruction while the uniform limited angle, non-adaptive acquisition algorithms, equipped with the same TV minimization solver achieve significantly lower image quality. Similar results are shown in Fig. 9 for the 'Shepp-Logan' phantom (see also the graphs in Fig. 10) and for the 'Zubal Head' in Fig. 11. We note that currently the running times of the adaptive acquisition Matlab simulations are about 7-10 times slower than the non-adaptive for the same number of line projections. This relates to the choice of M, the number of new line projections introduced at each iteration. So, for a given number of line projections n, the choice $M = 0.1n$, yields about 10 iterations, where the matrix $A^{(k)}$ contains about $0.1kn$, $k = 1, \ldots, 10$, rows. Solving these iterations

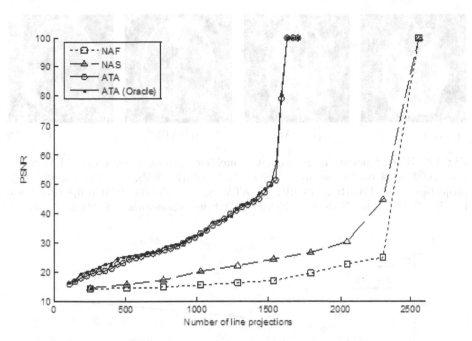

Fig. 10. Comparison of the adaptive acquisition algorithm (ATA) vs. the limited angle, non-adaptive methods of NAS and NAF on the Shepp-Logan image. PSNR=100 is actually a graphical cut-off line of perfect reconstruction (PSNR=∞).

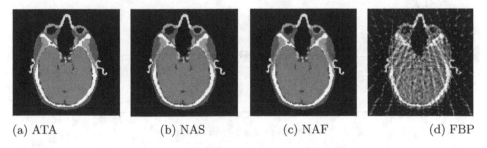

(a) ATA   (b) NAS   (c) NAF   (d) FBP

Fig. 11. Non-adaptive and adaptive acquisition on the 256×256 Zubal-Head. (a) ATA. 3834 line projections. Perfect reconstruction. (b) NAS. 4096 line projections. PSNR=31.16 dB. (c) NAF. 4096 line projections. PSNR=33.31 dB. (d) FBP. 5120 line projections. PSNR=17.67 dB.

is about 5.5 slower than solving the TV-minimization of order n only once. The rest of the running time of the adaptive method is spent on the Ridgelet analysis computations that are performed at each iteration.

Next, we show results with simulated low dose as in [10]. For a selected parameter of incident photon count γ_I, the simulated detected photon counts $\tilde{\gamma}$, were chosen as Poisson distributed random variables with mean equal to $\gamma_I e^{-p}$, where p is a noiseless line projection. The simulated noisy projection, $\tilde{p}$, is then

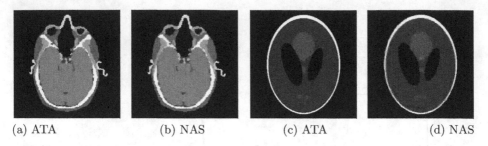

(a) ATA   (b) NAS   (c) ATA   (d) NAS

Fig. 12. Reconstruction from simulated incident photon counts (a) ATA. $\gamma_I =$ 1,000,000. 3834 projection lines. PSNR=33.44 dB.(b) NAS. $\gamma_I = 1,000,000$. 4096 projection lines. PSNR=29.26 dB. (c) ATA. $\gamma_I = 250,000$. 1630 projection lines. PSNR=37.58 dB. (d) NAS. $\gamma_I = 250,000$. 1792 projection lines. PSNR=25.30 dB.

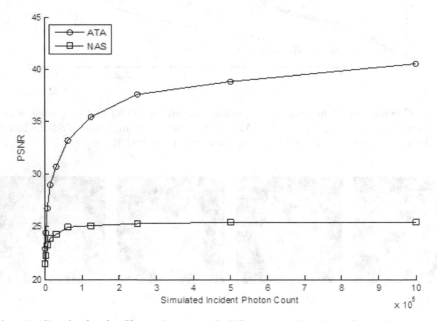

Fig. 13. Graph plot for Shepp-Logan with different simulated incident photon counts

determined by $\tilde{p} = -\log(\tilde{\gamma}/\gamma_I)$. This time, in our iterations, we used the right-hand side constraint in (1), which provides better regularity for noisy data. In Fig. 12 we see a comparison of adaptive and limited angle (non-adaptive) acquisitions using dose simulations. We see that the image quality of our adaptive is clearly higher for the same number of line integrals. In Fig. 13, we see a plot of the reconstructions at various levels of simulated dose levels.

5   Conclusion and Future Work

In this paper we proposed a mathematical model for adaptive Computed To-mography acquisition whose theoretical goal is to radically reduce dosage levels. We presented numerical simulations that demonstrate the potential of the math-ematical model of adaptive acquisition and compared our results to the state of the art non adaptive ones.

Our future research will focus on creating more realistic simulations to CT acquisition. We plan to enhance our algorithm to perform well on more realis-tic images and model more accurately adaptive low-dose radiation, beyond the simplistic model of the total number of line projections. The dose in a CT scan depends on the machine's flux intensity, with lower flux intensity implying lower dose, but higher Poisson-type noise in the detected measurements. We also plan to simulate true 3D scanning and add motion correction.

Lastly, in our work we used a form of the TVAL solver [37], adapted to our problem. It should be very interesting to test other TV solvers such as [25] and see if they (or a modified version of them) are better suited to the adaptive scheme.

References

1. Arias-Castro, E., Candès, E., Davenport, M.: On the fundamental limits of adaptive sensing. IEEE Trans Info. Theory 59, 472–481 (2013)
2. Barnes, E.: MBIR aims to outshine ASIR for sharpness, CT dose reduction. Article in AuntMinnie (May 18, 2010)
3. Brenner, D., Ellison, C.: Estimated risks of radiation-induced fatal cancer from Pediatric CT. American Journal of Roentgenology 176, 289–296 (2001)
4. Candès, E.: Ridgelets: Theory and applications. Ph.D. Thesis, Stanford University (1998)
5. Candès, E., Donoho, D.: New tight frames of Curvelets and optimal representations of objects with piecewise C^2 singularities. Comm. Pure App. Math. 57, 219–266 (2003)
6. Candès, E., Demanet, L., Donoho, D., Ying, L.: Fast Discrete Curvelet Transforms. Multiscale Modeling & Simulation 5, 861–899 (2006)
7. Candès, E., Romberg, J., Tao, T.: Robust uncertainty principles: Exact signal reconstruction from highly incomplete frequency information. IEEE Trans. Inf. Theory 52, 489–509 (2006)
8. Compressive Sensing Resources home page, http://dsp.rice.edu/cs
9. Courant, R.: Variational methods for the solution of problems with equilibrium and vibration. Bull. Amer. Math. Soc. 49, 1–23 (1943)
10. Delaney, A., Bresler, Y.: Globally Convergent Edge-Preserving Regularized Re-construction: An Application to Limited-Angle Tomography. IEEE Trans. Image Proc. 7, 204–221 (1998)
11. De Man, B., Basu, S.: Distance-driven projection and backprojection in three di-mensions. Phys. Med. Biol. 49, 2463–2475 (2004)
12. DeVore, R.: Nonlinear approximation. Acta Numerica 7, 50–150 (1998)

13. DeVore, R., Jawerth, B., Lucier, B.: Image compression through wavelet transform coding. IEEE Trans. Inf. Theory 38, 719–746 (1992)
14. Do, M., Vetterli, M.: The Finite Ridgelet transform for image representation. IEEE Trans. Image Proc. 12, 16–28 (2003)
15. Fortin, M., Glowinski, R.: Augmented Lagrangian Methods: Application to the Numerical Solution of Boundary Value Problems. North-Holland, Amsterdam (1983)
16. Frikel, J.: A new framework for sparse regularization in limited angle x-ray tomography. In: Proc. IEEE Conf. Biomedical Imaging, pp. 824–827 (2010)
17. Frikel, J.: Sparse regularization in limited angle tomography. ACHA 34, 117–141 (2013)
18. Hestenes, M.: Multiplier and gradient methods. Journal of Optimization Theory and Applications 4, 303–320 (1969)
19. Hsieh, J.: Computed Tomography, principles, design, artifacts and recent advances, 2nd edn. SPIE and Wiley-Interscience (2009)
20. Kolditz, D., Kyriakou, Y., Kalender, W.: Volume-of-interest (VOI) imaging in C-arm flat-detector CT for high image quality at reduced dose. Medical Physics 37, 2719–2730 (2010)
21. Kutyniok, G., Labate, D.: Construction of Regular and Irregular Shearlet Frames. J. Wavelet Theory and Appl. 1, 1–10 (2007)
22. Labate, D., Lim, W.-Q., Kutyniok, G., Weiss, G.: Sparse multidimensional representation using shearlets. Wavelets XI, 254–262 (2005); SPIE Proc. 5914
23. Li, C.: An Efficient Algorithm For Total Variation Regularization with Applications to the Single Pixel Camera and Compressive Sensing. Master's thesis, Rice University (September 2009)
24. Mallat, S.: A Wavelet Tour of Signal Processing, 3rd edn. The Sparse Way. Academic Press (2009)
25. Mao, Y., Fahimian, B., Osher, S., Miao, J.: Development and optimization of regularized tomographic reconstruction algorithms. IEEE Trans. Image Proc. 19, 1259–1268 (2010)
26. McCollough, C., Bruesewitz, M., Kofler, J.: CT Dose Reduction and Dose Management Tools: Overview of Available Options. Radiographics 26, 503–512 (2006)
27. Moore, J., Barret, H., Furenlid, L.: Adaptive CT for high-resolution, controlled-dose, region-of-interest imaging. In: IEEE Nucl. Sci. Symp. Conf., pp. 4154–4157 (2009)
28. Natterer, F.: The mathematics of computerized tomography. Classics in applied mathematics, vol. 32. SIAM (1986)
29. Nelson, R.: Thousands of new cancers predicted due to increased use of CT. Medscape news (December 17, 2009)
30. Powell, M.: A method for nonlinear constraints in minimization problems. In: Fletcher, R. (ed.) Optimization, pp. 283–298. Academic Press, London (1969)
31. Quinto, T.: Singularities of the X-ray transform and limited data tomography in $\mathbb{R}^2$ and $\mathbb{R}^3$. SIAM J. Math. Anal. 24, 1215–1225 (1993)
32. Ramani, S., Fessler, J.: Parallel MR image reconstruction using augmented Lagrangian methods. IEEE Trans. Med. Imag. 30, 694–706 (2011)
33. Rutt, B.: Scanned projection radiography using high speed computed tomographic scanning system. U.S Patent 4,573,179
34. Shuryak, I., Sachs, R., Brenner, D.: Cancer Risks after Radiation Exposure in Middle Age. J. National Cancer Institute 102, 1606–1609 (2010)

35. Thibault, J., Sauer, K., Bouman, C., Hsieh, J.: A three-dimensional statistical approach to improved image quality for multislice helical CT. Med. Phys. 34, 4526–4544 (2007)
36. Thibault, J., Sauer, K., Bouman, C., Hsieh, J.: Fast Model-Based X-Ray CT Reconstruction Using Spatially Nonhomogeneous ICD Optimization. IEEE Trans. Image Proc. 20, 161–175 (2011)
37. Li, C., Yin, W., Zhang, Y.: TVAL3 homepage, http://www.caam.rice.edu/~optimization/L1/TVAL3/
38. Zubal, I., Harrell, C., Smith, E., Rattner, Z., Gindi, G., Hoffer, P.: Computerized three-dimensional segmented human anatomy. Med. Phys. 21, 299–302 (1994)

Approximation of Implicit Blends by Canal Surfaces of Low Parameterization Degree

Michal Bizzarri and Miroslav Lávička

Department of Mathematics,
NTIS – New Technologies for the Information Society
Faculty of Applied Sciences, University of West Bohemia
Univerzitní 8, 301 00 Plzeň, Czech Republic
{bizzarri,lavicka}@kma.zcu.cz

Abstract. In this paper, we present a modified method for the computation of approximate rational parameterizations of implicitly given canal surfaces. The designed algorithm, which improves and completes a recent approach from [1], is mainly suitable for implicit blend surfaces of the canal-surface-type. Its main advantage is that it produces rational parameterizations of low bidegree $(7, 2)$. A distinguished feature of our approach is a combination of symbolic and numerical techniques yielding approximate topology-based cubic parameterizations of contour curves which are then applied to compute an approximate parameterization of the given canal surface.

1   Introduction

Blending is one of the most important operations in Computer-Aided (Geometric) Design. The main purpose of this operation is to generate one or more surfaces that create a smooth joint between the given shapes. Blending surfaces are necessary for rounding edges and corners of mechanical parts, or for smooth connection of separated objects. Thanks to its practical importance, blending has become an active research area in recent years. For overview of several blending techniques see e.g. [2,3,4] and references therein.

The existing approaches to the operation of blending are classified according to the type of surfaces which are used. In what follows, we would like to focus only on the implicit blend surfaces, offering a sufficient flexibility for designing blends. Important contributions for blending by implicitly given surfaces can be found in [5,6,7]. Several methods for constructing implicit blends were thoroughly investigated in [8,9,10,11].

In this paper, we will deal with the cases when implicit canal surfaces are used for the construction of a smooth transition between the primary shapes. Canal surfaces are defined as envelopes of one parameter families of spheres in 3-space, [1,12,13,14,15]. A special subclass of the canal surfaces are Dupin cyclides which can be defined as the envelopes of all spheres that touch three given spheres, see [16,17,18]. Cyclide blends between two cones were analysed in [19]. By generalizing the constructions of biarcs to Laguerre geometry it was presented how to construct blends between general canal surfaces using double-cyclide surfaces in [20].

M. Floater et al. (Eds.): MMCS 2012, LNCS 8177, pp. 34–48, 2014.

Generating shapes description in NURBS form has become a universal standard in technical applications since many years. However an exact rational parameterization does not exist for an arbitrary algebraic surface. Hence, suitable techniques producing (only) approximate parameterizations are often used to avoid these problems. The main aim of this paper is to show how to compute an approximate parameterization of the canal surface S given implicitly by the irreducible polynomial $f(x, y, z)$ obtained as a result of some blending technique. And as our approach yields 'only' approximate parameterizations, it can also be used for blends not being canal surfaces exactly but only approximately.

The technique presented in this paper improves the method introduced in [1], where an efficient algorithm for computing approximate parameterizations of canal surfaces was presented and studied. A main feature of the designed approach was a combination of symbolic and numerical techniques yielding approximate topology based parameterizations of contour curves, see [21], which are then applied to compute an approximate parameterization of the given canal surface. In what follows, we will suggest how to overcome the main drawback of the original method, i.e., a high rational bidegree of the obtained parameterization. Approximate rational canal surfaces obtained by the modified method from this paper possess bidegree $(7, 2)$. This is caused by the quality of the obtained $\mathbf{m}(t)$ which is now polynomial and cubic (in the original paper it was rational of degree 7).

The remainder of this paper is organized as follows. The next section summarizes several elementary facts concerning the canal surfaces, the critical points of algebraic curves and approximate parameterization techniques. Section 3 is devoted to the parameterization algorithm which produces an approximate parameterization of a given canal surface. After presenting some examples in Section 4, which illustrate the functionality of the designed method, we conclude this paper.

2   Preliminaries

A *canal surface* S is the envelope of the 1-parameter family of spheres whose centers trace the curve $\mathbf{m}(t)$ in $\mathbb{R}^3$ and possess the radii $r(t)$, i.e.,

$$F(t) = \left\| (x, y, z)^\top - \mathbf{m}(t) \right\|^2 - r(t)^2 = 0. \tag{1}$$

The curve $\mathbf{m}(t)$ is called the *spine curve* and $r(t)$ the radius function of S. The defining equations for the canal surface S are

$$F(t) = 0, \quad F'(t) = 0, \tag{2}$$

where F' is the derivative of F with respect to t. By eliminating the parameter t from (2) we arrive at the corresponding implicit equation $f(x, y, z) = 0$ of S. The equation $F'(t) = 0$ describes the plane perpendicular to the derivative vector $\mathbf{m}'(t)$. Thus the canal surface S contains a one parameter set of the so called *characteristic circles* $F(t) \cap F'(t)$, see Fig. 1. The envelope is real iff the condition $\|\mathbf{m}'(t)\| \geq \|r'(t)\|$ is fulfilled.

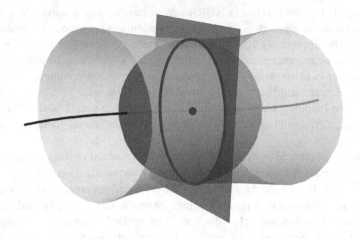

Fig. 1. A characteristic circle (black) of a canal surface

As proved in [13,14], any canal surface with a rational spine curve and a rational radius function possesses a rational parameterization. Later, a new method to study canal surfaces represented by rational or polynomial data using the technique of Pythagorean hodograph curves in the 4-dimensional Minkowski space was presented in [22,23]. Minimal rational parameterizations of canal surfaces were studied in [15].

The algorithm for computing parameterization of a canal surface $\mathcal{S}$ used in this paper (and also in [1,24]) is based on rotating a curve $\mathbf{c}(t)$ of $\mathcal{S}$ (different from the characteristic circles) around the tangents of the spine curve $\mathbf{m}(t)$. We obtain the parameterization of $\mathcal{S}$ in the form

$$\mathbf{s}(t,u) = \mathbf{m}(t) + \frac{(\varrho(u) + \mathbf{m}'(t)) \star (\mathbf{c}(t) - \mathbf{m}(t)) \star (\varrho(u) - \mathbf{m}'(t))}{(\varrho(u) + \mathbf{m}'(t)) \star (\varrho(u) - \mathbf{m}'(t))}, \qquad (3)$$

where $\varrho(u)$, $u \in \mathbb{R}$, is a rational function, the sums $\varrho(u) \pm \mathbf{m}'(t)$ of scalars and vectors are considered as quaternions, and $\star$ is the operation of quaternion multiplication

$$(a + \mathbf{a}) \star (b + \mathbf{b}) = ab - \mathbf{a} \cdot \mathbf{b} + a\mathbf{b} + b\mathbf{a} + \mathbf{a} \times \mathbf{b}, \qquad (4)$$

see [25,26] for more details about quaternions. Any rational choice of $\varrho(u)$ yields the rational parameterization of a canal surface – for the sake simplicity we choose $\varrho(u) = u$ for the low rational degree of $\mathbf{s}(t,u)$ in u, and $\varrho(u) = 2u/(1 - u^2)$ for a relatively uniform distribution of the t-parameter lines.

Let us emphasize that when using the spine curve and another curve for computing the rational parameterization of $\mathcal{S}$, we have to guarantee that their parameterizations are closely related. We say that the curve $\mathbf{c}(t)$ *corresponds in parameter* with the given curve $\mathbf{m}(t)$ in the interval I if it holds

$$\mathbf{c}'(t) \cdot (\mathbf{c}(t) - \mathbf{m}(t)) = 0, \quad \text{for all } t \in I. \qquad (5)$$

Hence, considering the spine curve $\mathbf{m}(t)$ and one curve $\mathbf{c}(t)$ on a given canal surface $\mathcal{S}$ which are corresponding in parameter we have ensured that the point $\mathbf{c}(t_0)$ lies at the

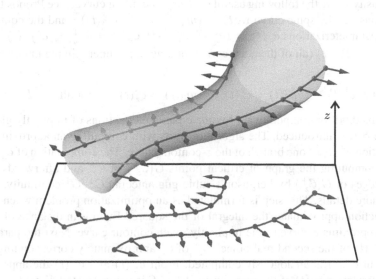

Fig. 2. z-contour curve (blue) together with its normal vectors on a canal surface

characteristic circle of S corresponding to the sphere centered at $\mathbf{m}(t_0)$ and with the radius $r(t_0)$, for all $t_0 \in I$.

Next, we can define by analogy a correspondence in parameter for two arbitrary curves $\mathbf{c}_1(t)$ and $\mathbf{c}_2(t)$ (different from the characteristic circles) on a canal surface S. We say that $\mathbf{c}_1(t)$ and $\mathbf{c}_2(t)$ *correspond in parameter* in the interval I if the points $\mathbf{c}_1(t_0)$ and $\mathbf{c}_2(t_0)$ lie on the same characteristic circle of S for all $t_0 \in I$. The points $\mathbf{c}_1(t_0)$ and $\mathbf{c}_2(t_0)$ are then called *the associated foot points*.

A prominent role among all curves on a given canal surface is played by the so called contour curves since the computation process becomes considerably simpler as we will see. An x-, y-, z-*contour curve* $\mathcal{C}_x$, $\mathcal{C}_y$, $\mathcal{C}_z$ of the canal surface S is the curve consisting of all the points at which the normals of S are orthogonal to the x-, y-, z-axis. Then, the x-contour curve $\mathcal{C}_x$ is determined by

$$f(x, y, z) = \frac{\partial f(x, y, z)}{\partial x} = 0 \tag{6}$$

and it is composed of two branches denoted by $\mathcal{C}_x^+$ and $\mathcal{C}_x^-$. Analogously, we can compute the y- and z-contour curves $\mathcal{C}_y$ and $\mathcal{C}_z$. Fig. 2 shows a canal surface with a z-contour curve.

Remark 1. The x-contour curves are given by (6) (and analogous expressions for $\mathcal{C}_y$ and $\mathcal{C}_z$), i.e., we consider them as the *complete intersections* of two algebraic surfaces. So we use the methods of algebraic geometry which work over an algebraically closed field (e.g. the field of complex numbers $\mathbb{C}$). However, problems originating in geometric modelling work especially over the reals. From this reason, in what follows we will consider only *real* branches of the contour curves (which, of course, do not have to exist for the chosen part of the canal surface). This is a limitation of the designed method.

One can easily prove the following useful property of contour curves, see Proposition 3.1 in [1]. Consider the spine curve $\mathbf{m}(t) = (m_1(t), m_2(t), m_3(t))^\top$ and the contour curves with parameterizations $\mathbf{c}_x(t) = (c_x^1, c_x^2, c_x^3)^\top(t)$, $\mathbf{c}_y(t) = (c_y^1, c_y^2, c_y^3)^\top(t)$ and $\mathbf{c}_z(t) = (c_z^1, c_z^2, c_z^3)^\top(t)$ (all of them are corresponding in parameter in the interval I). Then it holds

$$m_1(t) = c_x^1(t) \qquad m_2(t) = c_y^2(t) \qquad m_3(t) = c_z^3(t) \qquad \text{for all } t \in I. \qquad (7)$$

In [1] the method for computing approximate parameterizations of implicitly given canal surface $\mathcal{S}$ was introduced. The algorithm starts with computing an approximate parameterization $\mathbf{c}_z^+(t)$ of one branch of the z-contour curve. The computation of $\mathbf{c}_z^+(t)$ is based on computing the graph of critical points $G(\mathcal{C}_z^+)$ of $\mathcal{C}_z^+$ and afterwards replacing the edges of $G(\mathcal{C}_z^+)$ by Ferguson's cubic guaranteeing C^1/G^1 continuity, see [21,27] for more details. This step is formulated as an optimization problem when the objective function approximates the integral of the squared Euclidean distance of the constructed approximate curve to the implicitly given contour curve. Next the parameterization $\mathbf{c}_z^-(t)$ of the second real branch $\mathcal{C}_z^-$ of $\mathcal{C}_z$ approximately corresponding in parameter with $\mathbf{c}_z^+(t)$ is analogously computed. From $\mathbf{c}_z^+(t)$ and $\mathbf{c}_z^-(t)$ the approximate parameterization $\mathbf{m}(t)$ of the spine curve $\mathcal{M}$ of $\mathcal{S}$ is reconstructed. Finally rotating $\mathbf{c}_z^+(t)$ around the tangents of $\mathbf{m}(t)$ yields the parameterization $\mathbf{s}(u,t)$ of the canal surface $\mathcal{S}$, cf. (3).

Although the method from [1] is efficient, its main drawback lies in the fact that it produces parameterizations with high rational bidegree. This is caused by the quality of $\mathbf{m}(t)$ which is rational of degree 7. In this paper we improve the original technique such that $\mathbf{m}(t)$ will be polynomial and cubic.

3   The Parameterization Algorithm

In this section we design a method yielding the polynomial cubic approximations (corresponding in parameter) of contour and spine curves of implicitly given canal surface $\mathcal{S}$. Then, they will be used for computing an approximate parameterization of $\mathcal{S}$ having the rational degree $[7, 4]$ in t and $[2, 2]$ in u.

The idea of the method is based on the property (7). In particular when the parameterizations $\mathbf{c}_x(t) = (c_x^1, c_x^2, c_x^3)^\top(t)$, $\mathbf{c}_y(t) = (c_y^1, c_y^2, c_y^3)^\top(t)$ and $\mathbf{c}_z(t) = (c_z^1, c_z^2, c_z^3)^\top(t)$ (corresponding in parameter) of contour curves $\mathcal{C}_x, \mathcal{C}_y$ and $\mathcal{C}_z$ are known, then we can easily arrive at the parameteric description of the spine curve

$$\mathbf{m}(t) = \left(c_x^1(t), c_y^2(t), c_z^3(t)\right)^\top. \qquad (8)$$

Thus the problem of computing the approximation $\mathbf{m}(t)$ of the spine curve of a given canal surface is reduced to the problem of computing approximate parameterizations of the particular branches of contour curves, e.g. $\mathcal{C}_x^+, \mathcal{C}_y^+$ and $\mathcal{C}_z^+$. Moreover $\mathbf{m}(t)$ is polynomial if $\mathbf{c}_x^+(t), \mathbf{c}_y^+(t)$ and $\mathbf{c}_z^+(t)$ are polynomial and it has the degree which is equal to the maximum of the degrees of $c_x^1(t), c_y^2(t), c_z^3(t)$.

We proceed analogously as in Section 3.5 in [1]. First we construct the graphs of the critical points $G(\mathcal{C}_x), G(\mathcal{C}_y)$ and $G(\mathcal{C}_z)$ of the contour curves $\mathcal{C}_x, \mathcal{C}_y, \mathcal{C}_z$, respectively, such that each graph will contain the critical points of the corresponding contour

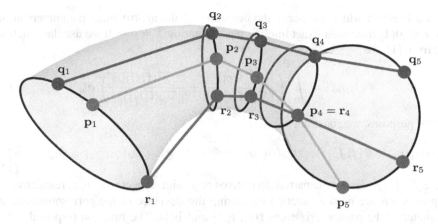

Fig. 3. One branch from each constructed graph of critical points of contour curves

curve and the foot points associated with the critical points of the two other contour curves as its vertices. The algorithm for constructing the graph of critical points having in addition some prescribed points as its vertices is described in Procedure 2 in [1]. In what follows it is sufficient to use only one branch from each constructed graph of critical points – in particular, we will use $G(\mathcal{C}_x^+), G(\mathcal{C}_y^+)$ and $G(\mathcal{C}_z^+)$.

Since the considered contour curves are related to canal surfaces originating in technical applications (i.e., with non-complicated topology), the individual graphs are the paths only. Thus, we can consider that $G(\mathcal{C}_x^+)$ is composed of the path $\mathbf{p}_1, \ldots, \mathbf{p}_k$, $G(\mathcal{C}_y^+)$ of the path $\mathbf{q}_1, \ldots, \mathbf{q}_k$, and $G(\mathcal{C}_z^+)$ of the path $\mathbf{r}_1, \ldots, \mathbf{r}_k$ and being simultaneously satisfied that $\mathbf{p}_i, \mathbf{q}_i$ and $\mathbf{r}_i$ are the associated foot points, see Fig. 3 for a particular example of the constructed graphs of critical points.

Remark 2. When reducing the graphs $G(\mathcal{C}_x), G(\mathcal{C}_y), G(\mathcal{C}_z)$ to $G(\mathcal{C}_x^+), G(\mathcal{C}_y^+), G(\mathcal{C}_z^+)$ it is convenient to omit from our further considerations the vertices which appeared in $G(\mathcal{C}_x), G(\mathcal{C}_y), G(\mathcal{C}_z)$ as the foot points associated to the critical points of $\mathcal{C}_x^-, \mathcal{C}_y^-, \mathcal{C}_z^-$.

Our goal is to replace the edges $\mathbf{p}_i, \mathbf{p}_{i+1} \in G(\mathcal{C}_x^+), \mathbf{q}_i, \mathbf{q}_{i+1} \in G(\mathcal{C}_y^+)$ and $\mathbf{r}_i, \mathbf{r}_{i+1} \in G(\mathcal{C}_z^+)$ by suitable Ferguson's cubics $\mathbf{f}_i(t), \mathbf{g}_i(t)$ and $\mathbf{h}_i(t)$, respectively, such that $\mathbf{f}_i(t)$ is an approximate parameterization of the corresponding segment of $\mathcal{C}_x^+, \mathbf{g}_i(t)$ of $\mathcal{C}_y^+$, $\mathbf{h}_i(t)$ of $\mathcal{C}_z^+$ and $\mathbf{f}_i(t), \mathbf{g}_i(t)$ and $\mathbf{h}_i(t)$ approximately correspond in parameter. This step will be formulated as an optimization process – when interpolating points $\mathbf{p}_i, \mathbf{p}_{i+1}$ and normalized tangent vectors $\mathbf{t}_{\mathbf{p}_i}, \mathbf{t}_{\mathbf{p}_{i+1}}$ at these points. We set the lengths of the tangent vectors at $\mathbf{p}_i, \mathbf{p}_{i+1}$ (input data to interpolation step) as free parameters α_i^1 and α_i^2, i.e.,

$$\mathbf{f}_i(t, \alpha_i^1, \alpha_i^2) = \mathbf{p}_i F_0 + \mathbf{p}_{i+1} F_1 + \alpha_i^1 \mathbf{t}_{\mathbf{p}_i} F_2 + \alpha_i^2 \mathbf{t}_{\mathbf{p}_{i+1}} F_3, \tag{9}$$

where F_i are the standard cubic Hermite polynomials. By analogy we consider $\mathbf{g}_i(t, \beta_i^1, \beta_i^2)$ and $\mathbf{h}_i(t, \gamma_i^1, \gamma_i^2)$. Now, we need to find such particular values of $\alpha_i^1, \alpha_i^2, \beta_i^1, \beta_i^2, \gamma_i^1, \gamma_i^2$ that the corresponding parameterizations will approximate the contour curves and simultaneously be approximately corresponding in parameter.

As a function which measures the deviation of the approximate parameterization $\mathbf{f}(t)$, $t \in [0, 1]$ from their exact implicit representation $f = g = 0$ we use the function (11) from [1] , i.e.,

$$\Phi(f, g, \mathbf{f}) = \int_0^1 \left(\frac{f^2(\mathbf{f}(t))}{\|\nabla f(\mathbf{f}(t))\|^2} + \frac{g^2(\mathbf{f}(t))}{\|\nabla g(\mathbf{f}(t)))\|^2} \right) dt. \tag{10}$$

For our purposes, we construct

$$\Phi_1(\alpha_1, \alpha_2) = \Phi(f, f_x, \mathbf{f}), \quad \Phi_2(\beta_1, \beta_2) = \Phi(f, f_y, \mathbf{g}), \quad \Phi_3(\gamma_1, \gamma_2) = \Phi(f, f_z, \mathbf{h}), \tag{11}$$

where f_x, f_y, f_z denote the partial derivatives of f with respect to x, y, z, respectively.

In addition, we need a function measuring the deviance of the correspondence in parameter of the parameterizations $\mathbf{f}(t)$, $\mathbf{g}(t)$ and $\mathbf{h}(t)$. The function responsible for the deviance of the correspondence in parameter of the curves $\mathbf{f}(t)$ and $\mathbf{g}(t)$ lying on a canal surface $\mathcal{S}$ defined by the polynomial $f(x, y, z)$ will be taken as the distance of the intersection points of the normal lines of $\mathcal{S}$ at $\mathbf{f}(t)$ and $\mathbf{g}(t)$ with the bisector plane of $\mathbf{f}(t)$ and $\mathbf{g}(t)$, respectively. In particular we start with constructing the normal lines of $\mathcal{S}$ at $\mathbf{f}(t)$ and $\mathbf{g}(t)$, i.e.,

$$\mathbf{n_f} = \mathbf{f} + s_1 \nabla f(\mathbf{f}) \quad \text{and} \quad \mathbf{n_g} = \mathbf{g} + s_2 \nabla f(\mathbf{g}), \quad s_1, s_2 \in \mathbb{R}. \tag{12}$$

The intersection points $\mathbf{x}_1$ and $\mathbf{x}_2$ of $\mathbf{n_f}$ and $\mathbf{n_g}$ with the bisector plane

$$(\mathbf{f} - \mathbf{g}) \cdot \frac{\mathbf{f} + \mathbf{g}}{2} - (\mathbf{f} - \mathbf{g}) \cdot \mathbf{x} = 0, \tag{13}$$

where $\mathbf{x} = (x, y, z)^\top$ will be of the form

$$\mathbf{x}_1 = \mathbf{f} + \frac{(\mathbf{f} - \mathbf{g}) \cdot (\mathbf{f} - \mathbf{g}) \nabla f(\mathbf{f})}{2(\mathbf{g} - \mathbf{f}) \cdot \nabla f(\mathbf{f})} \quad \text{and} \quad \mathbf{x}_2 = \mathbf{g} + \frac{(\mathbf{f} - \mathbf{g}) \cdot (\mathbf{f} - \mathbf{g}) \nabla f(\mathbf{g})}{2(\mathbf{f} - \mathbf{g}) \cdot \nabla f(\mathbf{g})}. \tag{14}$$

Hence the objective function measuring the deviance in parameter is as follows

$$\Psi(\mathbf{f}, \mathbf{g}, f) = \int_0^1 \|\mathbf{x}_1 - \mathbf{x}_2\|^2 \, dt. \tag{15}$$

Thus we arrive at two further objective functions:

$$\Psi_1(\alpha_1, \alpha_2, \beta_1, \beta_2) = \Psi(\mathbf{f}, \mathbf{g}, f), \quad \Psi_2(\alpha_1, \alpha_2, \gamma_1, \gamma_2) = \Psi(\mathbf{f}, \mathbf{h}, f). \tag{16}$$

To sum up, the global objective function will be of the form

$$\Upsilon(\alpha_i^1, \alpha_i^2, \beta_i^1, \beta_i^2, \gamma_i^1, \gamma_i^2) = w_1(\Phi_1 + \Phi_2 + \Phi_3) + w_2(\Psi_1 + \Psi_2) \tag{17}$$

for some weights w_i (in all presented examples in Section 4 we have chosen $w_1 = w_2 = 1$). In order to minimize Υ we used Newton iteration process since it allows us handle the integral, however any other optimization method could be used instead – for example it is acceptable to use Newton-Cotes integration formulas to dispose of the integral and afterwards minimize the function by an arbitrary optimization method. The whole method for computing approximate parameterizations of implicitly given canal surfaces is summarized in Algorithm 1.

Algorithm 1. Approximate parameterization of implicitly defined canal surface

INPUT: Defining polynomial $f(x, y, z)$ of a canal surface $\mathcal{S}$.

1: For each contour curve $\mathcal{C}_x, \mathcal{C}_y$ and $\mathcal{C}_z$ compute its critical points and the foot points associated to the critical points of two remaining contour curves;
2: Construct graphs $G(\mathcal{C}_x), G(\mathcal{C}_y)$ and $G(\mathcal{C}_z)$ of critical points having the critical and its associated foot points as its vertices;
3: Reduce the computed graphs to $G(\mathcal{C}_x^+), G(\mathcal{C}_y^+)$ and $G(\mathcal{C}_z^+)$ only, i.e., consider the graphs reflecting only one of the two branches for each contour curve. Each graph is composed of $k - 1$ edges (k vertices);
4: **for** each $i = 1, \ldots, k - 1$ **do**
5:    Construct Ferguson's cubic $\mathbf{f}_i(t, \alpha_i^1, \alpha_i^2)$ matching the points $\mathbf{p}_i, \mathbf{p}_{i+1}$ (the vertices of the i-th edge of $G(\mathcal{C}_x^+)$) and the tangent vectors $\alpha_i^1 \mathbf{t}_{\mathbf{p}_i}, \alpha_i^2 \mathbf{t}_{\mathbf{p}_{i+1}}$ (the tangent vectors of $\mathcal{C}_x^+$ at $\mathbf{p}_i, \mathbf{p}_{i+1}$, respectively);
6:    Construct Ferguson's cubic $\mathbf{g}_i(t, \beta_i^1, \beta_i^2)$ matching the points $\mathbf{q}_i, \mathbf{q}_{i+1}$ (the vertices of the i-th edge of $G(\mathcal{C}_y^+)$) and the tangent vectors $\beta_i^1 \mathbf{t}_{\mathbf{q}_i}, \beta_i^2 \mathbf{t}_{\mathbf{q}_{i+1}}$ (the tangent vectors of $\mathcal{C}_y^+$ at $\mathbf{q}_i, \mathbf{q}_{i+1}$, respectively);
7:    Construct Ferguson's cubic $\mathbf{h}_i(t, \gamma_i^1, \gamma_i^2)$ matching the points $\mathbf{r}_i, \mathbf{r}_{i+1}$ (the vertices of the i-th edge of $G(\mathcal{C}_z^+)$) and the tangent vectors $\gamma_i^1 \mathbf{t}_{\mathbf{r}_i}, \gamma_i^2 \mathbf{t}_{\mathbf{r}_{i+1}}$ (the tangent vectors of $\mathcal{C}_z^+$ at $\mathbf{r}_i, \mathbf{r}_{i+1}$, respectively);
8:    Minimize the objective function $\Upsilon(\alpha_i^1, \alpha_i^2, \beta_i^1, \beta_i^2, \gamma_i^1, \gamma_i^2)$ to get the particular lengths $\alpha_i^1, \alpha_i^2, \beta_i^1, \beta_i^2, \gamma_i^1, \gamma_i^2$ of the tangent vectors $\mathbf{t}_{\mathbf{p}_i}, \mathbf{t}_{\mathbf{p}_{i+1}}, \mathbf{t}_{\mathbf{q}_i}, \mathbf{t}_{\mathbf{q}_{i+1}}, \mathbf{t}_{\mathbf{r}_i}, \mathbf{t}_{\mathbf{r}_{i+1}}$, respectively, and obtain the cubics $\mathbf{f}(t), \mathbf{g}(t)$ and $\mathbf{h}(t)$ approximating the contour curves;
9:    Reconstruct the approximation of the spine curve of $\mathcal{S}$ in the form $\mathbf{m}_i(t) = (f_i^1(t), g_i^2(t), h_i^3(t))^\top$, where $t \in [0, 1]$;
10:    Rotate $\mathbf{f}_i(t)$ around the tangents of $\mathbf{m}_i(t)$ using (3). This process yields the approximate parameterization $\mathbf{s}_i(t, u)$ of the corresponding part of $\mathcal{S}$;
11:    Compute the deviance d_i of the approximate parameterization $\mathbf{s}_i(t, u)$ from the implicit equation $f(x, y, z) = 0$;
12: **end for**

OUTPUT: The piecewise approximate parameterization $\mathbf{s}_1(t, u), \ldots, \mathbf{s}_{k-1}(t, u)$, where $(t, u) \in [0, 1] \times \mathbb{R}$ such that the pieces $\mathbf{s}_i(t, u)$ have the deviance from f equal to d_i.

Remark 3. Instead of minimizing the objective function $\Upsilon(\alpha_i^1, \alpha_i^2, \beta_i^1, \beta_i^2, \gamma_i^1, \gamma_i^2)$, one can firstly compute an approximate parameterization $\mathbf{f}(t)$ of $\mathcal{C}_x^+$ and then construct two new objective functions considering the other contour curves $\mathcal{C}_y^+$ and $\mathcal{C}_z^+$. Thus, we have altogether three objective functions, the first one is

$$\Upsilon_1(\alpha_1, \alpha_2) = \Phi(f, f_x, \mathbf{f}). \tag{18}$$

Hence after minimizing $\Upsilon_1(\alpha_1, \alpha_2)$ the approximation $\mathbf{f}(t) = \mathbf{f}(t, \alpha_1, \alpha_2)$ of $\mathcal{C}_x^+$ is computed and we formulate another two objective functions concerning $\mathcal{C}_y^+$ and $\mathcal{C}_z^+$:

$$\Upsilon_2(\beta_1, \beta_2) = \Phi(f, f_y, \mathbf{g}) + \Psi(\mathbf{f}, \mathbf{g}, f), \tag{19}$$

and

$$\Upsilon_3(\gamma_1, \gamma_2) = \Phi(f, f_z, \mathbf{h}) + \Psi(\mathbf{f}, \mathbf{h}, f). \tag{20}$$

The objective functions Υ_1, Υ_2 and Υ_3 can be minimized separately, which is less complicated since we minimize (integrate/evaluate) less complicated functions having only two variables. On the other hand when computing the "fixed" parameterization $\mathbf{f}(t)$ first, the error is slightly bigger – our tests have shown that the error rises approximately ten times when using the simplified approach.

The deviance d_i of the approximately parameterized surface $\mathbf{s}_i(t, u)$ from the implicitly given surface $f(x, y, z) = 0$ (step 11 in Algorithm 1) can be computed by the following integral, see (24) in [1]

$$d_i = \int_{-\infty}^{\infty} \left(\int_0^1 \frac{f^2(\mathbf{s}_i(t, u))}{\|\nabla f(\mathbf{s}_i(t, u))\|^2} dt \right) du. \tag{21}$$

Remark 4. Algorithm 1 can be easily modified to perform the adaptive refinement similarly to the method presented in [1]. It is enough to increase the number of the edges of the constructed graphs when needed.

The most difficult part of Algorithm 1 consists in step 1. Firstly, the critical points of the contour curves need to be computed. The x-critical points of the curve $\mathcal{C}$ defined by $f = g = 0$ can be computed by solving the following system of equations:

$$f = g = \frac{\partial f}{\partial y} \cdot \frac{\partial g}{\partial z} - \frac{\partial f}{\partial z} \cdot \frac{\partial g}{\partial y} = 0. \tag{22}$$

The y- and z-critical points of $\mathcal{C}$ are computed analogously, see [27,21] for more details. Secondly, for each critical point the associated foot points have to be computed. Let us consider that the point $\mathbf{p}_1 \in \mathcal{C}_x^+$ is known. We shall find points $\mathbf{p}_2 \in \mathcal{C}_x^-$, $\mathbf{q}_1, \mathbf{q}_2 \in \mathcal{C}_y$ and $\mathbf{r}_1, \mathbf{r}_2 \in \mathcal{C}_z$ such that $\mathbf{p}_1, \mathbf{p}_2, \mathbf{q}_1, \mathbf{q}_2, \mathbf{r}_1, \mathbf{r}_2$ are the associated foot points, i.e., they lie on the same characteristic circle. Clearly, the associated foot point $\mathbf{p}_2$ has the same x-coordinate p_1^1 as $\mathbf{p}_1$. Thus we arrive at the points $\mathbf{p}_2^1, \ldots, \mathbf{p}_2^s$ by intersecting the curve $\mathcal{C}_x$ with the plane $x = p_1^1$; in particular we solve the system of non-linear equations

$$f(x, y, z) = \frac{\partial f(x, y, z)}{\partial x} = x - p_1^1 = 0. \tag{23}$$

Now, we have to choose the right point $\mathbf{p}_2 = \mathbf{p}_2^i$ from the set of points $\mathbf{p}_2^1, \ldots, \mathbf{p}_2^s$. For this we use the fact that the normal lines of $\mathcal{S}$ at the points $\mathbf{p}_1$ and $\mathbf{p}_2$ have to intersect at the point $\mathbf{m}$ such that $\|\mathbf{p}_1 - \mathbf{m}\| = \|\mathbf{p}_2 - \mathbf{m}\|$.

Thus, we have the points $\mathbf{p}_1 \in \mathcal{C}_x^+$ and $\mathbf{p}_2 \in \mathcal{C}_x^-$, i.e., the points lying on the same characteristic circle. By cutting the remaining critical curves $\mathcal{C}_y$ and $\mathcal{C}_z$ by the planes $y = m_2$ and $z = m_3$ ($\mathbf{m} = (m_1, m_2, m_3)^\top$ is the intersection point of the normal lines of $\mathcal{S}$ at the points $\mathbf{p}_1$ and $\mathbf{p}_2$) we arrive at the associated foot points $\mathbf{q}_1, \mathbf{q}_2 \in \mathcal{C}_y$ and $\mathbf{r}_1, \mathbf{r}_2 \in \mathcal{C}_z$, respectively. Finally, as in the previous step the "right" points need to be chosen as the points having the same distance from $\mathbf{m}$ as $\mathbf{p}_1$ and $\mathbf{p}_2$ have and all six lie in the same plane.

Remark 5. The only time-consuming parts of the presented algorithm (the same situation as for the method from [1]) inhere in the following steps: 1) The *computation of the critical points* (and the points on the contour curves corresponding to the critical points) which (in more complicated cases) have to be computed "only" numerically as the solution of the system of non-linear equations. 2) The *optimization process*, in which one has to (numerically) integrate and evaluate relatively complicated functions. Nevertheless, all presented examples were computed in seconds.

Proposition 1. *Algorithm 1 yields a G^1 continuous rational parameterization of the maximal rational degree $[7, 4]$ in t and $[2, 2]$ in u.*

Proof. In Algorithm 1, the parameterizations of contour curves are polynomial G^1 continuous parameterizations of degree 3 in t; so is the parameterization of the spine curve, cf. (7). Hence, using (3) we arrive at a G^1 continuous parameterization of the canal surface having the maximal rational degree $[7, 4]$ in t. The rational degree of the parameterization in u depends on the choice of the rational function $\varrho(u)$; the choice $\varrho(u) = u$ leads to the rational degree $[2, 2]$ in u. $\qquad\qquad\square$

Example 1. Let us consider the following cubic polynomial parameterizations of the contour curves:

$$\mathbf{c}_x^+ = \mathbf{c}_z^- = (t^3, 0, t)^\top \quad \text{and} \quad \mathbf{c}_y^- = (t^3, t, 0)^\top. \tag{24}$$

Then the parameterization of the spine curve (using (7)) is of the form

$$\mathbf{m} = (t^3, t, t)^\top. \tag{25}$$

It is easy to certify that formula (5) is fulfilled, i.e., the parameterizations $\mathbf{c}_x^+, \mathbf{c}_y^-$ and $\mathbf{c}_z^-$ are corresponding in parameter with $\mathbf{m}$ (and hence also mutually). Rotating $\mathbf{c}_x^+$ around the tangents of $\mathbf{m}$ (using formula (3)) yields the following rational parameterization of the corresponding canal surface

$$\mathbf{s}(t, u) = \left(\frac{9t^7 + t^3 \left(u^2 - 4\right) + 2tu}{9t^4 + u^2 + 2}, \frac{2 \left(9t^5 + t\right)}{9t^4 + u^2 + 2}, \frac{t \left(u - 3t^2\right)^2}{9t^4 + u^2 + 2} \right)^\top \tag{26}$$

having bidegree $(7, 2)$.

Remark 6. Let us note that the approximation of a given canal surface is usually needed in some prescribed region of interest $\mathfrak{R}$. To ensure this requirement it is enough to add to the set of vertices of the graphs $G(\mathcal{C}_x)$, $G(\mathcal{C}_y)$ and $G(\mathcal{C}_z)$ the intersection points of $\mathcal{C}_x, \mathcal{C}_y$ and $\mathcal{C}_z$ with $\mathfrak{R}$ and, of course, all the associated foot points.

4 Computed Examples

In this section we present some selected results obtained by applying the algorithm introduced in the previous section. The first theoretical example starts with the canal

Fig. 4. Implicitly given canal surface (yellow) together with the bounding planes (blue) from Example 2

surface given implicitly by the irreducible polynomial which was obtained e.g. by eliminating the parameter from (2). The second practical example presents the method which justifies the title of this paper, i.e., it shows the construction of an implicit blending surface which is then parameterized as a canal surface using the designed algorithm. We recall that our algorithm may be used also for the computed blends which look like a canal surface although they are not exactly surfaces of this type.

Example 2. We parameterize the implicit canal surface $\mathcal{S}$ given by the polynomial

$$\begin{aligned}
f = \ & 16x^6 + 48x^4y^2 + 16x^4z^2 - 160x^4z - 32x^4 - 288x^3yz - 288x^3y + 288x^3z + \\
& 288x^3 + 48x^2y^4 + 32x^2y^2z^2 - 104x^2y^2z - 280x^2y^2 + 144x^2yz + 1008x^2y - \\
& 128x^2z^3 + 480x^2z^2 - 168x^2z - 776x^2 - 288xy^3z + 360xy^3 + 288xy^2z + \\
& 72xy^2 - 256xyz^3 + 384xyz^2 + 672xyz - 1480xy + 256xz^3 - 384xz^2 - 672xz + \\
& 1048x + 16y^6 + 16y^4z^2 + 56y^4z - 59y^4 + 144y^3z - 180y^3 + 64y^2z^3 + 192y^2z^2 - \\
& 672y^2z + 334y^2 + 128yz^3 - 192yz^2 - 336yz + 524y + 256z^4 - 704z^3 + \\
& 240z^2 + 808z - 635
\end{aligned}$$

in the region bounded by the planes

$$p_1 : \ 8x + 4y - 12z + 57 = 0 \quad \text{and} \quad p_2 : \ 8x + 4y + 4z - 11 = 0. \tag{27}$$

In particular, we are interested in the part of $\mathcal{S}$ in the region fulfilling $p_1 > 0$ and $p_2 < 0$, see Fig. 4. First, we determine the contour curves, i.e., the curves defined by $f = f_x = 0$, $f = f_y = 0$ and $f = f_z = 0$. The next step is to compute the critical and bounding points of those curves, and to each of those points compute the associated foot points (the points lying on the same characteristic circles and on the other two contour curves). Then we construct the graph having those computed points as vertices and choose only one branch from each contour curve. Deleting redundant points (foot

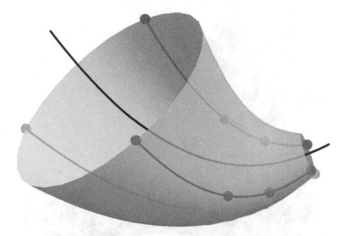

Fig. 5. The parameterized part of a implicitly given canal surface together with the spine and contour curves and the bounding, critical and its associated foot points from Example 2

points corresponding to the critical points on the second, not chosen, branches of the contour curves) yields the graph of critical points. In this particular example the graph of critical points is composed of the three following paths:

$$G\left(\mathcal{C}_x^+\right) = \left(\left(-3,1,\frac{9}{4}\right)^\top, \left(-1,1,\frac{1}{4}\right)^\top, (0,1,0)^\top, \left(1,1,\frac{1}{4}\right)^\top\right);$$

$$G\left(\mathcal{C}_y^+\right) = \left(\left(\frac{-34-15\sqrt{3}}{13}, -\frac{3}{2}, \frac{87-40\sqrt{3}}{52}\right)^\top, \left(-1,-\frac{1}{2},-\frac{5}{4}\right)^\top,\right.$$

$$\left.\left(\frac{1}{2},0,-\frac{\sqrt{3}}{2}\right)^\top, \left(\frac{7}{5},\frac{1}{2},-\frac{1}{20}\right)^\top\right);$$

$$G\left(\mathcal{C}_z^+\right) = \left(\left(-1,-3,\frac{9}{4}\right)^\top, \left(\frac{1}{5},-\frac{7}{5},\frac{1}{4}\right)^\top, \left(\frac{4}{5},-\frac{3}{5},0\right)^\top, \left(\frac{7}{5},\frac{1}{5},\frac{1}{4}\right)^\top\right).$$

Next we compute the parameterizations $\mathbf{f}(t)$, $\mathbf{g}(t)$ and $\mathbf{h}(t)$ of the chosen branches of the contour curves such that we replace the edges of the corresponding graph of critical points by Ferguson's cubics and minimize the objective functions (17). Finally, we reconstruct the approximation $\mathbf{m}(t)$ of the spine curve and by rotation of the one of the contour curves, e.g. $\mathbf{f}(t)$ around the tangents of the spine curve $\mathbf{m}(t)$ we obtain an approximate parameterization of the canal surface in the given region. The parameterized part of the canal surface with the contour and spine curves and the vertices of the graph of critical points is shown in Fig. 5. The error of the approximation, measured by integral (21), is less than $6 \cdot 10^{-5}$.

Fig. 6. Implicit blending surface (yellow) between cone and cylinder (blue) from Example 3

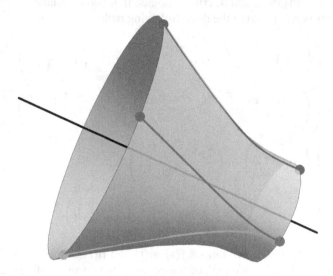

Fig. 7. The parameterized part of a blending canal surface together with the spine and contour curves and the bounding points from Example 3

Example 3. We compute an approximation of the blending surface $h = 0$ joining the cone $f_1 = 4x^2 + y^2 + 6yz - 4\sqrt{2}y + z^2 + 4\sqrt{2}z - 20$ (bounded by the plane $f_{10} = -y + z + 2\sqrt{2}$) and the cylinder $f_2 = 2x^2 + y^2 + 2yz + z^2 - 4$ (bounded by the plane

$f_{20} = -y + z - 3\sqrt{2}$) by a rational canal surface, see Fig. 6. Using the method from [9], the blending surface is of the form

$$h = (1 - u)f_1 f_{20}^{n+1} - u f_2 f_{10}^{n+1}, \tag{28}$$

where we used $u = 1/4$ and $n = 2$. We can check that the blending surface $h = 0$ is a canal surface by method presented in Remark 5.4 in [1]. None of the contour curves contains the critical points hence the approximate parameterization of the blending surface will be composed of one part only. The parameterized part of the blending canal surface with the contour and spine curves and the bounding points are shown in Fig. 7. The error of the approximation measured by (21) is less then $1.6 \cdot 10^{-2}$. Let us note that the error can be easily improved by an adaptive refinement.

Remark 7. We recall that the presented method approximated implicit surfaces from Examples 2 and 3 with errors about $6 \cdot 10^{-5}$ and $1.6 \cdot 10^{-2}$, respectively, whereas using the method from [1] (for the same canal surfaces) yields approximate canal surfaces with approximation errors $2 \cdot 10^{-5}$ and $4 \cdot 10^{-3}$, respectively. However, only one next step of adaptive refinement in the modified method is enough to overcome the errors given by the approach from [1].

5   Conclusion

In this paper we continued with the work started in [1]. We presented a modification and improvement of the method for computing approximate parameterizations of canal surfaces given implicitly. The designed algorithm is based on computing approximate topology-based parameterizations of spatial curves lying on the given canal surface. The distinguished feature of the obtained parameterizations is their low rational bidegree. The technique can be applied mainly to parameterizing implicit blends consisting of parts of canal (or canal-surface-like) surfaces.

Acknowledgments. The authors were supported by the European Regional Development Fund (ERDF), project "NTIS - New Technologies for the Information Society", European Centre of Excellence, CZ.1.05/1.1.00/02.0090. We thank to all referees for their valuable comments, which helped us to improve the paper.

References

1. Bizzarri, M., Lávička, M.: A symbolic-numerical method for computing approximate parameterizations of canal surfaces. Computer-Aided Design 44(9), 846–857 (2012)
2. Farin, G.: Curves and surfaces for CAGD: A practical guide. Morgan Kaufmann Publishers Inc., San Francisco (2002)
3. Hoschek, J., Lasser, D.: Fundamentals of computer-aided geometric design. AK Peters (1993)
4. Bastl, B., Jüttler, B., Lávička, M., Schulz, T.: Blends of canal surfaces from polyhedral medial surface transform representations. Computer-Aided Design Design 43(11), 1477–1484 (2011)

5. Hoffmann, C., Hopcroft, J.: Automatic surface generation in computer aided design. The Visual Computer 1(2), 92–100 (1985)
6. Hoffmann, C., Hopcroft, J.: Quadratic blending surfaces. Computer-Aided Design 18(6), 301–306 (1986)
7. Rockwood, A.P.: Displacement method for implicit blending surfaces in solid models 8, 279–297 (1989)
8. Hartman, E.: Blending of implicit surfaces with functional splines. Computer-Aided Design 22, 500–506 (1990)
9. Hartmann, E.: G^n-continuous connections between normal ringed surfaces. Computer Aided Geometric Design 18(8), 751–770 (2001)
10. Hartmann, E.: Implicit G^n-blending of vertices. Computer Aided Geometric Design 18(3), 267–285 (2001)
11. Zhou, P., Qian, W.-H.: Blending multiple parametric normal ringed surfaces using implicit functional splines and auxiliary spheres. Graphical Models 73(4), 87–96 (2011)
12. Dohm, M., Zube, S.: The implicit equation of a canal surface. Journal of Symbolic Computation 44, 111–130 (2009)
13. Peternell, M., Pottmann, H.: Computing rational parametrizations of canal surfaces. Journal of Symbolic Computation 23, 255–266 (1997)
14. Landsmann, G., Schicho, J., Winkler, F.: The parametrization of canal surfaces and the decomposition of polynomials into a sum of two squares. Journal of Symbolic Computation 32(1-2), 119–132 (2001)
15. Krasauskas, R.: Minimal rational parametrizations of canal surfaces. Computing 79, 281–290 (2007)
16. Dutta, D., Martin, R.R., Pratt, M.J.: Cyclides in surface and solid modeling. IEEE Computer Graphics and its Applications 13, 53–59 (1993)
17. Krasauskas, R., Mäurer, C.: Studying cyclides with Laguerre geometry. Computer Aided Geometric Design 17(2), 101–126 (2000)
18. Lávička, M., Vršek, J.: On the representation of Dupin cyclides in Lie sphere geometry with applications. Journal for Geometry and Graphics 13(2), 145–162 (2009)
19. Shene, C.-K.: Blending two cones with dupin cyclides. Computer Aided Geometric Design 15(7), 643–673 (1998)
20. Pottmann, H., Peternell, M.: Applications of Laguerre geometry in CAGD. Computer Aided Geometric Design 15, 165–186 (1998)
21. Bizzarri, M., Lávička, M.: A symbolic-numerical approach to approximate parameterizations of space curves using graphs of critical points. Journal of Computational and Applied Mathematics 242, 107–124 (2013)
22. Choi, H.I., Lee, D.S.: Rational parametrization of canal surface by 4 dimensional Minkowski Pythagorean hodograph curves. In: Proceedings of the Geometric Modeling and Processing 2000, GMP 2000, pp. 301–309. IEEE Computer Society, Washington, DC (2000)
23. Cho, H., Choi, H., Kwon, S.-H., Lee, D., Wee, N.-S.: Clifford algebra, Lorentzian geometry and rational parametrization of canal surfaces. Computer Aided Geometric Design 21, 327–339 (2004)
24. Bizzarri, M., Lávička, M.: Parameterizing rational offset canal surfaces via rational contour curves. Computer-Aided Design 45(2), 342–350 (2013)
25. Altmann, S.L.: Rotations, quaternions, and double groups. Dover Publications (2005)
26. Goldman, R.: Understanding quaternions. Graphical Models 73, 21–49 (2011)
27. Alcázar, J., Sendra, J.R.: Computation of the topology of real algebraic space curves. Journal of Symbolic Computation 39, 719–744 (2005)

Exploiting the Implicit Support Function
for a Topologically Accurate Approximation
of Algebraic Curves

Eva Blažková and Zbyněk Šír

Faculty of Mathematics and Physics, Charles University in Prague, Czech Republic
{eblazkova,sir}@karlin.mff.cuni.cz

Abstract. Describing the topology of real algebraic curves is a classical problem in computational algebraic geometry. It is usually based on algebraic techniques applied directly to the curve equation. We use the implicit support function representation for this purpose which can in certain cases considerably simplify this task. We describe possible strategies and demonstrate them on a simple example. We also exploit the implicit support function for a features-preserving approximation of the graph topologically equivalent to the curve. This contribution is meant as a first step towards an algorithm combining classical approaches with the dual description via the support function.

Keywords: algebraic curve, support function, critical points, approximation, trigonometric polynomial.

1 Introduction

Solution of many problems in Computer Aided Geometric Design depends on an approximation of a curve given by an implicitly defined bivariate polynomial with rational coefficients. It is very desirable to visualize the curve in any required precision, to find the number of components or to test to which component a given point belongs. All this information is fully contained in the planar graph topologically equivalent to the curve and whose vertices are points of the algebraic curve and edges correspond to regular arcs of the curve.

Known algorithms studying the topology of an algebraic curve have always two parts. First we find out the critical points and then we connect them appropriately. There are two main types of algorithms. The first type uses the same principle as the Cylindrical Algebraic Decomposition (CAD) algorithm, cf. [5, page 159]. The other approach is based on a subdivision of the given region.

Cylindrical Algebraic Decomposition based algorithms are usually divided into three phases: First find the x-coordinates of critical points of $\mathcal{C}$, then for each x_i compute the intersection points $P_{i,j}$ of $\mathcal{C}$ and the vertical line $x = x_i$ and finally for every $P_{i,j}$ determine the number of branches of $\mathcal{C}$ on the left and right and use this information to connect the points appropriately.

The main problem of these algorithms is the second phase, because the x-coordinates of the critical points are not necessarily rational numbers and therefore

M. Floater et al. (Eds.): MMCS 2012, LNCS 8177, pp. 49–67, 2014.

the polynomials $f(x_i, y)$ have non-rational coefficients. There are several methods to deal with this problem. In [12], Hong computes xy-parallel separating boxes of critical points with rational endpoints. Then he can count the branches in Phase 3 as roots of univariate polynomial with rational coefficients. In paper [7], the authors proposed a preprocessing - linear change of coordinate. The x-coordinate is transformed so that the curve is in generic position. When the curve is in generic position, the *Sturm-Habicht* sequence is used, a suitable generalization of polynomial remainder sequence, to derive the y-coordinates of critical points (Phase 2) as rational functions of their x-coordinate and also to deduce the multiplicity of the considered critical point. Another solution was given by paper [16] - they project critical points to three axes x, y and a random one. From these projections they can recover xy-parallel boxes with rational endpoints which separate the critical points. Paper [6] give the Bitstream Descartes algorithm (a variant of interval Descartes algorithm) as an efficient algorithm to isolate roots of a polynomial with non-rational coefficients. In contrast to all above algorithms, [13] replace the Sturm-Habicht sequence with a Gröbner basis and rational univariate representation, which ensure that we avoid working with polynomials with non-rational coefficients even in non-generic position.

The second type of algorithm is based on subdivision. The only certified algorithm (i.e. one which gives the correct output for every input) based on subdivision is [4]. This algorithm subdivides the region $\mathcal{D}$ into *regular regions* (the curve is smooth inside) and *regions with singular points*, which can be made sufficiently small. The topology inside the regions containing a singular point is recovered from the information on the boundary using the topological degree.

The main contribution of this paper consists in application of the (implicit) support function representation to the construction of the graph topologically equivalent to a given algebraic curve. We also consider the subsequent high precision approximation of the curve. The support function representation describes a curve as the envelope of its tangent lines, where the distance between the tangent line and the origin is specified by a function of the unit normal vector. This representation is one of the classical tools in the field of convex geometry [11]. In this representation offsetting and convolution of curves correspond to simple algebraic operations of the corresponding support functions. In addition, it provides a computationally simple way to extract curvature information [8]. Applications of this representation to problems from Computer Aided Design were foreseen in the classical paper [15] and developed in several recent publications, see e.g., [1–3, 9, 10, 14, 17, 18].

The remainder of this paper is organized as follows. Section 2 is devoted to basic definitions and results related to the (implicit) support function representation and to the topology of algebraic curves. Section 3 describes how the use of the implicit support function can contribute to the basic phases of determination of the topology of planar algebraic curves. Issues related both to the search for critical points and their connectivity are considered. In Section 4 we show how the support function representation can be exploited for an efficient approximation of segments of the curve connecting the critical points. In Section 5 we summarize our results in an algorithm and demonstrate it on a simple example.

2   Preliminaries

In this section we first recall the definitions and basic properties of the explicit and the implicit support functions. We also summarize concepts related to the determination of the topology of algebraic curves. In both cases we slightly extend standard approaches toward our goals.

2.1   Implicit Support Function Representation of Algebraic Curves

For an algebraic planar curve $\mathcal{C}$ we define its support function h as a (possibly multivalued) function defined on a subset of the unit circle

$$h : \mathbb{S}^1 \supset U \to \mathbb{R}^1$$

by which is any unit normal $\mathbf{n} = (n_1, n_2)$ associated with the distance(s) from the origin to the corresponding tangent line(s) of the curve.

As proved in [18] we can recover the curve $\mathcal{C}$ from h as the envelope of the system of lines $\{\mathbf{n} \cdot \mathbf{x} - h(\mathbf{n}) = 0 : \mathbf{n} \in U\}$. This envelope is locally parameterized via the formula

$$\mathcal{C}(\mathbf{n}) = h(\mathbf{n})\mathbf{n} + \nabla_{\mathbb{S}_1} h(\mathbf{n}) = h(\mathbf{n})\mathbf{n} + \dot{h}(\mathbf{n})\mathbf{n}^\perp , \tag{1}$$

where $\nabla_{\mathbb{S}_1}$ denotes the intrinsic gradient with respect to the unit circle, which is alternatively expressed using the derivative $\dot{h}(\mathbf{n})$ with respect to the arc-length and $\mathbf{n}^\perp$ is the clockwise rotation of $\mathbf{n}$ about the origin by the angle $\frac{\pi}{2}$.

For an algebraic curve $\mathcal{C}$ defined as the zero set of a polynomial $f(x, y) = 0$ we typically do not obtain an explicit expression of h but rather an implicit one, which is closely related to the notion of dual curve.

Definition 1. *Let $\mathcal{C}$ be a curve in projective plane. The* dual *of $\mathcal{C}$ is the Zariski closure of the set in the dual projective plane consisting of tangent lines of $\mathcal{C}$.*

The equation of the dual curve

$$D(h, \mathbf{n}) = 0 \tag{2}$$

can be computed by eliminating x and y from the following system of equations:

$$\mathbf{n} \cdot \left(\frac{\partial f}{\partial x}, \frac{\partial f}{\partial y} \right)^\perp = 0$$

$$\mathbf{n} \cdot (x, y) = h . \tag{3}$$

Definition 2. *The dual equation $D(h, \mathbf{n}) = 0$ together with the algebraic constraint $n_1^2 + n_2^2 = 1$ is called the* implicit definition of the support function h *or simply* the implicit support function.

If the partial derivative $\partial D/\partial h$ does not vanish at $(\mathbf{n}_0, h_0)$ then (2) implicitly defines the support function

$$\mathbf{n} \mapsto h(\mathbf{n})$$

in a certain neighborhood of $(\mathbf{n}_0, h_0) \in \mathbb{R}^3$.

The (implicit) support function is obviously a kind of dual representation which takes into account the Euclidean metric. It has many nice properties. Let us recall how it is affected by selected geometric operations, cf. [15, 18]:

(i) *translation* by a translation vector $\mathbf{v} \in \mathbb{R}^2$

$$h(\mathbf{n}) \mapsto \tilde{h}(\mathbf{n}) := h(\mathbf{n}) + \mathbf{v} \cdot \mathbf{n}$$
$$D(h, \mathbf{n}) = 0 \mapsto \tilde{D}(\tilde{h}(\mathbf{n}), \mathbf{n}) := D(h(\mathbf{n}) + \mathbf{v} \cdot \mathbf{n}, \mathbf{n}) = 0 \ ,$$

(ii) *rotation* by an orthogonal matrix $\mathbf{A} \in SO(2)$

$$h(\mathbf{n}) \mapsto \tilde{h}(\mathbf{n}) := h(\mathbf{A}\mathbf{n})$$
$$D(h(\mathbf{n}), \mathbf{n}) = 0 \mapsto \tilde{D}(\tilde{h}(\mathbf{n}), \mathbf{n}) := D(h(\mathbf{A}\mathbf{n}), \mathbf{n}) = 0 \ ,$$

(iii) *scaling* by a factor $\lambda \in \mathbb{R}$

$$h(\mathbf{n}) \mapsto \tilde{h}(\mathbf{n}) := \lambda h(\mathbf{n})$$
$$D(h(\mathbf{n}), \mathbf{n}) = 0 \mapsto \tilde{D}(\tilde{h}(\mathbf{n}), \mathbf{n}) := D(\lambda h(\mathbf{n}), \mathbf{n}) = 0 \ ,$$

(iv) *offseting* with a distance $\delta \in \mathbb{R}$

$$h(\mathbf{n}) \mapsto \tilde{h}(\mathbf{n}) := h(\mathbf{n}) + \delta$$
$$D(h(\mathbf{n}), \mathbf{n}) = 0 \mapsto \tilde{D}(\tilde{h}(\mathbf{n}), \mathbf{n}) := D(h(\mathbf{n}) + \delta, \mathbf{n}) = 0 \ .$$

Moreover, the support function representation is very suitable for describing the *convolution* $\mathcal{C}_3 = \mathcal{C}_1 \star \mathcal{C}_2$ of curves $\mathcal{C}_1, \mathcal{C}_2$ as this operation corresponds to the sum of the associated support functions $h_3 = h_1 + h_2$ and its implicit support function can be obtained by eliminating h_1, h_2 from the system of equations

$$D_1(h_1, \mathbf{n}) = 0, \quad D_2(h_2, \mathbf{n}) = 0 \quad \text{and} \quad h_3 = h_1 + h_2 \ ,$$

see [18, 14] for more details.

Another very useful property of the support function representation (especially in connection with G^2 Hermite interpolation problem) is that it can be efficiently used for describing the *signed curvature* of a given curve, cf. [18], in the form

$$\kappa = -\frac{1}{h + \ddot{h}} \ . \tag{4}$$

2.2   Topology of the Curve

We are given a real planar algebraic curve $\mathcal{C} = \{(x, y) \in \mathbb{R}^2 \mid f(x, y) = 0\}$ where $f \in \mathbb{Q}[x, y]$. We consider the problem of determining the topology of $\mathcal{C}$. The topology of $\mathcal{C}$ is usually described by a planar graph which can have vertices at infinity and which is topologically equivalent to the original curve.

Definition 3. *Let $\mathcal{C}$ be a curve and $\mathcal{G}$ be a planar graph (possibly with vertices at infinity). The curve $\mathcal{C}$ and a graph $\mathcal{G}$ are topologically equivalent if and only if they are isotopic as curves of Euclidean space, i.e., there exists a continuous map $H : \mathbb{R}^2 \times [0, 1] \to \mathbb{R}^2$, such that*

- *$H(x, t)$ is a homeomorphism for all $t \in [0, 1]$,*
- *$H(x, 0) = \mathrm{id}$,*
- *$H(\mathcal{C}, 1) = \mathcal{G}$.*

Consider a vertical line l moving from the left side ($x = -\infty$) to the right ($x = \infty$). At any position there is a finite number of intersections of l and $\mathcal{C}$. The number of intersections can change only when $\mathcal{C}$ has a *critical point* on this x-coordinate. To ensure that the graph $\mathcal{G}$ is topologically equivalent to $\mathcal{C}$ we have to include all critical points among vertices of $\mathcal{G}$. Namely

Definition 4. *Let $f(x, y) \in \mathbb{Q}[x, y]$ define the real algebraic curve*

$$\mathcal{C} = \{(x, y) \in \mathbb{R}^2 \mid f(x, y) = 0\} .$$

The point $(a, b) \in \mathcal{C}$ is called

- *x-critical point if $\frac{\partial f}{\partial x} = 0$, similarly we define y-critical point,*
- *singular point if $\frac{\partial f}{\partial x} = \frac{\partial f}{\partial y} = 0$,*
- *x-extremal point if $\frac{\partial f}{\partial x} = 0$ and $\frac{\partial f}{\partial y} \neq 0$, similarly we define y-extremal point.*

There are several methods to deal with the critical points. Our approach is related to the general scheme of Cylindrical Algebraic Decomposition (CAD) based algorithms. These algorithms are usually divided into three phases. In Phase 1 the x-coordinates of all the critical points of $\mathcal{C}$ are found. Using subresultant sequence, the discriminant $R(x)$ of f is computed. Then one determines the roots of $R(x)$ and obtain the x-coordinates $(x_i, 1 \leq i \leq n)$ of all critical points of $\mathcal{C}$. In Phase 2 for each x_i the intersection points $P_{i,j}$ of $\mathcal{C}$ and the vertical line $x = x_i$ are computed. These intersection points have as y-coordinates the roots of the polynomial $f(x_i, y)$. In Phase 3 the number of branches of $\mathcal{C}$ over every interval (x_i, x_{i+1}) is determined. It is the number of real roots of $f(x', y)$ for any x' from the given interval. Using this information it is possible to connect the points appropriately. In [7] a Phase 0 was proposed; a linear change of coordinate. The plane is sheared so that the curve is in generic position.

Definition 5. *The real algebraic curve C is in* generic position *if it satisfies the following conditions:*

- *the curve C has no vertical asymptotes*
- *on every vertical line $x = \alpha$, $\alpha \in \mathbb{R}$ is at most one critical point*

Obviously there are at most $\binom{c}{2}$ non-generic configurations, where c is a number of critical points. Therefore the change of coordinates is always possible.

3   The Topology of the Curve Using the Implicit Support Function

In this section we will discuss how the use of the implicit support function can contribute to the basic phases of determination of the topology of planar algebraic curves, see Section 2.2. We will handle certain issues related both to the search for critical points and to their connectivity.

3.1   Critical Points

When the critical points are determined we can profit from the use of the support function. We devote a paragraph to every type of critical points. As we will see the support function is particularly useful in the search for cusps, points with horizontal and vertical tangents and inflections. It can also provide interesting additional information allowing us to omit self-intersections from the list of critical points while preserving the accurate curve topology. On the other hand the determination of boundary points (for a curve studied within a box) is easier on the primary curve and therefore we omit them here. An efficient global strategy would therefore be based on a combination of the information about the primary curve and its support function.

Cusps. From the general theory of algebraic curves (see e.g., [19]) the cusps on C correspond to inflection points in the dual representation. Cusps are distinguished as points having infinite curvature. They can be quite easily determined from the support function due to (4). If only the implicit support function is available, a condition for cusps can be formulated as follows.

Proposition 1. *Let $D(h, \mathbf{n}) = 0$ be the implicit support function of the curve C. Then the cusps of C satisfy the following condition:*

$$
\begin{aligned}
hD_h^3 &- n_1^2(D_h^2 D_{n_2 n_2} + D_{hh} D_{n_2}^2 - 2D_h D_{hn_2} D_{n_2}) - n_1 D_h^2 D_{n_1} + \\
&+ n_2^2(D_h^2 D_{n_1 n_1} + D_{hh} D_{n_1}^2 - 2D_h D_{hn_1} D_{n_1}) - n_2 D_h^2 D_{n_2} + \\
&+ 2n_1 n_2(D_h D_{hn_2} D_{n_1} + D_h D_{hn_1} D_{n_2} + D_{hh} D_{n_1} D_{n_2} - D_h^2 D_{n_1 n_2}) = 0 ,
\end{aligned} \tag{5}
$$

where the subscripts denote corresponding partial derivatives.

Proof. Using (4) we get the necessary condition for cusps

$$h(\mathbf{n}) + \ddot{h}(\mathbf{n}) = 0 . \tag{6}$$

Let $\mathbf{n}(s) = (n_1(s), n_2(s))$ be a parametrization of the unit circle by arc-length s and suppose that we locally have $h(\mathbf{n}(s))$. Using the chain rule we get following derivatives:

$$\dot{h} = h_{n_1}\dot{n}_1 + h_{n_2}\dot{n}_2 = -h_{n_1}n_2 + h_{n_2}n_1 \tag{7}$$

$$\ddot{h} = h_{n_1 n_1}\dot{n}_1^2 + h_{n_1 n_2}\dot{n}_1 \dot{n}_2 + h_{n_1}\ddot{n}_1 + h_{n_2 n_2}\dot{n}_2^2 + h_{n_2 n_1}\dot{n}_1 \dot{n}_2 + h_{n_2}\ddot{n}_2 =$$
$$= h_{n_1 n_1}n_2^2 - h_{n_1 n_2}n_2 n_1 - h_{n_1}n_1 + h_{n_2 n_2}n_1^2 - h_{n_2 n_1}n_1 n_2 - h_{n_2}n_2 , \tag{8}$$

where the dot denotes the derivative with respect to arc length s and the subscript denotes the partial derivative. The second equality in (7) and in (8) is deduced using the equality $(\dot{n}_1, \dot{n}_2) = (-n_2, n_1)$.

The partial derivatives of h can be deduced from its implicit definition. For example:

$$\frac{\partial}{\partial n_1}D(h(\mathbf{n}), n_1, n_2) = D_{n_1}(h, n_1, n_2) + h_{n_1}D_h(h, n_1, n_2) = 0 .$$

And therefore

$$h_{n_1} = -\frac{D_{n_1}(h, n_1, n_2)}{D_h(h, n_1, n_2)} .$$

Similarly we can deduce all partial derivatives of h and substitute them into (8). That equation we substitue into (6) to get a necessary condition (5) for cusps in variable $\mathbf{n}$. □

In concrete computations the cusps will be found by simultaneously solving equation (6) and the fundamental equations (2) and $n_1^2 + n_2^2 - 1 = 0$. The primary points are fully defined by (1).

Extremal Points. Due to the dual nature of the (implicit) support function representation it is particularly easy to find the extremal points, as shown in the following

Lemma 1. *The x-extremal and y-extremal points have unit normal vectors $(\pm 1, 0)$ and $(0, \pm 1)$, respectively.*

Proof. From the definition it follows that $\frac{\partial f}{\partial x} = 0$ resp. $\frac{\partial f}{\partial y} = 0$. □

Corollary 1. *Let h be the support function implicitly defined by $D(h, \mathbf{n}) = 0$. The x and y-extremal points are the solutions of the polynomial equations in h*

$$D(h, (1, 0)) = 0 \quad and \quad D(h, (0, 1)) = 0 , \tag{9}$$

respectively.

Using the envelope formula (1) we can recover extremal points on the primary curve $\mathcal{C}$.

Inflection Points. Many algorithms for topologically exact description of algebraic curves do not consider inflection points. In the context of dual representations they however occur as natural splitting points. Indeed they simplify both the topology determination and subsequent approximation of individual segments.

Inflections are points where the normal vector changes its direction of movement as the point traverses the curve. Althought these can be found from the primary equation of the curve, this property is easily identified in the support function representation. Such points are of two types: the cusps and the t-extremal points of the support function, where t is the parameter on the unit circle. The first type corresponds to real inflection points, the second is the case of points at infinity. This leads to the following proposition:

Proposition 2. *Let $\mathcal{C}$ be an algebraic curve, let $t \mapsto \mathbf{n}(t)$ be a parametrization of the unit circle and consider the form $D(h, t) = 0$ of the implicit support function of $\mathcal{C}$. Then the inflection points of curve $\mathcal{C}$ are the t-critical points of the implicit support function which are neither isolated points nor self-intersections.*

We can identify the inflection points by counting the number of points of the curve on a line a little to the left and on a line a little to the right of the critical point.

Proposition 3. *Let $P = (x_0, y_0)$ be a point of the curve $\mathcal{C}$, $x_1, x_2 \in \mathbb{Q}$ and $I = [x_1, x_2]$ be an isolating interval of x_0, i.e., I does not contain other x-coordinate of x-critical point than x_0. The x-critical point P is an inflection point if and only if*

$$\sharp\{\alpha \in \mathbb{R} \mid f(x_1, \alpha) = 0\} \neq \sharp\{\alpha \in \mathbb{R} \mid f(x_0, \alpha) = 0\} \text{ or}$$
$$\sharp\{\alpha \in \mathbb{R} \mid f(x_0, \alpha) = 0\} \neq \sharp\{\alpha \in \mathbb{R} \mid f(x_2, \alpha) = 0\} ,$$

where $\sharp$ denotes the number of zeros counted with multiplicities.

Proof. We want to exclude self-intersections and isolated points, which are characterized by

$$\sharp\{\alpha \in \mathbb{R} \mid f(x_1, \alpha) = 0\} = \sharp\{\alpha \in \mathbb{R} \mid f(x_0, \alpha) = 0\} = \sharp\{\alpha \in \mathbb{R} \mid f(x_2, \alpha) = 0\} .$$

Self-intersections. Self-intersections are important features in standard algorithms for determination of the curve topology. The support function based approach however allows us to avoid the precise determination of self-intersections. From the dual point of view the two branches of the intersection are handled separately, but we can easily obtain geometrical bounds on the curve branches which certify existence and uniqueness of their intersections.

Definition 6. The tangent triangle $T(P_1, P_2)$ *is the triangle bounded by tangents at points P_1 and P_2 and by the segment $P_1 P_2$.*

Proposition 4. *Let C_k be a segment of the algebraic curve C connecting P_1, P_2 free of cusps, inflections and extremal points. Then C_k lies in the interior of the tangent triangle $T(P_1, P_2)$.*

Proof. Denote by t_1 and t_2 the tangent vectors at P_1 and P_2 respectively. Due to the fact that C is split at extremal points and cusps, the angle between t_1 and t_2 is at most $\frac{\pi}{2}$. Therefore the arc does not intersect itself and moreover the arc does not contain any cusp, because the curve is divided in cusps. Therefore the arc is smooth and from the implicit function theorem we can suppose that the explicit formula for given arc is $c(t)$. The vector $c''(t)$ can change its sign only at cusps and inflection points and therefore it has a constant sign on the arc. Without loss of generality we can suppose that it is positive, i.e., the arc is strictly convex. From the definition of convexity, the arc lies above both tangents and below the segment $P_1 P_2$. $\qquad\square$

Due to the previous proposition we can find the self-intersections of the curve as the non-empty intersections of envelope triangles of all arcs in which the curve is divided. This method give us the information about which pairs of arcs intersect and also the approximate positions of the self-intersections in the intersections of envelope triangles.

Proposition 5. *Let C_1 and C_2 be two simple curve segments. If their bounding triangles $T_1 = T(P_1, P_1')$ and $T_2 = T(P_2, P_2')$ intersect in the following way:*

- *The edge $P_1 P_1'$ intersects the edge $P_2 P_2'$,*
- *$P_1, P_1' \notin T_2$ and*
- *$P_2, P_2' \notin T_1$,*

then the segments have precisely one intersection and it lies in $T_1 \cap T_2$.

Proof. Existence of the intersection follows from the transversal intersection of the triangles. The uniqueness is ensured by the convexity of both curve segments within the bounding tangent triangles. $\qquad\square$

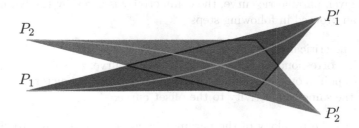

Fig. 1. Two simple curve segments and their tangent triangles. The intersection of segments lies inside the intersection of tangent triangles.

3.2   Connectivity of Critical Points

When we have determined the position of the critical points (Phase 1 of a general CAD based algorithm) we need to connect them appropriately. First we will study the general situation, when only the implicit formula of the curve is given. Then we describe the advantages of this approach when the given curve is an offset curve of a parametric curve.

Connectivity Based on Implicit Support Function. When we have the implicit support function of the curve, using the implicit function theorem we have also G^2 data at every point and we can profit from them. We describe some rules which the connected points have to satisfy:

1. *The difference of angles of tangents (normals) of two connected points is at most $\frac{\pi}{2}$.*
   This is because the curve is split at extremal points, cusps and inflection points.
2. *The sign of the second derivative at given point P determines in which half-plane given by the tangent line at P are the points connected to P. If the sign is negative, the points connected to P are in the same halfplane as the normal vector to C at P, if the sign is positive, they are in the other halfplane.*
   This rule follows immediately from the definition of convexity.

In many cases these two rules yield the connectivity of the given curve. If not, it seems that often we are able to determine the topology by subdividing (possibly several times) the maximal angle in rule 1, i.e, we add extra splitting points. For example, in the first iteration we add points with normal vector $\left(\pm\frac{1}{\sqrt{2}}, \pm\frac{1}{\sqrt{2}}\right)$. These points we can determine similarly to extremal points, see Section 3.1.

Additional Connectivity Information for Offsets. If this general approach turns out to be insufficient we can either use one of CAD based algorithms cited in Section 1 or exploit some additional properties of studied curves. Here we would like to emphasize that in the case when the curve under examination is an offset to a given parametric curve, the connectivity is given by the parametrization. We can proceed in following steps:

1. Determine critical points on the offset curve.
2. Find the corresponding points on the original curve.
3. Connect points on the original curve by decreasing parameter.
4. Apply the same connectivity to the offset curve.

In this way the topology of the parametric curve is transferred to the offset curve.

4 Implicit Support Function Based Approximation

The support function representation can be exploited for an efficient approximation of segments of the curve connecting the critical or inflection points. Compared to approximation in the primary space it can bring several advantages,
which will be discussed in this section.

Because we want to preserve features of the implicitly defined offsets and
convolutions, it is suitable to interpolate the critical points up to the second
order geometric data. Indeed, e.g. the cusps are distinguished by having infinite
curvature. Using the support function representation it is possible to perform
the G^2 Hermite interpolation by solving a system of linear equations [3]. The
interpolation of critical points can be combined with an optimization of the
approximation of the connection segments.

4.1 Approximation Space

A suitable space of implicit support functions must be fixed in order to perform
an efficient approximation.

Definition 7. *A set $\mathcal{A}$ of functions $h : \mathbb{S}^1 \to \mathbb{R}$ is called* a rational approximation space *if the following conditions hold:*

- *$\mathcal{A}$ is a real linear space of finite dimension.*
- *$\mathcal{A}$ is (as a set) invariant with respect to the rotations of $\mathbb{S}^1$.*
- *The curves with support functions from $\mathcal{A}$ are rational.*

Any segment of the primary algebraic curve will be approximated by a piece
of a parametric curve with support function $h \in \mathcal{A}$. If $\{a_i\}_{i=1}^n$ is a basis of $\mathcal{A}$
then

$$h(t) = \sum_{i=1}^n c_i a_i(t) ,$$

where c_i are free coefficients. The parametric segment $\mathbf{x}_i(t)$ is computed from
h via the envelope formula (1). Let us stress the fact, that in the definition of
approximation space we require that the resulting segments are rational. Their
union, which approximate the whole algebraic curve can therefore be represented
in the NURBS format.

It was shown in [18] that suitable subspaces of trigonometric polynomials
satisfy the three required conditions. In order to obtain a sufficient number of
degrees of freedom for G^2 Hermite interpolation we will from now on use the
trigonometric polynomials of degree 3:

$$\mathcal{A} = \mathrm{Span}\{1, \sin t, \cos t, \sin 2t, \cos 2t, \sin 3t, \cos 3t\} . \tag{10}$$

The main drawback of trigonometric polynomials is that they can not produce
curves with inflections (and interpolate zero curvature). For an accurate (G^2)
interpolation of inflections we plan to use other approximation spaces including

square roots of trigonometric polynomials and more generally implicitly defined multivalued support functions. Alternatively it is possible to approximate inflections only with G^1 precision.

4.2 G^2 Hermite Interpolation and Fixing Degrees of Freedom

G^2 Hermite interpolation with trigonometric polynomials is described in detail in [3]. We will extend this procedure to points with infinite curvature (cusps) and we will also discuss how to optimize the possible free degrees of freedom.

G^2 Hermite interpolation can efficiently be performed on the level of support function due to following

Proposition 6. *Let C be a planar curve with support function h, defined at least locally in a neighborhood of $\mathbf{n}_0$. If g is a function defined also in a neighborhood of $\mathbf{n}_0$ and satisfying*

$$g(\mathbf{n}_0) = h(\mathbf{n}_0), \qquad \dot{g}(\mathbf{n}_0) = \dot{h}(\mathbf{n}_0), \qquad \ddot{g}(\mathbf{n}_0) = \ddot{h}(\mathbf{n}_0) . \tag{11}$$

Then the corresponding curve $\mathbf{x}_g$ obtained via (1) interpolates the position of the point $C(n_0)$, its normal and its curvature.

Proof. Due to (1)

$$C(\mathbf{n}_0) = h(\mathbf{n}_0)\mathbf{n}_0 + \dot{h}(\mathbf{n}_0)\mathbf{n}_0^\perp = g(\mathbf{n}_0)\mathbf{n}_0 + \dot{g}(\mathbf{n}_0)\mathbf{n}_0^\perp = \mathbf{x}_g(\mathbf{n}_0) .$$

The two curves have also the common normal $\mathbf{n}_0$ at their common point. Finally they have also the same curvature

$$\kappa = -\frac{1}{h(\mathbf{n}_0) + \ddot{h}(\mathbf{n}_0)} = -\frac{1}{g(\mathbf{n}_0) + \ddot{g}(\mathbf{n}_0)}$$

due to (4).   □

A corollary of the previous proposition is that the G^2 Hermite interpolation in the curve space is thus reduced to the C^2 interpolation in the approximation space. The right hand sides of (11) will be obtained from $D(h, \mathbf{n})$ via implicit differentiation. Interpolation at any point thus imposes three linear conditions on coefficients c_i. More precisely, for $g(t) = \sum_{i=1}^{7} c_i a_i(t)$, an element of the approximation space (10), the conditions (11) has the following form

$$\sum_{i=1}^{7} c_i a_i(t) = h(\mathbf{n}_0), \quad \sum_{i=1}^{7} c_i a_i'(t) = \dot{h}(\mathbf{n}_0), \quad \sum_{i=1}^{7} c_i a_i''(t) = \ddot{h}(\mathbf{n}_0) , \tag{12}$$

where $t = \arctan\left(\frac{n_{01}}{n_{02}}\right)$. Matching the support function up to the second derivative also reproduces the cusps, which correspond to the case $h(\mathbf{n}_0) + \ddot{h}(\mathbf{n}_0) = 0$. This case, which is singular in the primary curve space, is completely regular from the point of view of the support function.

The interpolation of cusps and inflections is very important both for obtaining a low approximation error and for estimating the approximation error. In this case the error evaluates simply as the maximal error of the support function on the given interval.

Proposition 7. *Let h, g be two support functions defined on the interval $U = [\mathbf{n}_0, \mathbf{n}_1]$, such that*

$$g(\mathbf{n}_i) = h(\mathbf{n}_i), \qquad \dot{g}(\mathbf{n}_i) = \dot{h}(\mathbf{n}_i), \qquad i \in 0, 1 .$$

Suppose, that the corresponding curves $\mathbf{x}_h$, $\mathbf{x}_g$ are cusp-free on U. Then their Hausdorff distance corresponds to the error in support functions.

$$||\mathbf{x}_h - \mathbf{x}_g||_H = ||h - g||_\infty . \tag{13}$$

Proof. Due to boundary conditions and absence of singular points (cusps), the Hausdorff distance is realized by a common normal line to both curves. The distance of the points on this line is equal to the absolute value of the difference of the support functions. For a more formal proof see [18, Proposition 14]. $\square$

The approximation space can have a higher dimension than 6 and the remaining degrees of freedom can be used for minimizing the segment error. The two possible strategies are based on interpolation of some additional data and on minimizing some integral measure, respectively.

As we are using an approximation space (10) of dimension 7, after satisfying (12) for both boundary points, we are left one additional free parameter. In the following example we will use this parameter for interpolation of the support function value at the mid-normal

$$g(t') = \sum_{i=1}^{7} c_i a_i(t') = h\left(\frac{\mathbf{n}_0 + \mathbf{n}_1}{2}\right), \quad \text{for } t' = \arctan\frac{n_{01} + n_{11}}{n_{02} + n_{12}} \tag{14}$$

or alternatively to minimize the L_2 norm of the difference of supports. In this case every c_i is a function of the free parameter e used to minimize the quantity

$$||h(t) - g(t, e)||_\infty . \tag{15}$$

5   Algorithm and Example

In this section we summarize the previous results in an algorithm for topologically precise approximation of algebraic curves. We also demonstrate this algorithm on an example.

5.1   Algorithm Description

Algorithm 1 summarizes the process of determining the topology of an algebraic curve and the subsequent approximation of the curve.

Algorithm 1. Topologically accurate approximation of an algebraic curve

---

Input: Real algebraic curve $\mathcal{C}$ given as a zero set of a bivariate polynomial with rational coefficients $f(x, y) \in \mathbb{Q}[x, y]$

Output: Topologically accurate approximation of the curve $\mathcal{C}$.

1: Determine the support function h of $\mathcal{C}$.
2: Determine the cusps, extremal and inflection points in the implicit support function representation.
3: Find corresponding points on the primary curve.
4: Connect points.
5: Determine the self-intersections.
6: Approximate the support function of the segments by trigonometric polynomials.
7: Use envelope formula to find the approximation of $\mathcal{C}$.

In step 1 the implicit definition $D(h, \mathbf{n}) = 0$ of the support function is obtained by eliminating the variables x, y from (3). In the next step we determine the cusps - equation (5), the extremal points - equations (9) and the inflection points using Proposition 3. We get corresponding points in Step 3 from the envelope formula (1). Then we try to connect the points found in Step 3 using rules from Section 3.2. If this method fails we use a standard CAD based algorithm or additional information, e.g., the curve could be an offset of a known parametric curve, etc. As we have the connectivity of these points we can in Step 5 recover the self-intersections as the intersections of tangent triangles as shown in Proposition 5. The two steps - the approximation is described in Section 4.2.

5.2 Example

In order to demonstrate all features mentioned above, we will use them on the example of the offset at distance $-\frac{9}{10}$ to the ellipse given as the zero set of the bivariate polynomial $f(x, y) = x^2 + 4y^2 - 4$ and oriented by its outer normal.

Eliminating x and y from the system of equations (3)

$$x^2 + 4y^2 = 4 \ ,$$

$$-8yn_1 + 2xn_2 = 0 \ ,$$

$$xn_1 + yn_2 = h \ ,$$

we get the implicit definition of support function of f, $D(h, \mathbf{n}) = h^2 - 4n_1^2 - n_2^2 = 0$. The implicit support function of the offset at distance $-\frac{9}{10}$ is therefore easily evaluated as

$$D(h, \mathbf{n}) = \left(h - \frac{9}{10}\right)^2 - 4n_1^2 - n_2^2 = 0 \ .$$

The condition for cusps given by equation (5) becomes

$$h - \frac{30(n_1^2 - n_2^2)(10h - 9)^2 + 9000n_1^2 n_2^2}{(10h - 9)^3} = 0$$

Table 1. Cusps (C) and extremal points (E) of the offset curve at distance $-\frac{9}{10}$ to the ellipse $x^2 + 4y^2 - 4 = 0$

	type	h	h'	h''	$\mathbf{n}$	corresponding point
P_1	E	$-\frac{11}{10}$	0	$\frac{3}{2}$	$(1,0)$	$(-\frac{11}{10},0)$
P_2	C	-0.7441	0.9039	0.7441	$(0.7535, 0.6575)$	$(-1.155, 0.1918)$
P_3	E	$-\frac{1}{10}$	0	-3	$(0,-1)$	$(0, \frac{1}{10})$
P_4	C	-0.7441	-0.9039	0.7441	$(-0.7535, 0.6575)$	$(1.155, 0.1918)$
P_5	E	$-\frac{11}{10}$	0	$\frac{3}{2}$	$(-1,0)$	$(\frac{11}{10},0)$
P_6	C	-0.7441	0.9039	0.7441	$(-0.7535, -0.6575)$	$(1.155, -0.1918)$
P_7	E	$-\frac{1}{10}$	0	-3	$(0,1)$	$(0, -\frac{1}{10})$
P_8	C	-0.7441	-0.9039	0.7441	$(0.7535, -0.6575)$	$(-1.155, -0.1918)$

and has the 4 solutions listed in Table 1. We determine the extremal points by solving the equations

$$\left(h - \tfrac{9}{10}\right)^2 - 1 = 0 \quad \text{and} \quad \left(h - \tfrac{9}{10}\right)^2 - 4 = 0 \ .$$

These are also in Table 1.

These 8 points $P_1, P_2, \ldots, P_8$ divide the curve into 8 segments. The connectivity is found using rules from Section 3.2. We need only the rule 2, the value of h'' at P_1 is positive and therefore it have to be connected to points on the left from it - there are only two points P_2, P_8. Similarly P_5 is connected to P_4 and P_6. The value of h'' at P_3 is negative and therefore it is connected to points below it, i.e. P_2, P_4. And the same argument is used to connect P_7 to P_6 and P_8. The connectivity is on Fig. 2, left.

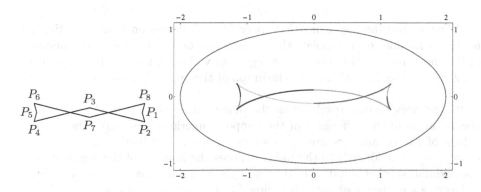

Fig. 2. Left: The graph topologically equivalent to the offset at distance $-\frac{9}{10}$ to the ellipse $x^2 + 4y^2 - 4 = 0$. Right: Its approximation by a spline curve composed of 8 arcs of trigonometric polynomials of degree 3.

For simplicity we use the approximation space of dimension 6

$$\mathcal{A} = \mathrm{Span}\{\sin t, \cos t, \sin 2t, \cos 2t, \sin 3t, \cos 3t\} \ .$$

Solving the system of linear equation (11) we interpolate every arc of the offset by an arc of trigonometric polynomial of degree 3. The resulting spline is on Fig. 2, right.

Table 2 shows the approximation error and its improvement (ratio of two consecutive errors). The error was obtained by sampling the Hausdorff distance, which is, due to Proposition 7, the maximal difference between the support functions. From the table it seems that the improvement of the error converge to 64, i.e., the approximation order is 6. The graphs of error for first few interpolation degrees are shown in Fig. 3.

Table 2. Errors of the interpolation of offset at distance $-\frac{9}{10}$ to the ellipse $x^2+4y^2-4 = 0$ by trigonometric spline coming as a solution of (11).

parts	error	improvement
8	$2.43023 \cdot 10^{-3}$	
16	$4.42354 \cdot 10^{-5}$	54.93871
32	$1.26347 \cdot 10^{-6}$	35.01110
64	$3.66130 \cdot 10^{-8}$	34.50865
128	$6.68349 \cdot 10^{-10}$	54.78136
256	$1.08374 \cdot 10^{-11}$	61.67045
512	$1.71052 \cdot 10^{-13}$	63.35748
1024	$2.64063 \cdot 10^{-15}$	64.77699
2048	$4.16170 \cdot 10^{-17}$	63.45069
4096	$6.48248 \cdot 10^{-19}$	64.19919

When we use the approximation space (10) of dimension 7 and use the last degree of freedom to interpolate the support function at mid-normal (condition (14)), the approximation error for 8 segments will decrease cca. 10 times (from $2.43023 \cdot 10^{-3}$ to $2.12534 \cdot 10^{-4}$). The graph of the approximation error is in Fig. 4, left.

We get very similar result when the degree of freedom is used to minimize the L_2 norm of the difference of the support functions, see (15). The optimal values of the parameters are $e_1 = e_4 = e_5 = e_8 = 2.9805$ and $e_2 = e_3 = e_6 = e_7 = -18.6308$, where the index denotes the number of the segment. The approximation error is $2.03011 \cdot 10^{-4}$ and the graph is shown in Fig. 4, right.

Every arc of the offset curve is enclosed in the tangent triangle due to Proposition 4. Therefore the curve yields a self-intersection only if there is a pair of triangles which have an intersection in way described in Proposition 5. From Fig. 5 we see that there are only two self-intersections and we also know their approximate position in the colored polygons. Using all this information we can construct the topologically equivalent graph to the given curve.

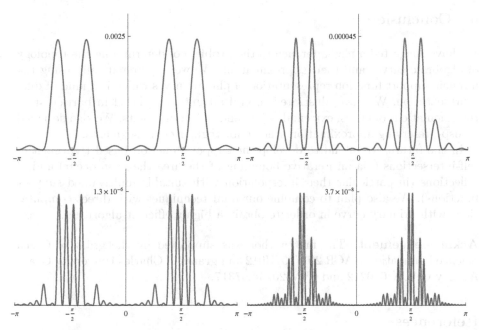

Fig. 3. The approximation error for 8, 16, 32 and 64 segments of spline in approximation space of dimension 6. The points where the error vanishes are the points in which we interpolate the curve.

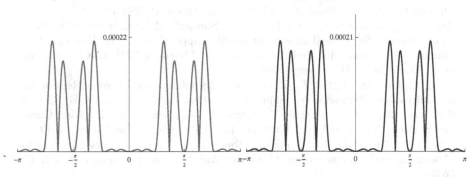

Fig. 4. The graph of the approximation error for 8 segments for different methods of fixing the degree of freedom: left the interpolation of support function at mid-normal, right: the minimization of the L_2 norm of the difference of the support functions.

Fig. 5. Every piece of the curve lies inside the envelope triangle. The self-intersections lies inside the intersection of these triangles.

6   Conclusion

We have suggested a new approach to the problem of determining the topology of algebraic curves and their approximation. We were systematically using the implicit support function representation of planar curves which is a kind of dual representation. We have illustrated several advantages related in particular to the calculations of cusps, extremal points and inflection points. We also designed a cusp-preserving approximation scheme for regular curve segments.

In the future, we intend to develop the support function based treatment of self-intersections (via an iterative bounding of the area they can occur) and of inflections (in particular their interpolation with suitable multivalued support functions). We also plan to combine our dual techniques with direct computations with primary curve in order to obtain a highly efficient algorithm.

Acknowledgement. The first author was supported by the grant of Czech Science Foundation GACR 201/09/H012 and grants of Charles University Grant Agency GAUK 640212 and SVV-2013-267317.

References

1. Aigner, M., Jüttler, B., Gonzalez-Vega, L., Schicho, J.: Parametrizing surfaces with certain special support functions, including offsets of quadrics and rationally supported surfaces. Journal of Symbolic Computation 44, 180–191 (2009)
2. Aigner, M., Gonzalez-Vega, L., Jüttler, B., Sampoli, M.L.: Computing isophotes on free-form surfaces based on support function approximation. In: Hancock, E.R., Martin, R.R., Sabin, M.A. (eds.) Mathematics of Surfaces XIII. LNCS, vol. 5654, pp. 1–18. Springer, Heidelberg (2009)
3. Bastl, B., Lávička, M., Šír, Z.: G^2 Hermite Interpolation with Curves Represented by Multi-valued Trigonometric Support Functions. In: Boissonnat, J.-D., Chenin, P., Cohen, A., Gout, C., Lyche, T., Mazure, M.-L., Schumaker, L. (eds.) Curves and Surfaces 2011. LNCS, vol. 6920, pp. 142–156. Springer, Heidelberg (2012)
4. Alberti, L., Mourrain, B., Wintz, J.: Topology and arrangements computation of semi-algebraic planar curves. Computer Aided Geometric Design 25, 631–651 (2008)
5. Basu, S., Pollack, R., Roy, M.-F.: Algorithms in Real Algebraic Geometry. Springer (2006)
6. Eigenwilling, A., Kerber, M., Wolpert, N.: Fast and Exact Geometric Analysis of Real Algebraic Plane Curves. In: Brown, C.W. (ed.) Proceedings of the 2007 International Symposium on Symbolic and Algebraic Computation (ISSAC 2007), pp. 151–158. ACM (2007)
7. Gonzalez-Vega, L., Necula, I.: Efficient topology determination of implicitly defined algebraic plane curves. Computer Aided Geometric Design 19, 719–749 (2002)
8. Gravesen, J.: Surfaces parametrised by the normals. Computing 79, 175–183 (2007)
9. Gravesen, J., Jütter, B., Šír, Z.: Approximating Offsets of Surfaces by using the Support Function Representation. In: Bonilla, L.L., Moscoso, M., Platero, G., Vega, J.M. (eds.) Progress in Industrial Mathematics at ECMI 2006. Mathematics in Industry, vol. 12, pp. 719–723. Springer (2007)

10. Gravesen, J., Jüttler, B., Šír, Z.: On rationally supported surfaces. Comput. Aided Geom. Design 5(4-5), 320–331 (2008)
11. Gruber, P.M., Wills, J.M.: Handbook of convex geometry. North–Holland, Amsterdam (1993)
12. Hong, H.: An efficient method for analyzing the topology of plane real algebraic curves. Mathematics and Computers in Simulation 42, 571–582 (1996)
13. Cheng, J., Lazard, S., Peñaranda, L., Pouget, M., Rouillier, F., Tsigaridas, E.: On the Topology of Planar Algebraic Curves. In: Proceedings of the 25th Annual Symposium on Computational Geometry, pp. 361–370. ACM, New York (2009)
14. Lávička, M., Bastl, B., Šír, Z.: Reparameterization of curves and surfaces with respect to convolutions. In: Dæhlen, M., Floater, M., Lyche, T., Merrien, J.-L., Mørken, K., Schumaker, L.L. (eds.) MMCS 2008. LNCS, vol. 5862, pp. 285–298. Springer, Heidelberg (2010)
15. Sabin, M.: A Class of Surfaces Closed under Five Important Geometric Operations. Technical Report VTO/MS/207, British Aircraft Corporation (1974), http://www.damtp.cam.ac.uk/user/na/people/Malcolm/vtoms/vtos.html
16. Seidel, R., Wolpert, N.: On the exact computation of the topology of real algebraic curves. In: Proceedings of 21st Annual Symposium on Computational Geometry, pp. 107–115. ACM, New York (2005)
17. Šír, Z., Bastl, B., Lávička, M.: Hermite interpolation by hypocycloids and epicycloids with rational offsets. Computer Aided Geometric Design 27, 405–417 (2010)
18. Šír, Z., Gravesen, J., Jüttler, B.: Curves and surfaces represented by polynomial support functions. Theoretical Computer Science 392, 141–157 (2008)
19. Walker, R.J.: Algebraic Curves. Springer (1978)

Non-regular Surface Approximation

Mira Bozzini[1], Licia Lenarduzzi[2], and Milvia Rossini[3]

[1] Dipartimento di Matematica e Applicazioni
Università di Milano Bicocca
via Cozzi 53 20125 Milano, Italy
`mira.bozzini@unimib.it`
[2] IMATI CNR
via Bassini 15, 20133 Milano, Italy
`licia@mi.imati.cnr.it`
[3] Dipartimento di Matematica e Applicazioni
Università di Milano Bicocca
via Cozzi 53 20125 Milano, Italy
`milvia.rossini@unimib.it`

Abstract. The aim of the paper is to provide a method for approximating non regular surfaces from a set of scattered data in a faithful way. The method we propose is effective and particularly well-suited for recovering geophysical surfaces with faults or drainage patterns. Some real examples will be presented.

Keywords: Discontinuity detection, fault, geophysical surfaces, gradient fault, recovering.

1 Introduction

In this paper we discuss the problem of recovering a non regular surface from a set of scattered data with large size N. By non regular surface, we mean that the function underlying the data or its gradient are discontinuous along a curve.

This topic is of great interest in many problems as, for instance, geophysical applications when one needs to describe the topography of seafloor surfaces, mountain regions and, more in general, the shape of geological entities. In all these cases the surfaces to be recovered may present faults (discontinuity curves for the function) and/or gradient faults (discontinuity curves for the gradient). See for instance [7], [8], [18].

The aim of the paper is to give an effective method to recover geophysical surfaces with non regular structures. In this case, for instance, we have digital elevation maps (DEMs) which are obtained by laser measurements (LIDAR).

Therefore, the problem is to approximate an unknown non-regular bivariate function $f(\mathbf{x})$, $\mathbf{x} \in \Omega \subset \mathbb{R}^2$ by a sample of scattered and noisy data of size N, large but not extra large, i.e. $N < 2^{16}$.

Let S denote the sample

$$S = \{(\mathbf{x}_i, \tilde{f}_i), \ i = 1, \dots, N\}; \tag{1}$$

M. Floater et al. (Eds.): MMCS 2012, LNCS 8177, pp. 68–87, 2014.

the point locations $X = \{\mathbf{x}_i \in \Omega \subset \mathbb{R}^2\}$ are scattered in $\Omega = [0,1]^2$, and the assigned values are such that

$$\tilde{f}_i = f(\mathbf{x}_i) + e_i, \quad i = 1, \ldots, N, \tag{2}$$

where

1. e_i are i.i.d random variables with expected value $E(e_i) = 0$ and unknown covariance matrix $C = \sigma^2 I$, being I the identity matrix of order N. We assume that the noise to signal ratio, $\sigma/\|f\|_2$, is small.
2. The function $f(\mathbf{x})$ or its gradient $\nabla f(\mathbf{x})$ are discontinuous across an unknown curve Γ of Ω and smooth in any neighborhood of Ω which does not intersect Γ.

In particular, we refer to:

i) geophysical surfaces with faults generated by tectonic movements that cause fractures in the ground following piecewise linear paths. For this reason, we assume that the fault Γ is a continuous piecewise linear curve;
ii) digital elevation maps of mountainous districts with valleys shaped like a (non symmetric) $\bigvee$. Usually, through the valley a river flows and, obviously, its drainage pattern is downhill. This means that we need to require monotonicity constraints on the approximation of the river $f(\Gamma)$. In this case, it makes sense to assume that Γ is a C^1 curve.

The problem is very complex because discretely defined surfaces, that exhibit such features, can not be recovered correctly without the knowledge of the discontinuity curve position and of the discontinuity type. Moreover it is not enough to recover Γ faithfully. We need also that the approximation $\hat{\Gamma}$ observes the partition of the sample given by Γ, otherwise the recovering will be poor near the discontinuity curve especially in the case of faults (see Fig. 3). In doing this, as we shall see later, it is fundamental to exploit the information i) and ii) given by the geophysical problem.

We need to solve three sub-problems:

1. To detect the position of the discontinuity curve and to say if it is a fault or a gradient fault (Section 2).
2. To approximate the discontinuity curve (Sections 3.1 and 4.2).
3. To recover the surface and to preserve the irregular structures (Sections 3.2 and 4.3).

Some real examples will be discussed in Section 5.

2   Detection and Classification of the Discontinuity Curve

The importance of detecting the discontinuities curves of a function, is evident also from the literature where we find several methods on the topic. In particular there is a wide literature about fault detection, often referred as edge detection (see for instance [1], [2], [6], [9],[13], [14], [15], [16], [17]). Many of these methods are based on wavelets and multiresolution techniques [1], [15], [16], [17] that are very popular in image processing where we have pixel (gridded) data with very large samples of size at least 2^{16}. On the other hand, we find only some strategy for detecting discontinuities in the gradient when gridded data are given [12], [5].

The technique that we use here is presented in [4]. For reader's convenience, we recall it shortly. It exploits the fact that when we have a set of exact data

$$F_n = \{(\mathbf{z}_{i_n,j_n}, f(\mathbf{z}_{i_n,j_n})) \quad i_n, j_n = 0, \ldots, n\}, \tag{3}$$

placed on a grid $G_n \subset \Omega$ with step-size $h_n = 1/n$

$$G_n = \{\mathbf{z}_{i_n,j_n} = (i_n h_n, j_n h_n), \quad i_n, j_n = 0, \ldots, n\}, \tag{4}$$

the centered differences Δ_n and the isotropic second order differences $\Delta_{1,n}^2$ applied to the data (3)

$$\Delta_n f(\mathbf{z}_{i_n,j_n}) = [\Delta_{x,n} f(\mathbf{z}_{i_n,j_n}), \Delta_{y,n} f(\mathbf{z}_{i_n,j_n})] \tag{5}$$
$$= [f(\mathbf{z}_{i_n+1,j_n}) - f(\mathbf{z}_{i_n-1,j_n}), f(\mathbf{z}_{i_n,j_n+1}) - f(\mathbf{z}_{i_n,j_n-1})], \tag{6}$$

$$\Delta_{1,n}^2 f(\mathbf{z}_{i_n,j_n}) = \sum_{l,r=-1}^{1} \gamma_{l,r} f(\mathbf{z}_{i_n+l,j_n+r}), \quad \gamma = \frac{1}{6} \begin{pmatrix} 1 & 4 & 1 \\ 4 & -20 & 4 \\ 1 & 4 & 1 \end{pmatrix}, \tag{7}$$

are able to characterize the grid points $\mathbf{z}_{i_n,j_n}$ near the curve Γ. Namely, let Q_{i_n,j_n} denote the square $[(i_n-1)h_n, (i_n+1)h_n] \times [(j_n-1)h_n, (j_n+1)h_n]$ and A_{i_n,j_n} the set of points at which the discontinuity curve intersects the horizontal, vertical and diagonal directions of the square Q_{i_n,j_n}. For simplicity, we assume that in any Q_{i_n,j_n}, Γ intersects each direction only one time at most. The possible intersection points are denoted as

$$\boldsymbol{\xi}_o = (\xi_{ox}, \xi_{oy}) = \Gamma \bigcap \overline{\mathbf{z}_{i_n-1,j_n} \mathbf{z}_{i_n+1,j_n}},$$

$$\boldsymbol{\xi}_v = (\xi_{vx}, \xi_{vy}) = \Gamma \bigcap \overline{\mathbf{z}_{i_n,j_n-1} \mathbf{z}_{i_n,j_n+1}},$$

$$\boldsymbol{\xi}_{d_1} = (\xi_{d_1x}, \xi_{d_1y}) = \Gamma \bigcap \overline{\mathbf{z}_{i_n-1,j_n-1} \mathbf{z}_{i_n+1,j_n+1}},$$

$$\boldsymbol{\xi}_{d_2} = (\xi_{d_2x}, \xi_{d_2y}) = \Gamma \bigcap \overline{\mathbf{z}_{i_n-1,j_n+1} \mathbf{z}_{i_n+1,j_n-1}}.$$

We indicate with

$$\text{jump}_x f|_{\boldsymbol{\xi}_o} = f(\xi_{ox}^+, j_n h_n) - f(\xi_{ox}^-, j_n h_n), \text{jump}_y f|_{\boldsymbol{\xi}_v} = f(i_n h_n, \xi_{vy}^+) - f(i_n h_n, \xi_{vy}^-),$$

and respectively with $\text{jump} f|_{Q_{i_n,j_n}}$ and $\text{jump} \nabla f|_{Q_{i_n,j_n}}$ a weighted average of the possible jumps of f and ∇f along the four directions in the square Q_{i_n,j_n}.

We have that (see [4]):

1. if Γ intersects $(\mathbf{z}_{i_n-1,j_n}, \mathbf{z}_{i_n+1,j_n})$ in $\boldsymbol{\xi}_o$

$$\Delta_{x,n} f(\mathbf{z}_{i_n,j_n}) = \begin{cases} \operatorname{jump}_x f|_{\boldsymbol{\xi}_o} + O(h_n), & \text{if } \Gamma \text{ is a fault} \\ O(h_n) & \text{if } \Gamma \text{ is a gradient fault}, \end{cases} \quad (8)$$

while

$$\Delta_{x,n} f(\mathbf{z}_{i_n,j_n}) = O(h_n) \quad \text{when} \quad \Gamma \bigcap (\mathbf{z}_{i_n-1,j_n}, \mathbf{z}_{i_n+1,j_n}) = \emptyset; \quad (9)$$

2. if Γ intersects $(\mathbf{z}_{i_n,j_n-1}, \mathbf{z}_{i_n,j_n+1})$ in $\boldsymbol{\xi}_v$

$$\Delta_{y,n} f(\mathbf{z}_{i_n,j_n}) = \begin{cases} \operatorname{jump}_y f|_{\boldsymbol{\xi}_v} + O(h_n), & \text{if } \Gamma \text{ is a fault} \\ O(h_n), & \text{if } \Gamma \text{ is a gradient fault} \end{cases} \quad (10)$$

while

$$\Delta_{y,n} f(\mathbf{z}_{i_n,j_n}) = O(h_n) \quad \text{when} \quad \Gamma \bigcap (\mathbf{z}_{i_n,j_n-1}, \mathbf{z}_{i_n,j_n+1}) = \emptyset; \quad (11)$$

3. if $A_{i_n,j_n} \neq \emptyset$ and at least one of its point is an interior point of Q_{i_n,j_n}

$$\Delta_{1,n}^2 f(\mathbf{z}_{i_n,j_n}) = \begin{cases} \operatorname{jump} f|_{Q_{i_n,j_n}} + O(h_n), & \text{if } \Gamma \text{ is a fault} \\ \operatorname{jump} \nabla f|_{Q_{i_n,j_n}} h_n + O(h_n^2), & \text{if } \Gamma \text{ is a gradient fault}, \end{cases} \quad (12)$$

otherwise

$$\Delta_{1,n}^2 f(\mathbf{z}_{i_n,j_n}) = O(h_n^2). \quad (13)$$

Now, given a sample of scattered data S of size N, we construct gridded pseudo-data

$$S_{G_n} = \{ (\mathbf{z}_{i_n,j_n}, \tilde{u}_{i_n,j_n}), \quad i_n, j_n = 0, \ldots, n \}. \quad (14)$$

Namely, we consider a suitable step-size $h_n = 1/n$, the associated grid (4) and an integer $n_0 << N$.

For each grid point, we indicate with $\mathcal{U}_{i_n,j_n}$ the circular neighborhood centered at $\mathbf{z}_{i_n,j_n}$ containing n_0 points $\mathbf{x}_k^{i_n,j_n} \in X$, with $\tilde{f}_k^{i_n,j_n}$ the corresponding sample values, and with μ_{i_n,j_n} its radius. We define $\tilde{u}_{i_n,j_n}$ to be the average

$$\tilde{u}_{i_n,j_n} = \frac{1}{n_0} \sum_{k=1}^{n_0} \tilde{f}_k^{i_n,j_n}. \quad (15)$$

It is worthwhile to remark that in this way we smooth the noise corrupting the data; in fact the random variables $\tilde{u}_{i_n,j_n}$ have expected values

$$E(\tilde{u}_{i_n,j_n}) = \frac{1}{n_0} \sum_{k=1}^{n_0} f(\mathbf{x}_k^{i_n,j_n})$$

and variances

$$Var(\tilde{u}_{i_n,j_n}) = \frac{\sigma^2}{n_0}.$$

By applying the discrete operators Δ_n, $\Delta^2_{1,n}$ to (14), we obtain the estimators

$$\Delta_{x,n}\tilde{u}_{i_n,j_n}, \quad \Delta_{y,n}\tilde{u}_{i_n,j_n}, \quad \Delta^2_{1,n}\tilde{u}_{i_n,j_n}, \quad i_n, j_n = 1, \ldots, n-1. \qquad (16)$$

In [4], it has been proven that if Γ is a fault, their expected values are

$$E(\Delta_{x,n}\tilde{u}_{i_n,j_n}) = \begin{cases} C^x_{i_n,j_n} + O(h_n), & \text{if } \Gamma \bigcap (\mathbf{z}_{i_n-1,j_n}, \mathbf{z}_{i_n+1,j_n}) \neq \emptyset \\ O(h_n), & \text{if } \Gamma \bigcap (\mathbf{z}_{i_n-1,j_n}, \mathbf{z}_{i_n+1,j_n}) = \emptyset, \end{cases} \qquad (17)$$

$$E(\Delta_{y,n}\tilde{u}_{i_n,j_n}) = \begin{cases} C^y_{i_n,j_n} + O(h_n), & \text{if } \Gamma \bigcap (\mathbf{z}_{i_n,j_n-1}, \mathbf{z}_{i_n,j_n+1}) \neq \emptyset \\ O(h_n), & \text{if } \Gamma \bigcap (\mathbf{z}_{i_n,j_n-1}, \mathbf{z}_{i_n,j_n+1}) = \emptyset, \end{cases} \qquad (18)$$

and

$$E(\Delta^2_{1,n}\tilde{u}_{i_n,j_n}) = \begin{cases} O(h_n^2), & \text{if } A_{i_n,j_n} = \emptyset \\ C^1_{i_n,j_n} + O(h_n), & \text{if } A_{i_n,j_n} \neq \emptyset \end{cases} \qquad (19)$$

being $C^x_{i_n,j_n}$, $C^y_{i_n,j_n}$, and $C^1_{i_n,j_n}$ constants depending on the discontinuity jumps of f at some points of Q_{i_n,j_n}.

Instead, if Γ is a gradient fault, we have that

$$E(\Delta_{x,n}\tilde{u}_{i_n,j_n}) = O(h_n), \qquad E(\Delta_{y,n}\tilde{u}_{i_n,j_n}) = O(h_n), \qquad (20)$$

and

$$E(\Delta^2_{1,n}\tilde{u}_{i_n,j_n}) = \begin{cases} O(h_n^2), & \text{if } A_{i_n,j_n} = \emptyset \\ D^1_{i_n,j_n} h_n + O(h_n^2), & \text{if } A_{i_n,j_n} \neq \emptyset, \end{cases} \qquad (21)$$

where $D^1_{i_n,j_n}$ is a constant depending on the discontinuity jumps of ∇f at some points of Q_{i_n,j_n}.

The variances of (16) are such that

$$Var(\Delta_{x,n}\tilde{u}_{i_n,j_n}) \leq \frac{2\sigma^2}{n_0}, \quad Var(\Delta_{y,n}\tilde{u}_{i_n,j_n}) \leq \frac{2\sigma^2}{n_0}, \qquad (22)$$

$$Var(\Delta^2_{1,n}\tilde{u}_{i_n,j_n}) \leq \frac{13\sigma^2}{n_0}. \qquad (23)$$

We now study the asymptotic behavior of (16). For $N \to \infty$ and $n \to \infty$, we consider a sequence of nested grids $G_{\bar{n}} \subset \cdots \subset G_n \subset G_{n+1} \ldots$, $n = 2^i\bar{n}$, $h_n = h_{\bar{n}}/2^i$ with $i = 0, 1, 2, \ldots$, and the associated sets

$$\widetilde{\Delta}_{x,n} = \{\Delta_{x,n}\tilde{u}_{i_n,j_n}, \quad i_n, j_n = 1, \ldots, n-1\}, \qquad (24)$$

$$\widetilde{\Delta}_{y,n} = \{\Delta_{y,n}\tilde{u}_{i_n,j_n}, \quad i_n, j_n = 1, \ldots, n-1\}, \qquad (25)$$

$$\widetilde{\Delta^2}_{1,n} = \{\Delta^2_{1,n}\tilde{u}_{i_n,j_n}, \quad i_n, j_n = 1, \ldots, n-1\}. \qquad (26)$$

Fixed a grid $G_{\hat{n}}$, we take the $\hat{j}_{\hat{n}}$th row (and respectively the $\hat{i}_{\hat{n}}$th column). Let $\boldsymbol{\eta}_{\hat{y}} = (x, \hat{y})$ be a point on the line $y = \hat{y}$ (and correspondingly let $\boldsymbol{\eta}_{\hat{x}} = (\bar{x}, y)$ be a point on the line $x = \hat{x}$). For $n > \hat{n}$, the sequence of nested grids G_n determines on $y = \hat{y}$ a sequence of nested intervals $I_{\eta_{\hat{y}},n} = [\mathbf{z}_{i_n-1,j_n}, \mathbf{z}_{i_n+1,j_n}]$ containing $\eta_{\hat{y}}$ (and correspondingly on $x = \hat{x}$ a sequence $I_{\eta_{\hat{x}},n} = [\mathbf{z}_{i_n,j_n-1}, \mathbf{z}_{i_n,j_n+1}]$ containing $\eta_{\hat{x}}$). We indicate with $\widetilde{\Delta}_{x,n}(\eta_{\hat{y}})$ and $\widetilde{\Delta}_{y,n}(\eta_{\hat{x}})$ the sequences of the centered differences of $\tilde{u}_{i_n,j_n}$ associated with $I_{\eta_{\hat{y}},n}$ and $I_{\eta_{\hat{x}},n}$.

Analogously, fixed a point η of Ω, we denote with $Q_{\eta,n} = [(i_n-1)h_n, (i_n+1)h_n] \times [(j_n-1)h_n, (j_n+1)h_n]$ a sequence of nested squares containing η and we indicate with $\widetilde{\Delta}^2_{1,n}(\eta)$ the sequence of the isotropic difference of $\tilde{u}_{i_n,j_n}$ associated with $Q_{\eta,n}$.

Asymptotically, the following results hold.

Proposition 1. *When $N \to \infty$, $h_n \to 0$, $n_0 \to \infty$ so that $\sqrt{N}h_n \to \infty$, $\mu_{i_n,j_n} \to 0$, $\mu_{i_n,j_n}n_0 \to \infty$, and $h_n n_0 \to \infty$, we have that in probability*

1. if $\eta_{\hat{y}} \in \Gamma$ and Γ is a fault curve

$$\frac{\widetilde{\Delta}_{x,n}(\eta_{\hat{y}})}{h_n} \to \infty \qquad (27)$$

while if $\eta_{\hat{y}} \in \Gamma$ and Γ is a gradient curve or $\eta_{\hat{y}} \notin \Gamma$

$$\frac{\widetilde{\Delta}_{x,n}(\eta_{\hat{y}})}{h_n} \to K^x_{\eta_{\hat{y}}}; \qquad (28)$$

2. if $\eta_{\hat{x}} \in \Gamma$ and Γ is a fault curve,

$$\frac{\widetilde{\Delta}_{y,n}(\eta_{\hat{x}})}{h_n} \to \infty \qquad (29)$$

while if $\eta_{\hat{x}} \in \Gamma$ and Γ is a gradient curve or $\eta_{\hat{x}} \notin \Gamma$

$$\frac{\widetilde{\Delta}_{y,n}(\eta_{\hat{x}})}{h_n} \to K^y_{\eta_{\hat{x}}} \qquad (30)$$

3. if $\eta \in \Gamma$ and Γ is a fault curve

$$\frac{\widetilde{\Delta}^2_{1,n}(\eta)}{h_n} \to \infty; \qquad (31)$$

4. if $\eta \in \Gamma$ and Γ is a gradient curve

$$\frac{\widetilde{\Delta}^2_{1,n}(\eta)}{h_n} \to C_\eta; \qquad (32)$$

5. if $\eta \notin \Gamma$

$$\frac{\widetilde{\Delta}^2_{1,n}(\eta)}{h_n} \to 0. \qquad (33)$$

In practice, we work with a given sample and consequently a fixed h_n. The previous results suggest how to perform the detection. First, we look for the position of the possible fault curve by considering and analyzing the quantities $\Delta_{x,n}\tilde{u}_{i_n,j_n}/h_n$ and $\Delta_{y,n}\tilde{u}_{i_n,j_n}/h_n$, $i_n, j_n = 1, \ldots, n-1$.

Proposition 1 allows us to say that they assume "big" values at the points $\mathbf{z}_{i_n,j_n}$ near the fault, otherwise they have "small" values with respect to the previous ones.

The set of possible fault points ("big values") can be detected by a classification method which separates the two different class of points ("big values", "small values"). For instance, this can be done by fixing threshold values depending on the range of $\Delta_{x,n}\tilde{u}_{i_n,j_n}/h_n$ and $\Delta_{y,n}\tilde{u}_{i_n,j_n}/h_n$.

We have to remark that, in the selected class, we can have also set of points for which $\frac{\Delta_{x,n}\tilde{u}_{i_n,j_n}}{h_n}$ and/or $\frac{\Delta_{y,n}\tilde{u}_{i_n,j_n}\tilde{u}_{i_n,j_n}}{h_n}$ are "big" because the constants $K^x_{\eta_{\hat{x}}}$ and/or $K^y_{\eta_{\hat{y}}}$ are large. These points correspond to high gradients.

Typically, high gradient points lie in a bidimensional region R of Ω and they do not follow the behavior of a curve of Ω. This allows us to discriminate them.

In Fig. 1, it is shown a function with a fault and a zone of high gradients (left). On the right, we can see the points that follows a curve and the region corresponding the the sharp variation of f. Here we have considered a sample of $N = 4900$ scattered points. In Fig. 2, we have considered the rapidly varying data corresponding to the seafloor surface of one of the deepest parts of the Tonga Trench [8]. On the left, it is shown a view of the Tonga Trench data set ($N = 8113$ gridded data) and on the right we can see that the detected points belong to a region and do not follow a curve. Hence they correspond to high gradients.

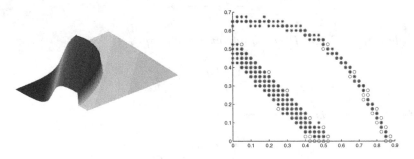

Fig. 1. $N = 4900$, $h_n = 1/40$, $\frac{\sigma}{\|f\|^2} \sim 0.015$

We indicate by

$$D_f = \{(\mathbf{z}_l^f, \tilde{u}_l^f), \, l = 1, \ldots, n_f\} \tag{34}$$

the points of S_{G_n} corresponding to the locations $\mathbf{z}_l^f$ detected as fault points. The unknown curve Γ is classified as fault.

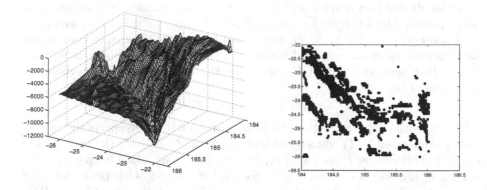

Fig. 2. Tonga example

In a second step, we look for the gradient fault. We consider the set $S_{G_n} \setminus D_f$. As before, we select from it the points at which $\frac{\Delta^2_{1,n} \tilde{u}_{i_n,j_n}}{h_n}$ takes "big values". We call

$$D_g = \{(\mathbf{z}^g_l, \tilde{u}^g_l), l = 1, \ldots, n_g\} \tag{35}$$

the points of S_{G_n} corresponding to the locations $\mathbf{z}^g_l$ detected as gradient fault points. The unknown curve Γ is classified as a gradient fault.

In both cases the distances between the detected points and the curve is less or equal to h_n

$$d(\mathbf{z}^f_l, \Gamma) \le h_n, \quad d(\mathbf{z}^g_l, \Gamma) \le h_n. \tag{36}$$

3 Recovering a Surface with a Fault

3.1 Approximation of the Fault Line Γ

According to assumption i) of the Introduction, the fault Γ is a C^0 piecewise linear curve.

Let us assume that the fault curve runs from west to east and that it divides Ω into the disjoint subsets Ω_N and Ω_S. It also separates the given sample S in two disjoint sub-samples S_N and S_S.

Having used the centered differences, the points $\mathbf{z}^f_l$, $l = 1, \ldots, n_f$, detect a stripe $\mathfrak{S}$ of Ω where most likely Γ lies.

The construction of the approximation $\hat{\Gamma}$ of Γ, starts with a first rough approximation that will be used to find the slope changes.

We construct a new set of points $\mathbf{z}^*_j \in \mathfrak{S}$ by associating to each $\mathbf{z}^f_l$ the average with the points $\mathbf{z}^f_i$ whose distance from $\mathbf{z}^f_l$ is less or equal to $\sqrt{2}h_n$ and, in order to obtain a behavior more close to the piecewise linear one, we apply again the average procedure to the new set with a distance less than h_n. Let

$$Z^* = \{\mathbf{z}^*_j(x^*_j, y^*_j) \quad j = 1, \ldots, n^*\}$$

be the so obtained set ordered with respect to $x_j^\star$ and purified by eventual coincident points. The set $Z^\star$ has a global piecewise linear behavior and we detect those points $\bar{\mathbf{z}}_i \in Z^\star$, $i = 1, \ldots, \bar{m}$, corresponding to slope changes by considering the locations where the second differences of the $Z^\star$ elements move away from zero. A first approximation of Γ is given by the linear spline $\hat{\Gamma}^1$ interpolating the points of the set

$$\bar{Z} = \{\mathbf{z}_1^\star, \bar{\mathbf{z}}_1, \bar{\mathbf{z}}_2 \ldots, \bar{\mathbf{z}}_{\bar{m}}, \mathbf{z}_{n^\star}^\star\}.$$

It is of crucial importance that the final approximation $\hat{\Gamma}$ respects the sample classification given by Γ. Namely, we indicate with $\hat{\Omega}_N$ and $\hat{\Omega}_S$ the two parts in which Ω is divided by $\hat{\Gamma}$, and with $\hat{S}_N$ and $\hat{S}_S$ the sample data having $\mathbf{x}_i \in \hat{\Omega}_N$ and $\mathbf{x}_j \in \hat{\Omega}_S$. We need that $S_N \equiv \hat{S}_N$ and $S_S \equiv \hat{S}_S$. Otherwise, we have undesired oscillations in the final recovering as shown on the left of Fig. 3. Here we have considered the test function (55) of example 1 (Section 5) where we approximate f by using $\hat{\Gamma}^1$ (on the right) which does not separate the sample points correctly.

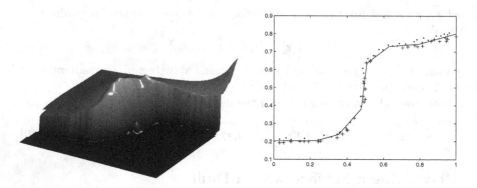

Fig. 3. Left: Approximation of (55). Right: the approximation $\hat{\Gamma}^1$ of Γ, + sample locations belonging to $\Omega_S \bigcap \mathfrak{S}$, · sample locations belonging to $\Omega_N \bigcap \mathfrak{S}$.

Obviously, a wrong classification can happen in $\mathfrak{S}$. Then it is necessary to establish whether a sample point with $\mathbf{x}_i \in \mathfrak{S}$ belongs to S_S or to S_N. For this purpose, we take the set S^1 of the points in S such that the distance $d(\mathbf{x}_i, \mathbf{z}_l^f) \leq h_n$ for some $\mathbf{z}_l^f \in D_f$.

By considering all the possible distances between the elements of S_1, and using standard algorithms of cluster analysis, it is possible to divide S^1 in two classes S_N^1 and S_S^1 whose locations $\mathbf{x}_i$ are in Ω_N and Ω_S respectively.

Now, by using S_N^1 and S_S^1, we verify whether $\hat{\Gamma}^1$ respects the classification, otherwise we recursively modify it segment by segment maintaining the continuity.

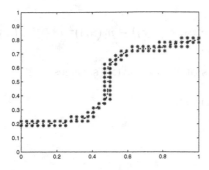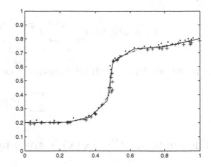

Fig. 4. Example 1 (Section 5). Left. $\star$: the detected points z_i^f, diamond: the points of Z^*. Right. $-$ the final approximation $\hat{\Gamma}$, $\cdot -$ the true fault Γ.

In Fig. 4, it is shown the final approximation $\hat{\Gamma}$ of Γ. The maximum error and the root mean least squares error computed on a grid of 200 point are $e_\infty = 0.046$ and $e_2 = 0.007$ respectively.

3.2   Recovering

The approximating surface $\tilde{s}(\mathbf{x})$ consists of two approximations

$$\tilde{s}_N(\mathbf{x}), \ \mathbf{x} \in \hat{\Omega}_N \quad \text{and} \quad \tilde{s}_S(\mathbf{x}), \ \mathbf{x} \in \hat{\Omega}_S.$$

Each of them is constructed by using the data points S_N and S_S respectively.

It is well-known that radial basis functions are a powerful tool when dealing with scattered data. Among the possible choices, we have considered the space spanned by the shifts of the C^1 Thin-Plate Spline (TPS)

$$v_2(x) = \frac{1}{16\pi} \|\mathbf{x}\|^2 \ln(\|\mathbf{x}\|^2)$$

because the surface to be recovered presents a non smooth behavior and because the TPS is a stable bases and enjoys scale invariance. Moreover, when recovering non-smooth features, the local approximation error is of the order of the scale squared.

With this choice, the approximation, in each sub-domain, is given by a convex linear combination of polyharmonic functions that locally satisfy to the least squares principle.

We fix a set of centers $Y = \{\mathbf{y}_i, \ i = 1, \ldots, N_0\}$ placed on a uniform grid of step-size $\bar{h} = k/\sqrt{N}$, (typically $k = 2, 3$.)

We indicate with $Y_N = Y \cap \Omega_N$ the N_{0N} centers in Ω_N and with $Y_S = Y \cap \Omega_S$ those in Ω_S, $|Y_S| = N_{0S}$.

For each center $\mathbf{y}_i \in Y_N$, we consider the local TPS approximations

$$\tilde{s}_i^N(\mathbf{x}) = \sum_{\mathbf{y}_j \in U_i} \tilde{c}_j^i v_2(\mathbf{x} - \mathbf{y}_j) + \tilde{p}_1^i(\mathbf{x}), \quad \mathbf{x} \in U_i \tag{37}$$

where

$$\tilde{c}_j^i = \arg\min_{c_j} \sum_{\mathbf{x}_l \in U_i} (\tilde{f}_l - \sum_{\mathbf{y}_j \in U_i} c_j^i v_2(\mathbf{x}_l - \mathbf{y}_j) - \tilde{p}_1^i(\mathbf{x}_l))^2. \tag{38}$$

The local set U_i is such to contain 9 centers of Y_N and $\tilde{p}_1^i(\mathbf{x})$ satisfies

$$\sum_{\mathbf{y}_j \in U_i} \tilde{c}_j^i \tilde{p}_1^i(\mathbf{y}_j) = 0.$$

The approximation $\tilde{s}_N(\mathbf{x})$ is given by

$$\tilde{s}_N(\mathbf{x}) = \sum_{i=1}^{N_{ON}} w_i(\mathbf{x})\tilde{s}_i^N(\mathbf{x}), \tag{39}$$

where $w_i(\mathbf{x})$ are interpolatory weights such that $\sum_{i=1}^{N_{ON}} w_i(\mathbf{x}) = 1$.

In the same way, considering the centers $\mathbf{y}_i \in Y_S$, we obtain $\tilde{s}_S(\mathbf{x})$. The final approximation is

$$\tilde{s}(\mathbf{x}) = \begin{cases} \tilde{s}_N(\mathbf{x}), & \text{if } \mathbf{x} \in \hat{\Omega}_N \\ \tilde{s}_S(\mathbf{x}), & \text{if } \mathbf{x} \in \hat{\Omega}_S. \end{cases} \tag{40}$$

4   Recovering of a Surface with a Gradient Fault

As said in Section 1, we mainly refer to digital elevation maps of mountain regions where we have gradient faults related to river flows behaving. Of course the procedure can be applied also to surfaces shaped like (non symmetric) $\wedge$ near the gradient fault.

4.1   Morphometric Hierarchy of the Branches of a River

In general the river flow, represented by the curve Γ, can be formed by two or more branches. The different branches are numbered hierarchically according to the following scheme. We define the branch order $\mathfrak{n}$ equal to one if it rises from a source and equal to $\mathfrak{n}+1$ if it rises in the confluence of branches of order $\mathfrak{n}$. If a branch of order $\mathfrak{n}$ meets a branch of order $\mathfrak{m} < \mathfrak{n}$, after the confluence the order remains $\mathfrak{n}$. By this hierarchical definition, it is possible to number the different branches Γ_b, of Γ with $b = 1, \ldots, B$ (see for instance Fig. 5 on the left) and then

$$\Gamma = \bigcup_{b=1}^{B} \Gamma_b.$$

The confluence points are individuated by visual inspection. This greatly reduces the possibility of taking wrong points.

Let R_j, $j = 1, \ldots, J$ denote the regions of Ω bounded by contiguous branches of Γ and ordered, for instance, clockwise starting from Γ_1 (see Fig. 5 on the right).

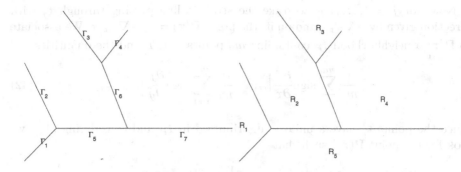

Fig. 5.

4.2   Approximation of the Gradient Fault

We consider the points $\mathbf{z}_i^g$ to construct a larger set of points $\mathcal{T}$ which mimic better the behavior of Γ. To this aim, we exploit the information given by the problem. Our surface has a gradient fault and the region containing it, is shaped like a (non symmetric) $\bigvee$.

Let us consider a point $\boldsymbol{\xi}$ on Γ and a local ball $B_{\boldsymbol{\xi}}$ centered at $\boldsymbol{\xi}$ which is divided by Γ in two disjoint parts $B_{\boldsymbol{\xi}}^N$ and $B_{\boldsymbol{\xi}}^S$. For any $\mathbf{x}_1 \in B_{\boldsymbol{\xi}}^N$, and $\mathbf{x}_2 \in B_{\boldsymbol{\xi}}^N$, we have that the gradient components at $\mathbf{x}_1$ and $\mathbf{x}_2$ have opposite signs

$$\frac{\partial f}{\partial x}\big|_{x_1}\, \frac{\partial f}{\partial x}\big|_{x_2} < 0 \quad \text{and} \quad \frac{\partial f}{\partial y}\big|_{x_1}\, \frac{\partial f}{\partial y}\big|_{x_2} < 0,$$

and the sign of each component of ∇f doesn't change in $B_{\boldsymbol{\xi}}^N$ and $B_{\boldsymbol{\xi}}^S$ respectively.

Then if we take p points ς_i with a uniform distribution in $B_{\boldsymbol{\xi}}$, we have

$$E\left(\frac{1}{p}\sum_{i=1}^{p} \text{sign}\frac{\partial f}{\partial x}\big|_{\varsigma_i}\right) = 0, \quad E\left(\frac{1}{p}\sum_{i=1}^{p} \text{sign}\frac{\partial f}{\partial y}\big|_{\varsigma_i}\right) = 0. \tag{41}$$

This simple remark, allows us to perform an algorithm to find $\mathcal{T}$.

We consider the points

$$T = \{\mathbf{t}_j, \quad j = 1,\ldots,n_T\}$$

of a grid of Ω with step-size $h < h_n$, whose distance from a one point $\mathbf{z}_i^g$, one at least, is less than a prescribed $H > h_n$.

We evaluate the signs of the gradient component at the points of T by applying the usual discretization of the gradient to the set $\{(\mathbf{t}_j, \tilde{s}_c(\mathbf{t}_j))\}_{j=1}^{n_T}$, where $\tilde{s}_c(\mathbf{x})$ is the Thin-Plate Spline approximation (45) that will be used also in the final recovering of f.

Now, for $j = 1, \ldots, n_T$, we take the straight line passing through $\mathbf{t}_j$ with direction given by $-\nabla s|_{\mathbf{t}_j}$, and on it, the point $\mathbf{P}(\tau) = \mathbf{t}_j - \nabla s|_{\mathbf{t}_j} \tau$. We associate to $\mathbf{P}(\tau)$ a neighborhood I_P containing m_0 points $\mathbf{t}_i^j \in T$ and the quantities

$$\frac{1}{m_0} \sum_{i=1}^{m_0} \text{sign} \frac{\partial f}{\partial x}\Big|_{\mathbf{t}_i^j}, \quad \frac{1}{m_0} \sum_{i=1}^{m_0} \text{sign} \frac{\partial f}{\partial y}\Big|_{\mathbf{t}_i^j}. \tag{42}$$

Since the points $\mathbf{t}_i^j$ have a uniform distribution in I_P and considering (41), we look for the point $\mathbf{P}(\bar{\tau}^j)$ such that

$$\left| \frac{1}{m_0} \sum_{i=1}^{m_0} \text{sign} \frac{\partial f}{\partial x}\Big|_{\mathbf{t}_i^j} \right| \leq \varepsilon(m_0), \quad \left| \frac{1}{m_0} \sum_{i=1}^{m_0} \text{sign} \frac{\partial f}{\partial y}\Big|_{\mathbf{t}_i^j} \right| \leq \varepsilon(m_0), \tag{43}$$

where $\varepsilon(m_0) \to 0$ when $m_0 \to \infty$.

The approximation $\hat{\Gamma}$ of Γ is obtained by fitting in the least squares sense the points $\mathcal{T} = \{\mathbf{P}(\bar{\tau}^j) \in \Omega, \ j = 1, \ldots, n_T\}$, by the shifts of a multiquadric basis having small variable parameter chosen, as suggested in [3], in order to have a shape preserving curve.

We indicate with $\hat{\Gamma}$ the union of the different branches $\hat{\Gamma}_b$ each defined in an interval I_b

$$\hat{\Gamma} = \bigcup_{b=1}^{B} \hat{\Gamma}_b. \tag{44}$$

4.3   Recovering of the Surface

The approximation $\tilde{s}(\mathbf{x})$ of $f(\mathbf{x})$ is obtained by a blending of two surfaces $\tilde{s}_c(\mathbf{x})$ and $\tilde{s}_v(\mathbf{x})$

$$\tilde{s}(\mathbf{x}) = (1 - w(\mathbf{x}))\tilde{s}_c(\mathbf{x}) + w(\mathbf{x})\tilde{s}_v(\mathbf{x}).$$

$\tilde{s}_c(\mathbf{x})$ is a C^1 approximation obtained on the whole domain Ω that over-smooths near $\hat{\Gamma}$, while $\tilde{s}_v(\mathbf{x})$ is a C^0 function constructed in a sheath $\mathcal{G}$ containing $\hat{\Gamma}$ that approximates the behavior of the unknown surface near Γ.

The weight $w(\mathbf{x})$ is constructed in order to localize strongly the recovering near the gradient fault.

Construction of $\tilde{s}_c(\mathbf{x})$. The smooth approximation is obtained, as before (Section 3.2), considering again the set of centers $Y = \{\mathbf{y}_i, \ i = 1, \ldots, N_0\}$ placed on the uniform grid of step-size $\bar{h}$ in Ω.

For $i = 1, \ldots, N_0$ we consider the local approximations

$$\tilde{s}_i(\mathbf{x}) = \sum_{\mathbf{y}_j \in U_i} \tilde{c}_j^i v_2(\mathbf{x} - \mathbf{y}_j) + \tilde{p}_1^i(\mathbf{x}), \quad \mathbf{x} \in U_i$$

where

$$\tilde{c}_j^i = \arg \min_{c_j} \sum_{\mathbf{x}_l \in U_i} (\tilde{f}_l - \sum_{\mathbf{y}_j \in U_i} c_j^i v_2(\mathbf{x}_l - \mathbf{y}_j) - \tilde{p}_1^i(\mathbf{x}_l))^2,$$

the local set U_i is such to contain 9 centers of Y and $\tilde{p}_1^i(\mathbf{x})$ satisfies

$$\sum_{\mathbf{y}_j \in U_i} \tilde{c}_j^i \tilde{p}_1^i(\mathbf{y}_j) = 0.$$

The smooth approximation is given by

$$\tilde{s}_c(\mathbf{x}) = \sum_{i=1}^{N_0} w_i^c(\mathbf{x}) \tilde{s}_i(\mathbf{x}), \tag{45}$$

where $w_i^c(\mathbf{x})$ are interpolatory weights such that $\sum_{i=1}^{N_0} w_i^c(\mathbf{x}) = 1$.
$\tilde{s}_c(\mathbf{x})$ belongs to $C^1(\Omega)$ and does not recover correctly the discontinuity while it provides a good approximation away from Γ.

Construction of $\tilde{s}_v(\mathbf{x})$. The construction of the C^0 approximation $\tilde{s}_v(\mathbf{x})$ requires to define a region $\mathcal{G}$ that we call sheath in which $\tilde{s}_v(\mathbf{x})$ is defined.

Firstly, by an iterative procedure, we find a uniform partition $\{t_{b,i}\}_{i=0}^{n_b}$ on each interval I_b so that the linear spline s_b connecting the corresponding points on $\hat{\Gamma}_b$, satisfies

$$\|s_b - \hat{\Gamma}_b\|_\infty < \delta(N), \tag{46}$$

where $\delta(N)$ is a prefixed tolerance such that $\delta(N) \to 0$ and $N\delta(N) \to \infty$ when $N \to \infty$.

We now consider the line

$$\tilde{\Gamma} = \bigcup_{b=1}^{B} s_b, \tag{47}$$

and we indicate with $\hat{R}_j$, $j = 1, \ldots, J$, the regions bounded by contiguous linear splines.

Secondly, we find a sheath containing $\tilde{\Gamma}$.

We indicate with $\overline{s_b(t_{b,i})s_b(t_{b,i+1})}$ a generic segment of s_b and with $\mu_{b,i}$ its measure. We consider the circle $C_{i,b}$ of radius $r_{b,i} > \mu_{b,i}/2$ centered at $\frac{s_b(t_{b,i}) + s_b(t_{b,i+1})}{2}$. The linear spline s_b cuts $C_{i,b}$ in two parts: one belonging to the region $\hat{R}_j$ and the other to the region $\hat{R}_{\bar{j}}$. We indicate with $C_{i,b}^j$ the part of $C_{i,b}$ in $\hat{R}_j$ and with $C_{i,b}^{\bar{j}}$ the part in $\hat{R}_{\bar{j}}$.

Let $\mathcal{G}_j$ be the union, with respect to i and b, of all the portions $C_{i,b}^j$ belonging to $\hat{R}_j$, i.e

$$\mathcal{G}_j = \bigcup_{i,b} C_{i,b}^j, \quad C_{i,b}^j \subset \hat{R}_j.$$

The sheath $\mathcal{G}$ is

$$\mathcal{G} = \bigcup_{j=1}^{J} \mathcal{G}_j.$$

Thirdly, for any circle $C_{i,b}$ we construct the two linear least squares polynomials $\hat{p}_{i,b}^j$ and $\hat{p}_{i,b}^{\bar{j}}$ related to the sample data with locations respectively in

$C_{i,b}^{j}$ and in $C_{i,b}^{\bar{j}}$. We impose that they assume the same values on the segment $\overline{s_b(t_{b,i})s_b(t_{b,i+1})}$, that is

$$\hat{p}_{i,b}^{j}(\mathbf{x}) = \hat{p}_{i,b}^{\bar{j}}(\mathbf{x}) \quad \mathbf{x} \in \overline{s_b(t_{b,i})s_b(t_{b,i+1})}. \tag{48}$$

For each $\mathcal{G}_j$, $j = 1, \ldots, J$, we define the function

$$\tilde{s}_{j,v}(\mathbf{x}) = \sum_{i,b} w_{i,j}^{v}(\mathbf{x})\hat{p}_{i,b}^{j}(\mathbf{x}), \quad \mathbf{x} \in \mathcal{G}_j, \tag{49}$$

where $w_{i,j}^{v}$ are C^1 interpolatory weights with compact support restricted to $\mathcal{G}_j$ and such that $\sum_{i,b} w_{i,j}^{v}(\mathbf{x}) = 1$.

The C^0 approximation of f in $\mathcal{G}$ is given by

$$\tilde{s}_{v}(\mathbf{x}) = \bigcup_{j=1}^{J} \tilde{s}_{j,v}(\mathbf{x}). \tag{50}$$

Let us remak that when the curve $f(\Gamma)$ on the surface is a drainage pattern, $f(\Gamma)$ is monotone. Generally, the curve $\tilde{s}_v(\tilde{\Gamma})$ will not exhibit such behavior.

In fact, as remarked in the fault case, it could happen that, when we compute the least squares polynomials $\hat{p}_{i,b}^{j}$ ($\hat{p}_{i,b}^{\bar{j}}$), we take functional values $\tilde{f}_i$ that we should not take because they stay in $C_{i,b}^{j}$ ($C_{i,b}^{\bar{j}}$) but by opposite parts with respect to Γ. This causes small oscillations in $\tilde{s}_v(\tilde{\Gamma})$ that globally shows a monotone behavior.

We eliminate the undue oscillations by computing a linear spline subject to monotonicity constraints.

Namely, for any $b = 1, \ldots, B$, we consider on I_b a new uniform partition $\{\zeta_{b,i}\}_{i=0}^{m_b}$ where $m_b = \lceil n_b/k_b \rceil$, $k_b > 1$. On each subinterval $[\zeta_{b,i}, \zeta_{b,i+1}]$, we consider M uniform nodes at which we evaluate s_b. We indicate with $A_b = \{\mathbf{a}_l, l = 1, \ldots, m_b M\}$ such set of values.

We compute $\tilde{s}_v(\mathbf{x})$ at the points $\mathbf{a}_j$. Having fixed the centers $\{s_b(\zeta_{b,i})\}_{i=0}^{m_b}$, we construct the linear least squares spline $L_b(s_b(t))$, which minimizes

$$\sum_{\mathbf{a}_l \in A_b} |\tilde{s}_v(\mathbf{a}_l) - L_b(\mathbf{a}_l)|^2 \tag{51}$$

subject to the monotonicity constraints and to the continuity constraint between the different branches. Let

$$L^m = \bigcup_{b=1}^{B} L_b. \tag{52}$$

We evaluate L^m at the points $s_b(t_{i,b})$, $i = 1, \ldots, n_b$, $b = 1, \ldots, B$ and as before we compute the linear least squares polynomials $\tilde{p}_{i,b}^{j}$ subject to the constraints

$$\tilde{p}_{i,b}^{j}(s_b(t_{b,i})) = L^m(s_b(t_{b,i})), \quad i = 1, \ldots, n_b, \quad b = 1, \ldots, B.$$

Again, using the same weights used in (49), we compute

$$\tilde{s}^m_{j,v}(\mathbf{x}) = \sum_{i,b} w^v_{i,j}(\mathbf{x}) \tilde{p}^j_{i,b}(\mathbf{x}), \quad \mathbf{x} \in \mathcal{G}_j, \quad j = 1, \ldots, J. \tag{53}$$

Hence the approximation $\tilde{s}_v$ is given by

$$\tilde{s}_v(\mathbf{x}) = \bigcup_{j=1}^{J} \tilde{s}^m_{j,v}(\mathbf{x}) \tag{54}$$

which is C^0 in $\mathcal{G}$ and monotone along $\tilde{\Gamma}$.

5 Numerical Results

In this section we start by giving two examples related to faulted surfaces both taken from [8]. In the former example, the data are taken from a test function while the latter is relevant to real data. Then we show a real example with gradient faults where the data come from a digital elevation map of a mountainous region (Piemonte-Italy) with drainage patterns.

5.1 Example 1

We consider $N = 1600$ scattered points $\mathbf{x}_i$ with a uniform distribution in $\Omega = [0,1]^2$, and the sample S where

$$f(x,y) = \frac{1}{(3x - \frac{7}{2})^2 + (3y - \frac{7}{2})^2} + g(x,y), \tag{55}$$

and

$$g(x,y) = \begin{cases} 0.35 + \exp\left(-(3x - 3/2)^2 - (3y - 3/2)^2\right), & \text{if } y \geq \Gamma(x) \\ 0, & \text{otherwise,} \end{cases}$$

being $\Gamma(x)$ the piecewise linear function connecting the points $(0, 0.2)$, $(0.2, 0.2)$, $(0.35, 0.225)$, $(0.42, 0.3)$, $(0.48, 0.4)$, $(0.49, 0.53)$, $(0.5, 0.65)$, $(0.65, 0.725)$, $(0.8, 0.75)$, $(1.0, 0.8)$.

The approximation of $\Gamma(x)$ has been discussed in Section 3.1 (Fig. 4). In Fig. 5 we can see the recovered surface that has been computed when choosing $N_0 = 16 \times 16$ centers $\mathbf{y}_j \in Y$. The errors computed with respect to the sample data are $e_2 = 0.0035$, and $e_\infty = 0.0744$; instead, for the recovering of Fig. 3 we have $e_2 = 0.0514$ and $e_\infty = 1.0757$.

5.2 Example 2

This example is based on real data coming from a region of the Pyrénées (France) [8].

We have considered $N = 10000$ gridded data. The first approximation $\hat{\Gamma}^1$ of the fault fullfills the classification of the sample. In Fig. 7 we can see the data on the left and the recovering on the right. The error with respect to the given data is $e_\infty = 0.0029m$.

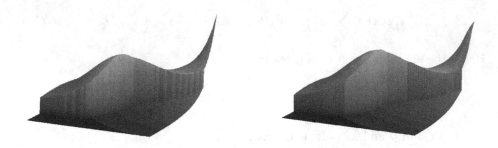

Fig. 6. Left: the true surface. Right: The approximation.

Fig. 7. Left: The data. Right: The recovered surface

5.3   Example 3

The final example is relevant to a DEM of a mountainous region situated in Piemonte. We have $N = 14080 = 110 \times 128$ gridded data shown in Fig. 8.

Fig. 8. The Piemonte data

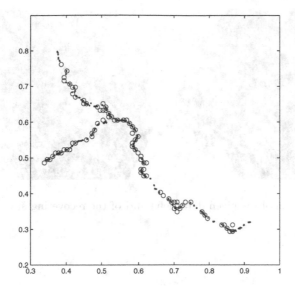

Fig. 9. Circles: the points $\mathbf{z}_l^g$. Dots: the points of $\mathcal{T}$.

Fig. 10. The final recovering

With the procedure of Section 2 we have detected the fault gradient points $\mathbf{z}_l^g$ that are shown with circles in Fig. 9. In this figure we can see also the points (dots) of $\mathcal{T}$ obtained as described in Section 4.2.

To construct the approximation $\tilde{s}_c(\mathbf{x})$, we have taken $N_0 = 12100$ gridded knots. The final recovering is shown in Fig. 10. The quality of reproduction is good. This is also confirmed by the errors computed with respect to the given data. In fact $e_\infty = 18m$ and $e_2 = 0.55m$, while the maximum relative error is $E_\infty = 0.009m$. We have a good approximation also near the drainage curves where the maximum relative error is 0.029.

Fig. 11. Zoom of the given data (left) and of the recovering $\tilde{s}(\mathbf{x})$ (right)

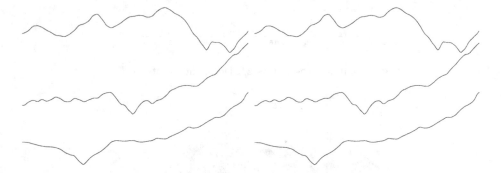

Fig. 12. Left. Data set sections. Right. Sections of $\tilde{s}(\mathbf{x})$.

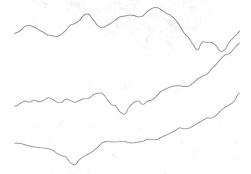

Fig. 13. Sections of $\tilde{s}_c(\mathbf{x})$

To complete the discussion, we show on the left of Fig. 12 three sections of the data set relevant to the low part of the drainage pattern. On the right, we can see that the sections approximated by $\tilde{s}(\mathbf{x})$ are rendered well. On the contrary, as expected, the same sections of $\tilde{s}_c(\mathbf{x})$ exhibit a lack of shape reproduction (Fig. 13).

References

1. Aràndiga, F., Cohen, A., Donat, R., Dyn, N., Matei, B.: Approximation of piecewise smooth functions and images by edge-adapted (ENO-EA) nonlinear multiresolution techniques. Appl. Comput. Harmon. Anal. 24(2), 225–250 (2008)
2. Archibald, R., Gelb, A., Yoon, J.: Polynomial fitting for edge detection in irregularly sampled signals and images. SIAM J. Numer. Anal. 43(1), 259–279 (2005)
3. Bozzini, M., Lenarduzzi, L., Schaback, R.: Adaptive Interpolation by Scaled Multiquadrics. Adv. Comput. Math. 16, 375–387 (2002)
4. Bozzini, M., Rossini, M.: Detection of faults and gradient faults from scattered data with noise. In: Vigo-Aguiar, J., Alonso, P., Oharu, S., Venturino, E., Wade, B. (eds.) Proceedings of the 2009 International Conference on Computational and Mathematical Methods in Science and Engineering, pp. 189–200 (2009)
5. Cates, D., Gelb, A.: Detecting derivative discontinuity locations in piecewise continuous functions from Fourier spectral data. Numer. Algorithms 46(1), 59–84 (2007)
6. Crampton, A., Mason, J.C.: Detecting and approximating fault lines from randomly scattered data. Numer. Algorithms 39(1-3), 115–130 (2005)
7. Fua, P.: Fast, Accurate and Consistent Modeling of Drainage and Surrounding Terrain. Int. J. of Computer Vision 26, 215–234 (1998)
8. Gout, C., Le Guyader, C., Romani, L., Saint-Guirons, A.G.: Approximation of Surfaces with Fault(s) and/or Rapidly Varying Data, Using a Segmentation Process, D^m-Splines and the Finite Element Method. Numer. Alg. 48, 67–92 (2008)
9. Gutzmer, T., Iske, A.: Detection of Discontinuities in Scattered Data Approximation. Numerical Algorithms 16, 155–170 (1997)
10. Hand, D.J.: Discrimination and Classification. Wiley (1981)
11. Lazzaro, D., Montefusco, L.: Radial Basis Functions for the Multivariate Interpolation of Large Scattered Data Sets. J. Comput. Appl. Maths. 140, 521–536 (2002)
12. López de Silanes, M.C., Parra, M.C., Torrens, J.J.: Vertical and oblique fault detection in explicit surfaces. J. Comput. Appl. Math. 140, 559–585 (2002)
13. López de Silanes, M.C., Parra, M.C., Torrens, J.J.: On a new characterization of finite jump discontinuities and its application to vertical fault detection. Math. Comput. Simulation 77(2-3), 247–256 (2008)
14. Parra, M.C., López de Silanes, M.C.: Some results obtained in the study of fault detection from scattered data. In: Seventh Zaragoza-Pau Conference on Applied Mathematics and Statistics, Spanish, Jaca. (2001); Monogr. Semin. Mat. García de Galdeano, vol. 27, pp. 483–490. Univ. Zaragoza, Zaragoza (2003)
15. Rossini, M.: Irregularity Detection from Noisy Data in One and Two Dimension. Numer. Algor. 16, 283–301 (1997)
16. Rossini, M.: 2D- Discontinuity Detection from Scattered Data. Computing 61, 215–234 (1998)
17. Rossini, M.: Detecting discontinuities in two dimensional signal sampled on a grid. Journal of Numerical Analysis, Industrial and Applied Mathematics 4(3-4), 203–215 (2009)
18. Solé, A., Caselles, V., Sapiro, G., Aràndiga, F.: Morse Description and Geometric Encoding of Digital Elevation Maps. IEEE Trans. on Image Processing 13, 1245–1262 (2004)

Piecewise Rational Parametrizations
of Canal Surfaces

Heidi E.I. Dahl

SINTEF ICT, Department of Applied Mathematics
Forskningsveien 1, 0373 Oslo, Norway
heidi.dahl@sintef.no

Abstract. Canal surfaces, as envelopes of one-parameter families of
spheres, correspond to curves in Minkowski space. We show that the
continuity properties of a canal surface are inherited from the continuity
properties of the associated curve, i.e., two curves joined with G^1 or G^2
continuity in Minkowski space correspond to two canal surfaces joined
with the same level of continuity. We also describe an algorithm for minimal bi-degree rational parametrizations of patches on canal surfaces, and
show how this can be used to parametrize piecewise rational corner and
edge blends.

1   Introduction

In Computer Aided Design (CAD) complex shapes are constructed from a small
set of simple primitives. To a large extent, and in particular in the design of
mechanical parts, these primitive shapes are planes, the natural quadrics, and
rolling ball blends between them. The natural quadrics (spheres, and right circular cylinders and cones) are rational surfaces with rational offsets, however,
a rolling ball blend between two natural quadrics is not necessarily rational. In
current CAD systems they are therefore constructed by approximation in all but
the simplest cases, e.g., where the blend is a patch on a cylinder or torus.

Although shape accuracy is important in current CAD systems there is no
requirement that adjacent surfaces match exactly, so gaps within fine tolerances
are allowed. However, with the introduction of Isogeometric Analysis (IGA) (see,
e.g., [1]) this changes, as in Finite Element Analysis (FEA) adjacent elements
are required to match exactly. As a result of this, there is a renewed interest
in exact rational parametrizations of curves and surfaces. The parametrization
degrees of these exact surfaces will necessarily be higher than for approximative
blends, but this disadvantage is offset by the possibility of constructing watertight patchworks of rational surfaces where the limiting curves of two adjacent
patches match exactly. For rolling ball blends, we can then construct an exact
rational parametrization of the blending surface, such that its limiting curves
are contained in the two original surfaces.

Rolling ball blends are patches on *canal surfaces*, which are defined as envelopes of one-parameter families of spheres. Such a family can be described as

M. Floater et al. (Eds.): MMCS 2012, LNCS 8177, pp. 88–111, 2014.

a curve in the 4-dimensional Minkowski space $\mathbb{R}^{3,1}$, where a point $(s; r)$ corresponds to a sphere with centre s and radius r (the sign of the radius determining the orientation of the sphere). The properties of a canal surface can therefore be completely defined in terms of the properties of the corresponding curve in $\mathbb{R}^{3,1}$.

In this paper we will examine the differential geometry of canal surfaces. In particular, we will show that if two curves in Minkowski space meet with G^1 or G^2 continuity (extending the differential geometry of curves in $\mathbb{R}^3$ to $\mathbb{R}^{3,1}$), the join of the two canal surface inherits the same degree of continuity.

In a previous paper (see [2]), we have classified and parametrized rational fixed radius rolling ball blends of pairs of natural quadrics, and established a lower bound on the bi-degree of their parametrizations. By a slight modification of the parametrization algorithm, it extends to variable radius rolling ball blends. Using piecewise rational rolling ball blends of edges and corners, we extend the range of configurations that can be blended rationally at a relatively low degree, and give designers an added flexibility in creating rational blends.

We start by deriving the algorithm for minimal bi-degree rational parametrizations of canal surfaces in Section 2. In Section 3 we extend the differential geometry of $\mathbb{R}^3$ to $\mathbb{R}^{3,1}$, and in Section 4 we describe properties of canal surfaces in terms of the properties of the associated curves in $\mathbb{R}^{3,1}$. Finally, in Section 5 we demonstrate two approaches to corner blends using piecewise rational rolling ball blends, applying the parametrization algorithm from Section 2 and the continuity results from the following sections.

2 Rational Parametrizations of Canal Surfaces

A canal surface is defined as the envelope of the family of spheres

$$f(t) = (s(t); r(t)) \in \mathbb{R}^{3,1}. \tag{1}$$

The curve $s(t)$ traced by the centres of the spheres is called the *spine curve* and $r(t)$ the *radius function* of the canal surface. A canal surface has a parametrization on the form

$$F(t, u) = s(t) + r(t)N(t, u) \tag{2}$$

where the isoparametric curve $F_t(u)$ for a given t is a circle, known as a *characteristic circle*. In fact, $N_t(u)$ is a circle on the unit sphere. An algorithm for minimal bi-degree rational parametrizations of $N(t, u)$ can be found in [3]. Thus if $f(t)$ is rational, we can also construct a rational parametrization of the canal surface $F(t, u)$ (this is proved in, e.g., Theorem 5.1 of [4]). Furthermore, $N(t, u)$ is the unit normal vector of the canal surface, so $F(t, u)$ is a rational surface with a rational unit normal vector field, i.e., it is a Pythagorean Normal (PN) surface. PN surfaces have rational offsets, so by considering canal surfaces corresponding to rational curves in $\mathbb{R}^{3,1}$, we obtain a class of rational rolling ball blends that have some of the advantages of the natural quadrics.

Remark 1. There are non-rational curves in $\mathbb{R}^{3,1}$ whose canal surfaces can be parametrized rationally. However, for these rational parametrizations the rationality of the unit normal vector field is no longer automatic.

2.1   Arcs of Circles on the Unit Sphere

The parametrization of canal surfaces has thus been reduced to the parametrization of circles on the unit sphere, and the parametrization of rolling ball blends to the parametrization of the isoparametric arcs of circles $N_t(u)$ on the unit sphere.

In [2], we considered this problem for *pipe surfaces*, i.e., canal surfaces with constant radius function. As well as closed expressions for the minimal bidegree rational parametrizations of the fixed radius blends, in the cases where such a blend exists, we presented a general parametrization algorithm for patches on pipe surfaces. Only a slight modification of the algorithm is necessary to extend it to variable radius blends.

The parametrization algorithm is a simplification of the results presented in [3], where minimal degree parametrizations are described both for patches and for the complete canal surface. As mentioned above, the decomposition of the parametrization has reduced the problem to the parametrization of a specific arc of circle on the unit sphere. A first naive approach would be to use the inverse stereographic projection to send the arc of circle to a line segment or an arc of circle in the plane $\mathbb{R}^2$ and then parametrize the image. However, this would result in a parametrization of a relatively high degree, as in most cases we parametrize a curve of degree 2 in the plane before projecting it back onto the sphere. This raises the question of whether we can find a projection that sends arcs of circles on the unit sphere onto line segments, and which increases the parametrization degree as little as possible.

The answer is *generalized stereographic projection* $\boldsymbol{P_S}$. In [5], $\boldsymbol{P_S}$ is defined as a map from $\mathbb{RP}^3$ to the unit sphere, and in [3] this is reformulated, by identifying $\mathbb{R}^4$ with $\mathbb{C}^2$, as the *universal rational parametrization* of the unit sphere:

$$\boldsymbol{P_S}\left(\boldsymbol{U}\right) = \left(U_0\overline{U_0} + U_1\overline{U_1}, 2Re\left(U_0\overline{U_1}\right), 2Im\left(U_0\overline{U_1}\right), U_0\overline{U_0} - U_1\overline{U_1}\right)^T \tag{3}$$

where $\boldsymbol{U} = (U_0, U_1) \in \mathbb{C}^2$. This expression is homogeneous: $\boldsymbol{P_S}\left(\lambda\boldsymbol{U}\right) = |\lambda|^2\boldsymbol{P_S}\left(\boldsymbol{U}\right)$, so the generalized stereographic projection can be interpreted as a map from the complex projective line $\mathbb{CP}^1$ to the unit sphere in $\mathbb{R}^3$ (after projection into affine coordinates).

Remark 2. If we restrict the domain of the generalized stereographic projection to the unit sphere in $\mathbb{R}^4$, $\boldsymbol{P_S}$ is called the *Hopf map*.

We consider a *line* in $\mathbb{CP}^1$ as the interpolation of two complex projective points using a real parameter. Then $\boldsymbol{P_S}$ sends a line in $\mathbb{CP}^1$ onto a circle on the unit sphere. More importantly, any circle $\mathscr{C}$ on the unit sphere can be lifted onto a line $\mathscr{L}$ in its preimage in $\mathbb{CP}^1$ so that $\boldsymbol{P_S}(\mathscr{L}) = \mathscr{C}$.

Remark 3. One of the advantages of the generalized stereographic projection is that unlike the stereographic projection, it does not have a distinguished point on the sphere where it is not defined. However, the choice of lifting distinguishes one point. For arcs of circles ending in this point, we need to choose a different lifting, e.g., by a circular permutation of the coordinates.

The algorithms in [2] and [3] describe how to lift the endpoints $\alpha(t)$ and $\beta(t)$ of an arc of circle on the unit sphere to $X(t)$ and $Y(t)$ in $\mathbb{CP}^1$. The lifting of a point $x(t) = (x_0(t), x_1(t), x_2(t), x_3(t)) \in \mathbb{RP}^3$ is defined as

$$L(x(t)) = \left(U_0, U_0 \frac{x_0(t) - x_3(t)}{x_1(t) + \mathrm{i}x_2(t)} \right), \quad U_0 = gcd\left(x_0(t) + x_3(t), x_1(t) + \mathrm{i}x_2(t) \right) \tag{4}$$

(see Equation 10 in [3]). They then parametrize a line segment between $X(t)$ and $Y(t)$ in such a way that the projection of the line segment onto the unit sphere, using the generalized stereographic projection, is the original arc. As the generalized stereographic projection is also the universal rational parametrization of the sphere, if we minimize the parametrization degree of the line in $\mathbb{CP}^1$, its projection on the unit sphere has minimal parametrization degree.

Remark 4. Given two points $X, Y \in \mathbb{CP}^1$, the line between them is unique up to a complex scalar λ. The choice of λ determines which arc of circle between $P_S(X)$ and $P_S(Y)$ the line segment is projected onto. As part of the parametrization algorithm, we therefore have to determine the correct *lifting coefficient* λ.

The parametrization of arcs of circles on the unit sphere is summarized in the following algorithm:

Algorithm 1. *An arc of circle on the unit sphere is parametrized by executing the following steps:*

1. *Lift the endpoints $\alpha(t), \beta(t) \in \mathbb{RP}[t]^3$, $i = 1, 2$, to $X(t), Y(t) \in \mathbb{CP}^1[t]$.*
2. *Determine the lifting coefficient $\lambda(t)$, and factorize $\lambda(t) = \lambda_0(t)\lambda_1(t)$ as evenly as possible in order to distribute the increase in degree across $X(t)$ and $Y(t)$.*
3. *Parametrize the line segment $(1 - u)\lambda_0(t)X(t) + u\lambda_1(t)Y(t)$, $u \in [0, 1]$ between $X(t)$ and $Y(t)$.*
4. *Project onto the unit sphere using the generalized stereographic projection P_S.*

Steps 1, 3, and 4 are identical to the corresponding steps in the algorithm for the parametrization of pipe surfaces (Alg. 13, [2]). However, while the isoparametric circles on pipe surfaces correspond to large circles on the unit sphere, this is not the case for canal surfaces in general. We therefore have to adjust λ accordingly.

The lifting coefficient λ is defined in [3] as the unique solution of a set of linear equations. In [2] we reduced this to a single equation for the case of pipe surfaces. We can make similar simplifications for canal surfaces in general.

Lemma 1. *The system of linear complex equations determining the lifting coefficient $\lambda = \lambda_0\overline{\lambda_1} \in \mathbb{C}[t]$ has a unique solution up to multiplication by a real number. If $[\alpha(t)] \neq -[\beta(t)]$, where $[\alpha(t)] = (\alpha_1/\alpha_0, \alpha_2/\alpha_0, \alpha_3/\alpha_0)$ is the projection from $\mathbb{RP}^3$ to $\mathbb{R}^3$, then*

$$\lambda = \frac{(\alpha_0 + \alpha_3)(\beta_0 + \beta_3)(B_0 + B_3) - (\alpha_1 + \mathrm{i}\alpha_2)(\beta_1 - \mathrm{i}\beta_2)(B_0 - B_3)}{X_0\overline{Y_0}}. \tag{5}$$

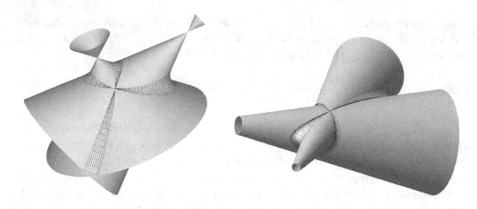

Fig. 1. Rolling ball blends of two cones

Otherwise the two points on the unit sphere are diametrically opposite, i.e.,
$\boldsymbol{\alpha}(t) = (\alpha_0, \alpha_1, \alpha_2, \alpha_3)$, $\boldsymbol{\beta}(t) = (-\alpha_0, \alpha_1, \alpha_2, \alpha_3)$, *and*

$$\lambda = \frac{\mathrm{i}(\alpha_0 + \alpha_3)(B_0 - B_3) - (\alpha_1 + \mathrm{i}\alpha_2)(\mathrm{i}B_1 + B_2)}{X_0 \overline{Y_0}} \tag{6}$$

Here $X_0 = gcd(\alpha_0 + \alpha_3, \alpha_1 + \mathrm{i}\alpha_2)$, $\overline{Y_0} = gcd(\beta_0 + \beta_3, \beta_1 - \mathrm{i}\beta_2)$, *and*

$$\boldsymbol{B} = (B_0, B_1, B_2, B_3) = \delta(t)(\dot{r}(t), \dot{s}_1(t), \dot{s}_2(t), \dot{s}_3(t)) \tag{7}$$

where $\delta(t)$ *is the common denominator of* $\dot{r}(t)$, $\dot{s}_1(t)$, $\dot{s}_2(t)$, *and* $\dot{s}_3(t)$.

Remark 5. The arc of circle is contained in the intersection of the unit sphere with the plane $\boldsymbol{B}.\boldsymbol{x} = 0$, $\boldsymbol{x} = (x_0, x_1, x_2, x_3)$.

Proof. In [3], λ is given as the unique solution of a set of four complex linear equations. The main challenge in deriving (5) is the size of the expressions, so we will only give an outline of the calculations here.

Starting from the four complex equations, we consider their real and imaginary components. We solve two of these real equations for the real and imaginary parts of λ. We can then show that the remaining equations are equivalent. The expression for λ is then simplified exploiting the fact that the endpoints $[\boldsymbol{\alpha}(t)]$ and $[\boldsymbol{\beta}(t)]$ are contained in the intersection of the plane $\boldsymbol{B}.\boldsymbol{x} = 0$ with the unit sphere. Eliminating any real factors, we arrive at (5).

This completes the parametrization algorithm for arcs of circles on the unit sphere.

2.2   Parametrizing Variable Radius Rolling Ball Blends

The first step in constructing a parametrization of a blend, such as the two cone-cone blends in Figure 1, is to parametrize the *touching curves*: the curves

Fig. 2. Linear family generated by two spheres in oriented contact (left), and the rolling ball in oriented contact with a cone (right)

the rolling ball trace on the two natural quadrics, i.e., the limiting curves of the patch on the canal surface. Consider two natural quadrics, and a variable radius rolling ball blend between them. Let $\boldsymbol{f}(t) \in \mathbb{R}^{3,1}$ be the curve in Minkowski space corresponding to the blend. How we determine the touching curves in the original surfaces depends on the type of the natural quadric.

To find the touching curve on a sphere, recall that the sphere and the rolling ball is in *oriented contact* for any t, i.e., the two spheres are tangent, and their unit normal vectors coincide at the touching point. Equivalently, the two corresponding points $\boldsymbol{p}$ and $\boldsymbol{q} \in \mathbb{R}^{3,1}$ are at zero distance, measured using the *Minkowski metric*

$$\|\boldsymbol{p} - \boldsymbol{q}\| = \sqrt{\langle \boldsymbol{p} - \boldsymbol{q}, \boldsymbol{p} - \boldsymbol{q} \rangle}, \tag{8}$$

where

$$\langle \boldsymbol{v}, \boldsymbol{v}' \rangle = v_1 v_1' + v_2 v_2' + v_3 v_3' - v_4 v_4' \tag{9}$$

is the *Minkowski scalar product*. We use the notation $(\_, \_)$ and $|\_|$ for the Euclidean scalar product and metric, respectively. The Minkowski metric measures the tangential distance between two oriented spheres. If we consider the linear family of spheres generated by the sphere and the rolling ball, the touching point is the member of the family with zero radius. To find the touching curve on the sphere, it therefore suffices to solve a linear equation, and we find that if $\boldsymbol{f}(t)$ is rational, the degree of the touching curve on the sphere is the same as the degree of $\boldsymbol{f}(t)$.

To find the touching curve on a cone or cylinder, recall that they are envelopes of linear families of spheres. Furthermore, for a given t the rolling ball is in oriented contact with exactly one sphere in the family (see Figure 2, right). We find the touching curve by the same procedure as in [2], Lemma 4, (noting that r is now a function in t), first solving a linear equation to find the tangent sphere, and then solving a second linear equation to find the touching point. From the explicit expressions in [2], we see that if $\boldsymbol{f}(t)$ is rational, the degree of the touching curve is double the degree of $\boldsymbol{f}(t)$.

The second step in the construction of the parametrization is to determine the Gaussian images $\boldsymbol{\alpha}$ and $\boldsymbol{\beta}$ of the two touching curves, i.e., the curves limiting the patch corresponding to the blend on the unit sphere. The touching curves $\boldsymbol{T}_0(t)$ and $\boldsymbol{T}_1(t)$ are the isoparametric curves of $\boldsymbol{F}(t, u)$ for $u = 0$ and $u = 1$, and accordingly their Gaussian images are respectively $\boldsymbol{N}_0(t)$ and $\boldsymbol{N}_1(t)$. This gives us the expression for $\boldsymbol{\alpha}$ and $\boldsymbol{\beta}$:

$$[\boldsymbol{\alpha}(t)] = \boldsymbol{N}_0(t) = \frac{\boldsymbol{T}_0(t) - \boldsymbol{s}(t)}{r(t)}, \quad [\boldsymbol{\beta}(t)] = \boldsymbol{N}_1(t) = \frac{\boldsymbol{T}_1(t) - \boldsymbol{s}(t)}{r(t)}, \tag{10}$$

where $[\boldsymbol{\alpha}(t)] = (\alpha_1/\alpha_0, \alpha_2/\alpha_0, \alpha_3/\alpha_0)$ is the projection from $\mathbb{RP}^3$ to $\mathbb{R}^3$.

And finally, after parametrizing the arc of circle between $\boldsymbol{\alpha}$ and $\boldsymbol{\beta}$ using Algorithm 1, we put together the components of the parametrization using (2):

$$\boldsymbol{F}(t, u) = \boldsymbol{s}(t) + r(t) \left[\boldsymbol{P}_{\mathcal{S}} \left((1 - u) \lambda_0 \boldsymbol{X} + u \lambda_1 \boldsymbol{Y} \right) \right]. \tag{11}$$

Combining these steps, we arrive at the following algorithm for minimal bidegree parametrizations of variable radius rolling ball blends:

Algorithm 2. *Consider two surfaces, and a canal surface containing a rolling ball blend between them corresponding to a rational curve $\boldsymbol{f}(t) = (\boldsymbol{s}(t); r(t)) \in \mathbb{R}^{3,1}$. The rolling ball blend is parametrized by calculating:*

1. *The touching curves $\boldsymbol{T}_0(t)$ and $\boldsymbol{T}_1(t)$ of the blend.*
2. *Their Gaussian images $\boldsymbol{\alpha}(t)$ and $\boldsymbol{\beta}(t)$.*
3. *The liftings $\boldsymbol{X}(t)$ and $\boldsymbol{Y}(t) \in \mathbb{CP}^1[t]$.*
4. *The lifting coefficient $\lambda(t) = \lambda_0(t)\lambda_1(t)$.*
5. *The line segment $(1 - u)\lambda_0(t)\boldsymbol{X}(t) + u \lambda_1(t)\boldsymbol{Y}(t)$, $u \in [0, 1]$.*
6. *Its generalized stereographic projection onto the unit sphere*

$$\boldsymbol{N}(t, u) = \left[\boldsymbol{P}_{\mathcal{S}} \left((1 - u) \lambda_0(t) \boldsymbol{X}(t) + u \lambda_1(t) \boldsymbol{Y}(t) \right) \right].$$

7. *The parametrization of the blend $\boldsymbol{F}(t, u) = \boldsymbol{s}(t) + r(t)\boldsymbol{N}(t, u)$.*

If the degree of the parametrization of the line in $\mathbb{CP}^1$ is minimized, this parametrization of the variable radius rolling ball blend is of minimal bi-degree $(n, 2)$.

2.3   Rational Blends of the Natural Quadrics

In the previous section, Algorithm 2 was constructed under the assumption that the canal surface containing the blend is known. Given two surfaces, determining which curves $\boldsymbol{f}(t) \in \mathbb{R}^{3,1}$ correspond to blends between them is a separate question.

Consider the cones $\mathscr{C}$ and $\mathscr{C}' \subset \mathbb{R}^3$ corresponding to a lines $\mathscr{L}$ and $\mathscr{L}' \subset \mathbb{R}^{3,1}$, and a rolling ball blend $\boldsymbol{F}(t, u)$ between them. At any point the rolling ball is in oriented contact with both cones, so the curve $\boldsymbol{f}(t)$ in Minkowski space is at zero distance to the lines. This means that it is contained in the *isotropic quadric* of the cone $\mathscr{C}$ (see [2]), i.e., the hypersurface in $\mathbb{R}^{3,1}$ of points at zero distance to $\mathscr{L}$, or equivalently the spheres in oriented contact with the cone. The curve corresponding to a blend between $\mathscr{C}$ and $\mathscr{C}'$ is therefore contained in the 2-dimensional intersection of their isotropic quadrics: the bisector surface of $\mathscr{L}$ and $\mathscr{L}'$ in Minkowski space. This surface is toric, so we may apply the results of e.g. [6] to construct spline curves in the bisector surface in $\mathbb{R}^{3,1}$ corresponding to rational blends of the two cones.

Another approach is to consider the hyperplane sections of the bisector surface. This is the approach used for fixed radius blends, where the bisector surface is intersected with the hyperplane $x_4 = R$. In [2], our classification of the configurations of natural quadrics that admit a rational fixed radius blends is based on the fact that there are exactly two types of surfaces where all hyperplane sections are rational: rational ruled surfaces and the Steiner surface (see [7]). Thus for these configurations, any hyperplane section of the bisector surface is a rational curve that corresponds to a rational blend of the two cones.

In the rest of the paper, we will move on from this question to consider how we can join curves in Minkowski space, and canal surfaces in $\mathbb{R}^3$, with G^1 and G^2 continuity. In particular, we want to determine how the curvature properties at the join of two curves in $\mathbb{R}^{3,1}$ are reflected in the curvature properties at the join of the corresponding canal surfaces in $\mathbb{R}^3$. Our main interest is to construct piecewise rational canal surfaces with a given degree of continuity at the joins, but the constructions below are valid as long as the parametrization $\boldsymbol{F}(t, u)$ of the canal surface is respectively once and twice differentiable at the join for G^1 and G^2 continuity.

3   Differential Geometry of Curves in $\mathbb{R}^{3,1}$

In the previous section, we described how the parametrization $\boldsymbol{F}(t, u)$ of a canal surface can be reduced to the parametrization of a circle $\boldsymbol{N}(t, u)$ on the unit sphere, and how this circle can be parametrized rationally. For the purposes of differential geometry, however, it is convenient to choose a non-rational parametrization of $\boldsymbol{N}(t, u)$, to avoid the increase in degree from the differentiation of rational expressions. To ensure that $\boldsymbol{F}(t, u)$ is non-degenerate, i.e., that the envelope of the family of spheres is real, we require that $\|\dot{\boldsymbol{f}}(t)\|^2 > 0$.

A convenient parametrization is based on the Frenet frame of the spine curve:

$$\boldsymbol{N}_t(\theta) = \frac{\dot{r}}{\nu}\boldsymbol{t} + \sqrt{1 - \left(\frac{\dot{r}}{\nu}\right)^2}\,(\cos\theta\,\boldsymbol{n} + \sin\theta\,\boldsymbol{b}) \tag{12}$$

where $\boldsymbol{t}$, $\boldsymbol{n}$, and $\boldsymbol{b}$ are the unit tangent, principal normal, and bi-normal vectors of the spine curve $\boldsymbol{s}$ at t, and $\nu = |\dot{\boldsymbol{s}}(t)|$. The non-degeneracy condition ensures that $\nu > \dot{r}$, so the parametrization is well defined as long as the Frenet frame exists.

Remark 6. The choice of the Frenet frame is motivated by the existence of the Frenet equations expressing the derivatives $\dot{\boldsymbol{t}}$, $\dot{\boldsymbol{n}}$, and $\dot{\boldsymbol{b}}$ in terms of $\boldsymbol{t}$, $\boldsymbol{n}$, and

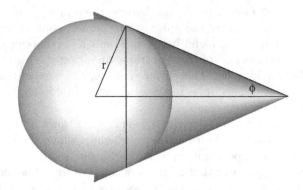

Fig. 3. The sphere $\boldsymbol{f}_t$ with its tangent cone

b. However, there are curves that do not have well defined Frenet frames in all points. An alternative choice of frame may, e.g., be rotation minimizing frames, which have relations with its derivatives similar to the Frenet equations. We expect that the result would be similar using this frame, however this will be the subject of further study. For now, the results in the following sections are valid if the Frenet frames of the two spine curves are well defined and identical at the join.

The parametrization in (12) can be constructed by recalling that each sphere $\boldsymbol{f}_t$ contributes a characteristic circle $\boldsymbol{F}_t(\theta)$ to the envelope. Consider the tangent line of the curve $\boldsymbol{f}(t) \in \mathbb{R}^{3,1}$, i.e., the line through $\boldsymbol{f}(t)$ with direction vector $\dot{\boldsymbol{f}}(t)$. The envelope of this linear one-parameter family of spheres is the *tangent cone* of the canal surface at $\boldsymbol{f}_t$, which is tangent to $\boldsymbol{f}_t$ along the characteristic circle. If ϕ is the half angle of the tangent cone we find that $\sin \phi = \frac{\dot{r}}{\nu}$, which gives us the above parametrization of $\boldsymbol{N}(t, \theta)$ (see Figure 3).

Remark 7. A parametrization similar to (12) has already been used in the investigation of the analytic and algebraic properties of canal surfaces (see, e.g., [8]), where the focus have been on canal surfaces whose spine curves are parametrized by arc length, i.e., where $\nu = |\dot{\boldsymbol{s}}(t)| = 1$. However, it has been shown that it is impossible to parametrize a space curve, other than a straight line, by rational functions of its arc length (see [9]), so for a canal surface with rational spine we need to consider the general case. Fortunately, this does not significantly increase the complexity of the resulting expressions.

3.1 G^1 Continuity of Curves in $\mathbb{R}^{3,1}$ and Canal Surfaces

By the definition of the envelope, the tangent planes of the sphere $\boldsymbol{f}_t$ coincide with the tangent planes of the canal surface along the characteristic circle. This gives us a first result relating the continuity of the curve in Minkowski space with the continuity of the canal surface:

Theorem 3. *If two curves $f(t)$ and $g(t) \in \mathbb{R}^{3,1}$ are joined with G^1 continuity, so are the corresponding canal surfaces $F(t, u)$ and $G(t, u) \in \mathbb{R}^3$.*

Proof. G^1 continuity is also called *tangent continuity*. The two curves $f(t)$ and $g(t) \in \mathbb{R}^{3,1}$, are tangent continuous in the point $f(t_0) = g(t_0)$ if the tangent lines of f and g coincide at $t = t_0$. Thus the two canal surfaces $F(t, u)$ and $G(t, u)$ have the same tangent cone and characteristic circle at $t = t_0$. They therefore have the same unit normal vector field $N(t_0, u)$ along the characteristic circle, i.e., the tangent planes of the two adjacent surfaces coincide: the two canal surfaces are joined with tangent continuity.

For rolling ball blends, G^1 continuity may be considered sufficient, as it is the level of continuity of the blend with the original surfaces. However, existing approximate rational blends have internal G^2 continuity, so we need at least to consider if this is achievable with exact piecewise rational blends. Determining the conditions for a G^2 join of canal surfaces will therefore be the main focus of the rest of this paper.

3.2 G^2 Continuity of Curves in $\mathbb{R}^{3,1}$

G^2 continuity is also known as *curvature continuity*. In $\mathbb{R}^3$, two curves are joined with curvature continuity if their osculating circles coincide at the join, i.e., if their unit tangent and principal normal vectors t and n, and curvature κ coincide. This can be extended to curves in $\mathbb{R}^{3,1}$ (see [10]).

The unit tangent vector of a curve $s(t) \in \mathbb{R}^3$ is defined as

$$t = \frac{\dot{s}}{|\dot{s}|}. \tag{13}$$

We generalize this to a curve $f(t)$ in $\mathbb{R}^{3,1}$ using the Minkowski scalar product to normalize the vector

$$t_m = \frac{\dot{f}}{\|\dot{f}\|}. \tag{14}$$

Remark 8. In $\mathbb{R}^3$, the only vector with zero length is the vector $(0, 0, 0)$. In Minkowski space this is no longer the case. We therefore have to be careful when normalizing vectors. However, the non-degeneracy condition we stated earlier requires that $\|\dot{f}\| > 0$, so the Minkowski unit tangent vector is well defined for curves corresponding to non-degenerate canal surfaces.

To find the unit principal normal vector n of $s(t)$, we derive $\dot{s} = |\dot{s}|t$, using the Frenet equation $\dot{t} = \kappa|\dot{s}|n$, and substitute the expression for t in (13). This gives us

$$n = \frac{|\dot{s}|^2\ddot{s} - (\dot{s}, \ddot{s})\dot{s}}{\kappa|\dot{s}|^4} \tag{15}$$

where, taking the norm on both sides, we find the curvature

$$\kappa = \frac{\||\dot{s}|^2\ddot{s} - (\dot{s}, \ddot{s})\dot{s}\|}{|\dot{s}|^4}. \tag{16}$$

Fig. 4. The spine curve $s(t)$ with its Frenet frame and osculating circle (left), and the canal surface $F(t, \theta)$ with its tangent and principal normal cones (right)

Generalizing (15) and (16) to $\mathbb{R}^{3,1}$, we define the unit principal normal vector of a curve $f(t)$ in Minkowski space in terms of the Minkowski scalar product:

$$n_m = \frac{\|\dot{f}\|^2 \ddot{f} - \langle \dot{f}, \ddot{f} \rangle \dot{f}}{\kappa_m \|\dot{f}\|^4}, \tag{17}$$

and its *Minkowski curvature*

$$\kappa_m = \frac{\|\|\dot{f}\|^2 \ddot{f} - (\dot{f}, \ddot{f}) \dot{f}\|}{\|\dot{f}\|^4}. \tag{18}$$

Figure 4 shows the spine curve $s(t) \in \mathbb{R}^3$ with its Frenet frame and osculating circle, and the associated canal surface $F(t, \theta)$ with tangent and principal normal cones generated by its Minkowski tangent and principal normal vectors t_m and n_m.

A similar generalization can be constructed for the first and second unit binormal vectors (see [10]).

Remark 9. A point $p \in \mathbb{R}^3$ can be considered a sphere of zero radius, corresponding to the point $(p; 0) \in \mathbb{R}^{3,1}$. Thus a curve $s(t) \in \mathbb{R}^3$ can be identified with the curve $(s(t); 0)$ in Minkowski space. For this curve, we then find that $t_m = (t; 0)$, $n_m = (n; 0)$, and $\kappa = \kappa_m$.

Two curves $f(t)$ and $g(t) \in \mathbb{R}^{3,1}$ are joined with G^2 or *curvature continuity* if t_m, n_m, and κ_m coincide at the join. We now want to prove the G^2 analogue of Theorem 3, i.e., that if two curves in Minkowski space are joined with G^2 continuity, then so are the associated canal surfaces.

4 Differential Geometry of Canal Surfaces

The principal curvatures κ_1 and κ_2 in a point on a surface in $\mathbb{R}^3$ are defined as the maximum and minimum curvatures of its normal sections. The curvature of a surface is completely defined by κ_1, κ_2, and its principal curvature directions, as by Euler's theorem they determine the curvature of any normal section of the surface. κ_1 and κ_2 can be calculated as the eigenvalues, and the principal curvature directions as the eigenvectors, of the shape operator of the surface.

4.1   The Principal Curvatures of a Canal Surface

The shape operator S of a surface can be expressed in terms of the coefficients of its first and second fundamental forms. In order to compare principal curvatures, we therefore start by calculating these. The calculations are straightforward but large, so we will not show them here. The Maple file containing the details of the calculations can be obtained by application to the author.

Let $\boldsymbol{f}(t) = (\boldsymbol{s}(t); r(t))$ be a curve in $\mathbb{R}^{3,1}$. The coefficients of the first fundamental form of $\boldsymbol{f}(t)$ are:

$$E = \nu^2 \kappa^2 \left(\left(\Lambda_1 (1 + r\Lambda_2) - X \right)^2 + r^2 \left(\Lambda_1 \tau + \frac{\dot{r}}{\nu} Y \right)^2 \right) \tag{19}$$

$$F = r^2 \nu \kappa^2 \left(\tau \Lambda_1 + \frac{\dot{r}}{\nu} Y \right) \Lambda_1 \tag{20}$$

$$G = r^2 \kappa^2 \Lambda_1^2 \tag{21}$$

The coefficients of the second fundamental form of $\boldsymbol{f}(t)$ are:

$$e = \left(r \left(\frac{\tau}{\kappa} \frac{\dot{r}}{\nu} - \kappa \Lambda_1 Y \right)^2 + \frac{\Lambda_1 (1 + 2r\Lambda_2)}{r} (X - r\Lambda_1 \Lambda_2) - r \left(1 - \Lambda_1^2 \Lambda_2^2 + \frac{\tau^2}{\kappa^2} \right) \right) \nu^2 \kappa^2 \tag{22}$$

$$f = -r\kappa^2 \nu \Lambda_1 \left(\tau \Lambda_1 + \frac{\dot{r}}{\nu} Y \right) \tag{23}$$

$$g = -r\kappa^2 \Lambda_1^2 \tag{24}$$

where $r = r(t)$, κ and τ are respectively the curvature and torsion of the spine curve $\boldsymbol{s}(t)$, $X = r \cos \theta$ and $Y = \sin \theta$, and

$$\Lambda_1 = \frac{1}{\kappa} \sqrt{1 - \left(\frac{\dot{r}}{\nu} \right)^2}, \quad \Lambda_2 = \frac{\dot{\nu}\dot{r} - \ddot{r}\nu}{\left(1 - \left(\frac{\dot{r}}{\nu} \right)^2 \right) \nu^3}. \tag{25}$$

The shape operator S of a surface can be defined in terms of these six coefficients

$$S = \frac{1}{EG - F^2} \begin{pmatrix} eG - fF & fG - gF \\ fE - eF & gE - fF \end{pmatrix}. \tag{26}$$

Inserting (19 - 24) and simplifying, we arrive at the shape operator of a canal surface

$$S = \begin{pmatrix} \frac{r\Lambda_1 \Lambda_2 - X}{r(\Lambda_1(1+r\Lambda_2) - X)} & \frac{\nu\tau\Lambda_1 + \dot{r}Y}{r(\Lambda_1(1+r\Lambda_2) - X)} \\ 0 & \frac{1}{r} \end{pmatrix}. \tag{27}$$

The shape operator S is triangular, so its eigenvalues are the coefficients along the diagonal. The principal curvatures of a canal surface are therefore

$$\kappa_1 = \frac{r\Lambda_1\Lambda_2 - X}{r\left(\Lambda_1\left(1 + \Lambda_2 r\right) - X\right)}, \quad \kappa_2 = \frac{1}{r}. \tag{28}$$

The second principal curvature κ_2 is the curvature of the sphere f_t, and the associated curvature line is the characteristic circle. Two canal surfaces joined with G^1 continuity therefore automatically have one coinciding principal curvature and coinciding directions of principal curvature. What remains to be shown, in order to have G^2 continuity, is that the remaining principal curvature κ_1 is the same for the two canal surfaces.

4.2 G^2 Continuous Canal Surfaces

To simplify the expressions, we assume that the Frenet frames of the two spine curves coincide. We then need to solve the equation

$$\frac{r\Lambda_1\Lambda_2 - X}{r\left(\Lambda_1\left(1 + \Lambda_2 r\right) - X\right)} = \frac{r\Lambda_1'\Lambda_2' - X}{r\left(\Lambda_1'\left(1 + \Lambda_2'r\right) - X\right)} \tag{29}$$

for any $X = r\cos\theta$, i.e., for any θ, as we want to ensure $\kappa_1 = \kappa_1'$ at any point along the characteristic circle.

Lemma 2. *Two canal surfaces $F(t, u)$ and $F'(t, u) \in \mathbb{R}^3$ are joined with G^2 continuity if they have the same Λ_1 and Λ_2.*

Proof. Expanding (29), we arrive at

$$r\Lambda_1\Lambda_1'\left(\Lambda_2 - \Lambda_2'\right) + \left(\Lambda_1 - \Lambda_1'\right)X = 0, \quad \forall X. \tag{30}$$

Eliminating the two coefficients of the monomials in X proves the lemma, as the non-degeneracy condition requires $\Lambda_1\Lambda_1' \neq 0$.

By examining Λ_1 and Λ_2 further, we arrive at the conjectured theorem, relating the level of continuity of the canal surface to the level of continuity of the associated curve in Minkowski space

Theorem 4. *If two curves $f(t)$ and $f'(t) \in \mathbb{R}^{3,1}$ are joined with G^2 continuity, so are the corresponding canal surfaces $F(t, u)$ and $F'(t, u) \in \mathbb{R}^3$.*

Proof. G^2 continuity of curves in $\mathbb{R}^{3,1}$ is defined by coinciding unit tangent t_m, principal normal n_m, and Minkowski curvature κ_m. We want to demonstrate that this implies that Λ_1 and Λ_2 are identical for the two canal surfaces.
  The vectors t_m and n_m span a 2-dimensional plane in $\mathbb{R}^{3,1}$, whose restriction to the first three coordinates is the plane spanned by t and $n \in \mathbb{R}^3$. Thus if the two curves f and $f' \in \mathbb{R}^{3,1}$ have coinciding t_m and n_m, their spines s and $s' \in \mathbb{R}^3$ have coinciding t and n.

If the unit tangent vectors t_m coincide at the join, the two canal surfaces have the same tangent cone, thus

$$\kappa \Lambda_1 = \sqrt{1 - \left(\frac{\dot{r}}{\nu}\right)^2} = \kappa' \Lambda_1', \tag{31}$$

as $\frac{\dot{r}}{\nu} = \sin\varphi$ where φ is the half angle of the tangent cone. In the plane spanned by t_m and n_m, we find the vector $(n; -\kappa \Lambda_1^2 \Lambda_2)$. As $n = n'$, we get $\kappa \Lambda_1^2 \Lambda_2 = \kappa' \Lambda_1'^2 \Lambda_2'$, i.e.,

$$\frac{\Lambda_2}{\kappa} = \frac{\Lambda_2'}{\kappa'}. \tag{32}$$

The Minkowski curvature can be reformulated as

$$\frac{\kappa_m}{\kappa} = \frac{\sqrt{1 - \Lambda_1^2 \Lambda_2^2}}{\kappa^2 \Lambda_1^2} = \frac{\sqrt{1 - \left(1 - \left(\frac{\dot{r}}{\nu}\right)^2\right)\frac{\Lambda_2^2}{\kappa^2}}}{\left(1 - \left(\frac{\dot{r}}{\nu}\right)^2\right)} = \frac{\kappa_m'}{\kappa'} \tag{33}$$

so if $\kappa_m = \kappa_m'$, we get $\kappa = \kappa'$, $\Lambda_1 = \Lambda_1'$, and $\Lambda_2 = \Lambda_2'$. Applying Lemma 2, this proves the theorem.

Remark 10. Theorems 3 and 4 confirm the conjecture in Remark 2.1 of [11] for contact of order 1 and 2, i.e., joins of G^1 and G^2 continuity.

4.3 G^2 Continuity with the End Sphere

In some cases, for example when constructing corner blends (see Section 5.2), we want a segment of canal surface to be G^2 continuous with its end sphere. The principal curvatures of the sphere are $\kappa_1 = \kappa_2 = 1/r$, so to calculate the conditions for G^2 continuity with the sphere we need to solve $\kappa_1 = 1/r$, i.e.,

$$\frac{\Lambda_1}{r\left(\Lambda_1\left(1 + r\Lambda_2\right) - X\right)} = 0 \tag{34}$$

Theorem 5. *A segment of canal surface is joined with G^2 continuity to its end sphere if either $\|\dot{f}\| = 0$ or $|\dot{s}| = 0$.*

Proof. By inserting the expressions for Λ_1 and Λ_2 in (34), we arrive at a rational expression with numerator

$$\left(\nu^2 - \dot{r}^2\right)\nu = \|\dot{f}\|^2 |\dot{s}| = 0. \tag{35}$$

This gives us two cases for G^2 continuity.

The first case in the theorem, $\|\dot{f}\| = 0$, was initially excluded to avoid degeneracies in the canal surface. Now consider what happens when we allow this at the end of the segment of canal surface. The characteristic circle parametrized in (12) is reduced to a single point, so we are in fact closing the canal surface, and the surface is then by construction G^2 at the endpoint.

The second case for G^2 continuity with the end sphere is $\nu = 0$. Considering the non-degeneracy condition $\nu^2 > \dot{r}^2$, we need to consider the limit value of $\dot{r}/\nu$ to determine the characteristic circle at the end sphere.

Remark 11. Consider a rational Bézier curve $\boldsymbol{f}(t)$ in $\mathbb{R}^{3,1}$, with control points $\{\boldsymbol{p}_0, \dots, \boldsymbol{p}_n\}$ (corresponding to *control spheres* of the canal surface). The associated canal surface is closed at $t = 0$ if $\|\boldsymbol{p}_0 - \boldsymbol{p}_1\| = 0$, i.e., if the two first control spheres are in oriented contact. The canal surface is joined with its end sphere at G^2 continuity if $\boldsymbol{p}_0 = \boldsymbol{p}_1$, i.e., the first control sphere is double.

Corollary 1. *Let $\boldsymbol{f}(t)$ and $\boldsymbol{g}(t)$ be two Bézier curves in $\mathbb{R}^{3,1}$ with control points respectively $\{\boldsymbol{p}_{-m}, \dots, \boldsymbol{p}_0\}$ and $\{\boldsymbol{p}_0, \dots, \boldsymbol{p}_n\}$. If*

1. *$\boldsymbol{p}_{-1}, \boldsymbol{p}_0, \boldsymbol{p}_1$ are collinear,*
2. *$\boldsymbol{p}_{-2}, \boldsymbol{p}_{-1}, \boldsymbol{p}_0, \boldsymbol{p}_1, \boldsymbol{p}_2$ span a 2-dimensional plane, and*
3. *$\boldsymbol{p}_0$ is a double control point for both curves,*

then the associated canal surfaces $\boldsymbol{F}(t, u)$ and $\boldsymbol{G}(t, u)$ are joined with G^2 continuity.

Proof. If $\boldsymbol{p}_{-1}, \boldsymbol{p}_0, \boldsymbol{p}_1$ are collinear, then $\boldsymbol{f}(t)$ and $\boldsymbol{g}(t)$ have the same tangent line, i.e., the same $\boldsymbol{t}_m$. The control points $\{\boldsymbol{p}_{-2}, \boldsymbol{p}_{-1}, \boldsymbol{p}_0\}$, and $\{\boldsymbol{p}_0, \boldsymbol{p}_1, \boldsymbol{p}_2\}$ span the 2-dimensional tangent/principal normal plane, so if the two planes coincide, the two curves have the same $\boldsymbol{n}_m$ as well. When $\boldsymbol{p}_0$ is a double control point, the canal surfaces are G^2 continuous with the common control sphere along the same characteristic circle, thus they are G^2 continuous with each other.

4.4   The Osculating Cyclide

For a curve $\boldsymbol{s}(t) \in \mathbb{R}^3$, a geometric interpretation of its curvature κ at a point $\boldsymbol{s}_t$ is that its inverse, the radius of curvature $r_\kappa = \frac{1}{\kappa}$, is the radius of the circle best approximating the curve close to $\boldsymbol{s}_t$. This circle, called the *osculating circle*, lies in the plane spanned by the unit tangent vector $\boldsymbol{t}$ and the unit principal normal vector $\boldsymbol{n}$. This interpretation can be generalized to Minkowski space.

Theorem 6. *If at a point $\boldsymbol{f}_t$ on a curve $\boldsymbol{f}(t) \in \mathbb{R}^{3,1}$ the unit tangent and principal normal vectors $\boldsymbol{t}_m$ and $\boldsymbol{n}_m$ are well defined, then there exists a unique pseudo-Euclidean (PE) circle $\boldsymbol{c}_t(u)$ in the 2-dimensional plane spanned by $\boldsymbol{t}_m$ and $\boldsymbol{n}_m$, tangent to the curve at $\boldsymbol{f}_t$, and with the same Minkowski curvature κ_m.*

Proof. A PE circle in $\mathbb{R}^{3,1}$ is uniquely defined by three finite points (see [12]). Equivalently, in the plane spanned by $\boldsymbol{t}_m$ and $\boldsymbol{n}_m$, a PE circle is uniquely defined by the point $\boldsymbol{f}_t$, the unit tangent vector $\boldsymbol{t}_m$ and the Minkowski curvature κ_m. This gives us the uniqueness of the osculating PE circle. Its existence will be shown by construction later in this section.

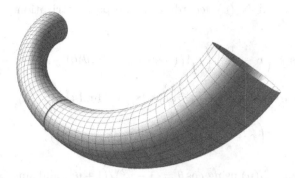

Fig. 5. A canal surface (left) and its osculating cyclide (right)

A *Dupin cyclide* is the canal surface associated with a PE circle in Minkowski space. As a curve in $\mathbb{R}^{3,1}$ has a unique osculating PE circle, a canal surface has a unique *osculating cyclide*, see Figure 5.

Theorem 7. *Along a characteristic circle, a canal surface is G^2 continuous with its osculating cyclide.*

Proof. We showed in Theorem 4 that if two curves in $\mathbb{R}^{3,1}$ are joined with G^2 continuity, so are the associated canal surfaces. By construction, the osculating PE circle is G^2 continuous with the curve, which proves the theorem.

Thus we can always extend a canal surface with G^2 continuity by joining it with its osculating cyclide.

In Section 4.1 we derived the principal curvatures κ_1 and κ_2 at a given point $\boldsymbol{p}$ on a surface. The two spheres with radii $1/\kappa_1$ and $1/\kappa_2$ in oriented contact with the surface at $\boldsymbol{p}$ are called its *osculating spheres*, and give us a geometric interpretation of the principal curvatures. A definition of a G^2 continuous join of two surfaces is that both tangent planes, osculating spheres, and principal curvature directions coincide along the join.

For the canal surface corresponding to the curve $\boldsymbol{f}(t) = (\boldsymbol{s}(t); r(t)) \in \mathbb{R}^{3,1}$, $\boldsymbol{f}_t$ is the osculating sphere corresponding to the principal curvature $\kappa_2 = 1/r(t)$. Along the characteristic circle $\boldsymbol{F}_t(u)$, the second osculating sphere $\widehat{\boldsymbol{c}}_t(u)$ is in oriented contact with $\boldsymbol{f}_t$. It is therefore the member with radius $1/\kappa_1$ of the parabolic family of spheres generated by $\boldsymbol{f}_t$ and $(\boldsymbol{F}_t(u); 0)$:

$$v\boldsymbol{f}_t + (1 - v)(\boldsymbol{F}_t(u); 0). \tag{36}$$

Solving the linear equation $vr(t) + 0 = 1/\kappa_1$ for v, we find that the osculating spheres along the characteristic circle $\boldsymbol{F}_t(u)$ are parametrized by

$$\widehat{\boldsymbol{c}}_t(u) = \left(\boldsymbol{F}_t(u) - \frac{1}{\kappa_1}\boldsymbol{N}_t(u); \frac{1}{\kappa_1} \right). \tag{37}$$

Expanding $\boldsymbol{F}_t(u)$ and $\boldsymbol{N}_t(u)$ according to the parametrization in (12), we can write

$$\widehat{\boldsymbol{c}}_t(u) = \left(\boldsymbol{s} + \left(r - \frac{1}{\kappa_1}\right)\left(\kappa\Lambda_1\left(\cos(\theta)\boldsymbol{n} + \sin(\theta)\boldsymbol{b}\right) - \frac{r'}{\nu}\boldsymbol{t}\right); \frac{1}{\kappa_1}\right). \tag{38}$$

Substituting the expression for κ_1 from (28), we find that

$$\left(r - \frac{1}{\kappa_1}\right) = -r\Lambda_1/\left(r\Lambda_1\Lambda_2 - \cos(\theta)\right). \tag{39}$$

By reparametrizing $\widehat{\boldsymbol{c}}_t(u)$ using $\cos\theta = (1-u^2)/(1+u^2)$ and $\sin\theta = 2u/(1+u^2)$, we see that for each t, the curve is quadratic in u. From [13], we know that this is in fact a PE circle, whose envelope is the *Dupin necklace* of the canal surface. However, since we know the explicit rational parametrization of the curve, we can show this directly.

Theorem 8. *The envelope of the family of osculating spheres $\widehat{\boldsymbol{c}}_t(u)$ is a Dupin cyclide.*

Proof. Dupin cyclides are characterized as the only surfaces which are canal surfaces with respect to two distinct one-parameter families of spheres. Considering the curve $\widehat{\boldsymbol{c}}_t(u) \in \mathbb{R}^{3,1}$, we can assume without loss of generality that the point $(\boldsymbol{s}; r)$ is located at the origin. The canal surface corresponding to $\widehat{\boldsymbol{c}}_t(u)$ is a Dupin cyclide iff there exists a second curve $\boldsymbol{c}_t(u)$ in $\mathbb{R}^{3,1}$ such that any point $(x_t, x_n, x_b; x_r) \in \boldsymbol{c}_t(u)$ is at zero distance from $\widehat{\boldsymbol{c}}_t(u)$ (here the variables x_t, x_n, x_b correspond to the Frenet frame of the spine and x_r to a fourth unit vector for the radius dimension):

$$\|\widehat{\boldsymbol{c}}_t(u) - (x_t, x_n, x_b; x_r)\|^2 = 0, \quad \forall u. \tag{40}$$

This is a quartic rational expression in u, so in order to eliminate it for any u, the five monomial coefficients in the numerator have to be identically zero. Two of them give the implicit equations of two hyperplanes:

$$x_b = 0, \quad \frac{\dot{r}}{\nu}x_t - \kappa\,\Lambda_1^2\Lambda_2 x_n - x_r = 0. \tag{41}$$

The remaining equations are equivalent, and gives us the implicit equation of the curve $\boldsymbol{c}_t(u)$ in the 2-dimensional intersection of the two hyperplanes:

$$\left(\kappa x_t + \frac{\dot{r}\Lambda_2 x_n}{\nu}\right)^2 + \left(\kappa_m\kappa\Lambda_1 x_n - \frac{1}{\kappa_m\Lambda_1}\right)^2 = \frac{1}{(\kappa_m\Lambda_1)^2} \tag{42}$$

The existence of this second curve proves the theorem.

Corollary 2. *The curve $\boldsymbol{c}_t(u)$ is the osculating PE circle of the curve $\boldsymbol{f}(t)$, and its envelope (the Dupin collar) the osculating cyclide of the canal surface.*

Proof. To prove the corollary, we need to show that $c_t(u)$ is contained in the 2-dimensional plane spanned by t_m and n_m, and that its Minkowski curvature is κ_m. The curve lies in the intersection of the two hyperplanes in (41), which both pass through the origin. It therefore suffices to show that t_m and n_m are contained in their intersection.

The vector t_m is collinear with $(t; \dot{r}/\nu) = (1, 0, 0; \dot{r}/\nu)$. This point satisfies both equations in (41), so t_m is contained in the intersection of the two hyperplanes.

When we consider the vector n_m, we can remove the component collinear with $(t; \dot{r}/\nu)$. What remains is proportional to $(n; -\kappa \Lambda_1^2 \Lambda_2) = (0, 1, 0; -\kappa \Lambda_1^2 \Lambda_2)$. This proves that n_m is also contained in the intersection of the two hyperplanes, and as t_m and n_m are not collinear, they span the 2-dimensional intersection.

The Minkowski curvature of $c_t(u)$ is found, e.g., by parametrizing (42) as a general conic, and then applying (18). We find that its curvature at the origin is indeed κ_m.

Remark 12. From the above proof we see that the restriction of the 2-dimensional plane in $\mathbb{R}^{3,1}$ spanned by t_m and n_m to the first three coordinates is the plane in $\mathbb{R}^3$ spanned by t and n. And though n is not the restriction of n_m, it is uniquely determined given t_m and n_m.

The Corollary proves the existence of the osculating PE circle, concluding the proof of Theorem 6.

4.5   Additional Properties of Canal Surfaces

Knowing the expressions for the two principal curvatures of a canal surface, we can easily calculate several other properties.

For example, the Gaussian curvature K is defined as

$$K = \kappa_1 \kappa_2 = \frac{\Lambda_1 \Lambda_2 - \cos\theta}{r\left(\Lambda_1\left(1 + \Lambda_2 r\right) - r\cos\theta\right)}. \tag{43}$$

We can then easily prove the following theorem:

Theorem 9. *The only developable regular canal surfaces are cones and cylinders.*

Proof. At regular points, the Gaussian curvature of a developable surface is identically zero. Inserting the expressions for Λ_1 and Λ_2 into (43), the numerator is linear in $\cos\theta$ giving us two expressions to eliminate: $\|\dot{f}\|^2 \nu^2 \kappa = 0$ and $\|\dot{f}\|(\nu \dot{r} - \ddot{r}\nu) = 0, \forall t$.

We have assumed $\|\dot{f}\| > 0$ except at the end of the canal surface, and $\nu \equiv 0$ gives us a canal surface with only one point as its spine. The remaining case is $\kappa = 0, \forall t$, which means that the spine is linear and ν is constant, so $\dot{\nu} = 0$. To eliminate the second expression, we therefore need $\ddot{r} = 0$, i.e., r is linear. Thus the only developable regular canal surfaces have linear spines and radius functions: they are cones and cylinders.

We can also calculate the mean curvature H of the canal surface:

$$H = \frac{\kappa_1 + \kappa_2}{2} = \frac{\Lambda_1\left(\frac{1}{2} + r\Lambda_2\right) - r\cos\theta}{r\left(\Lambda_1\left(1 + \Lambda_2 r\right) - r\cos\theta\right)}. \tag{44}$$

The first fundamental form determines whether the canal surface has any local self-intersections, as local self-intersections occur when $EG - F^2 = 0$. For a canal surface this gives us a condition on the radius function

$$|r| < \left|\frac{\Lambda_1}{\cos\theta - \Lambda_1\Lambda_2}\right|, \quad \forall t, \theta. \tag{45}$$

When the radius function is constant, i.e., for pipe surfaces, this is reduced to

$$|r| < \frac{1}{|\kappa\cos\theta|}, \tag{46}$$

which should be true for any θ. Since $\min_\theta 1/|\kappa\cos\theta| = 1/|\kappa|$, the pipe surface has no local self-intersections if

$$|r| < \frac{1}{|\kappa|} = |r_\kappa|, \quad \forall t. \tag{47}$$

Theorem 10. *As long as its radius is less than the minimal radius of curvature of the spine, a pipe surface has no local self-intersections.*

5   Applications: Piecewise Rational Corner Blends

To demonstrate the applications of the conditions for G^1 and G^2 continuity of canal surfaces, we present two constructions of blends of a three sided corner. The constructions can be generalized to certain n-sided corners, and to corners whose faces are patches on natural quadrics.

5.1   Sequential Corner Blends

Consider a three sided corner, whose edges are blended by patches on cylinders. If the radii of the three cylinders are equal, the associated lines in Minkowski space intersect in a point. Then a patch on the sphere corresponding to the point of intersection gives us a blend of the corner, which is G^1 continuous with the edge blends (Figure 6, left).

If only two of the radii are equal, we can blend the corner with a patch on a torus, again with G^1 continuity with the edge blend (Figure 6, right). In current CAD systems, these corner blends are implemented exactly. When the radii of the three cylinders are all different, however, current CAD systems have to resort to approximative blends. By constructing the blend sequentially, we can apply the fixed radius blend parametrization algorithm from [2] to construct a piecewise rational blend of the corner with internal G^1 continuity.

Fig. 6. Three sided corner blended with a patch on the sphere (left) and torus (right)

In this sequential construction, one of the edges of the corner is distinguished, and its blend is extended into the corner blend. We start by blending the other two edges (in Figure 7 we blend the edges with the largest radii). The remaining edge is then composed of three rational pieces:

1. a line segment: the intersection of the two faces,
2. an arc of circle: the intersection of the largest cylindrical edge blend with the opposing face, and
3. a segment of a rational quartic curve: the intersection of two cylinders tangent in a single point.

When constructing a blend with fixed radius R, we find its spine by intersecting the R-offsets of the original surfaces. And in this offset corner (Figure 7, left), the spine of the blend of the remaining edge (including the corner) is similarly composed of a line segment, an arc of circle, and a segment of a rational quartic. These three curves are joined with G^1 continuity, and the radius function of the blend is constant, so as a consequence of Theorem 3 the resulting composite edge-corner blend is also internally G^1 continuous.

When constructing the rational parametrization of the blend, we parametrize each of the components separately according to the closed formulae provided in [2].

5.2 Spherical Corner Blends

The spherical corner blend is a generalization of the case where the three cylindrical edge blends have the same radius. Consider the three curves in $\mathbb{R}^{3,1}$ corresponding to the three edge blends of the corner. Only in rare cases will they intersect in a single point. In the general case, we construct transitional curves from given points on the edge blend curves, to a common point. This will give us three transitional edge blends, and if we choose the common point properly,

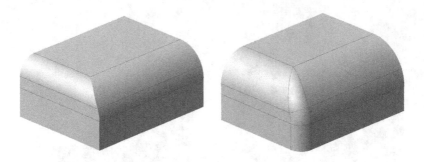

Fig. 7. Sequential corner blend

it corresponds to a vertex sphere (Figure 8, right) tangent to all three faces, on which we determine a three-sided patch closing the composite corner blend.

The spheres tangent to all three faces constitutes a linear one-parameter family of spheres, and the associated cone is inscribed in the corner. This gives us one degree of freedom in the choice of a vertex sphere.

The spine of an edge blend lies in the bisector of the adjoining faces, which is a plane through the edge. Any rational spine can therefore be parametrized as a planar rational Bézier curve. The radius of the blend is the distance of the spine to the two adjoining faces. If we, for simplicity, assume that the vertex of the corner is in the origin and the unit normal vectors of the three faces are e_1, e_2, and e_3, then the radius function of the edge blend with spine s of the face with normal vector e_i is $s.e_i$. Thus any rational curve in the bisector plane will give us a rational edge blend. The construction of edge blends can thus be reduced to the construction of rational Bézier curves in the bisector plane of the adjoining faces, and ultimately to the construction of control polygons with associated weights.

For a spherical corner blend, the construction of a transitional edge blend is therefore reduced to the construction of a transitional control polygon connecting the spine curve of the edge blend to the centre of the vertex sphere. Let p_0 be the endpoint of the edge spine curve, p_n the centre of the vertex sphere, and $\{p_1, \ldots, p_{n-1}\}$ the control points in between.

If we want the corner blend to be internally G^1 continuous, it is sufficient that p_1 is on the tangent line at the end of the edge spine curve, as the join with the vertex sphere is G^1 continuous by construction.

In order to achieve internal G^2 continuity, we have to apply the results from Sections 4.2 and 4.3. Recalling Remark 11, in order to have a G^2 join with the vertex sphere, we choose $p_{n-1} = p_n$ in order to have a double control point at the vertex sphere.

In $\mathbb{R}^{3,1}$, the bisector surface of the two hyperplanes corresponding to the two adjoining faces is a 2-dimensional plane. This constrains any curve corresponding to a blend to the plane spanned by t_m and n_m. If the corner blend is internally G^1 continuous, the remaining condition to ensure internal G^2 continuity is that

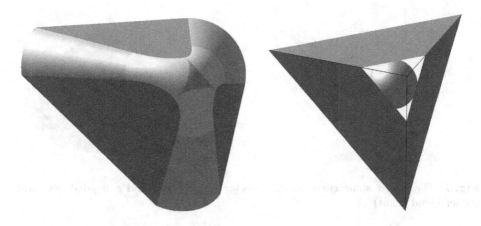

Fig. 8. Spherical corner blend of a three sided corner

the spines of the edge blend and the transitional edge blend must have the same curvature κ at p_0. For a rational Bézier curve, the curvature at an endpoint is

$$\kappa = \frac{w_0 w_2}{w_1^2} \frac{n-1}{n} \frac{h}{a^2} \tag{48}$$

where w_i are the weights, n the degree of the spine curve, a the length of the first edge of the control polygon, and h the height of the third control point above the line containing the first edge. Given the curvature of the spine curve of the edge blend at p_0, requiring G^2 continuity therefore gives only one additional constraint in our choice of control polygon and weights.

Summarizing the requirements for a spherical corner blend with internal G^2 continuity:

Theorem 11. *A spherical corner blend is internally G^2 continuous if the control points $\{p_0, \ldots, p_n\}$ and weights $\{w_0, \ldots, w_n\}$ of the transitional edge blends satisfy the following conditions:*

1. *p_1 is on the tangent line at the end of the edge spine curve,*
2. *$p_{n-1} = p_n$, and*
3. *$\kappa = \frac{w_0 w_2}{w_1^2} \frac{n-1}{n} \frac{h}{a^2}$ where κ is the curvature at the end of the edge spine curve, $a = |p_1 - p_0|$, and h the distance of p_2 from the line spanned by p_0 and p_1.*

The transitional edge blends of two adjacent edges meet in at least one point: the point where the vertex sphere touch their common face, i.e., the corner of the triangular patch on the vertex sphere. In order to avoid an overlap of adjacent transitional edge blends at the vertex sphere, we impose an additional condition on their control polygons. For a three sided corner, the penultimate control point p_{n-1} has to be in outside the triangle defined by p_n, the origin (the vertex of the corner), and $2R e_i$, if the vertex sphere corresponds to the point $(p_n; R)$

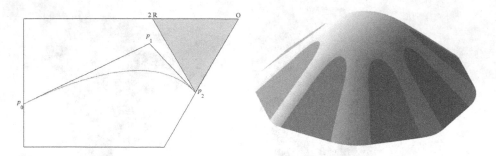

Fig. 9. The Bézier spine curve in the bisector plane (left), and a n-sided spherical corner blend (right)

and the edge has direction vector e_i (Figure 9, left). A similar condition can be formulated for n-sided corners (Figure 9, right).

6   Conclusions

We have shown that, as envelopes of one-parameter families of spheres, canal surfaces inherit their geometric continuity properties from the associated curve in Minkowski space $\mathbb{R}^{3,1}$: if two curves are joined with G^1 or G^2 continuity, so are the corresponding canal surfaces. By extending the differential geometry of curves in $\mathbb{R}^3$ to $\mathbb{R}^{3,1}$, we find that if a curve has well defined unit tangent and principal normal vectors in a point, it has a unique osculating PE circle, and the corresponding canal surface has a unique osculating cyclide. The osculating PE circle is G^2 continuous with the curve, so the osculating cyclide is G^2 continuous with the canal surface.

Rational curves in $\mathbb{R}^{3,1}$ correspond to rational canal surfaces, which can be parametrized using Algorithm 2 with a minimal bi-degree $(n, 2)$.

Combining these two independent results we can construct piecewise rational rolling ball blends of edges and corners of patchworks of planes and natural quadrics, with internal G^1 and G^2 continuity.

Acknowledgements This paper has been partially financed by the Marie-Curie Initial Training Network SAGA (ShApes, Geometry, Algebra), FP7-PEOPLE contract PITN-GA-2008-214584.

References

1. Cottrell, J.A., Hughes, T.J.R., Bazilevs, Y.: Isogeometric analysis: Toward integration of CAD and FEA. John Wiley and Sons (2009)
2. Dahl, H.E.I., Krasauskas, R.: Rational fixed radius rolling ball blends between natural quadrics. Computer-Aided Geometric Design 29, 691–706 (2012)

3. Krasauskas, R.: Minimal rational parametrizations of canal surfaces. Computing 79(2-4), 281–290 (2007)
4. Peternell, M., Pottmann, H.: Computing rational parametrizations of canal surfaces. Journal of Symbolic Computation 23(2-3), 255–266 (1997)
5. Dietz, R., Hoschek, J., Jüttler, B.: An algebraic approach to curves and surfaces on the sphere and on other quadrics. Computer Aided Geometric Design 10(3-4), 211–229 (1993)
6. Krasauskas, R., Kazakevičiūté, M.: Universal rational parametrizations and spline curves on toric surfaces. In: Dokken, T., Jüttler, B. (eds.) Computational Methods for Algebraic Spline Surfaces, pp. 213–231 (2005)
7. Moore, E.H.: Algebraic surfaces of which every plane-section is unicursal in the light of n-dimensional geometry. American Journal of Mathematics 10(1), 17–28 (1887)
8. Xu, Z., Feng, R., Sun, J.-G.: Analytic and algebraic properties of canal surfaces. Journal of Computational and Applied Mathematics 195(1-2), 220–228 (2006)
9. Farouki, R.T., Sakkalis, T.: Rational space curves are not "unit speed". Computer Aided Geometric Design 24(4), 238–240 (2007)
10. Yilmaz, S., Turgut, M.: On the differential geometry of the curves in Minkowski space-time I. Int. J. Contemp. Math. Sciences 3(27), 1343–1349 (2008)
11. Peternell, M., Pottmann, H.: Applications of Laguerre geometry in CAGD. Computer Aided Geometric Design 15(2), 165–186 (1998)
12. Krasauskas, R., Mäurer, C.: Studying cyclides with Laguerre geometry. Computer Aided Geometric Design 17(2), 101–126 (2000)
13. Bartoszek, A., Langevin, R., Walczak, P.G.: Special canal surfaces of S^3. Bulletin of the Brazilian Mathematical Society 42(2), 301–320 (2010)

Macro-element Hierarchical Riesz Bases

Oleg Davydov and Wee Ping Yeo

Department of Mathematics and Statistics, University of Strathclyde,
26 Richmond Street, Glasgow G1 1XH, Scotland
{oleg.davydov,weeping.yeo}@strath.ac.uk

Abstract. We show that a nested sequence of C^r macro-element spline spaces on quasi-uniform triangulations gives rise to hierarchical Riesz bases of Sobolev spaces $H^s(\Omega)$, $1 < s < r + \frac{3}{2}$, and $H_0^s(\Omega)$, $1 < s < \sigma + \frac{3}{2}$, $s \notin \mathbb{Z} + \frac{1}{2}$, as soon as there is a nested sequence of Lagrange interpolation sets with uniformly local and bounded basis functions, and, in case of $H_0^s(\Omega)$, the nodal interpolation operators associated with the macro-element spaces are boundary conforming of order σ. In addition, we provide a brief review of the existing constructions of C^1 Largange type hierarchical bases.

Keywords: Hierarchical bases, Riesz bases, macro-elements, bivariate splines, Jackson inequality, Bernstein inequality.

1 Introduction

Smooth macro-element spaces are among most practically useful spaces of piecewise polynomial splines in two and three space dimensions, see [20]. They are available on arbitrary polygonal domains and possess stable local bases and hence full approximation order. Some of them are refinable and therefore suitable for the multiresolution analysis [5–7, 9, 11–15, 17, 26, 27], with applications in particular to multilevel methods in numerical partial differential equations and surface modelling.

Given a sequence of nested spline spaces $S_0 \subset S_1 \subset \cdots \subset S_n \subset \cdots$, and corresponding nested interpolation sets $\Xi_0 \subset \Xi_1 \subset \cdots \subset \Xi_n \subset \cdots$ with Lagrange bases $\{B_\xi^{(n)}\}_{\xi \in \Xi_n}$, *hierarchical bases* are obtained from the appropriately re-scaled functions

$$B_\xi^{(n)}, \quad \xi \in \Xi_n \setminus \Xi_{n-1}, \qquad n = 0, 1, \ldots \ (\Xi_{-1} := \emptyset).$$

The most famous example is given by the piecewise linear basis functions (hat functions), where the hierarchical basis is used for the multilevel preconditioning of the discretised second order elliptic equations [31]. The effectiveness of this method is related to the Riesz basis (or "stability") property of this hierarchical basis in the Sobolev spaces $H^s(\Omega)$ and $H_0^s(\Omega)$, $1 < s < \frac{3}{2}$. For elliptic equations of forth order, stability in $H^2(\Omega)$ and $H_0^2(\Omega)$ is needed, and this can be achieved by C^1 hierarchical bases [12] that are Riesz bases in the range $1 < s < \frac{5}{2}$. In fact,

M. Floater et al. (Eds.): MMCS 2012, LNCS 8177, pp. 112–134, 2014.
© Springer-Verlag Berlin Heidelberg 2014

as noted in [21], bases with stability in $H^s(\Omega)$ with as large as possible range of s is advantageous, in particular when an elliptic operator includes parts of different order. Moreover, a good preconditioning effect is expected when s corresponding to a given variational problem lies in the central part of the stability interval.

In this paper we study general conditions for the nested sequences of macro-element spline spaces to give rise to Riesz bases in $H^s(\Omega)$ and $H_0^s(\Omega)$. The main results (see Theorem 5) show that the stability range $1 < s < r + \frac{3}{2}$ in $H^s(\Omega)$ is guaranteed for refinable C^r macro-elements on quasi-uniform triangulations in $\mathbb{R}^2$ if the Lagrange bases $\{B_\xi^{(n)}\}_{\xi \in \Xi_n}$ are uniformly local and bounded, and the nodal bases of the macro-element spaces are also uniformly bounded. Moreover, the same stability range (up to the half-integer values) is obtained in $H_0^s(\Omega)$ if the macro-element nodal (Hermite) interpolation operators Π_n are *boundary conforming* of order r in the sense that for any function f vanishing on the boundary of Ω together with its derivatives up to order r, the interpolants $\Pi_n f$ have the same property.

The paper is organised as follows. In Section 2 we list some auxiliary results on K-functionals, interpolation spaces and Sobolev spaces $H^s(\Omega)$ and $H_0^s(\Omega)$. Section 3 is devoted to Bernstein and Jackson inequalities for bivariate splines, including the Bernstein inequality in $H^s(\Omega)$ for spline spaces possessing stable local bases, and error bounds for the macro-element nodal interpolation of functions in Sobolev spaces of integer order. General results on hierarchical bases of Lagrange type are given in Section 4, whereas C^1 macro-element spaces where such bases are known are reviewed in Section 5. In particular, we verify that the sequence of nested triangulations suggested in [12] is quasi-uniform.

Throughout we employ the usual notation $a \lesssim b$ and $a \sim b$ to indicate that the inequality (respectively, the double inequality) includes bounding constants which are not of interest. The parameters on which these constants may depend are either explicitly mentioned or clear from the context.

2   Preliminaries

We denote by $W_p^k(\Omega)$, $k \in \mathbb{N}$, $1 \le p \le \infty$, the usual Sobolev spaces on a bounded Lipschitz domain Ω. The space $C^k(\Omega) \subset W_\infty^k(\Omega)$ consists of all k times continuously differentiable functions f on the closure of Ω, with $\|f\|_{C^k(\Omega)} = \|f\|_{W_\infty^k(\Omega)}$. The space $W_2^k(\Omega)$ is also denoted by $H^k(\Omega)$, with $H^0(\Omega) := L_2(\Omega)$. It is a Hilbert space with inner product

$$\langle f, g \rangle_{H^k(\Omega)} = \langle f, g \rangle_{L_2(\Omega)} + \sum_{|\alpha|=k} \left\langle \frac{\partial^\alpha f}{\partial x^\alpha}, \frac{\partial^\alpha g}{\partial x^\alpha} \right\rangle_{L_2(\Omega)},$$

where $\alpha = (\alpha_1, \ldots, \alpha_n) \in \mathbb{Z}_+^n$ is a multi-index, with $|\alpha| := \alpha_1 + \cdots + \alpha_n$.

Let X and $Y \subset X$ be two Hilbert spaces with norms $\|\cdot\|_X$ and $\|\cdot\|_Y = \|\cdot\|_X + |\cdot|_Y$, respectively, where $|\cdot|_Y$ is a seminorm. The K-functional is defined for each $f \in X$ and $t > 0$ by

$$K_{XY}(f, t) := \inf_{g \in Y} \|f - g\|_X + t|g|_Y,$$

or equivalently (see [22, Remark 4.8]) by the same expression with $|g|_Y$ replaced by $\|g\|_Y$.

One of the key properties of the K-functional is the following Jackson type inequality.

Lemma 1. *Let S be linear subspace of X. Suppose that for some $t > 0$,*

$$\inf_{s \in S} \|g - s\|_X \le t|g|_Y, \quad \text{for all} \quad g \in Y.$$

Then for any $f \in X$,

$$\inf_{s \in S} \|f - s\|_X \le K_{XY}(f, t).$$

Proof. Indeed,

$$\inf_{s \in S} \|f - s\|_X \le \inf_{g \in Y} \inf_{s \in S} (\|f - g\|_X + \|g - s\|_X) \le K_{XY}(f, t)$$

if the assumption holds.   □

According to the K-method [1], the interpolation space $[X, Y]_\theta$, $0 < \theta < 1$, consists of all $f \in X$ for which the functional

$$|f|_{\theta;K} = \left(\int_0^\infty \left(t^{-\theta} K_{XY}(f, t) \right)^2 \frac{dt}{t} \right)^{1/2} \tag{1}$$

is finite. Given any $\alpha > 1$, by splitting the domain of integration $(0, \infty)$ into the intervals $(\alpha^{-n-1}, \alpha^{-n})$, $n = 0, 1, \dots$, and $(1, \infty)$, and using standard properties of the K-functional, it is easy to show that

$$|f|_{\theta;K} \sim \left(\sum_{n=0}^\infty \left[\alpha^{n\theta} K_{XY}(f, \alpha^{-n}) \right]^2 \right)^{1/2}, \tag{2}$$

where the constants of equivalence depend only on θ and α.

The k-*th modulus of smoothness* of $f \in L_p(\Omega)$, $0 < p \le \infty$, is defined by

$$\omega_k(f, t)_p = \sup_{|\delta| < t} \|\Delta_\delta^k f\|_{L_p(\Omega_{k\delta})},$$

where $|\delta|$ denotes the Euclidean length of $\delta \in \mathbb{R}^n$, $\Omega_{k\delta} := \{x \in \Omega : x + j\delta \in \Omega, \ j = 0, \dots, k\}$, and

$$(\Delta_\delta^k f)(x) := \sum_{j=0}^k \binom{k}{j} (-1)^{k-j} f(x + j\delta), \quad x \in \mathbb{R}^n,$$

is the usual difference operator. By [27, Theorem 1], the modulus of smoothness is equivalent to the K-functional,

$$\omega_k(f, t)_2 \sim K_{L_2, H^k}(f, t^k), \qquad t > 0. \tag{3}$$

Therefore, in view of Lemma 1, error bounds for functions in Sobolev spaces immediately lead to Jackson type estimates in terms of the modulus of smoothness.

The Sobolev spaces $H^s(\Omega)$ of a fractional order $s > 0$ can be defined as interpolation spaces

$$H^s(\Omega) = \big[L_2(\Omega),\, H^k(\Omega)\big]_\theta,$$

where $s = k\theta$, k integer, $0 < \theta < 1$. In view of (1) and (3),

$$|f|_{H^s(\Omega)} \sim \left(\int_0^\infty (t^{-s}\omega_k(f,t)_2)^2 \frac{dt}{t} \right)^{1/2}. \tag{4}$$

Let $C_c^\infty(\Omega)$ be the linear space of all infinitely differentiable functions on Ω with compact support contained in Ω. We use $H_0^s(\Omega)$ to denote the closure of $C_c^\infty(\Omega)$ in $H^s(\Omega)$. It is well known [19] that $C_c^\infty(\Omega)$ is dense in $H^s(\Omega)$ if and only if $s \leq \frac{1}{2}$. If $s > \frac{1}{2}$ and the boundary of Ω is smooth, then $H_0^s(\Omega)$ is a proper subspace of $H^s(\Omega)$ given by

$$H_0^s(\Omega) = \left\{ u \in H^s(\Omega) : \frac{\partial^\alpha u}{\partial x^\alpha} = 0 \text{ on } \partial\Omega, \text{ for all } 0 \leq |\alpha| < s - \frac{1}{2},\, \alpha \in \mathbb{Z}^n \right\},$$

see [19, Theorem 11.5]. Hence, $H_0^s(\Omega) = H^s(\Omega)$ if $s \leq \frac{1}{2}$ and $H_0^s(\Omega) = H^s(\Omega) \cap H_0^{s_0}(\Omega)$, where $s_0 = \lceil s - \frac{1}{2} \rceil$ if $s > \frac{1}{2}$. According to [19, Theorem 11.6] the spaces $H_0^s(\Omega)$ of fractional order $s \notin \mathbb{Z} + \frac{1}{2}$ can be obtained from the integer order spaces $H_0^k(\Omega)$, $k > s$, by interpolation

$$H_0^s(\Omega) = \big[L_2(\Omega), H_0^k(\Omega)\big]_\theta, \quad \theta = \frac{s}{k}, \quad s \notin \mathbb{Z} + \frac{1}{2}. \tag{5}$$

For $s \in \mathbb{Z} + \frac{1}{2}$ a description of the interpolation spaces $H_{00}^s(\Omega) := \big[L_2(\Omega), H_0^k(\Omega)\big]_\theta$, $\theta = \frac{s}{k}$, can be found in [19, Theorem 11.7].

For a domain $\Omega \subset \mathbb{R}^2$ with piecewise smooth boundary in the sense of [16, p. 34], which includes the case of Lipschitz polygonal domains, the interpolation property (5) has been shown in [32]. As shown in [16], $H_0^s(\Omega)$, $s \notin \mathbb{Z} + \frac{1}{2}$, in this case coincides with the space $\tilde{H}^s(\Omega)$ of all those functions $f \in H^s(\Omega)$ whose extension to $\mathbb{R}^2$ by zero belongs to $H^s(\mathbb{R}^2)$. See also [2] for (5) in the case of a bounded Lipschitz domain in any space dimensions and integer s.

3   Bernstein and Jackson Inequalities for Bivariate Splines

We first recall standard definitions from the theory of bivariate piecewise polynomial splines, see [20] for more details.

Let Ω be polygonal domain in $\mathbb{R}^2$ and Δ a finite collection of (closed) triangles whose union coincides with Ω. We assume that the intersection of any two triangles in Δ is empty, or a common vertex, or a common edge of them. Then Δ is a *triangulation* of Ω. The length of an edge e of Δ is denoted by $|e|$. Let ξ be the set of all edges of Δ. The maximum length of the edges of Δ, denoted by $h = h_\Delta = \sup_{e \in \mathcal{E}} |e|$, is called the *diameter* or *mesh size* of Δ. We denote the smallest angle of the triangles $T \in \Delta$ by β_Δ, and set

$$\gamma_\Delta = \min\{\operatorname{diam} T : T \in \Delta\}/h_\Delta.$$

A family of triangulations is called *regular* if $\beta_\Delta \geq \beta > 0$ for every Δ in the family. A regular family is said to be *quasi-uniform* if $\gamma_\Delta \geq \gamma > 0$ for every Δ.

For any positive integer d, let $S_d(\Delta)$ denote the space of all piecewise polynomials of degree d with respect to Δ. In other words, $s \in S_d(\Delta)$ if and only if, on each triangle $T \in \Delta$, s agrees with a polynomial in $\mathbb{P}_d$, the space of all bivariate polynomials of total degree at most d. For any $r = 0, 1, \ldots, d-1$, let

$$S_d^r(\Delta) := S_d(\Delta) \cap C^r(\Omega)$$

be the space of all piecewise polynomials of degree d and smoothness r with respect to Δ.

Let $\{s_1, \ldots, s_N\}$ be a basis for a linear space $S \subset S_d(\Delta)$. We say that the basis is *m-local* if for each $i = 1, \ldots, N$ there is a triangle $T_i \in \Delta$ such that $\operatorname{supp} s_i \subset \operatorname{star}^m(T_i)$. Here $\operatorname{star}^k(T) := \operatorname{star}(\operatorname{star}^{k-1}(T))$ for $k \geq 2$, where if U is the union of a cluster of triangles, then $\operatorname{star}(U) = \operatorname{star}^1(U)$ is the union of all triangles in Δ that have a non-empty intersection with U. A basis is called *local* if it is m-local for some m.

Suppose that $\{\lambda_1, \ldots, \lambda_N\} \subset S^*$ is the dual basis, that is,

$$\lambda_i s_j = \begin{cases} 1, & i = j, \\ 0, & \text{otherwise.} \end{cases}$$

A basis $\{s_1, \ldots, s_N\}$ for $S \subset S_d(\Delta)$ is said to be a *stable local basis* [8] if for an integer m and positive constants C_1, C_2,

(a) $\{s_1, \ldots, s_N\}$ is m-local,
(b) $|\lambda_i s| \leq C_1 \|s\|_{L_\infty(\operatorname{star}^m(T_i))}$ for all $s \in S$, $i = 1, \ldots, N$, and
(c) $\|s_i\|_{L_\infty(\Omega)} \leq C_2$, $i = 1, \ldots, N$.

Any stable local basis is L_p-*stable* for all $1 \leq p \leq \infty$ after appropriate renorming, that is, for any $\alpha = (\alpha_1, \ldots, \alpha_N) \in \mathbb{R}^N$,

$$k_1 C_2^{-1} \|\alpha\|_{l_p} \leq \left\| \sum_{i=1}^N \alpha_i \frac{s_i}{|\operatorname{supp} s_i|^{1/p}} \right\|_{L_p(\Omega)} \leq k_2 C_1 \|\alpha\|_{l_p}, \quad 1 \leq p \leq \infty,$$

where k_1, k_2 are some constants depending only on p, r, d and m, and $|M|$ denotes the area of a set $M \subset \mathbb{R}^2$.

3.1 Bernstein Inequality

Functions in subspaces of $S_d^r(\Delta)$ possessing a stable local basis satisfy a Bernstein type inequality in the norm of $H^s(\Omega)$ for all $0 < s < r + \frac{3}{2}$.

Theorem 1 (Bernstein Inequality). *Suppose that* $S \subset S_d^r(\Delta)$ *has a stable local basis* $\{\phi_i\}_{i \in I}$. *Then for any* $f \in S$,

$$\|f\|_{H^s(\Omega)} \lesssim h_\Delta^{-s} \|f\|_{L_2(\Omega)}, \qquad 0 < s < r + \frac{3}{2}, \tag{6}$$

where the bounding constant depends only on $s, r, d, \beta_\Delta, \gamma_\Delta$ *and the parameters* m, C_1, C_2 *of the stable local basis.*

Under slightly different assumptions on S, a proof of the Bernstein inequality can be found in [27], see also [29]. We provide a proof based on the following lemma.

Lemma 2 ([17, Lemma 2.2]). *Let* $f \in S_d^r(\Delta)$. *Then* $f \in H^s(\Omega)$ *for all* $s < r + \frac{3}{2}$, *and*

$$\|f\|_{H^s(\Omega)} \lesssim h_\Delta^{-s} \|f\|_{L_2(\Omega)}, \qquad 0 < s < r + \frac{3}{2}, \tag{7}$$

where the bounding constant depends only on $s, r, d, \beta_\Delta, \gamma_\Delta$ *and the number of triangles* $T \in \Delta$ *in the support of* f.

Proof (of Theorem 1). Since $\{\phi_i\}_{i \in I}$ is a stable local basis, the functions $\psi_i = |\operatorname{supp} \phi_i|^{-1/2} \phi_i$, $i \in I$, form an L_2-stable basis for S. In particular $\|\psi_i\|_{L_2(\Omega)} \leq M$, where M depends only on the parameters m, C_1, C_2 of the stable local basis.

Let $f = \sum_{i \in I} c_i \psi_i$, so that $\|f\|_{L_2(\Omega)}^2 \sim \sum_{i \in I} |c_i|^2$. Choose an integer $k > s$. Since the basis $\{\psi_i\}_{i \in I}$ is m-local, it is not difficult to see that

$$I_k(f, \delta)^2 \lesssim \sum_{i \in I} |c_i|^2 I_k(\psi_i, \delta)^2, \quad \text{where } I_k(f, \delta) := \|\Delta_\delta^k f\|_{L_2(\Omega_{k\delta})}, \quad \delta \in \mathbb{R}^2.$$

Hence,

$$\omega_k(f, t)_2^2 \lesssim \sum_{i \in I} |c_i|^2 \omega_k(\psi_i, t)_2^2,$$

and by (4),

$$|f|_{H^s(\Omega)}^2 \lesssim \sum_{i \in I} |c_i|^2 \int_0^\infty (t^{-s} \omega_k(\psi_i, t)_2)^2 \frac{dt}{t} \sim \sum_{i \in I} |c_i|^2 |\psi_i|_{H^s(\Omega)}^2.$$

By applying Bernstein inequality (7) to the locally supported functions ψ_i and using the L_2-stability of the basis $\{\psi_i\}_{i \in I}$, in particular, uniform L_2-boundedness of ψ_i, we obtain

$$\|f\|_{H^s(\Omega)}^2 \lesssim h_\Delta^{-2s} \sum_{i \in I} |c_i|^2 \|\psi_i\|_{L_2(\Omega)}^2 \lesssim h_\Delta^{-2s} \sum_{i \in I} |c_i|^2 \lesssim h_\Delta^{-2s} \|f\|_{L_2(\Omega)}^2,$$

which completes the proof. $\qquad\qquad\qquad\qquad\qquad\qquad\qquad\qquad\qquad\qquad\qquad$ $\square$

3.2 Jackson Inequality for Macro-element Spline Spaces

We restrict our attention to the macro-element spaces, see [20, Section 5.10], because of the availability of boundary conforming interpolation operators that allow appropriate treatment of subspaces with zero boundary conditions.

Recall that a linear functional λ is called a *nodal functional* provided λf is a scalar multiple of the value of f or its (directional) partial derivative at some point $\eta = \eta(\lambda) \in \mathbb{R}^2$, that is $\lambda f = \gamma \frac{\partial^{\nu+\mu} f}{\partial \sigma^\nu \partial \tau^\mu}(\eta)$, for suitable $\nu, \mu \in \mathbb{Z}_+$, $\eta \in \Omega$, unit vectors σ, τ, and a scaling coefficient $\gamma \in \mathbb{R}$. The number $\kappa(\lambda) = \nu + \mu$ is called the *order* of λ.

A collection $\mathcal{N} = \{\lambda_i\}_{i=1}^N$ is called a *nodal determining set* for a spline space $S \subset S_d(\Delta)$ if every $s \in S$ is $\kappa(\lambda)$ times continuously differentiable at $\eta(\lambda)$, and $\lambda s = 0$ for all $\lambda \in \mathcal{N}$ implies $s \equiv 0$. $\mathcal{N}$ is called a *nodal minimal determining set* (NMDS) for S if there is no smaller nodal determining set. In other words, $\mathcal{N}$ is an NMDS if it is a basis for the dual S^* of S. Let $\{s_i\}_{i=1}^N$ be the basis of S dual to $\mathcal{N}$, called the *nodal basis*.

We will work with spaces of splines that are defined on triangulations $\Delta_R = \bigcup_{K \in \Delta} K_R$ obtained from a given partition Δ of Ω into polygonal cells K by applying some refinement process to each $K \in \Delta$. Examples are provided by Clough-Tocher and Powell-Sabin splits of the triangles of a triangulation Δ of Ω. We assume that each K is star-shaped with respect to a disk. We denote by χ_K the *chunkiness parameter* $\operatorname{diam} K / \rho_{\max}$ of K, where $\rho_{\max}$ is the maximum radius of disks with respect to which K is star-shaped [3, Section 4.3]. Recall that χ_K is bounded in terms of the minimum angle of K if K is a triangle. We set $\chi_\Delta := \max_{K \in \Delta} \chi_K$.

For each cell $K \in \Delta$, we define

$$\mathcal{N}_K = \{\lambda \in \mathcal{N} : \eta(\lambda) \in K\}.$$

We call $S \subset S_d(\Delta_R)$ a *macro-element space* provided there is a NMDS $\mathcal{N}$ for S such that for each $K \in \Delta$, $S|_K$ is uniquely determined from the values $\{\lambda s\}_{\lambda \in \mathcal{N}_K}$. It is easy to see that the support of a basis function s_i in a macro-element space is contained in the union of all $K \in \Delta$ containing $\eta(\lambda_i)$. For each $\lambda_i \in \mathcal{N}$, we choose the scaling coefficient γ to be equal to $\gamma_i = \operatorname{diam}(T_i)^{\kappa(\lambda_i)}$, where $T_i \in \Delta_R$ is a triangle containing $\eta(\lambda_i)$. Note that $\operatorname{diam}(T_i) \sim \operatorname{diam}(T')$ for any other triangle $T' \in \Delta_R$ sharing a vertex with T_i, with the constant of equivalence depending only on β_{Δ_R}, see [20, Section 4.7], and $\operatorname{diam}(T_i) \sim \operatorname{diam}(K)$, where $T_i \subset K \in \Delta$, and the constant of equivalence depends only on β_{Δ_R} and $\nu_{\Delta_R} := \max_{K \in \Delta} |K_R|$. Then by Markov inequality [20, Theorem 2.32] $|\lambda_i s| \leq C_1 \|s\|_{L_\infty(T_i)}$ for any $s \in S$, where C_1 depends only on d, $\kappa(S) := \max_i \kappa(\lambda_i)$ and β_{Δ_R}. It follows that $\{s_i\}_{i=1}^N$ is a stable local basis for S with parameters depending only on d, $\kappa(S)$, β_{Δ_R} and ν_{Δ_R} as soon as $\|s_i\|_{L_\infty(\Omega)} \leq C_2$, $i = 1, \ldots, N$, for some constant C_2.

The interpolation operator $\Pi : C^{\kappa(S)}(\Omega) \to S$ is defined by

$$\Pi f = \sum_{i=1}^N \lambda_i(f) s_i. \tag{8}$$

By the duality of the basis functions s_i, it is clear that $\Pi s = s$ for all $s \in S$. In particular, Π reproduces polynomials of degree at most k if $\mathbb{P}_k \subset S$. The definition of the macro-element space implies that the *local interpolation operators* $\Pi_K : C^{\kappa(S)}(K) \to S|_K$,

$$\Pi_K f = \sum_{i:\,\eta(\lambda_i)\in K} \lambda_i(f)s_i$$

satisfy $\Pi_K f = (\Pi f)|_K$ for $f \in C^{\kappa(S)}(\Omega)$.

We say that the interpolation operator Π is *boundary conforming* of order σ if the homogeneous boundary conditions of order σ are preserved by the interpolant, that is, if

$$\frac{\partial^{\nu+\mu} f}{\partial x^\nu \partial y^\mu} = 0 \quad \text{on } \partial\Omega, \qquad \text{for all} \quad \nu, \mu \geq 0, \quad \nu + \mu \leq \sigma,$$

implies

$$\Pi f \in S_{0,\sigma} := \left\{ s \in S : \frac{\partial^{\nu+\mu} s}{\partial x^\nu \partial y^\mu} = 0 \text{ on } \partial\Omega, \text{ for all } \nu, \mu \geq 0, \, \nu + \mu \leq \sigma \right\}.$$

The proof of the following version of the Jackson inequality follows the scheme used in [3, Section 4.4], where it is proved for *finite elements*, thus making an assumption of *affine equivalence* of the spaces $S|_K$, $K \in \Delta$. In place of affine equivalence, we only assume that the nodal basis is uniformly bounded, see (9).

Theorem 2 (Jackson Inequality). *Let $S \subset S_d^r(\Delta_R)$ be a macro-element space such that $\mathbb{P}_k \subset S$ for some $1 \leq k \leq d$, and $\kappa(S) \leq k-1$. Assume that its nodal basis $\{s_i\}_{i=1}^N$ satisfies*

$$\|s_i\|_{L_\infty(\Omega)} \leq C_2, \qquad i = 1, \ldots, N. \tag{9}$$

Then for every $f \in H^{k+1}(\Omega)$,

$$\|f - \Pi f\|_{H^\nu(\Omega)} \leq C h_\Delta^{k+1-\nu} |f|_{H^{k+1}(\Omega)}, \qquad \nu = 0, \ldots, \min\{r, k\} + 1, \tag{10}$$

where C depends only on d, β_{Δ_R}, ν_{Δ_R}, χ_Δ and C_2.

Proof. Recall that by Sobolev embedding theorem any function $f \in H^{k+1}(\Omega)$ belongs (after possible modification on a set of zero measure) to $C^{k-1}(\Omega)$. This implies that Πf is well defined for all $f \in H^{k+1}(\Omega)$, and $f - \Pi f \in H^{r+1}(\Omega)$ since $S_d^r(\Delta_R) \subset H^{r+1}(\Omega)$.

Given any $K \in \Delta$, we define

$$\hat{K} := \left\{ \frac{x}{\operatorname{diam}(K)} : x \in K \right\}.$$

Then $\operatorname{diam} \hat{K} = 1$ and hence $|\hat{K}| \leq \pi/4$. For any function g defined on K we set $\hat{g}(y) := g(\operatorname{diam}(K)y)$, $y \in \hat{K}$. The functions

$$\hat{s}_i := \widehat{s_i|_K}, \qquad \text{for all } i \text{ such that } \lambda_i \in \mathcal{N}_K,$$

form a basis for the spline space $\hat{S}_K := \{\hat{s} : s \in S|_K\}$ on $\hat{K}$, with its dual basis given by the linear functionals $\hat{\lambda}_i(\hat{g}) := \lambda_i(g)$, $g \in C^{k-1}(K)$. Since $\mathrm{diam}(T_i) \sim \mathrm{diam}(K)$, we have

$$\hat{\lambda}_i \hat{g} = \mathrm{diam}(T_i)^{\nu+\mu} \frac{\partial^{\nu+\mu} g}{\partial\sigma^\nu \partial\tau^\mu}(\eta) \sim \frac{\partial^{\nu+\mu}\hat{g}}{\partial\sigma^\nu \partial\tau^\mu}(\mathrm{diam}(K)^{-1}\eta),$$

and it follows that

$$|\hat{\lambda}_i(g)| \leq \hat{C}_1 \|g\|_{C^{k-1}(\hat{K})}, \quad g \in C^{k-1}(\hat{K}), \quad \lambda_i \in \mathcal{N}_K, \tag{11}$$

where $\hat{C}_1$ depends only on β_{Δ_R}, ν_{Δ_R} and d. Note that by Sobolev inequality [3, Section 4.3],

$$\|g\|_{C^{k-1}(\hat{K})} \lesssim \|g\|_{H^{k+1}(\hat{K})}, \quad g \in H^{k+1}(\hat{K}) \subset C^{k-1}(\hat{K}), \tag{12}$$

where the bounding constant depends only on k and the chunkiness parameter $\chi_{\hat{K}}\,(=\chi_K)$.

We define the interpolation operator $\Pi_{\hat{K}} : C^{k-1}(\hat{K}) \to \hat{S}_K$ by

$$\Pi_{\hat{K}} g := \sum_{i:\,\lambda_i \in \mathcal{N}_K} \hat{\lambda}_i(g)\hat{s}_i.$$

By (9) we get

$$\|\hat{s}_i\|_{L_2(\hat{K})} \leq \frac{\sqrt{\pi}}{2} \|\hat{s}_i\|_{L_\infty(\hat{K})} \leq \frac{\sqrt{\pi}}{2} C_2,$$

which in view of the Bernstein inequality (7) leads to

$$\|\hat{s}_i\|_{H^{r+1}(\hat{K})} \leq \hat{C}_2, \tag{13}$$

where $\hat{C}_2$ depends only on d, r, β_{Δ_R}, $|K_R|$ and C_2.

The inequalities (11) and (13) imply that the operator $\Pi_{\hat{K}} : C^{k-1}(\hat{K}) \to H^{r+1}(\hat{K})$ is uniformly bounded, i.e.,

$$\|\Pi_{\hat{K}} g\|_{H^{r+1}(\hat{K})} \leq \hat{C}_3 \|g\|_{C^{k-1}(\hat{K})}, \tag{14}$$

where the constant $\hat{C}_3$ depends only on $\hat{C}_1$, $\hat{C}_2$, d and $|K_R|$. Indeed, let $g \in C^{k-1}(\hat{K})$. Then $\Pi_{\hat{K}} g \in \hat{S}_K \subset W_\infty^{r+1}(\hat{K}) \subset H^{r+1}(\hat{K})$. Clearly, $|\mathcal{N}_K|$ does not exceed a constant C' depending only on d and $|K_R|$. In view of (11) and (13),

$$\|\Pi_{\hat{K}} g\|_{H^{r+1}(\hat{K})} \leq \sum_{i:\,\lambda_i \in \mathcal{N}_K} |\hat{\lambda}_i(g)|\,\|\hat{s}_i\|_{H^{r+1}(\hat{K})} \leq C'\hat{C}_1\hat{C}_2 \|g\|_{C^{k-1}(\hat{K})}.$$

We now show that for every $K \in \Delta$ and $g \in H^{k+1}(K)$,

$$|g - \Pi_K g|_{H^\nu(K)} \lesssim \mathrm{diam}(K)^{k+1-\nu}|g|_{H^{k+1}(K)}, \quad 0 \leq \nu \leq \min\{r,k\}+1, \tag{15}$$

where the constant in the bound depends only on d, β_{Δ_R}, ν_{Δ_R}, χ_K and C_2. If $g \in H^{k+1}(K)$, then $\hat{g} \in H^{k+1}(\hat{K})$ and, by the Bramble-Hilbert lemma [3, Section 4.3] there exists a polynomial $p \in \mathbb{P}_k$ such that

$$\|\hat{g} - p\|_{H^\ell(\hat{K})} \lesssim |\hat{g}|_{H^{k+1}(\hat{K})}, \qquad 0 \le \ell \le k+1, \tag{16}$$

where the bounding constant depends only on k and the chunkiness parameter $\chi_{\hat{K}} (= \chi_K)$. Let $m = \min\{r, k\}$. Since $\Pi_{\hat{K}} p = p$, we have by (14), (12) and (16),

$$\begin{aligned}
\|\hat{g} - \Pi_{\hat{K}}\hat{g}\|_{H^{m+1}(\hat{K})} &\le \|\hat{g} - p\|_{H^{m+1}(\hat{K})} + \|\Pi_{\hat{K}}(p - \hat{g})\|_{H^{m+1}(\hat{K})} \\
&\lesssim \|\hat{g} - p\|_{H^{k+1}(\hat{K})} + \|p - \hat{g}\|_{C^{k-1}(\hat{K})} \\
&\lesssim \|\hat{g} - p\|_{H^{k+1}(\hat{K})} + \|p - \hat{g}\|_{H^{k+1}(\hat{K})} \\
&\lesssim |\hat{g}|_{H^{k+1}(\hat{K})},
\end{aligned}$$

and (15) follows since

$$|g - \Pi_K g|_{H^\nu(K)} = \operatorname{diam}(K)^{1-\nu}|\hat{g} - \Pi_{\hat{K}}\hat{g}|_{H^\nu(\hat{K})},$$

$$|\hat{g} - \Pi_{\hat{K}}\hat{g}|_{H^\nu(\hat{K})} \le \|\hat{g} - \Pi_{\hat{K}}\hat{g}\|_{H^{m+1}(\hat{K})}, \text{ and}$$

$$|\hat{g}|_{H^{k+1}(\hat{K})} = \operatorname{diam}(K)^k |g|_{H^{k+1}(K)}.$$

The estimate (10) follows from (15) because

$$\|f - \Pi f\|_{H^\nu(\Omega)}^2 = \sum_{K\in\Omega}\sum_{i=0}^{\nu}|f|_K - \Pi_K f|_K|_{H^i(K)}^2, \quad |f|_{H^{k+1}(\Omega)}^2 = \sum_{K\in\Omega}|f|_K|_{H^{k+1}(K)}^2$$

and $h_\Delta = \max_{K\in\Delta}\operatorname{diam}(K)$. $\qquad\square$

Note that the estimate

$$\inf_{g\in S}\|f - g\|_{H^\nu(\Omega)} \le Ch_\Delta^{k+1-\nu}|f|_{H^{k+1}(\Omega)}, \quad f \in H^{k+1}(\Omega),$$

can be obtained by using quasi-interpolation operators for any spline spaces S with a stable local basis, see [20] or [10]. Even though Theorem 2 is only applicable to macro-element spaces, its importance for the results below about Riesz bases in $H_0^s(\Omega)$ is that it leads to the estimate

$$\inf_{g\in S_{0,\sigma}}\|f - g\|_{H^\nu(\Omega)} \le Ch_\Delta^{k+1-\nu}|f|_{H^{k+1}(\Omega)}, \quad f \in H_0^{k+1}(\Omega), \tag{17}$$

as soon as the interpolation operator Π is boundary conforming of some order $\sigma \le r$, which is normally the case for the macro-elements.

Corollary 1. *In addition to the assumptions of Theorem 2 suppose that the interpolation operator Π is boundary conforming of order $\sigma \le r$. Then the estimate (17) holds for all $\nu = 0, \ldots, \min\{r, k\} + 1$, where C depends only on d, β_{Δ_R}, ν_{Δ_R}, χ_Δ and C_2.*

4   General Theory of Hierarchical Riesz Bases

Recall that a basis $\{\phi_n\}_{n=1}^\infty$ for a Hilbert space H is said to be a *Riesz basis* if for any $c \in \ell_2$,

$$\Big\| \sum_{n=1}^\infty c_n \phi_n \Big\|_H \sim \Big(\sum_{n=1}^\infty c_n^2 \Big)^{1/2}.$$

Suppose that S_n, $n = 0, 1, 2 \ldots$, is a *nested sequence* of finite dimensional subspaces of a Hilbert space H, that is

$$S_0 \subset S_1 \subset \ldots \subset S_n \subset \ldots \quad n = 0, 1, 2, \ldots. \tag{18}$$

We assume that $\cup_{n=0}^\infty S_n$ is dense in H and set $S_{-1} := \{0\}$. Then every element $f \in H$ can be represented as a convergent series $\sum_{n=0}^\infty f_n$ in H with $f_n \in S_n$. For $n = 0, 1, 2, \ldots$, let P_n be a linear projection from S_n onto S_{n-1}, and let W_n be the complement space, that is, $P_n(W_n) = \{0\}$ and $S_n = S_{n-1} + W_n$. In particular, $W_0 = S_0$.

We will use the following general result about construction of Riesz bases for certain subspaces of H using stable bases of W_n. More standard statements of this type are usually restricted to the case when the projectors P_n are uniformly bounded, see e.g. [6].

Theorem 3 ([18]). *Assume that for some $v > 0$ and $\rho > 1$,*

$$\|P_{n+1} \cdots P_m f\|_H \lesssim \rho^{v(m-n)} \|f\|_H, \qquad f \in S_m, \tag{19}$$

for all $m, n = 0, 1, 2, \ldots$ with $n < m$. Let $s > v$ and let H_s be a linear subspace of H which itself is a Hilbert space with norm $\| \cdot \|_{H_s}$ satisfying

$$\|f\|_{H_s} \sim \inf_{f_n \in S_n : f = \sum_{n=0}^\infty f_n} \Big(\sum_{n=0}^\infty \big[\rho^{ns} \|f_n\|_H \big]^2 \Big)^{1/2}, \qquad f \in H_s. \tag{20}$$

Suppose that for each $n = 0, 1 \ldots$, $W_n \subset H_s$ and there is a stable basis $\{\phi_k^{(n)}\}_{k \in K_n}$ for W_n in the sense that

$$\Big\| \sum_{k \in K_n} c_k \phi_k^{(n)} \Big\|_H \sim \Big(\sum_{k \in K_n} c_k^2 \Big)^{1/2}, \tag{21}$$

with constants of equivalence independent of n. Then $\bigcup_{n=0}^\infty \{\rho^{-ns} \phi_k^{(n)}\}_{k \in K_n}$ is a Riesz basis for H_s.

Assumption (20) of Theorem 3 can often be verified with the help of the following theorem. Although it can be derived from more general results in e.g. [4, 22] (see also [28]), we provide here a short and self-contained proof based on arguments similar to those in [27, Theorem 6] and [7, Corollary 5.2].

Theorem 4. *Let H and $H' \subset H$ be Hilbert spaces with norms $\|\cdot\|_H$ and $\|\cdot\|_{H'} = \|\cdot\|_H + |\cdot|_{H'}$, where $|\cdot|_{H'}$ is a seminorm. Suppose that for some $\alpha > 1$ and $0 < \lambda < 1$ nested finite dimensional linear subspaces $S_n \subset H$ satisfy the Jackson inequality*

$$\inf_{s \in S_n} \|f - s\|_H \lesssim \alpha^{-n} |f|_{H'}, \quad f \in H', \qquad n = 0, 1, \ldots, \tag{22}$$

and the Bernstein inequality in the norm $\|\cdot\|_{\lambda;K}$ of the interpolation space $[H, H']_\lambda$,

$$\|s\|_{\lambda;K} \lesssim \alpha^{n\lambda} \|s\|_H, \qquad s \in S_n. \tag{23}$$

Then for any $0 < \theta < \lambda$,

$$\|f\|_{\theta;K} \sim \inf_{f_n \in S_n : f = \sum_{n=0}^{\infty} f_n} \Big(\sum_{n=0}^{\infty} \big[\alpha^{n\theta} \|f_n\|_H\big]^2 \Big)^{1/2}, \quad f \in [H, H']_\theta, \tag{24}$$

where the constants of equivalence depend only on α, the difference $\lambda - \theta$ and the bounding constants in (22) and (23).

Proof. Recall from (2) that

$$\|f\|_{\theta;K} \sim \|f\| := \|f\|_H + \Big(\sum_{n=0}^{\infty} \big[\alpha^{n\theta} K_{H,H'}(f, \alpha^{-n})\big]^2 \Big)^{1/2}.$$

We will show that $\|f\| \sim \|f\|^*$, where $\|f\|^*$ denotes the right hand side of (24).

We first prove that $\|f\|^* \lesssim \|f\|$. Let $f \in H$. It follows from (22) by Lemma 1 that there exists a sequence of elements $f_n \in S_n$ such that

$$\|f - f_n\|_H \lesssim K_{H,H'}(f, \alpha^{-n}), \qquad n = 0, 1, \ldots.$$

Then

$$\|f_n - f_{n-1}\|_H \leq \|f_n - f\|_H + \|f_{n-1} - f\|_H \lesssim K_{H,H'}(f, \alpha^{-n}), \quad n \geq 1,$$

and $\|f_0\|_H \lesssim \|f\|_H + K_{H,H'}(f, 1)$. If $\|f\| < \infty$, then $\|f - f_n\|_H \to 0$ when $n \to \infty$ and hence

$$f = \sum_{n=0}^{\infty} (f_n - f_{n-1}), \qquad f_{-1} = 0,$$

where $f_n - f_{n-1} \in S_n$ since $S_{n-1} \subset S_n$, which implies

$$\|f\|^* \leq \Big(\sum_{n=0}^{\infty} \big[\alpha^{n\theta} \|f_n - f_{n-1}\|_H\big]^2 \Big)^{1/2} \lesssim \|f\|.$$

We now proceed to showing the opposite inequality $\|f\| \lesssim \|f\|^*$. Let $f = \sum_{n=0}^{\infty} f_n$ with some $f_n \in S_n$. By (23) we have for $t \in [\alpha^{-(j+1)}, \alpha^{-j}]$,

$$K_{H,H'}(f_n, t)^2 \leq K_{H,H'}(f_n, \alpha^{-j})^2 \lesssim \alpha^{-2\lambda j} |f_n|_{\lambda,K}^2 \lesssim (t\alpha^n)^{2\lambda} \|f_n\|_H^2. \tag{25}$$

Let $0 < \theta < \lambda$. Then

$$\sum_{j=0}^{\infty} \alpha^{2\theta j} K_{H,H'}(f, \alpha^{-j})^2 \leq 2(A + B),$$

where

$$A = \sum_{j=0}^{\infty} \alpha^{2j\theta} \left(\sum_{n=0}^{j} K_{H,H'}(f_n, \alpha^{-j}) \right)^2, \quad B = \sum_{j=0}^{\infty} \alpha^{2j\theta} \left(\sum_{n=j+1}^{\infty} K_{H,H'}(f_n, \alpha^{-j}) \right)^2.$$

By (25) and Cauchy-Schwarz inequality,

$$A \lesssim \sum_{j=0}^{\infty} \alpha^{2j\theta} \left(\sum_{n=0}^{j} \alpha^{(n-j)\lambda} \|f_n\|_H \right)^2 = \sum_{j=0}^{\infty} \alpha^{2j(\theta-\lambda)} \left(\sum_{n=0}^{j} \alpha^{n(\lambda-\theta)} \alpha^{n\theta} \|f_n\|_H \right)^2$$

$$\leq \sum_{j=0}^{\infty} \alpha^{2j(\theta-\lambda)} \sum_{n=0}^{j} \alpha^{n(\lambda-\theta)} \sum_{n=0}^{j} \alpha^{n(\lambda-\theta)} \alpha^{2n\theta} \|f_n\|_H^2.$$

Since

$$\sum_{n=0}^{j} \alpha^{n(\lambda-\theta)} = \frac{\alpha^{(j+1)(\lambda-\theta)} - 1}{\alpha^{(\lambda-\theta)} - 1} \leq \frac{\alpha^{(\lambda-\theta)}}{\alpha^{(\lambda-\theta)} - 1} \cdot \alpha^{j(\lambda-\theta)},$$

we get

$$A \lesssim \sum_{j=0}^{\infty} \alpha^{-j(\lambda-\theta)} \sum_{n=0}^{j} \alpha^{n(\lambda-\theta)} \alpha^{2n\theta} \|f_n\|_H^2 = C_1 \sum_{n=0}^{\infty} \alpha^{2n\theta} \|f_n\|_H^2,$$

where $C_1 = \sum_{k=0}^{\infty} \alpha^{-k(\lambda-\theta)} = \frac{1}{1-\alpha^{-(\lambda-\theta)}}$.

The bound $K_{H,H'}(f_n, \alpha^{-j}) \leq \|f_n\|_H$ and the Cauchy-Schwarz inequality imply

$$B \leq \sum_{j=0}^{\infty} \alpha^{2j\theta} \left(\sum_{n=j+1}^{\infty} \|f_n\|_H \right)^2 = \sum_{j=0}^{\infty} \alpha^{2j\theta} \left(\sum_{n=j+1}^{\infty} \alpha^{\frac{-n\theta}{2}} \alpha^{\frac{-n\theta}{2}} \alpha^{n\theta} \|f_n\|_H \right)^2$$

$$\leq \sum_{j=0}^{\infty} \alpha^{2j\theta} \sum_{n=j+1}^{\infty} \alpha^{-n\theta} \sum_{n=j+1}^{\infty} \alpha^{-n\theta} \alpha^{2n\theta} \|f_n\|_H^2$$

$$= \frac{\alpha^{-\theta}}{1-\alpha^{-\theta}} \sum_{n=1}^{\infty} \sum_{j=0}^{n-1} \alpha^{(j-n)\theta} \alpha^{2n\theta} \|f_n\|_H^2 \leq C_2 \sum_{n=0}^{\infty} \alpha^{2n\theta} \|f_n\|_H^2,$$

where $C_2 = \frac{\alpha^{-\theta-1}}{(1-\alpha^{-\theta})(1-\alpha^{-1})}$. Combining the above estimates for A and B yields $\|f\| \lesssim \|f\|^*$. $\square$

We will use Theorems 3 and 4 with $H = L_2(\Omega)$ and $H_s = H^s(\Omega)$ or $H_0^s(\Omega)$, where $\Omega \subset \mathbb{R}^2$ is an arbitrary polygonal domain, and $\{S_n\}_{n=0}^{\infty}$ is a nested sequence of macro-element spline spaces.

A sequence of triangulations $\{\Delta_n\}_{n=0}^{\infty}$ of Ω is said to be *nested* if each Δ_{n+1} is a refinement of Δ_n, that is Δ_{n+1} is obtained from Δ_n by subdividing the triangles of Δ_n. Then obviously $S_d^r(\Delta_{n+1}) \subset S_d^r(\Delta_n)$, so that $\{S_d^r(\Delta_n)\}_{n=0}^{\infty}$ is a nested sequence of spaces. However, certain subspaces $S_n \subset S_d^r(\Delta_n)$ may also be nested, see for example [9, 11, 13].

Recall that a sequence of triangulations $\{\Delta_n\}_{n=0}^{\infty}$ of Ω is *regular* if the minimum angle of all Δ_n remains bounded below by a positive constant $\beta > 0$ independent of n, and the triangulations Δ_n are quasi-uniform in the sense that there exist constants $\rho > 1$ and $c_1, c_2 > 0$ independent of n such that

$$c_1 \rho^{-n} \leq \operatorname{diam} T \leq c_2 \rho^{-n}, \qquad T \in \Delta_n. \tag{26}$$

Parameter ρ will be called the *refinement factor* of $\{\Delta_n\}_{n=0}^{\infty}$.

Recall that a finite set $\Xi \subset \Omega$ is said to be a *Lagrange interpolation set* for a finite dimensional linear space S of functions on Ω if $\#\Xi = \dim S$ and for each $\xi \in \Xi$ there is a unique function $B_\xi \in S$ satisfying $B_\xi(\eta) = \delta_{\xi,\eta}$ for all $\xi, \eta \in \Xi$, where $\delta_{\xi,\eta} = 1$ if $\xi = \eta$ and $\delta_{\xi,\eta} = 0$ otherwise. The set $\{B_\xi\}_{\xi \in \Xi}$ is a basis for S called the *Lagrange basis*.

A sequence of Lagrange interpolation sets $\{\Xi_n\}_{n=0}^{\infty}$ for the corresponding spaces S_n is said to be *nested* if

$$\Xi_0 \subset \Xi_1 \subset \ldots \subset \Xi_n \subset \ldots. \tag{27}$$

We are ready to formulate the main result of the paper.

Theorem 5. *Let $\{S_n\}_{n=0}^{\infty}$ be a nested sequence of spaces $S_n \subset S_d^r(\Delta_n)$, $r \geq 0$, with respect to a regular nested sequence of triangulations $\{\Delta_n\}_{n=0}^{\infty}$ of a polygonal domain $\Omega \subset \mathbb{R}^2$, with refinement factor $\rho > 1$, and let $\{\Xi_n\}_{n=0}^{\infty}$ be a nested sequence of Lagrange interpolation sets for the spaces S_n, with the corresponding Lagrange basis $\{B_\xi^{(n)}\}_{\xi \in \Xi_n}$ for S_n. Assume that the bases $\{B_\xi^{(n)}\}_{\xi \in \Xi_n}$ are uniformly local and bounded, that is they are m-local and satisfy $\|B_\xi^{(n)}\|_{L_\infty(\Omega)} \leq M$, $\xi \in \Xi_n$, for some m, M independent of n.*

(a) Assume that the spaces S_n satisfy the Jackson inequality

$$\inf_{g \in S_n} \|f - g\|_{L_2(\Omega)} \lesssim \rho^{-n(k+1)} |f|_{H^{k+1}(\Omega)}, \quad f \in H^{k+1}(\Omega), \tag{28}$$

For some $k \in \mathbb{N}$ with $r < k \leq d$. Then for any $s \in (1, r + \frac{3}{2})$ the set

$$\mathcal{B}_s := \bigcup_{n=0}^{\infty} \{\rho^{n(1-s)} B_\xi^{(n)}\}_{\xi \in \Xi_n \setminus \Xi_{n-1}}$$

is a Riesz basis for $H^s(\Omega)$.

(b) Moreover, if the spaces S_n, $n = 0, 1, \ldots$, satisfy the homogeneous boundary conditions of order $\sigma \leq r$, that is

$$\frac{\partial^{\nu+\mu} g}{\partial x^\nu \partial y^\mu} = 0 \text{ on } \partial\Omega, \text{ for all } \nu, \mu \geq 0, \ \nu + \mu \leq \sigma, \quad g \in S_n,$$

and (28) *holds for all* $f \in H_0^{k+1}(\Omega)$ *rather than for all* $f \in H^{k+1}(\Omega)$, *then* $\mathcal{B}_s$ *is a Riesz basis for* $H_0^s(\Omega)$ *if* $s \in (1, \sigma + \frac{3}{2}) \setminus (\mathbb{Z} + \frac{1}{2})$.

Proof. Under the assumptions of the theorem, the bases $\{B_\xi^{(n)}\}_{\xi \in \Xi_n}$ are stable and local in the sense of the definition in Section 3. Since $\operatorname{diam}(T) \sim \rho^{-n}$, $T \in \Delta_n$, the bases $\{\rho^n B_\xi^{(n)}\}_{\xi \in \Xi_n}$ are L_2-stable, which implies

$$\Big\| \sum_{\xi \in \Xi_n} c_\xi B_\xi^{(n)} \Big\|_{L_2(\Omega)} \sim \rho^{-n} \Big(\sum_{\xi \in \Xi_n} c_\xi^2 \Big)^{1/2}, \tag{29}$$

for any real numbers c_ξ, with constants of equivalence independent of n.

Let $0 < s < r + \frac{3}{2}$. We choose a number $\bar{s}$ such that $s < \bar{s} < r + \frac{3}{2}$. By Theorem 1, since the spaces S_n possess stable local bases, we obtain the Bernstein inequality

$$\|g\|_{H^s(\Omega)} \lesssim \rho^{n\bar{s}} \|g\|_{L_2(\Omega)}, \qquad g \in S_n.$$

By Theorem 4, applied with $\alpha = \rho^{k+1}$, $\lambda = \bar{s}/(k+1) < 1$ and $\theta = s/(k+1)$, we see that under the assumptions of part (a) condition (20) of Theorem 3 is satisfied for $H = L_2(\Omega)$, $H' = H^{k+1}(\Omega)$ and $H_s = H^s(\Omega) = [L_2(\Omega), H^{k+1}(\Omega)]_\theta$. Similarly, under the assumptions of part (b) condition (20) follows from Theorem 4 with $H = L_2(\Omega)$, $H' = H_0^{k+1}(\Omega)$ and $H_s = [L_2(\Omega), H_0^{k+1}(\Omega)]_\theta$.

We now verify the other assumptions of Theorem 3. The density of $\cup_{n=0}^\infty S_n$ in $H = L_2(\Omega)$ follows from the Jackson inequality (28) since both $H^{k+1}(\Omega)$ and $H_0^{k+1}(\Omega)$ are dense in $L_2(\Omega)$. Furthermore, let $I_n : C(\Omega) \to S_n$, $n = 0, 1, \ldots$, be the Lagrange interpolation operator

$$I_n f := \sum_{\xi \in \Xi_n} f(\xi) B_\xi^{(n)}.$$

We set $P_n := I_{n-1}|_{S_n}$, $n \geq 1$, and $P_0 := 0$. Then $P_n : S_n \to S_{n-1}$ is a linear projection, and, in view of the nestedness (27) of $\{\Xi_n\}_{n=0}^\infty$, we have $P_{n+1} \cdots P_m = I_n|_{S_m}$ for all $m > n$. Let $g \in S_m$ and $h := P_{n+1} \cdots P_m g$. Then $g = \sum_{\xi \in \Xi_m} g(\xi) B_\xi^{(m)}$ and $h = \sum_{\xi \in \Xi_n} g(\xi) B_\xi^{(n)}$. By (29) and (27) we obtain

$$\|h\|_{L_2(\Omega)}^2 \lesssim \rho^{-2n} \sum_{\xi \in \Xi_n} |g(\xi)|^2 \leq \rho^{-2n} \sum_{\xi \in \Xi_m} |g(\xi)|^2$$

$$\lesssim \rho^{2(m-n)} \|g\|_{L_2(\Omega)}^2,$$

which implies (19) with $H = L_2(\Omega)$ and $v = 1$. Because of (27) the sets

$$\{\rho^n B_\xi^{(n)} : \xi \in \Xi_n \setminus \Xi_{n-1}\}, \quad n = 0, 1, \ldots \quad (\Xi_{-1} = 0)$$

form L_2-stable bases for the complement spaces W_n. Since $W_n \subset S_n \subset H^s(\Omega)$ for all $s < r + \frac{3}{2}$ by Theorem 1, an application of Theorem 3 with $v = 1$ completes the proof of part (a). Under the assumptions of part (b) it is easy to see that $S_n \subset \tilde{H}^s(\Omega) = H_0^s(\Omega)$ for all $s < \sigma + \frac{3}{2}$, $s \notin \mathbb{Z} + \frac{1}{2}$, and Theorem 3 implies that $\mathcal{B}_s$ is a Riesz basis for $[L_2(\Omega), H_0^{k+1}(\Omega)]_{s/(k+1)}$ for all $1 < s < \sigma + \frac{3}{2}$. The statement of part (b) follows in view of the description (5) of these interpolation spaces in Section 2. □

Note that in the case $r = \sigma = 0$ the condition (20) of Theorem 3 for $H = L_2(\Omega)$ and $H_s = H_0^s(\Omega)$, $s < \frac{3}{2}$, can be verified with the help of [25, Corollary 3] without using interpolation spaces.

The argumentation of Theorem 5 for $\Omega \subset \mathbb{R}^d$ would lead to the Riesz basis for $H^s(\Omega)$ with the expectable range $\frac{d}{2} < s < r + \frac{3}{2}$. Indeed, (29) then holds with $\rho^{-\frac{dn}{2}}$ replacing ρ^{-n}, and hence Theorem 3 is applicable with $v = \frac{d}{2}$.

The standard C^0 piecewise linear hierarchical basis [31] is, after appropriate scaling, a Riesz basis of $H^s(\Omega)$ $s \in (1, \frac{3}{2})$ in two dimensions, see [21]. Clearly, Theorem 5 applies to this case, where the triangulations Δ_n are obtained by the uniform refinement of an initial triangulation of Ω, $\rho = 2$, S_n is either $S_1^0(\Delta_n)$ (for $H^s(\Omega)$) or its subspace $\{s \in S_n : s|_{\partial\Omega} = 0\}$ (for $H_0^s(\Omega)$), and Ξ_n is either the set of all vertices of Δ_n or the set of all interior vertices, respectively. The Jackson inequality (28) for $k = 1$ follows from Theorem 2 since $S_1^0(\Delta_n)$ are macro-element spaces with uniformly bounded basis functions, $\mathbb{P}_1 \subset S_1^0(\Delta_n)$, and the interpolation operator Π is boundary confirming of order $\sigma = 0$.

In the next section we provide a brief review of the existing constructions of C^1 Lagrange type hierarchical Riesz bases for Sobolev spaces $H^s(\Omega)$, $s \in (1, \frac{5}{2})$, and $H_0^s(\Omega)$, $s \in (1, \frac{3}{2}) \cup (\frac{3}{2}, \frac{5}{2})$. Note that C^1 hierarchical bases of Hermite type are also known [5, 26]. They form Riesz bases for $H^s(\Omega)$, $s \in (2, \frac{5}{2})$.

5 C^1 Lagrange Hierarchical Riesz Bases for Sobolev Spaces

Spline spaces $S_n \subset S_d^r(\Delta_n)$ and Lagrange interpolation sets Ξ_n satisfying the hypotheses of Theorem 5 give rise to hierarchical Riesz bases for $H^s(\Omega)$, $s \in (1, r + \frac{3}{2})$, respectively $H_0^s(\Omega)$, $s \in (1, \sigma + \frac{3}{2}) \setminus (\mathbb{Z} + \frac{1}{2})$. However, specific constructions are only available for $r = 0, 1$. In this section we review such constructions of the spaces S_n in the case $r = 1$. We do not describe the corresponding sets Ξ_n as they are quite technical, and the interested reader is instead referred to the original literature.

5.1   Piecewise Cubics on Triangulated Quadrangulations

The first construction of C^1 Lagrange hierarchical bases has been suggested in [12], where the nested spline spaces are the macro element spaces of C^1 piecewise cubic polynomials on the triangulations (see [20, Section 6.5]) obtained by adding two diagonals to the quadrilaterals of a *checkerboard* quadrangulation of any polygonal domain, which means that all interior vertices of the quadrangulation are of degree 4 and quadrilaterals can be coloured black and white in such a way that any two quadrilaterals sharing an edge have opposite colours. The corresponding nodal basis satisfies (9) with a constant C_2 dependent only on the minimum angle of the triangles $T \in \Delta_R$ and the interpolation operator Π is boundary conforming of order 1.

Nested spaces are obtained by the *triadic refinement* of the quadrilaterals and their subtriangles illustrated in Figures 1 and 2. More precisely, Let $Q = \langle v_1, v_2, v_3, v_3 \rangle$ be a quadrilateral and let $p_1 = 1/3(2v_1 + v_2)$, $p_2 = 1/3(v_1 + 2v_2)$, $p_3 = 1/3(2v_2 + v_3)$, $p_4 = 1/3(v_2 + 2v_3)$, $p_5 = 1/3(2v_3 + v_4)$, $p_6 = 1/3(v_3 + 2v_4)$, $p_7 = 1/3(2v_4 + v_1)$, $p_8 = 1/3(v_4 + 2v_1)$, $p_9 = 1/3(v_1 + 2\bar{v})$, $p_{10} = 1/3(v_2 + 2\bar{v})$, $p_{11} = 1/3(v_3 + 2\bar{v})$, $p_{12} = 1/3(v_4 + 2\bar{v})$, where $\bar{v}$ is the point of intersection of the diagonals of Q. The refinement is obtained by connecting the points p_1 and p_8 to p_9, p_2 and p_3 to p_{10}, p_4 and p_5 to p_{11}, p_6 and p_7 to p_{12}, and finally connecting the points $p_9, p_{10}, p_{11}, p_{12}$ together, as shown in Figure 1. Each of the 9 quadrilaterals is subdivided into 4 triangles by its diagonals as in Figure 2.

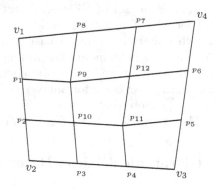

Fig. 1. A triadic refinement $\Diamond_Q$ of a quadrilateral Q

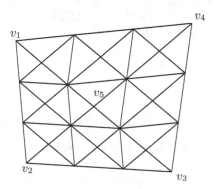

Fig. 2. The triangulation Δ_Q of $\Diamond_Q$

Given an initial quadrangulation $\Diamond_0$ of Ω, this method generates a sequence of successively refined quadrangulations $\Diamond_0, \Diamond_1, \ldots, \Diamond_n, \ldots$, and triangulations $\Delta_0, \Delta_1, \ldots, \Delta_n, \ldots$, and the nested macro-element spaces are $S_n = S_3^1(\Delta_n)$. While the nestedness of the sequence of triangulations $\{\Delta_n\}_{n=0}^{\infty}$ is obvious, its regularity, which has not been fully addressed in [12], follows from Proposition 1 below. For the nested sequence of Lagrange interpolation sets $\{\Xi_n\}_{n=0}^{\infty}$ described in [12] all assumptions of Theorem 5 (b) are satisfied, with $r = \sigma = 1$, $k = 3$ and $\rho = 3$, which leads to a Riesz basis for $H_0^s(\Omega)$, $s \in (1, \frac{3}{2}) \cup (\frac{3}{2}, \frac{5}{2})$.

Proposition 1. *Each triangle $T \in \Delta_n$, $n \geq 2$, is similar to a triangle in Δ_1 with the scaling factor $\frac{1}{3^{n-1}}$.*

Proof. Consider the quadrangulation $\Diamond_Q$ of a quadrilateral Q obtained by the triadic refinement. It is easy to see that the quadrilateral $\langle p_9, p_{10}, p_{11}, p_{12} \rangle$ is similar to the parent quadrilateral $Q = \langle v_1, v_2, v_3, v_4 \rangle$, whereas $\langle p_1, p_2, p_{10}, p_9 \rangle$ is a parallelogram with side length $\frac{1}{3}$ of the size of the parent edge $\langle v_1, v_2 \rangle$, see Figure 1. Three other children of Q in similar position are also parallelograms.

Let Δ_Q be the triangulation of $\Diamond_Q$ shown in Figure 2. We observe that there are 8 different types of similar triangles in Δ_Q as shown in Figure 3. The triangles of types $1, 2, 3$ and 4 are similar to their parent triangles (obtained from Q by splitting along its diagonals) with the coefficient $\frac{1}{3}$. The triangles of types $5, 6, 7$ and 8 will be referred to as "median" triangles because each of them has a side parallel to the median of its parent triangle and of length $\frac{2}{3}$ of that median, as illustrated in Figure 4, where the section $\langle v_1, v_2, v_5 \rangle$ of the triangulations Δ_Q of Figure 2 is shown separately.

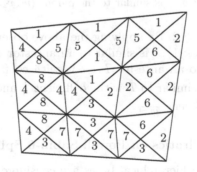

Fig. 3. Eight types of similar triangles in Δ_Q

We now apply the next refinement step and look at the median subtriangle $\langle a, b, c \rangle$ of the median triangle in Δ_Q as shown in Figure 5. We note that the dotted line $\langle q_1, q_2 \rangle$ is of length $\frac{2}{3}$ of the side $\langle p_1, p_9 \rangle$ of the parent which is parallel to the median $\langle p_1, p_2 \rangle$ of the grandparent. Hence the median of the median triangle $\langle a, b, c \rangle$ is of length $\frac{1}{4} \times \frac{2}{3} \times \frac{2}{3} = \frac{1}{9}$ of the median $\langle m, v_5 \rangle$ of the grandparent $\langle v_1, v_2, v_5 \rangle$. Therefore, the median subtriangle $\langle a, b, c \rangle$ is similar to the grandparent $\langle v_1, v_2, v_5 \rangle$ with coefficient $\frac{1}{9}$.

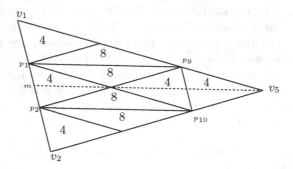

Fig. 4. The triangle $\langle v_1, v_2, v_5 \rangle$, its median $\langle m, v_5 \rangle$ and 9 children

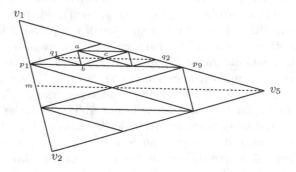

Fig. 5. The triangle $\langle a, b, c \rangle$ is similar to the parent $\langle v_1, v_2, v_5 \rangle$ with coefficients $\frac{1}{9}$

Let $T \in \Delta_n$, with $n \geq 2$. By applying the above observations recursively, we have two following cases: 1) T is similar to an ancestor $\tilde{T} \in \Delta_1$ with coefficient $\frac{1}{3^{n-1}}$. 2) T is similar to an ancestor $\hat{T} \in \Delta_0$ with coefficient $\frac{1}{3^n}$. But $\hat{T}$ has a child $\tilde{T} \in \Delta_1$ which is similar to $\hat{T}$ with coefficient $\frac{1}{3}$ and this implies that T is similar to $\tilde{T}$ with coefficient $\frac{1}{3^{n-1}}$. □

5.2 Piecewise Quadratics on Powell-Sabin-6 Splits

C^1 piecewise quadratic hierarchical bases are considered in [23]. Here, an initial checkerboard quadrangulation of Ω is first turned into a triangulation by adding one diagonal of each quadrilateral, and then each triangle is subdivided using a Powell-Sabin-6 (PS-6) split. To obtain a nested sequence of triangulations $\{\Delta_n\}_{n=0}^\infty$, a triadic refinement of the PS-6 split [30] is performed, see Figure 6. The nested spline spaces S_n are the C^1 piecewise quadratic Powell-Sabin macro-elements [20, Section 6.3]. Lagrange interpolation sets Ξ_n with the required properties are selected using a scheme which can be seen as a specific realisation of the interpolation method described in [24]. It is shown in [23] that this construction leads to a Riesz basis for $H^s(\Omega)$, $1 < s < \frac{5}{2}$, under the assumption that the triangulation sequence $\{\Delta_n\}_{n=0}^\infty$ is regular. Indeed, in this case Theorem 5 is applicable with $r = \sigma = 1$ and $k = 2$. We note however that this assumption does not seem easy to verify unless Δ_0 is a uniform triangulation, in which case $\rho = 3$. It is an open question whether an arbitrary polygonal domain Ω admits an initial triangulation such that the sequence of triangulations obtained by the triadic refinement of its PS-6 split is regular.

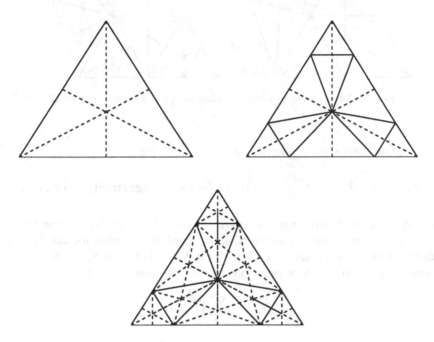

Fig. 6. The triadic refinement of the PS-6 split: A new vertex is placed at the position of the interior point in the PS-6 split and two new vertices on each edge.

5.3   Piecewise Quadratics on Mixed PS-6/PS-12 Splits

In our paper [14] we construct C^1 piecewise quadratic hierarchical bases on arbitrary polygonal domains using nested sequences of triangulations and spline spaces introduced in [17]. Beginning with an arbitrary triangulation Δ_0 of Ω, a nested sequence of triangulations $\{\Delta_n\}_{n=0}^{\infty}$ is obtained by the standard uniform refinement, where the middle points of edges are connected to each other. An edge of Δ_n is said to be regular if it is shared by two triangles that form a parallelogram. Clearly, all boundary edges are irregular, but an interior edge may only be irregular if it overlaps a part of an edge of Δ_0. Furthermore, let Δ_n^* be the triangulation obtained by subdividing each triangle $T \in \Delta_n$ using the Powell-Sabin-6 split if all edges of T are regular, or the Powell-Sabin-12 split [20, Section 6.4] otherwise. For both PS-6 and PS-12 splits the central vertex is chosen at the barycentre of the triangle and the edge splitting vertices are at the midpoints of the edges. Then $\{\Delta_n^*\}_{n=0}^{\infty}$ is also a nested sequence of triangulations, as illustrated in Figure 7. It is obviously regular, with refinement factor $\rho = 2$.

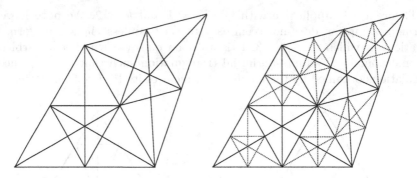

Fig. 7. An example to illustrate that Δ_{n+1}^* is a refinement of Δ_n^*

The spline spaces are defined by

$$S_n = \left\{ s \in S_2^1(\Delta_n^*) : \frac{\partial s}{\partial e^{\perp}}\Big|_e \text{ is linear for each irregular edge } e \text{ of } \Delta_n \right\},$$

where $\frac{\partial s}{\partial e^{\perp}}$ denotes the normal derivative of s on e. It is easy to see that $\{S_n\}_{n=0}^{\infty}$ are nested macro-element spaces, their interpolation operators are boundary conforming of order 1, and $\mathbb{P}_2 \subset S_n$, $n = 0, 1, \ldots$. Let $P_n : S_n \to S_{n-1}$ be the orthogonal projector with respect to the inner product defined by

$$(f, g) = \sum_{e \in \mathcal{E}_n} (f, g)_e,$$

where $\mathcal{E}_n$ is the set of all edges of Δ_n and, for $e = \langle v_1, v_2 \rangle$,

$$(f, g)_e := \frac{1}{2^{2n}} \left[f(v_1)g(v_1) + \left(f(v_1) + \frac{1}{4}\frac{\partial f}{\partial e}(v_1) \right)\left(g(v_1) + \frac{1}{4}\frac{\partial g}{\partial e}(v_1) \right) \right.$$
$$\left. + f(v_2)g(v_2) + \left(f(v_2) - \frac{1}{4}\frac{\partial f}{\partial e}(v_2) \right)\left(g(v_2) - \frac{1}{4}\frac{\partial g}{\partial e}(v_2) \right) \right].$$

It is shown in [17] that the projectors P_n satisfy (19) with

$$v = \log_2\left(\frac{2(1 + \sqrt{13})}{3} \right) \approx 1.618,$$

and thus lead to a construction of Riesz bases in $H^s(\Omega)$ for $v < s < \frac{5}{2}$.

In [14] we present a construction of nested Lagrange interpolation sets for S_n and their subspaces with homogeneous boundary conditions of order 1, which leads to a Riesz basis for $H^s(\Omega)$, $s \in (1, \frac{5}{2})$ and $H_0^s(\Omega)$, $s \in (1, \frac{3}{2}) \cup (\frac{3}{2}, \frac{5}{2})$, by applying Theorem 5 with $r = \sigma = 1$, $k = 2$ and $\rho = 2$.

Acknowledgements. This research has been supported in part by the grant UBD/PNC2/2/RG/1(159) from Universiti Brunei Darussalam.

References

1. Bergh, J., Löfström, J.: Interpolation Spaces: An Introduction. Springer, New York (1976)
2. Bramble, J.: Interpolation between Sobolev spaces in Lipschitz domains with an application to multigrid theory. Math. Comp. 64, 1359–1365 (1995)
3. Brenner, S.C., Scott, L.R.: The Mathematical Theory of Finite Element Methods, 3rd edn. Springer, New York (2008)
4. Butzer, P.L.: Approximationsprozesse und Interpolationsmethoden. Bibliographisches Institut, Mannheim (1968)
5. Dahmen, W., Oswald, P., Shi, X.-Q.: C^1-hierarchical bases. J. Comput. Appl. Math. 51, 37–56 (1994)
6. Dahmen, W.: Multiscale analysis, approximation and interpolation spaces. In: Chui, C.K., Schumaker, L.L. (eds.) Approximation Theory VIII. Wavelets and Multilevel Approximation, vol. 2, pp. 47–88. World Scientific Publishing (1995)
7. Dahmen, W.: Stability of multiscale transformations. J. Fourier. Anal. Appl. 2, 341–361 (1996)
8. Davydov, O.: Stable local bases for multivariate spline spaces. J. Approx. Theory 111, 267–297 (2001)
9. Davydov, O.: Locally stable spline bases on nested triangulations. In: Chui, C.K., Schumaker, L.L., Stöckler, J. (eds.) Approximation Theory X: Wavelets, Splines, and Applications, pp. 231–240. Vanderbilt University Press (2002)
10. Davydov, O.: Smooth finite elements and stable splitting. Berichte "Reihe Mathematik" der Philipps-Universität Marburg, Marburg (April 2007); An adapted version of this article has appeared as Section 4.2.6 "Smooth FEs on polyhedral domains" of the book K. Böhmer, Numerical Methods for Nonlinear Elliptic Differential Equations: A Synopsis. Oxford University Press, Oxford, (2010)
11. Davydov, O., Petrushev, P.: Nonlinear approximation from differentiable piecewise polynomials. SIAM J. Math. Anal. 35, 708–758 (2003)
12. Davydov, O., Stevenson, R.: Hierarchical Riesz bases for $H^s(\Omega)$, $1 < s < \frac{5}{2}$. Constr. Approx. 22, 365–394 (2005)
13. Davydov, O., Yeo, W.P.: Refinable C^2 piecewise quintic polynomials on Powell-Sabin-12 triangulations. J. Comput. Appl. Math. 240, 62–73 (2013)
14. Davydov, O., Yeo, W.P.: C^1 piecewise quadratic hierarchical bases (in preparation)
15. Goodman, T., Hardin, D.: Refinable multivariate spline functions. In: Jetter, K., et al. (eds.) Topics in Multivariate Approximation and Interpolation, pp. 55–83. Elsevier (2006)
16. Grisvard, P.: Elliptic Problems in Nonsmooth Domains. Pitman (1985)
17. Jia, R.Q., Liu, S.T.: C^1 Spline wavelets on triangulations. Math. Comp. 77, 287–312 (2007)
18. Jia, R.Q., Zhao, W.: Riesz bases of wavelets and applications to numerical solutions of elliptic equations. Math. Comp. 80, 1525–1556 (2011)
19. Lions, J.L., Magenes, E.: Non-homogeneous Boundary Value Problems and Applications, vol. I. Springer, Berlin (1972)
20. Lai, M.J., Schumaker, L.L.: Spline Functions on Triangulations. Cambridge University Press, Cambridge (2007)
21. Lorentz, R., Oswald, P.: Multilevel finite element Riesz bases for Sobolev spaces. In: Bjorstad, P., et al. (eds.) Domain Decomposition Methods in Science and Engineering: 9th International Conference, pp. 178–187. Domain Decomposition Press, Bergen (1997)

22. Luther, U.: Representation, interpolation, and reiteration theorems for generalized approximation spaces. Annali di Matematica 182, 161–200 (2003)
23. Maes, J., Bultheel, A.: C^1 hierarchical Riesz bases of Lagrange type on Powell-Sabin triangulations. J. Comp. Appl. Math. 196, 1–19 (2006)
24. Nürnberger, G., Zeilfelder, F.: Local Lagrange interpolation on Powell-Sabin triangulations and terrain modelling. In: Haussmann, W., Jetter, K., Reimer, M. (eds.) Recent Progress in Multivariate Approximation, pp. 227–244. Birhäuser (2001)
25. Oswald, P.: On function spaces related to finite element approximation theory. Z. Anal. Anwendungen. 9, 43–64 (1990)
26. Oswald, P.: Hierarchical conforming finite element methods for the biharmonic equation. SIAM J. Numer. Anal. 29, 1610–1625 (1992)
27. Oswald, P.: Multilevel Finite Element Approximation: Theory and Applications. B.G. Teubner, Stuttgart (1994)
28. Oswald, P.: Frames and space splittings in Hilbert spaces (1997) (manuscript), http://www.faculty.jacobs-university.de/poswald/bonn1.pdf
29. Oswald, P.: Multilevel frames and Riesz bases in Sobolev spaces (1998) (manuscript), http://www.faculty.jacobs-university.de/poswald/bonn2.pdf
30. Vanraes, E., Windmolders, J., Bultheel, A., Dierckx, P.: Automatic construction of control triangles for subdivided Powell-Sabin splines. Computer Aided Geometric Design 21, 671–682 (2004)
31. Yserentant, H.: On the multi-level splitting of finite element spaces. Numer. Numer. Math. 49, 379–412 (1986)
32. Zolesio, J.-L.: Interpolation d'espaces de Sobolev avec conditions aux limites de type mêlé. C. R. Acad. Sc. Paris, Série A 285, 621–624 (1977)

Generalized Metric Energies for Continuous Shape Deformation

Janick Martinez Esturo[1,2], Christian Rössl[1], and Holger Theisel[1]

[1] Visual Computing Group, University of Magdeburg, Germany
[2] Max Planck Institute for Computer Science, Saarland University, Germany

Abstract. High quality deformations of planar and volumetric domains are central to many computer graphics related problems like modeling, character animation, and non-rigid registration. Besides common "as-rigid-as-possible" approaches the class of nearly-isometric deformations is highly relevant to solve this kind of problems. Recent continuous deformation approaches try to find planar first order nearly-isometric deformations by integrating along approximate Killing vector fields (AKVFs). In this work we derive a generalized metric energy for deformation vector fields that has close-to-isometric AKVFs as a special case and additionally supports close-to-length-preserving, close-to-conformal as well as close-to-equiareal deformations. Like AKVF-based deformations we minimize nonlinear energies to first order using efficient linear optimizations. Our energy formulation supports nonhomogeneous as well as anisotropic behavior and we show that it is applicable to both planar and volumetric domains. We apply energy specific regularization to achieve smoothness and provide a GPU implementation for interactivity. We compare our approach to AKVF-based deformations for the planar case and demonstrate the effectiveness of our method for the $2d$ and $3d$ case.

Keywords: Shape Deformation, Isometry, Vector Field.

1 Introduction

Persistent shape deformation is a classic problem in computer graphics and design. Even though numerous approaches haven been developed in the previous decades, is is still an important and active area of research. Applications for planar shape deformations include, e. g., image warping and cartoon animation. Deformation of $3d$ shapes is used in classic domains like in engineering for shape modeling or to create animations in the media industry, but also, e. g., for data registration in medical applications.

A recent trend is the development of continuous *nearly isometric* methods [21,29]. These deformations should preserve distances and, as a result, angles and area as much as possible. Intuitively, isometry is a good measure for the quality of a deformation: while the shape should accurately satisfy the constraints defining the deformation, it should not unnecessarily stretch or bend. Hence, near-isometric deformations yield intuitive and high quality results.

M. Floater et al. (Eds.): MMCS 2012, LNCS 8177, pp. 135–157, 2014.
© Springer-Verlag Berlin Heidelberg 2014

However, this does not come for free! Roughly speaking, high quality near-isometric deformations come for the price of solving *nonlinear* problems. This is a major issue especially for interactive applications, which are typical in computer graphics and are mandatory for interactive design. There is a competition with more efficient linear methods (see, e.g., [5]), which are based on simpler, often approximated differential quantities. It is well known that linear methods fail to handle isometry: most approaches either cope well with translations or with rotations – but not with both simultaneously. Also, there is no guarantee that the deformation does not induce local folds or self-intersections. We arrive at the conclusion that both, linear and nonlinear methods, have their own right to co-exist in shape deformation frameworks: the user has the choice between fast linear methods at the cost of sacrificing quality, and high quality nonlinear methods that are significantly more expensive to compute.

The user has to pay a certain price – higher computation times or smaller data sets – and therefore expects benefits from nonlinear methods. These include not only geometric properties of the deformation but also other important criteria related to usability. In summary, the computation of near-isometric shape deformations should fulfill a number of requirements, which make their computation a challenging problem:

- The isometric deformation problem is nonlinear. Nevertheless, computation must be *effective* and robust to guarantee a unique global optimum. In addition, computation must be *efficient* enough to enable real-time response to user input.
- Deformations must *interpolate* constraints, which can be defined for *any* point of the shape. Approximate satisfaction of "soft constraints" can be tolerated only if arbitrarily small tolerances are possible in principle.
- Ideally, the user can – globally and locally – attenuate isometry such that continuous blends from angle preservation to area preservation are possible. Anisotropic behavior is an additional design parameter for the user.
- Deformations must be *smooth* in a sense that the energy or metric error is distributed smoothly over the shape. In particular, the error must not concentrate near positional constraints.
- The discrete deformations must be independent of the particular partition of the shape or the domain. This implies *resolution/tessellation invariance*.
- Ideally, the formulation of the solution should be same for the 2d and the 3d case. This alleviates implementation.

So far, we are not aware of any isometry-preserving shape deformation method that meets all of the above design goals. In this paper, we present a new integral approach to continuous shape deformation that fulfills *all* requirements. Our approach is more general – but not more complicated – than previous methods. We define a generalized metric energy that has flows as minimizer that determine near-isometric, near-conformal, and to some extent near-equiareal deformations. In particular, we show that the recently proposed planar deformations based on as-Killing-as-possible vector fields (AKVF) [29] constitute a special case of our energy.

The derivation of error measures used in our method is neither based on the popular as-rigid-as-possible (ARAP) approaches nor on the recently used notion of discrete Killing fields. In contrast to iterative energy optimization required for ARAP, which converges to local optima only, our method is non-iterative. Instead, deformation is a time-dependent function, and we optimize for its derivative w.r.t. time and solve an initial value problem.

Our method can easily be integrated into existing tools. It is applicable to triangle meshes in $2d$ as well as to tetrahedral meshes in $3d$. It shares the common intuitive user interface where few points are fixed and few points act as handles, which can be dragged along paths in the domain by the user. In addition, the user can control metric properties of the deformation: we provide a single scalar parameter to obtain combinations of near-isometric and near-conformal deformations on a continuous scale. This parameter can be given globally as a single scalar or locally as a scalar field over the shape. Local anisotropic behavior is achieved by incorporating varying anisotropic energy norms.

2 Background and Related Work

The deformation of a shape consists of a map from the original shape to the deformed shape. Isometric maps preserve distances, which is equivalent to simultaneously preserving angles (conformal maps) and area (equiareal maps). For a rigorous introduction of the differential geometry of such maps we refer to [6]. Related to shape deformation is parametrization of surfaces, i.e., finding a map between a surface in $3d$ and a planar domain. Naturally, isometry is a desired property for such maps; a pioneering approach is the construction of most isometric parametrizations [14]. Liu et al. [24] present hybrid parametrizations that interpolate locally rigid or local similarity transformations, which is similar to our generalized framework for continuous deformations. In the following, we consider only shape deformation methods. For a discussion of parametrization methods we refer to the survey of Hormann et al. [15]. We restrict our review of related work to nonlinear deformation methods. For a review of linear methods and a discussion of differences to nonlinear methods we refer to the survey [5].

A popular approach to isometry preservation is to restrict deformations locally to rigid transformations, i.e., translation and rotation. (Reflection is undesired.) This leads to the notion of the well-established as-rigid-as-possible (ARAP) maps, which where initially introduced for shape interpolation [1] and later applied for shape deformation [17,31]. Until today, there have emerged numerous extensions like [3,32,20,8], to mention just a few. ARAP approaches minimize a nonlinear energy expressing rigidity subject to constraints like fixed and displaced points. The classic approach consists in an iterative algorithm, which repeatedly estimates local rotations to build the global deformation until convergence. There are also alternative nonlinear deformation energies that enforce rigidity in form of, e. g., the rest energy of coupled rigid prisms [4]. Independent of the energy and the particular numerical scheme, the deformation is obtained as the minimizer of a particular energy in the shape coordinates at a singular point in time: we refer to such methods as *single step* methods.

In contrast, *continuous* methods evolve the deformation over time: the energy minimizer at each infinitesimal time step determines the gradient of the deformation, and the final deformation is obtained as the solution of an ordinary differential equation. From a technical point of view, the iterative solvers for minimizing nonlinear energies are, roughly speaking, replaced by a numerical ODE integration method. The latter is a standard numerical problem that is well-understood and that can be solved efficiently and reliably. In addition, the mapping to the deformed shape is guaranteed to be locally bijective if the deformation gradient does not vanish, i.e., if the deformation flow does not contain critical points, and hence the deformations does not show local fold-overs. The main benefit of the continuous methods, however, consists in the fact that finding deformation gradients is a linear problem for near-isometric deformations.

Isometry preservation is guaranteed for integration of exact Killing vector fields, see, e.g., [2]. Kilian et al. [21] approximate Killing vector fields for interpolation in a shape space, which yields deformations of a 2-manifold that is embedded in 3-space. Note that they compare this isometry preserving approach to continuous deformations based on the ARAP concept, which yields a related but different deformation class. Martinez et al. [25] extend their discretization towards tessellation independence and smoothness. Heeren et al. [13] use physical discrete shell energies to construct time-discrete geodesics in a different shape space. Solomon et al. [29] introduce the notion of as-Killing-as-possible (AKVF) deformations in planar domains. In contrast to the above approaches, they ensure smoothness by a post-process rather than by a regularization term, and instead of a standard ODE solver, they use planar holomorphic curves as a predictor to construct the trajectories. They obtain high quality deformations, which they compare to various other planar shape deformation methods. In summary, their results suggest that it is more than worthwhile and often preferable to consider near-isometric shape deformations.

Funck et al. [11] developed a remarkably different approach to continuous $3d$ shape deformation, which preserves volume by integration of divergence-free vector fields. Continuous deformation can also be obtained by fitting continuous shape manifolds to key frames [7].

There are various alternative methods for planar and volumetric shape deformation. One prominent class of methods is based on generalized barycentric coordinates, e.g., [16,19,23,18,33,34]. Besides isometry here is also a demand for conformal maps, which are produced by none of the above methods.

3   Continuous Metric Energies

In this section we introduce continuous deformations formally and derive energy terms that determine isometric and conformal deformations in $2d$ and $3d$. The section concludes with a generalized formulation of an integral energy that determines a one-parameter family of continuous deformations, which includes near-isometric, near-conformal, and close-to-equiareal deformations.

(a) Near-isometric.   (b) Near-conformal.

Fig. 1. $2D$ Deformation Examples. On the straight strip some vertices were fixed ($\bullet$) while some vertices were moved ($\circ$) (all models have the same scale). The deformations are generated by AMAP (8) and ACAP (12) vector fields. Note the approximate length preservation in (a), and the preservation of angles and the area deviation in (b).

3.1 Continuous Deformations

A continuous deformation is a time-dependent map $\mathbf{f} : \Omega_d^0 \times \mathbb{R} \to \mathbb{R}^d$ with $\Omega_d^0 \subseteq \mathbb{R}^d$, i.e., a time-dependent map from a domain Ω_d^0 to $\mathbb{R}^d$. We primarily consider the important dimensions $d = 2$ and $d = 3$ in this work but also provide some generalizations for higher dimensions. Let $\mathcal{X}_0 \subseteq \Omega_d^0$ be a point set defining some initial shape. Then the deformed shape at time t is expressed as the image $\mathbf{f}(\mathcal{X}_0, t)$. We use the short notation $\Omega_d = \mathbf{f}(\Omega_d^0, t)$ for the deformed domain at the current time t, which is clear from the context.

We define the velocity of $\mathbf{f}$ as the vector field $\mathbf{v}(\mathbf{x}, t) = \frac{d}{dt}\mathbf{f}(\mathbf{x}, t)$. Then $\mathbf{f}(\mathbf{x}, t)$ can be reconstructed from $\mathbf{v}$ by solving the initial value problem

$$\frac{d}{dt}\mathbf{x}(t) = \mathbf{v}(\mathbf{x}, t) \quad \text{with } \mathbf{x}(0) = \mathcal{X}_0 .$$

In the following, we derive conditions on $\mathbf{v}$ that lead to near-isometric and near-conformal maps $\mathbf{f}$. The conditions are characterized as the minimizers of certain energy terms w.r.t. interpolation constraints on $\mathbf{v}$. Figure 1 shows examples for deformations that were determined by this kind of vector fields.

3.2 Characteristic Deformations

For a single step (i. e., not time-dependent) deformation $\mathbf{f} : \Omega_d^0 \to \mathbb{R}^d$ with deformation gradient $\mathbf{D} = \nabla \mathbf{f}$ the first fundamental form $\mathcal{I}$ of $\mathbf{f}$ has the particularly simple form

$$\mathcal{I} = \mathbf{D}^{\mathrm{T}}\mathbf{D} .$$

Therefore, the *singular values* σ_i of $\mathbf{D}$ are square roots of the *eigenvalues* λ_i of $\mathcal{I}$. Then the following equivalent local properties of the deformation map can be shown (see, e. g., the work of Floater and Hormann in the context of parameterizations for the case of $d = 2$ [10]):

1. $\mathbf{f}$ is isometric $\Leftrightarrow$ $\mathcal{I} = \mathbf{I}$ $\Leftrightarrow$ $\lambda_i = 1$ $\Leftrightarrow$ $\sigma_i = 1,$ (1)

2. $\mathbf{f}$ is conformal $\Leftrightarrow$ $\mathcal{I} = \mu\,\mathbf{I}$ $\Leftrightarrow$ $\dfrac{\lambda_i}{\lambda_j} = 1$ $\Leftrightarrow$ $\dfrac{\sigma_i}{\sigma_j} = 1,$ (2)

3. $\mathbf{f}$ is equiareal $\Leftrightarrow$ $\det \mathcal{I} = 1$ $\Leftrightarrow$ $\prod_{i=1}^{d} \lambda_i = 1$ $\Leftrightarrow$ $\prod_{i=1}^{d} \sigma_i = 1 .$ (3)

Note that surface parameterizations can be regarded as deformations between $2d$ and $3d$, and in this work we consider the instantaneous deformation energy.

When deformations are parameterized by time t (i.e., we have continuous $\mathbf{f}(\mathbf{x}, t)$ and $\mathbf{D}(\mathbf{x}, t)$) these properties can be differentiated in order to obtain defining conditions on the vector field of the continuous deformation. Specifically, we apply the matrix algebra described by Minka [27] to obtain matrix derivatives w.r.t. t. They define the differential $d\mathbf{y}(\mathbf{x})$ to be the part of $\mathbf{y}(\mathbf{x}+d\mathbf{x}) - \mathbf{y}(\mathbf{x})$ that is linear in $d\mathbf{x}$. Differentials are obtained by iteratively applying a set of differentiation rules. After transformation into canonical form the matrix derivative can directly be read off.

3.3   Isometric Energies

Exact isometric deformations that fulfill all user constraints are not always possible. Therefore, measures for the *deviation* from isometry are required and we continue to present two possible models: Killing and metric energies.

Killing Energy. The matrix derivative of the isometry property is obtained by deducing the differential of (1), which gives

$$d\mathbf{D}^{\mathrm{T}}\mathbf{D} + \mathbf{D}^{\mathrm{T}}\,d\mathbf{D} = \mathbf{0}$$

using the product rule $d(\mathbf{A}\,\mathbf{B}) = d\mathbf{A}\,\mathbf{B} + \mathbf{A}\,d\mathbf{B}$ and $d\mathbf{I} = \mathbf{0}$. This equality has to hold for every time t of the continuous deformation. Specifically, for $t = 0$ we have $\mathbf{D}(\mathbf{x}_0, 0) = \mathbf{I}$ and by using $d\mathbf{D} = \mathbf{J}^{\mathrm{T}}$, where $\mathbf{J}$ is the Jacobian of the tangent vector field of $\mathbf{f}$, we obtain

$$\mathbf{J}^{\mathrm{T}} + \mathbf{J} = \mathbf{0} \tag{4}$$

as the condition for $\mathbf{f}$ to be isometric expressed in the vector field of the continuous deformation. Equation (4) corresponds to the constraint that exact isometric deformations are generated by infinitesimal rotations, since the symmetric part of their Jacobian, which is skew-symmetric then, vanishes.

The L^2 deviation of (4) over a domain Ω_d

$$E_{\mathrm{AKVF}}(\mathbf{v}) = \int_{\Omega_d} \left|\left|\mathbf{J}^{\mathrm{T}} + \mathbf{J}\right|\right|_F^2 \, d\mathbf{x} \tag{5}$$

is called *Killing energy* with the Frobenius norm $||\cdot||_F$. It is used by Solomon et al. [29] for the case $d = 2$ to define *as-Killing-as-possible* vector fields $\mathbf{v}$ that minimize E_{AKVF} and which therefore generate near-isometric planar deformations. Higher dimensional cases ($d > 2$) are also well-defined. Note that the Jacobian is linear in the unknown vector fields $\mathbf{v}$ as differentiation is a linear operation, i.e., there exists a gradient operator $\mathbf{G}$ on Ω_d with $\mathbf{J} = \mathbf{G}\,\mathbf{v}$. Therefore, the energy (5) is quadratic in $\mathbf{v}$ and the corresponding variational optimization of (5) leads to a linear system that can efficiently be solved for the optimal vector field.

Metric Energy. The classic Killing energy (5) uses the Frobenius norm of (4) to measure deviation from isometry. We propose a related energy that measures another form of deviation from isometry that is not based on a L^2 deviation of (4). Informally spoken, our energy directly observes an infinitesimally small line segment and measures change of length under an infinitesimal integration step in $\mathbf{v}$. This is done for all possible infinitesimal segments, i.e., we integrate the (squared) change of length over all possible directions. We call this energy *metric* as distance variations are measured explicitly. We start with the derivation of the $2d$ case followed by the $3d$ case.

In order to measure the variation of length under integration in $\mathbf{v}$ we consider a line segment $\mathcal{S}$ between points $\mathbf{x}_0$ and $\mathbf{x}_1 = \mathbf{x}_0 + r_1 \mathbf{r}_1$ for a unit direction $\mathbf{r}_1$ and segment length r_1. The flow of $\mathcal{S}$ in $\mathbf{v}$ is given as $\mathbf{x}_0'(h) = \mathbf{x}_0 + \int_0^h \mathbf{v}(\mathbf{x}_0'(s))ds$ and $\mathbf{x}_1'(h) = \mathbf{x}_1 + \int_0^h \mathbf{v}(\mathbf{x}_1'(s))ds$. This induces the quadratic length variation

$$d^l(h) = ||\mathbf{x}_1 - \mathbf{x}_0||^2 - ||\mathbf{x}_1'(h) - \mathbf{x}_0'(h)||^2 .$$

Since we are interested in instantaneous variations (i.e., the length variation of an infinitesimal small line segment during an infinitesimal small integration) only, we consider the limit

$$d_0^l(\mathbf{r}_1) = \lim_{h \to 0, \, r_1 \to 0} \frac{d^l(h)}{r_1^2 \, h} = \frac{\partial^3 d^l(h)}{\partial r_1^2 \, \partial h} .$$

d_0^l measures the instantaneous quadratic length variation for the direction $\mathbf{r}_1$. We obtain the pointwise quadratic isometric energy $e_{\text{METR}}(\mathbf{x}_0, \mathbf{v})$ at $\mathbf{x}_0$ by considering all possible line segment directions given by $\mathbf{r}_1(\alpha) = (\cos(\alpha), \sin(\alpha))^{\text{T}}$:

$$e_{\text{METR}}(\mathbf{x}_0, \mathbf{v}) = \frac{1}{2\pi} \int_0^{2\pi} d_0^l(\mathbf{r}_1(\alpha))^2 \, d\alpha . \tag{6}$$

It can be shown that (6) has the following closed form solution that depends only on the Jacobian of $\mathbf{v}$[1]:

$$e_{\text{METR}}(\mathbf{x}_0, \mathbf{v}) = u_x^2 + v_y^2 + \frac{1}{2}(u_y + v_x)^2 + \frac{1}{2}(u_x + v_y)^2$$

$$= c \left(||\mathbf{J} + \mathbf{J}^{\text{T}}||_F^2 + 2\,(\text{Tr}\,\mathbf{J})^2 \right) . \tag{7}$$

Here $\mathbf{J} = \left[\begin{smallmatrix} u_x & u_y \\ v_x & v_y \end{smallmatrix} \right]$ denotes the Jacobian of $\mathbf{v}$ at $\mathbf{x}_0$, $\text{Tr} \cdot$ is the trace of a matrix, and c is a constant factor. The total *metric energy* of a vector field $\mathbf{v}$ on Ω_2 is now given by

$$E_{\text{METR}}(\mathbf{v}) = \int_{\Omega_2} e_{\text{METR}}(\mathbf{x}, \mathbf{v}) \, d\mathbf{x} . \tag{8}$$

We call vector fields that minimize this energy *as-metric-as-possible* (AMAP) vector fields. Figure 1(a) shows examples for deformations that were determined by this kind of vector fields.

---

[1] The derivation of equivalence is lengthy but consists only of basic algebraic transformations and therefore is omitted in the paper. — We provide derivations in form of Maple scripts for *all closed form solutions of integrals* in this submission with the additional material.

We derive a similar energy for $d = 3$ dimensions using the same ansatz as above for $d = 2$. Again we take the integral over all possible configurations of an infinitesimal integration step of an infinitesimally small line segment between two points $\mathbf{x}_0$ and $\mathbf{x}_1 = \mathbf{x}_0 + r_1\,\mathbf{r}_1$. The main difference to the $2d$ case is that angles in the plane now have to be replaced by solid angles. For the spherical parametrization of the unit direction $\mathbf{r}_1(\alpha, \beta) = (\cos(\alpha)\cos(\beta)\,,\sin(\alpha)\cos(\beta)\,,\sin(\beta)\,)^{\mathrm{T}} \in \mathbb{R}^3$ we obtain the pointwise quadratic metric energy as the integral

$$e_{\mathrm{METR3D}}(\mathbf{x}_0, \mathbf{v}) = \frac{1}{4\pi} \int_0^{2\pi} \int_{-\frac{\pi}{2}}^{\frac{\pi}{2}} \cos(\beta)\; d_0^l(\mathbf{r}_1(\alpha, \beta))^2 \, \mathrm{d}\beta \, \mathrm{d}\alpha \;, \tag{9}$$

which again has the closed form solution

$$e_{\mathrm{METR3D}}(\mathbf{x}_0, \mathbf{v}) = c\left(\left\| \mathbf{J} + \mathbf{J}^{\mathrm{T}} \right\|_F^2 + 2\left(\mathrm{Tr}\,\mathbf{J} \right)^2 \right) . \tag{10}$$

Interestingly the factors of (7) and (10) only differ in the constant c, although their dimensions differ. The total $3d$ metric energy is then obtained as

$$E_{\mathrm{METR3D}}(\mathbf{v}) = \int_{\Omega_3} e_{\mathrm{METR3D}}(\mathbf{x}, \mathbf{v}) \, \mathrm{d}\mathbf{x} \;.$$

We again call the minimizers of this energy *as-metric-as-possible* vector fields. In the following we will use the terms METR and METR3D synonymously whenever the context is clear.

3.4   Conformal Energy

The differential of (2) is given by

$$\mathrm{d}\mathbf{D}^{\mathrm{T}}\mathbf{D} + \mathbf{D}^{\mathrm{T}}\,\mathrm{d}\mathbf{D} = \mathrm{d}\mu\,\mathbf{I} \;.$$

We again evaluate it at $t = 0$, and by setting $\mathrm{d}\mu = \alpha$ we obtain

$$\mathbf{J}^{\mathrm{T}} + \mathbf{J} = \alpha\,\mathbf{I}$$

as the condition for the continuous deformation $\mathbf{f}$ to be conformal. Note that here α is an additional degree of freedom stating the fact that instantaneous uniform scaling is conformal for every scaling factor.

We derive a pointwise energy / energy density e_{CONF} that measures the L^2 deviation of this conformality condition. The construction of the energy holds for any dimension d from which important two and three-dimensional special cases can be obtained:

$$\begin{aligned}
e_{\mathrm{CONF}} &= \left\| \mathbf{J}^{\mathrm{T}} + \mathbf{J} - \alpha\,\mathbf{I} \right\|_F^2 \\
&= \mathrm{Tr}\!\left(\left(\mathbf{J}^{\mathrm{T}} + \mathbf{J} - \alpha\,\mathbf{I} \right)^{\mathrm{T}} \left(\mathbf{J}^{\mathrm{T}} + \mathbf{J} - \alpha\,\mathbf{I} \right) \right) \\
&= \mathrm{Tr}\!\left(\left(\mathbf{J}^{\mathrm{T}} + \mathbf{J} \right)^{\mathrm{T}} \left(\mathbf{J}^{\mathrm{T}} + \mathbf{J} \right) \right) + \mathrm{Tr}\!\left(-2\alpha \left(\mathbf{J}^{\mathrm{T}} + \mathbf{J} \right) + \alpha^2 \mathbf{I} \right) \\
&= \left\| \mathbf{J}^{\mathrm{T}} + \mathbf{J} \right\|_F^2 - 2\alpha\,\mathrm{Tr}\!\left(\mathbf{J}^{\mathrm{T}} + \mathbf{J} \right) + d\,\alpha^2 \\
&= \left\| \mathbf{J}^{\mathrm{T}} + \mathbf{J} \right\|_F^2 - 4\alpha\,\mathrm{Tr}\,\mathbf{J} + d\,\alpha^2
\end{aligned} \tag{11}$$

This energy formulation still depends on the scaling factor α. To obtain an expression that is independent of this parameter we consistently set it to the value that minimizes the value of the energy. That is, we solve $\nabla_\alpha e_{\text{CONF}} = 0$ for α, which gives $\alpha = \frac{2}{d} \operatorname{Tr} \mathbf{J}$. Inserting this result into (11) we obtain

$$e_{\text{CONF}} = \left|\left|\mathbf{J}^{\mathsf{T}} + \mathbf{J}\right|\right|_F^2 - \frac{4}{d}(\operatorname{Tr} \mathbf{J})^2$$

for the general d-dimensional pointwise conformal energy in the vector field of the continuous deformation. The total *conformal energy* of the vector field is then given by

$$E_{\text{CONF}}(\mathbf{v}) = \int_{\Omega_d} \left|\left|\mathbf{J}^{\mathsf{T}} + \mathbf{J}\right|\right|_F^2 - \frac{4}{d}(\operatorname{Tr} \mathbf{J})^2 \; d\mathbf{x} \; .$$

Again, this energy is quadratic in the vector field. We call vector fields minimizing this energy *as-conformal-as-possible* (ACAP). The important low-dimensional special cases are

$$E_{\text{CONF2D}}(\mathbf{v}) = \int_{\Omega_2} \left|\left|\mathbf{J}^{\mathsf{T}} + \mathbf{J}\right|\right|_F^2 - 2\,(\operatorname{Tr} \mathbf{J})^2 \; d\mathbf{x} \quad \text{and} \tag{12}$$

$$E_{\text{CONF3D}}(\mathbf{v}) = \int_{\Omega_3} \left|\left|\mathbf{J}^{\mathsf{T}} + \mathbf{J}\right|\right|_F^2 - \frac{4}{3}\,(\operatorname{Tr} \mathbf{J})^2 \; d\mathbf{x} \; .$$

See Figure 1(b) for example deformations that were determined by these vector fields.

3.5   Equiareal Energy

In order to obtain the condition on the vector field for the continuous deformation to be equiareal, we differentiate (3) using the differentiation rule $d \det \mathbf{A} = \det \mathbf{A} \operatorname{Tr}(\mathbf{A}^{-1} d\mathbf{A})$:

$$\begin{aligned}
d \det(\mathbf{D}^{\mathsf{T}}\mathbf{D}) &= 2\,(d \det \mathbf{D})\,\det \mathbf{D} \\
&= 2\left(\det \mathbf{D} \operatorname{Tr}(\mathbf{D}^{-1} d\mathbf{D})\right) \det \mathbf{D} \\
&= \operatorname{Tr}\!\left(2\,(\det \mathbf{D})^2 \, \mathbf{D}^{-1} d\mathbf{D}\right)
\end{aligned} \tag{13}$$

Evaluating (13) at $t = 0$ and using $d\mathbf{D} = \mathbf{J}$ the equiareal condition on the vector field simplifies to

$$\operatorname{Tr} \mathbf{J} = 0,$$

which states that the vector field has to be divergence free as $\operatorname{Tr} \mathbf{J} = \nabla \cdot \mathbf{v}$. The corresponding L^2 pointwise equiareal energy $e_{\text{EQUIA}} = (\nabla \cdot \mathbf{v})^2$ yields the total equiareal energy

$$E_{\text{EQUIA}}(\mathbf{v}) = \int_{\Omega_d} (\nabla \cdot \mathbf{v})^2 \; d\mathbf{x} \; .$$

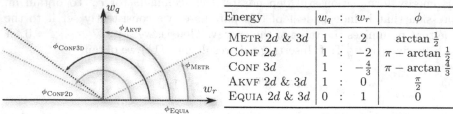

Energy	w_q : w_r	ϕ
Metr $2d$ & $3d$	1 : 2	$\arctan \frac{1}{2}$
Conf $2d$	1 : -2	$\pi - \arctan \frac{1}{2}$
Conf $3d$	1 : $-\frac{4}{3}$	$\pi - \arctan \frac{4}{3}$
Akvf $2d$ & $3d$	1 : 0	$\frac{\pi}{2}$
Equia $2d$ & $3d$	0 : 1	0

Fig. 2. Energy Parameter Domain. The different energies obtained from the general metric energy $E_\phi(\mathbf{v})$ are linear subspaces in the visualized domain of weights w_q and w_r, i.e., every pair of weights in a subspace yields the same energy minimizer. However, energies may not have unique minimizers, like the Equia energy in the limit.

3.6   A Generalized Family of Energies

In the following we relate the near-isometric, near-conformal, and near-equiareal energies to derive a generalized energy. This is a one-parameter family of energies that determine smooth blends between the different types of deformation.

We define

$$q(\mathbf{x}) := \left\| \mathbf{J}(\mathbf{x}) + \mathbf{J}(\mathbf{x})^{\mathrm{T}} \right\|_F^2 \quad \text{and} \quad r(\mathbf{x}) := (\mathrm{Tr}\,\mathbf{J}(\mathbf{x}))^2 .$$

Then all energy densities introduced so far can be expressed as linear combinations of $q(\mathbf{x})$ and $r(\mathbf{x})$. Uniform scaling of such an energy does not change the minimizing vector field. Therefore, we can describe all energies as a one-parameter family of *generalized metric energies* depending on ϕ:

$$E_\phi(\mathbf{v}) = \int_{\Omega_{2/3}} w_q(\phi)\, q(\mathbf{x}) \; + \; w_r(\phi)\, r(\mathbf{x})\, \mathrm{d}\mathbf{x}$$

with the weights $w_q(\phi) := \sin(\phi)$ and $w_r(\phi) := \cos(\phi)$ having specific ratios. In $2d$, ϕ can vary in the interval $]0, \pi - \arctan \frac{1}{2}]$, while in $3d$, ϕ varies in $]0, \pi - \arctan \frac{4}{3}]$. Then the parameter of the isometric energies is given by $\phi = \phi_{\mathrm{AKVF}} = \frac{\pi}{2}$, resp. $\phi = \phi_{\mathrm{METR}} = \arctan \frac{1}{2}$, and the minimizers of $E_{\phi_{\mathrm{AKVF}}}(\mathbf{v})$ and $E_{\mathrm{AKVF}}(\mathbf{v})$, resp. $E_{\phi_{\mathrm{METR}}}(\mathbf{v})$ and $E_{\mathrm{METR}}(\mathbf{v})$ are equal. Furthermore, the conformal energy is given by $\phi = \phi_{\mathrm{CONF2D}} = \pi - \arctan \frac{1}{2}$ in $2d$ and by $\phi = \phi_{\mathrm{CONF3D}} = \pi - \arctan \frac{4}{3}$ in $3d$, respectively. The equiareal energy is recovered for $\phi = \phi_{\mathrm{EQUIA}} = 0$. We note that volume preservation is not a sufficient condition for uniquely defining $\mathbf{v}$. However, adding a small amount of q to E_ϕ (i.e., choosing ϕ slightly above zero) gives unique solutions corresponding to near-equiareal deformations.

Figure 2 illustrates different choices of ϕ. Note that for $\phi > \frac{\pi}{2}$, E_ϕ contains negative quadratic terms. However, due to the definition of conformal energy density (11) as a squared matrix norm, it is guaranteed that E_ϕ is non-negative as long as $\phi \le \pi - \arctan \frac{1}{2}$ ($2d$) and $\phi \le \pi - \arctan \frac{4}{3}$ ($3d$), and that a unique minimizer exists.

Anisotropic Energies. The energy formulations presented so far are isotropic as distortions are measured in every direction in an uniform way. We model anisotropic behavior by replacing the isotropic Frobenius norm with an anisotropic norm $||\cdot||_{\mathbf{B}}^2$ defined by a rank-2 tensor field of symmetric positive definite matrices $\mathbf{B}$: $||\mathbf{A}||_{\mathbf{B}}^2 = \mathrm{Tr}(\mathbf{A}^{\mathrm{T}}\mathbf{B}\mathbf{A})$. For example, the pointwise energy (11) then becomes

$$
\begin{aligned}
e_{\mathrm{CONF}} &= ||\mathbf{J}^{\mathrm{T}} + \mathbf{J} - \alpha\,\mathbf{I}||_{\mathbf{B}}^2 \\
&= ||\mathbf{J}^{\mathrm{T}} + \mathbf{J}||_{\mathbf{B}}^2 + \mathrm{Tr}\big(-2\alpha\,(\mathbf{J}^{\mathrm{T}}\mathbf{B} + \mathbf{B}\,\mathbf{J})\big) + \mathrm{Tr}\big(\alpha^2\,\mathbf{B}\big) \\
&= ||\mathbf{J}^{\mathrm{T}} + \mathbf{J}||_{\mathbf{B}}^2 - 4\alpha\,\mathrm{Tr}(\mathbf{B}\,\mathbf{J}) + \gamma\,\alpha^2 \\
&= ||\mathbf{J}^{\mathrm{T}} + \mathbf{J}||_{\mathbf{B}}^2 - \frac{4}{\gamma}(\mathrm{Tr}(\mathbf{B}\,\mathbf{J}))^2 \;,
\end{aligned}
$$

where we have set $\gamma = \mathrm{Tr}\,\mathbf{B}$ and used the identity $\mathrm{Tr}(\mathbf{A}\,\mathbf{B}) = \mathrm{Tr}(\mathbf{B}\,\mathbf{A})$ together with the solution of $\nabla_\alpha\,e_{\mathrm{CONF}} = 0$, which is $\alpha = \frac{2}{\gamma}\,\mathrm{Tr}(\mathbf{B}\,\mathbf{J})$. In the special case of $\mathbf{B} = \mathbf{I}$ the isotropic case is recovered as then $||\cdot||_{\mathbf{B}} \equiv ||\cdot||_F$ and $\gamma = d$.

4   Discrete Setting

Let $\mathcal{P} = (\mathcal{V}, \mathcal{T}, \mathbf{x})$ be a partition of Ω_d (at a particular time t) with vertices $\mathcal{V}$ and cells $\mathcal{T}$ (triangles for $d = 2$ and tetrahedra for $d = 3$). Furthermore, let $m = |\mathcal{V}|$ denote the number of vertices, and $\mathbf{x}_i \in \mathbb{R}^d$ with $i \in \mathcal{V}$ denote vertex positions. We express a vector field $\mathbf{v}$ as piecewise linear functions on $\mathcal{P}$: $\mathbf{v}$ is given as nodal values $\mathbf{v}_i$, $i \in \mathcal{V}$; we write $\mathbf{v}$ as the a single column vector $\mathbf{v} = (\mathbf{v}_1{}^{\mathrm{T}}, \ldots, \mathbf{v}_m{}^{\mathrm{T}})^{\mathrm{T}} \in \mathbb{R}^{dm}$. Its piecewise constant Jacobian field is given as matrices $\mathbf{J}_c$ on cells $c \in \mathcal{T}$.

Energy Minimization. In the discrete setting, E_ϕ is a quadratic form in the unknown vector field: $E_\phi(\mathbf{v}) = \mathbf{v}^{\mathrm{T}}\mathbf{E}_\phi\,\mathbf{v}$. The matrix $\mathbf{E}_\phi \in \mathbb{R}^{dm \times dm}$ is the symmetric positive definite sparse matrix defining E_ϕ. With the Jacobians being constant on each cell, the coefficients of $\mathbf{E}_\phi$ are the sum of matrices $\mathbf{E}_\phi^c$ that capture the local error E_ϕ^c on cell c as

$$
\begin{aligned}
E_\phi^c(\mathbf{v}) &= \int_{\Omega_c} w_q(\phi)\,q(\mathbf{x}) + w_r(\phi)\,r(\mathbf{x})\,\mathrm{d}\mathbf{x} \\
&= V_c\left(w_q\,\big|\big|\mathbf{J}_c + \mathbf{J}_c{}^{\mathrm{T}}\big|\big|_F^2 + w_r\,(\mathrm{Tr}\,\mathbf{J}_c)^2\right) \quad = \mathbf{v}_c{}^{\mathrm{T}}\mathbf{E}_\phi^c\,\mathbf{v}_c\;.
\end{aligned}
$$

Here the vector $\mathbf{v}_c \in \mathbb{R}^6$ $(d = 2)$ or resp. $\mathbf{v}_c \in \mathbb{R}^{12}$ $(d = 3)$ is the concatenation of velocities of the vertices of c, $\mathbf{J}_c$ is the constant Jacobian on c, and V_c is the volume of the cell, triangle area or tetrahedral volume, which weights the constant expressions during integration over the discrete domain $\mathcal{P}$.

We use interpolation constraints on the flow $\mathbf{v}$. This means that the user prescribes trajectories $\gamma_k(t)$ that define the flow of some vertices $k \in \mathcal{V}$. This yields conditions $\mathbf{v}_k(t) = \frac{\mathrm{d}}{\mathrm{d}t}\gamma_k(t)$ as the flow along the trajectories is defined by their tangents. Not that this includes the special case of "fixed" vertices for which the trajectory is a constant domain point with $\mathbf{v}_k(t) \equiv \mathbf{0}$.

The vector field $\hat{\mathbf{v}}$ minimizing the energy is given as the solution to the linear system $\nabla E_\phi(\mathbf{v}) = \mathbf{0}$ subject to these constraints. In Section 5 we discuss how to setup $\mathbf{E}_\phi$ and solve the arising linear systems efficiently.

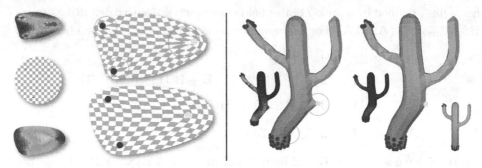

Fig. 3. Smoothness Energy. In $2d$ (left) and $3d$ (right) minimizers of the generalized metric energy are discontinuous near user constraints (fixed • and handle • vertices) leading to the highlighted local discontinuous deformations (small images show metric distortions, see Section 6). The smoothness energy term yields smooth vector fields and therefore smooth deformations. Note that both interior (left) and boundary vertices (right) can be constrained.

Enforcing Smoothness. The derived energies do not enforce smoothness of the solution. This means that even though we obtain a minimizer the residual energy is not distributed smoothly over the domain. In particular, this leads to high concentration of metric error near constrained vertices. Lipman observes this effect for finite ARAP deformations [22]. This problem was also already discussed by Solomon et al. [29]. Their solution consists in a post-process: they solve an additional linear system that diffuses the error to construct smooth vector fields. This consequences: firstly, an additional solving step is required, and secondly, the previously defined constraints can only be satisfied approximately. During time-integration, this approach leads to significant drift from the user defined trajectories. Moreover, we show that the total resulting deformation error increases unnecessarily (see Section 6).

We take a different approach based on regularization. We define smoothness as the local first order energy variation. This way local deformation errors vary smoothly and do not concentrate, e. g., only at the constrained vertices. The local energy of a cell depends on its constant local Jacobian, i.e., there is variation only on the cell boundaries. Let $c_i, c_j \in \mathcal{T}$ be two neighboring cells with Jacobians $\mathbf{J}_i, \mathbf{J}_j$, and local energy parameters ϕ_i, ϕ_j, respectively. As E_ϕ^c depends only on $\mathbf{J}_c$ we obtain the integrated variation $E_S^{i,j}$ for the pair (c_i, c_j) as

$$E_S^{i,j}(\mathbf{v}, \phi) = B_{i,j} \left\| 4 \left(\mathbf{D}_{i,j}^q + \mathbf{D}_{i,j}^{q\,\mathrm{T}} \right) + 2 \operatorname{Tr} \mathbf{D}_{i,j}^r \, \mathbf{I} \right\|_F^2$$

with $\mathbf{D}_{i,j}^s = w_{s_j} \mathbf{J}_j - w_{s_i} \mathbf{J}_i$ and $w_{s_i} = w_s(\phi_i)$ for $s \in \{q, r\}$, and $B_{i,j}$ denotes the length of the common edge of adjacent triangles $(d = 2)$ or the area of the common triangle of adjacent tetrahedra $(d = 3)$. See Appendix A for a derivation. Note that $E_S^{i,j}$ is quadratic in $\mathbf{v}$. The total discrete smoothness energy is then given by the sum over all adjacent cells

$$E_S(\mathbf{v}, \phi) = \sum_{i,j \in \mathcal{T}\,\mathrm{adjacent}} E_S^{i,j}(\mathbf{v}, \phi) \, .$$

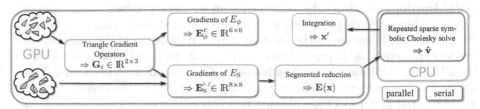

Fig. 4. GPU Pipeline. We use the GPU to setup linear systems and perform vector field integration. The linear systems are solved on the CPU using an efficient sparse solver. Operations marked (•) are performed in parallel on the GPU.

This energy has the quadratic form $E_S = \mathbf{v}^T \mathbf{E}_S \mathbf{v}$ that acts as a regularization term in a weighted total energy in the deformation vector field

$$E(\mathbf{v}, \phi) = E_\phi(\mathbf{v}) + \lambda E_S(\mathbf{v}, \phi) .$$

Its quadratic form is $E(\mathbf{v}, \phi) = \mathbf{v}^T (\mathbf{E}_\phi + \lambda \mathbf{E}_S) \mathbf{v} := \mathbf{v}^T \mathbf{E} \mathbf{v}$. Hence, we compute a smooth minimizer of E_ϕ by solving $\nabla \hat{\mathbf{v}}^T \mathbf{E} \hat{\mathbf{v}} = \mathbf{0}$. We use a factor of $\lambda = 0.1$ in all our examples. Figure 3 illustrates the effect of using the regularization term E_S in two and three dimensions. Note that smoothness of the vector field is preserved for handles in the interior as well as on the boundary of the domain.

Shape Integration. We are left with the problem of solving an ODE numerically: we solve $\frac{\mathrm{d}}{\mathrm{d}t}\mathbf{x}_i(t) = \hat{\mathbf{v}}(\mathbf{x}_i(t), t)$ with initial vertex positions $\mathbf{x}_i(0)$, $i \in \mathcal{V}$, using a standard ODE solver. For every evaluation of the vector field the energy minimizing flow $\hat{\mathbf{v}}$ is computed from the current shape configuration.

5 Implementation

Modeling Metaphor. In contrast to finite deformation methods, continuous deformations require velocities as boundary constraints (cf. [29,21]). There are various ways to prescribe velocities. In the simplest case they are provided as zero vectors for fixed vertices. Translations can be modeled by constant velocities, rotations can be expressed by linear flows. A fairly general and intuitive approach is the definition of a space-time curve that acts acts as trajectory, i.e., velocity along the curve is defined as the tangent vector. It is easy to extend this approach to define a laminar "bundle" of trajectories that are defined by the Frenet frame of a single curve [12].

In addition to constraints, the user can model nonhomogeneous energies by changing the scalar parameter ϕ and the tensor field $\mathbf{B}$. This can be done globally or locally per cell, e.g., by a spatial blend (see Figure 7). From the users point of view, near-isometric deformations often behave similarly to stiff real materials, while near-conformal deformations often exhibit strong scaling components towards smaller and larger area.

GPU Implementation. We use the GPU to accelerate certain steps of our deformation algorithm. Figure 4 provides an overview, with matrix dimensions given for the $2d$ case. In summary, the setup of the linear system and the integration of vertices are performed in parallel on the GPU, and the sparse system is solved on the CPU. First, all triangle gradient operators $\mathbf{G}_i$ are computed in parallel. These are required to compute the energy terms E_ϕ and E_S. Then the energy gradients are computed in parallel by exploiting symmetry for each cell and for each pair of adjacent cells. The results are summed by a parallel segmented reduction operation to give the final linear system. The sparse system is downloaded to CPU memory, where it is solved using a state-of-the-art sparse Cholesky solver that uses a precomputed symbolic factorization and an approximate minimum degree preordering to reduce fill-in [9]. In our experiments this direct system solve is up to four times faster than solving the linear system on the GPU using an iterative sparse solver. Compared to a pure CPU implementation using the GPU is up to three times faster. This is because the cost for system setup are significant as multiple systems need to be solved during integration. Finally, shape integration along the optimal flow $\hat{\mathbf{v}}$ is performed on the GPU. We use a standard fourth-order Runge-Kutta integrator with adaptive step size control.

6   Analysis and Results

Energy Comparison. We evaluate the angle and volume quality of deformations using the following error terms

$$F_{angle}^{2D} = \sum_{c\in T_2} \rho_c \left(\frac{\sigma_c^1}{\sigma_c^2} + \frac{\sigma_c^2}{\sigma_c^1} - 2 \right) \qquad F_{angle}^{3D} = \sum_{c\in T_3} \rho_c \left(\sum_{(j,k)\in P_3} \left(\frac{\sigma_c^j}{\sigma_c^k} + \frac{\sigma_c^k}{\sigma_c^j} \right) - 6 \right)$$

$$F_{area}^{2D} = \sum_{c\in T_2} \rho_c \left(\sigma_c^1 \sigma_c^2 + \frac{1}{\sigma_c^1 \sigma_c^2} - 2 \right) \quad F_{volume}^{3D} = \sum_{c\in T_3} \rho_c \left(\sigma_c^1 \sigma_c^2 \sigma_c^3 + \frac{1}{\sigma_c^1 \sigma_c^2 \sigma_c^3} - 2 \right).$$

These errors are established in the literature (see, e.g., [29]) and are based on Equations (1-3). Here σ_c^j is the jth singular value of the Jacobian of the map of triangle or tetrahedra c, $\rho_c = V_c/\sum_{j\in T} V_j$ with triangle area or tetrahedral volume V_c, and $P_3 = \{(1,2),(2,3),(3,1)\}$. To measure metric errors we introduce the error terms

$$F_{metric}^{2D} = \sum_{c\in T_2} \rho_c \left(\left(\sigma_c^1 - 1\right)^2 + \left(\sigma_c^2 - 1\right)^2 - \frac{1}{4}\left(\sigma_c^1 - \sigma_c^2\right)^2 \right) \qquad (14)$$

$$F_{metric}^{3D} = \sum_{c\in T_3} \rho_c \left(\sum_{j=1}^{3} (\sigma_c^j - 1)^2 - \frac{1}{5} \sum_{(j,k)\in P_3} (\sigma_c^j - \sigma_c^k)^2 \right) \qquad (15)$$

that are the weighted sum of solutions of integrals of the form of (6) and (9). The difference to the previous derivation of energies is that the integrand is no pointwise infinitesimal quadratic length variation but the pointwise finite quadratic length variation induced by the map and integrated along all possible directions. In the optimal case all error terms are zero.

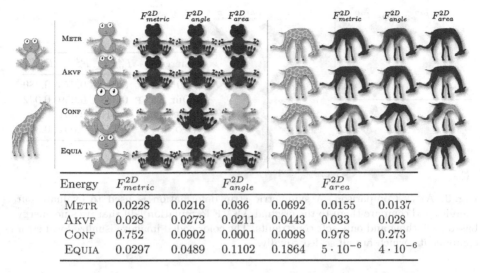

Energy	F^{2D}_{metric}		F^{2D}_{angle}		F^{2D}_{area}	
METR	0.0228	0.0216	0.036	0.0692	0.0155	0.0137
AKVF	0.028	0.0273	0.0211	0.0443	0.033	0.028
CONF	0.752	0.0902	0.0001	0.0098	0.978	0.273
EQUIA	0.0297	0.0489	0.1102	0.1864	$5 \cdot 10^{-6}$	$4 \cdot 10^{-6}$

Fig. 5. $2D$ Energy Evaluation. Two initial models (left) are deformed using the same boundary constraints for the different energy types. The plots visualize color coded local errors. The CONF error plots for the frog are downscaled to 75% size. The table gives total errors for each method and each model (frog left / giraffe right column).

Figure 5 shows error values and error visualizations for two planar deformations. For the EQUIA results we used $\phi = \arctan 2^{-9}$. The METR energy generating near-isometric vector fields achieves lowest metric and area distortions at the cost of change of angle. Deformations based on the AKVF energy show better angle preservation compared to METR, but they also show greater errors in length and area variation. Almost no angle distortion is introduced by ACAP vector fields based on the CONF energy, however, this is at the cost of area errors. The opposite is true for the EQUIA deformation that introduces almost no area error but instead a large angular error. The experiment confirms that the parameter ϕ corresponds to balance between metric and area preservation on the one side and angular preservation on the other side (cf. Figure 2). No deformation can preserve all properties at the same time.

In Figure 6 we compare our energies (including AKVF) to the original method in [29] that uses "soft" handle constraints and achieves smoothness by a error diffusion. Note that the softly constrained vertices drifted significantly. To compensate for this effect and for a fair comparison, the constraints were selected such that the trajectories of all handles ( ) end in the (optimally) fixed soft handles (●) after the same integration time. The two AKVF and the METR results look visually similar, however, all three error values indicate that our AKVF approach using a problem dependent smoothing term achieves deformation of lower error.

Figure 10 shows frames of the animation of a volumetric mesh. The wings of the eagle were deformed symmetrically using three-dimensional AMAP and ACAP deformations as well as with an EQUIA vector field with $\phi = \arctan 2^{-9}$.

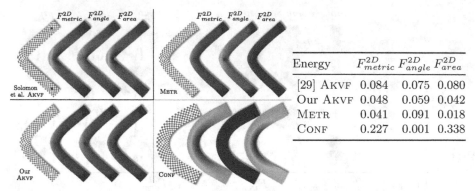

Energy	F^{2D}_{metric}	F^{2D}_{angle}	F^{2D}_{area}
[29] AKVF	0.084	0.075	0.080
Our AKVF	0.048	0.059	0.042
METR	0.041	0.091	0.018
CONF	0.227	0.001	0.338

Fig. 6. AKVF Comparison. A symmetric strip deformation is used to evaluate our energies and compare them to the original AKVF formulation [29] that uses no energy based smoothing and only soft constraints. The color coded images visualize local error components, which are all scaled equally.

Fig. 7. Left: A nonhomogeneous parameter ϕ is given as a scalar field in form of a blend from ϕ_{METR} on the left side of the frog's domain to ϕ_{CONF} on the right. Deformation constraints are defined symmetrically on both sides of the model. Right: An isotropic deformation compared to a deformation of anisotropic material with a locally "stiffer" axis direction. Equal deformation constraints were applied in both cases together with ϕ_{AKVF}.

Again, the conformal energy trades volumetric error for angle preservation while the isometric energy has better length and volume preservation properties at the expense of angular distortion. Best volume preservation but also most angular distortions is achieved by the equiareal deformation. This is also reflected in the error values of the animation steps (I) and (III) given in the table. Note that we also included the AKVF errors of the same $3d$ deformation, which is not shown.

Energy Parameter. Figure 7 (left) shows an example where a nonhomogeneous parameter ϕ is prescribed as a scalar field on the domain. In the example we use a spatial blend from near-isometric (left side) to conformal (right side). Defining symmetric constraints shows the nonhomogeneous effect of ϕ.

Fig. 8. $2D$ Deformation Examples. The triangulated models in the box were deformed using AMAP and ACAP vector fields.

In Figure 7 (right) we demonstrate the effect of using an anisotropic material compared to an isotropic one. Specifically, we define a region in the center of the strip that is "stiffer" along one prescribed axis modeled by a corresponding tensor field $\mathbf{B}$. The material modification leads to two near-isometric deformations of different characteristics for the same boundary constraints.

Independence of Tessellation. The discretization of our energies are integrated measures on the discretized domain. We expect that the resulting deformations are independent of the partition, i.e., of the tessellation, as long as there are enough degrees of freedom available to represent the constrained deformation. This is confirmed by all our experiments (see the adjacent Figure for an example).

Modeling Results. Figure 8 shows initial $2d$ shapes and two deformed versions using AMAP and ACAP deformations. The model size ranges from 5k to 11k vertices and the modeling time was below four minutes in every example. More $2d$ examples are shown in Figures 1(a), 1(b), and 5. Besides the animation in Figure 10 we show further tetrahedral deformations in Figure 9. Again we have the initial shapes together with AMAP and ACAP deformations. The meshes contain between $1,500$ to 5000 vertices. Again modeling time of an inexperienced user ranges from a few seconds to a few minutes. None of our tests suffered from stability issues, not even for extreme deformations. In particular, we didn't observe local folds or flips. This is due to the fact that the energy minimizing vector field generally does not vanish.

Timings. The following tables lists timings of our approach for the smallest and largest models, $2d$ and $3d$, respectively. We measured the time for the initial fac-

Fig. 9. 3D Deformation Examples. The tetrahedral models in the box were deformed using AMAP and ACAP vector fields to give the shown results.

torization of the linear system (t_1), system setup time (t_2), and time to solve the system (t_3), and the total time T to perform ten consecutive integration steps. Compared to a sole CPU implementation our parallel system setup using the GPU is up to three times faster even though the system has to be transferred to the CPU before solving it. Timings were measured on an

Model $(|\mathcal{T}|,|\mathcal{V}|)$	t_1(ms)	t_2(ms)	t_3(ms)	T(s)
Toucan (5.6k, 9.6k)	230	6	40	1.9
Cat (11k, 18k)	642	13	68	3.8
Octopus (1.5k, 5k)	168	19	51	2.9
Teapot (5k, 16k)	424	31	118	6.3

AMD Phenom II 955 quad-core CPU with 3.2GHz clock speed equipped with a NVIDIA GTX 560 Ti GPU with 2 GB of memory. Our approach is interactive for reasonably sized models. However, as is true for most solvers of nonlinear measures, also has much higher computational costs compared to linear methods. Please see also the accompanying video.

7   Discussion

Most existent geometrically-motivated approaches either optimize for near-isometric [17,31,21,29] or for near-conformal deformations [33,34]. In contrast to this our generalization combines both extrema in an integral formulation, and it can be applied the same way in $2d$ and in $3d$!

The results in Solomon et al. [29] indicate that their AKVF approach yields deformations of superior quality compared to related methods. It seems that for many shape deformation tasks near-isometry is the desired property. Our approach not only is able to reproduce their results but it shows even better behavior, getting even closer to isometric maps. At the same time our method is less complex and more efficient as we achieve smoothness by a regularization, enable true interpolation constraints and use standard ODE solvers.

Our main feature, however, is the ability to control the deformation by the parameter ϕ from conformal to equiareal with AKVF and our definition of near-

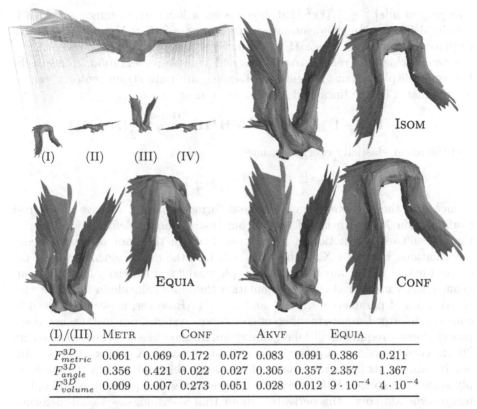

(I)/(III)	METR		CONF		AKVF		EQUIA	
F_{metric}^{3D}	0.061	0.069	0.172	0.072	0.083	0.091	0.386	0.211
F_{angle}^{3D}	0.356	0.421	0.022	0.027	0.305	0.357	2.357	1.367
F_{volume}^{3D}	0.009	0.007	0.273	0.051	0.028	0.012	$9 \cdot 10^{-4}$	$4 \cdot 10^{-4}$

Fig. 10. 3D Eagle Deformation. A tetrahedral model of an eagle (top left, with instantaneous vector field) was deformed in an animation of steps I-IV. The closeups show intermediate steps for different energies. Note the greatest volume of the conformal deformation.

isometric deformations in between. To the best of our knowledge, this approach is the first that provides such range of deformations in a single and concise mathematical framework. Our METR energies are an alternative way to measure deviation from isometry. In a direct comparison to AKVF deformations these new near-isometric energies show a better area preservation at the expense of a slightly higher angle deviation. We also note that close-to-equiareal deformations have not been studied thoroughly in the literature. Even though it is well known that these maps are not uniquely defined, for $\phi \to 0$ we get close to this limit, and even though the condition of system matrices degrades we obtain meaningful results.

Relation to Linear Elasticity. Our geometrically-motivated energy formulation can be related to physically-based theory of linear elasticity (see, e. g., [26]). This formalism assumes that a rest configuration with material coordinates $\mathbf{X}$ is deformed by a displacement field $\mathbf{u} = \mathbf{x} - \mathbf{X}$ into a deformed shape $\mathbf{x}$. The deformation results in an isotropic internal potential deformation energy $\psi = \mu||\epsilon||_F^2 + \frac{\lambda}{2}\operatorname{Tr}\epsilon^2$ that depends on a local strain tensor ϵ, which is usually defined using the deformation gradient tensor $\mathbf{F} = \nabla_{\mathbf{X}}\mathbf{x} = \mathbf{H} + \mathbf{I}$ with displacement gradient tensor $\mathbf{H} = \nabla_{\mathbf{X}}\mathbf{u}$. Here μ and λ are the physical Lamé constants, which are related to stiffness and volume preservation, respectively. For small displacement gradients the Lagrangian finite strain tensor ϵ can be approximated by the linearized small strain tensor

$$\epsilon := \frac{1}{2}(\mathbf{F}^\mathsf{T}\mathbf{F} - \mathbf{I}) = \frac{1}{2}(\mathbf{H} + \mathbf{H}^\mathsf{T} + \mathbf{H}^\mathsf{T}\mathbf{H}) \overset{||\mathbf{H}||_F \ll 1}{\approx} \frac{1}{2}(\mathbf{H} + \mathbf{H}^\mathsf{T}) \ ,$$

and the linear elasticity energy becomes

$$\psi = \frac{\mu}{4}\left|\left|\mathbf{H} + \mathbf{H}^\mathsf{T}\right|\right|_F^2 + \frac{\lambda}{2}\operatorname{Tr}\mathbf{H}^2 \ .$$

It measures the potential energy of the deformed shape $\mathbf{x}$ relative to the rest configuration $\mathbf{X}$, which is different to our instantaneous deformation energies that doesn't use the notion of a rest post. Still, in the limit of instantaneous deformations, i.e., $\mathbf{x} \to \mathbf{X}$, we have $\mathbf{H} \to \mathbf{J}$, i.e., the displacement gradient becomes the vector field Jacobian. Both the physical linear elasticity model and our geometrically motivated energy formulation therefore coincide in this case with the relation of parameters $w_q = \frac{\mu}{4}$ and $w_r = \frac{\lambda}{2}$. However, as we don't need to consider deformed shapes in different coordinate systems, our instantaneous approach doesn't require artificial regularization method like corotational elasticity [28] to correct artifact of diverging coordinate systems $\mathbf{X}$ and $\mathbf{x}$. Additionally, our instantaneous approach is unconditionally stable and we can therefore apply standard explicit ODE solvers for integration and require no, e.g., implicit integration. Moreover, this derivation shows that as-Killing-as-possible deformations [29] can be regarded as a geometric special case of physically-based linear elasticity that describes near-isometric materials. Additionally, in this work we provide the parameters for materials that show near-conformal behavior, which might not always give physically plausible results.

Limitations and Future Work. Nonlinear methods are expensive. Although we use a parallelized GPU implementation it is impossible to outperform linear methods in terms of computation time. This is a general drawback, and the user must decide if the additional cost is worthwhile. Still, all shown examples were modeled interactively.

Until now we consider only space deformations. So far, there is no extension to the explicit deformation of surfaces that are embedded in $3d$ space. This is because the vector field Jacobians capture only the tangential components of the vector field. They do not measure variations normal to the surface. For the same reason approximate Killing vector fields are, until now, considered only tangentially for triangle meshes [30]. We believe that this is an interesting direction for future research direction.

So far we consider only the initial value problem for path constrained deformation. It is much harder to solve the boundary value problem to find an energy minimizing path between two poses, e.g., for interpolation between poses (cf. [21,13]). We would like to use our generalized energy in such settings.

We furthermore want to study the application of generalized energies for parametrization applications to allow locally varying conformal / equiareal parametrization. Moreover, the eigen-spectrum of the energy might allow a multiresolution (in the parameter ϕ) segmentation of shapes.

8   Conclusions

In this paper we introduce a novel generalized metric energy for continuous shape deformation. We obtain near-isometric and near-conformal deformations by integration of as-isometric-as-possible and as-conformal-as-possible vector fields. Our approach works for two and three dimensions, we have applied it for deformations of triangular and tetrahedral meshes. For the discretization of the energy we have introduced a first order smoothness criterion based on the energy itself that guarantees vector field differentiability at handle vertices. Our implementation uses the GPU to achieve interactivity and we support nonhomogeneous and anisotropic behavior.

A   Smoothness Regularization

Given a generalized pointwise energy

$$e_\phi := w_q \left|\left| \mathbf{J} + \mathbf{J}^T \right|\right|_F^2 + w_r \left(\mathrm{Tr}\, \mathbf{J} \right)^2$$

the derivative w.r.t. $\mathbf{J}$ is given by

$$
\begin{aligned}
\frac{\partial}{\partial \mathbf{J}} e_\phi &= \frac{\partial}{\partial \mathbf{J}}\; w_q \left|\left| \mathbf{J} + \mathbf{J}^T \right|\right|_F^2 + w_r \left(\mathrm{Tr}\, \mathbf{J} \right)^2 \\
&= \frac{\partial}{\partial \mathbf{J}}\; w_q\, \mathrm{Tr} \left(\mathbf{J} + \mathbf{J}^T \right)^T \left(\mathbf{J} + \mathbf{J}^T \right) + w_r \left(\mathrm{Tr}\, \mathbf{J} \right)^2 \\
&= \frac{\partial}{\partial \mathbf{J}}\; w_q \left(2\, \mathrm{Tr}\, \mathbf{J}^T \mathbf{J} + 2\, \mathrm{Tr}\, \mathbf{J}\mathbf{J} \right) + w_r\, \mathrm{Tr}\, \mathbf{J}\, \mathrm{Tr}\, \mathbf{J} \\
&= 4\, w_q\, \mathbf{J} + 4\, w_q\, \mathbf{J}^T + w_r \left(\mathbf{I}\, \mathrm{Tr}\, \mathbf{J} + \mathrm{Tr}\, \mathbf{J}\, \mathbf{I} \right) \\
&= 4\, w_q \left(\mathbf{J} + \mathbf{J}^T \right) + 2\, w_r\, \mathrm{Tr}\, \mathbf{J}\, \mathbf{I} \,.
\end{aligned}
$$

Given two neighboring cells $c_i, c_j \in \mathcal{T}$ with local energy parameters ϕ_i, ϕ_j, we enforce smoothness by minimizing the variation of derivatives along a common edge ($n = 2$) or face ($n = 3$) $\mathcal{B}_{i,j} = \Omega_i \cap \Omega_j$, i.e., we regularize by minimizing smoothness energies of the form

$$
\begin{aligned}
E_S^{i,j}(\mathbf{v}, \phi) &= \int_{\mathcal{B}_{i,j}} \left|\left| \frac{\partial}{\partial \mathbf{J}_j} e_{\phi_j} - \frac{\partial}{\partial \mathbf{J}_i} e_{\phi_i} \right|\right|_F^2 \, \mathrm{d}\mathbf{x} \\
&= B_{i,j} \left|\left| \frac{\partial}{\partial \mathbf{J}_j} e_{\phi_j} - \frac{\partial}{\partial \mathbf{J}_i} e_{\phi_i} \right|\right|_F^2 \\
&= B_{i,j} \left|\left| 4 \left(\mathbf{D}_{i,j}^q + \mathbf{D}_{i,j}^{q\,T} \right) + 2\, \mathrm{Tr}\, \mathbf{D}_{i,j}^r\, \mathbf{I} \right|\right|_F^2
\end{aligned}
$$

with $\mathbf{D}_{i,j}^s = w_{s_j}\, \mathbf{J}_j - w_{s_i}\, \mathbf{J}_i$ and $w_{s_i} = w_s(\phi_i)$ for $s \in \{q, r\}$, and common edge length or face area $B_{i,j} = |\mathcal{B}_{i,j}|$. The integral can be simplified this way as the Jacobians are constant on each cell.

References

1. Alexa, M., Cohen-Or, D., Levin, D.: As-rigid-as-possible shape interpolation. In: Proc. SIGGRAPH, pp. 157–164 (2000)
2. Ben-Chen, M., Butscher, A., Solomon, J., Guibas, L.: On discrete killing vector fields and patterns on surfaces. In: Proc. SGP, pp. 1701–1711 (2010)
3. Ben-Chen, M., Weber, O., Gotsman, C.: Variational harmonic maps for space deformation. ACM Trans. Graph. 28(3), 34:1–34:11 (2009)
4. Botsch, M., Pauly, M., Gross, M., Kobbelt, L.: Primo: coupled prisms for intuitive surface modeling. In: Proc. SGP, pp. 11–20 (2006)
5. Botsch, M., Sorkine, O.: On linear variational surface deformation methods. IEEE TVCG 14(1), 213–230 (2008)
6. do Carmo, M.P.: Differential Geometry of Curves and Surfaces. Prentice–Hall (1976)
7. Cashman, T.J., Hormann, K.: A continuous, editable representation for deforming mesh sequences with separate signals for time, pose and shape. Comput. Graph. Forum 31(2), 735–744 (2012)
8. Chao, I., Pinkall, U., Sanan, P., Schröder, P.: A simple geometric model for elastic deformations. ACM Trans. Graph. 29(4), 1–38 (2010)
9. Chen, Y., Davis, T.A., Hager, W.W., Rajamanickam, S.: Algorithm 887: Cholmod, supernodal sparse cholesky factorization and update/downdate. ACM Trans. Math. Softw. 35(3), 1–22 (2008)
10. Floater, M.S., Hormann, K.: Surface parameterization: a tutorial and survey. In: Advances in Multiresolution for Geometric Modelling, pp. 157–186. Springer (2005)
11. von Funck, W., Theisel, H., Seidel, H.P.: Vector field based shape deformations. ACM Trans. Graph. 25(3), 1118–1125 (2006)
12. von Funck, W., Theisel, H., Seidel, H.P.: Explicit control of vector field based shape deformations. In: Proc. Pacific Graphics, pp. 291–300 (2007)
13. Heeren, B., Rumpf, M., Wardetzky, M., Wirth, B.: Time-discrete geodesics in the space of shells. Comp. Graph. Forum 31(5), 1755–1764 (2012)
14. Hormann, K., Greiner, G.: MIPS: An efficient global parametrization method. In: Curve and Surface Design 1999, pp. 153–162. Vanderbilt Press (2000)
15. Hormann, K., Polthier, K., Sheffer, A.: Mesh parameterization: Theory and practice. In: Proc. SIGGRAPH Asia (2008)
16. Hormann, K., Floater, M.S.: Mean value coordinates for arbitrary planar polygons. ACM Trans. Graph. 25(4), 1424–1441 (2006)
17. Igarashi, T., Moscovich, T., Hughes, J.F.: As-rigid-as-possible shape manipulation. ACM Trans. Graph. 24(3), 1134–1141 (2005)
18. Jacobson, A., Baran, I., Popović, J., Sorkine, O.: Bounded biharmonic weights for real-time deformation. ACM Trans. Graph. 30(4), 1–78 (2011)
19. Joshi, P., Meyer, M., DeRose, T., Green, B., Sanocki, T.: Harmonic coordinates for character articulation. ACM Trans. Graph. 26(3) (2007)
20. Karni, Z., Freedman, D., Gotsman, C.: Energy-based image deformation. In: Proc. SGP, pp. 1257–1268 (2009)
21. Kilian, M., Mitra, N.J., Pottmann, H.: Geometric modeling in shape space. ACM Trans. Graph. 26(3), 1–64 (2007)
22. Lipman, Y.: Bounded distortion mapping spaces for triangular meshes. ACM Trans. Graph. 31(4), 108:1–108:13 (2012)
23. Lipman, Y., Levin, D., Cohen-Or, D.: Green coordinates. ACM Trans. Graph. 27(3), 78:1–78:10 (2008)

24. Liu, L., Zhang, L., Xu, Y., Gotsman, C., Gortler, S.J.: A local/global approach to mesh parameterization. In: Proc. SGP, pp. 1495–1504 (2008)
25. Martinez Esturo, J., Rössl, C., Theisel, H.: Continuous deformations by isometry preserving shape integration. In: Boissonnat, J.-D., Chenin, P., Cohen, A., Gout, C., Lyche, T., Mazure, M.-L., Schumaker, L. (eds.) Curves and Surfaces 2011. LNCS, vol. 6920, pp. 456–472. Springer, Heidelberg (2012)
26. McAdams, A., Zhu, Y., Selle, A., Empey, M., Tamstorf, R., Teran, J., Sifakis, E.: Efficient elasticity for character skinning with contact and collisions. ACM Trans. Graph. 30(4), 1–37 (2011)
27. Minka, T.P.: Old and new matrix algebra useful for statistics. Tech. rep., MIT (2001)
28. Müller, M., Dorsey, J., McMillan, L., Jagnow, R., Cutler, B.: Stable real-time deformations. In: Proc. SCA, pp. 49–54 (2002)
29. Solomon, J., Ben-Chen, M., Butscher, A., Guibas, L.: As-killing-as-possible vector fields for planar deformation. Comput. Graph. Forum 30(5), 1543–1552 (2011)
30. Solomon, J., Ben-Chen, M., Butscher, A., Guibas, L.: Discovery of intrinsic primitives on triangle meshes. Comput. Graph. Forum 30(2), 365–374 (2011)
31. Sorkine, O., Alexa, M.: As-rigid-as-possible surface modeling. In: Proc. SGP, pp. 109–116 (2007)
32. Sýkora, D., Dingliana, J., Collins, S.: As-rigid-as-possible image registration for hand-drawn cartoon animations. In: Proc. NPAR, pp. 25–33 (2009)
33. Weber, O., Ben-Chen, M., Gotsman, C., Hormann, K.: A complex view of barycentric mappings. Comput. Graph. Forum 30(5), 1533–1542 (2011)
34. Weber, O., Myles, A., Zorin, D.: Computing extremal quasiconformal maps. Comput. Graph. Forum 31(5), 1679–1689 (2012)

Least Squares Fitting of Harmonic Functions Based on Radon Projections

Irina Georgieva[1], Clemens Hofreither[2], and Rumen Uluchev[3]

[1] Institute of Mathematics and Informatics, Bulgarian Academy of Sciences
Acad. G. Bonchev St., Bl. 8, 1113 Sofia, Bulgaria
irina@math.bas.bg
[2] Institute of Information and Communication Technologies,
Bulgarian Academy of Sciences
Acad. G. Bonchev St., Bl. 25A, 1113 Sofia, Bulgaria
chofreither@numa.uni-linz.ac.at
[3] Department of Mathematics and Informatics, University of Transport
158 Geo Milev St., Sofia 1574, Bulgaria
rumenu@vtu.bg

Abstract. Given the line integrals of a harmonic function on a finite set of chords of the unit circle, we consider the problem of fitting these Radon projections type of data by a harmonic polynomial in the unit disk. In particular, we focus on the overdetermined case where the amount of given data is greater than the dimension of the polynomial space. We prove sufficient conditions for existence and uniqueness of a harmonic polynomial fitting the data by using least squares method. Combining with recent results on interpolation with harmonic polynomials, we obtain an algorithm of practical application. We extend our results to fitting of more general mixed data consisting of both Radon projections and function values. We perform a comparative numerical study of the least-squares approach with two other reconstruction methods for the case of noisy data.

Keywords: multivariate interpolation, Radon transform, harmonic polynomials, least-squares fitting.

1 Introduction

There are many important problems in medicine, geophysics, biology, materials science, radiology, oceanography, and other sciences, where information about processes can only be obtained by nondestructive testing methods. Among the most successful techniques for reconstruction of objects with non-homogeneous density are tomographic imaging methods. Johann Radon and his results on the Radon transform [23] later to be named after him laid the mathematical foundation for this approach.

From the mathematical point of view, the problem is to recover a multivariate function using information given as line integrals of the unknown function. This problem has been intensively studied since the 1960s by different approaches

M. Floater et al. (Eds.): MMCS 2012, LNCS 8177, pp. 158–171, 2014.
© Springer-Verlag Berlin Heidelberg 2014

[17,5,6,7,15,19,20,24,16] and continues to find a lot of applications. Various reconstruction algorithms have been developed: filtered backprojection, iterative reconstruction, direct methods, etc., and some are based on the inverse Radon transform (see [21] and the bibliography therein).

Another class of methods for function reconstruction use direct interpolation by multivariate polynomials [20,14,1,4,11,12,13,10]. Many results along these lines are due to a research group founded by Prof. Borislav Bojanov which studies approximation problems using Radon projections type of data (see also [2,3,22]). A key question in this approach is how to construct a regular set of line segments, i.e., in what manner to choose chords of the unit circle so that there exists a unique polynomial of a corresponding degree with preassigned Radon projections over the configuration of chords.

To improve the approximation accuracy and to reduce the amount of input data required as well as the computational effort, one could try to incorporate some characteristic about the function to be recovered into approximation methods. According to this concept, interpolation of a harmonic function by harmonic polynomials based on Radon projections was studied in [9], where tools from symbolic computation were used, and in [8], where an analytical proof in a more general setting was given.

In the present paper, we continue the investigation of approximating harmonic functions using Radon projections type of data. In particular, we focus on the overdetermined case where the amount of data is greater than the dimension of the polynomial space. We use a least-squares method to determine a harmonic polynomial which fits the given data.

It turns out that the least-squares fitting problem and the interpolation problem are closely related. In [12], it was shown for the non-harmonic case that existence and uniqueness of the least-squares fitting polynomial relies on a regularity property of a subset of the scheme of chords.

With a similar proof technique, we derive sufficient conditions for existence and uniqueness of the least-squares harmonic polynomial, making use of recent results on interpolation with harmonic polynomials. We also consider fitting more general mixed data consisting of both Radon projections and function values. A reconstruction algorithm is developed and tested and numerical results are presented in the last section.

2   Preliminaries and Related Work

Let $D \subset \mathbb{R}^2$ denote the open unit disk and ∂D the unit circle. By $I(\theta, t)$ we denote a chord of the unit circle at angle $\theta \in [0, 2\pi)$ and distance $t \in (-1, 1)$ from the origin (see Figure 1). The chord $I(\theta, t)$ is parameterized by

$$s \mapsto (t \cos \theta - s \sin \theta, t \sin \theta + s \cos \theta)^\top, \quad \text{where } s \in (-\sqrt{1 - t^2}, \sqrt{1 - t^2}).$$

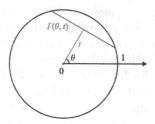

Fig. 1. The chord $I(\theta, t)$ of the unit circle

Definition 1. *Let $f(x, y)$ be a real-valued bivariate function in the unit disk D. The Radon projection $\mathcal{R}_\theta(f; t)$ of f in direction θ is defined by the line integral*

$$\mathcal{R}_\theta(f; t) := \int_{I(\theta, t)} f(\mathbf{x})\, d\mathbf{x} = \int_{-\sqrt{1-t^2}}^{\sqrt{1-t^2}} f(t\cos\theta - s\sin\theta, t\sin\theta + s\cos\theta)\, ds.$$

Johann Radon [23] showed in 1917 that a differentiable function f is uniquely determined by the values of its Radon transform,

$$f \mapsto \{\mathcal{R}_\theta(f; t) : -1 \le t \le 1,\ 0 \le \theta < \pi\}.$$

Further works in this area are due to John [18], Solmon [24], and others.

2.1 Interpolation and Fitting by Bivariate Polynomials

A fundamental problem in our investigations is to recover a polynomial using a finite number of values of its Radon transform. Essentially, this may be viewed as a bivariate interpolation problem where the usual function values are replaced by means over chords of the unit circle.

Let $\Pi_n^2 = \left\{ \sum_{i+j \le n} a_{ij} x^i y^j : a_{ij} \in \mathbb{R} \right\}$ denote the space of real bivariate polynomials of total degree at most n. This space has dimension $\binom{n+2}{2}$. Assume that a set $\mathcal{I} = \{I_m = I(\theta_m, t_m) : m = 1, \ldots, \binom{n+2}{2}\}$ of chords of ∂D is given. Furthermore, to each chord $I_m \in \mathcal{I}$ a given value $\gamma_m \in \mathbb{R}$ is associated. Then, the aim is to find a polynomial $p \in \Pi_n^2$ such that

$$\mathcal{R}_{\theta_m}(p, t_m) = \int_{I_m} p(\mathbf{x})\, d\mathbf{x} = \gamma_m \qquad \forall I_m \in \mathcal{I}. \tag{1}$$

If this interpolation problem has a unique solution for every choice of values $\Gamma = \{\gamma_m, : m = 1, \ldots, \binom{n+2}{2}\}$, then the scheme $\mathcal{I}$ of chords is called *regular*.

The question of how to construct such regular schemes has been extensively studied. The first general result was given by Marr [20] in 1974, who proved that the set of chords connecting $n + 2$ equally spaced points on the unit circle is regular for Π_n^2. A more general result for $\mathbb{R}^d$ and general convex domains was published by Hakopian [14] in 1982.

Different families of regular schemes of chords of the unit circle were constructed by Bojanov and Georgieva [1], Bojanov and Xu [4], Georgieva and Ismail [11], Georgieva and Uluchev [12], A mixed regular scheme which incorporates Radon projections and function values at points on the unit circle was proposed by Georgieva, Hofreither, and Uluchev [10].

Georgieva and Uluchev [12] considered a least-squares fitting problem for the overdetermined case of Radon projections type of data with algebraic polynomials and proved existence and uniqueness of the fitting polynomial. The proof was based on the above cited previous interpolation results. Moreover, this least-squares fitting was extended to mixed type data of Radon projections and function values.

2.2 Interpolation by Harmonic Polynomials

If we know *a priori* that the function to be interpolated is harmonic, then it seems natural to work in the space $\mathcal{H}_n$ of real bivariate harmonic polynomials of total degree at most n, which has dimension $2n + 1$. Analogous to (1), we prescribe chords $\mathcal{I} := \{I(\theta_i, t_i) : \theta_i \in [0, \pi), \ t_i \in (-1, 1)\}_{i=1}^{2n+1}$ of the unit circle and associated given values $\Gamma = \{\gamma_i\}_{i=1}^{2n+1}$, and wish to find a harmonic polynomial $p \in \mathcal{H}_n$ such that

$$\mathcal{R}_{\theta_i}(p, t_i) = \int_{I(\theta_i, t_i)} p(\mathbf{x}) \, dx = \gamma_i, \qquad i = 1, \ldots, 2n + 1. \tag{2}$$

Again we call $\mathcal{I}$ *regular* if the interpolation problem (2) has a unique solution for all given values Γ. The conditions in (2) can be equivalently rewritten as a system of linear equations for the coefficients of p. Thus, $\mathcal{I}$ is regular if and only if the matrix of this system is nonsingular. In the following, we present one family of such regular schemes.

We use the following basis of the space of harmonic polynomials $\mathcal{H}_n$,

$$h_0(x, y) = 1,$$
$$h_{2k-1}(x, y) = \mathrm{Re}(x + iy)^k, \quad h_{2k}(x, y) = \mathrm{Im}(x + iy)^k, \quad k = 1, \ldots, n,$$

with representation in polar coordinates

$$h_0(r, \theta) = 1,$$
$$h_{2k-1}(r, \theta) = r^k \cos(k\theta), \quad h_{2k}(r, \theta) = r^k \sin(k\theta), \quad k = 1, \ldots, n.$$

Every harmonic polynomial p of degree less than or equal to n can be expanded in this basis,

$$p = \sum_{k=0}^{2n} p_k h_k,$$

where p_k are real coefficients.

The following result, which gives a closed formula for Radon projections of the basis harmonic polynomials can be considered a harmonic analogue to the famous

Marr's formula [20]. A special case of this harmonic version was first derived using tools from symbolic computation [9]. Later, Georgieva and Hofreither [8] have given an analytic proof in a more general setting.

Lemma 1. *The Radon projections of the basis harmonic polynomials $h_k, k \in \mathbb{N}$, are given by*

$$\mathcal{R}_\theta(h_{2k-1}, t) = \int_{I(\theta, t)} h_{2k-1}(\mathbf{x})\, d\mathbf{x} = \frac{2}{k+1}\sqrt{1 - t^2} U_k(t) \cos(k\theta),$$

$$\mathcal{R}_\theta(h_{2k}, t) = \int_{I(\theta, t)} h_{2k}(\mathbf{x})\, d\mathbf{x} = \frac{2}{k+1}\sqrt{1 - t^2} U_k(t) \sin(k\theta),$$

where $\theta \in \mathbb{R}$, $t \in (-1, 1)$ and $U_k(t)$ is the k-th degree Chebyshev polynomial of second kind.

The above lemma plays a crucial role in proving regularity of a particular family of schemes $\mathcal{I}$ of chords.

Theorem 1 (Existence and uniqueness [9,8]). *The interpolation problem (2) has a unique solution for any set of chords $\mathcal{I} = \{I(\theta_i, t_i)\}_{i=1}^{2n+1}$ with*

$$0 \le \theta_1 < \theta_2 < \ldots < \theta_{2n+1} < 2\pi$$

and with constant distances $t_i = t \in (-1, 1)$ such that t is not a zero of any Chebyshev polynomial of the second kind $U_1, \ldots, U_n$.

See Figure 2 for some examples of (regular) schemes which satisfy the conditions of the above theorem, and one which does not and is in fact not regular.

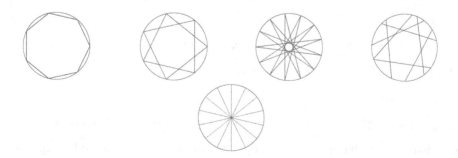

Fig. 2. *Top:* Some admissible schemes according to Theorem 1. *Bottom:* A scheme which does not satisfy the assumptions of Theorem 1 since $t = 0$ is a root of every Chebyshev polynomial of odd degree. This scheme is not regular.

3   Least-squares Fitting

Here we deal with the problem of fitting some given Radon projections of a harmonic function by a harmonic polynomial in the overdetermined case where the amount of data is greater than the dimension of the polynomial space. A least-squares method is used to determine a harmonic polynomial which fits the given data. The problem of least-squares fitting of Radon projections was first considered for the case of algebraic polynomials by Marr [20].

3.1   Radon Projections Type of Data

Let a set $\mathcal{I} := \left\{ I(\theta_i, t_i) : \theta_i \in [0, \pi), \ t_i \in (-1, 1) \right\}_{i=1}^{N}$ of N distinct chords of the unit circle ∂D, be given, and let $\Gamma := \left\{ \gamma_i \right\}_{i=1}^{N}$ be the Radon projections of a harmonic function u along these chords, i.e.,

$$\mathcal{R}_{\theta_i}(u, t_i) = \gamma_i, \qquad i = 1, \dots, N.$$

We regard the set of chords $\mathcal{I}$ and the set of values Γ generally as *data*. Finally, by $\Lambda := \left\{ \lambda_i \right\}_{i=1}^{N}$ we denote a set of positive real numbers which we consider to be *weights* related to the corresponding Radon projections.

The least squares fitting problem is formulated as follows.

Given data $\mathcal{I}$ and Γ, and weights Λ, find a polynomial $p \in \mathcal{H}_n$, $N > 2n + 1$, such that

$$\sum_{i=1}^{N} \lambda_i \left(\mathcal{R}_{\theta_i}(p, t_i) - \gamma_i \right)^2 \to \min. \tag{3}$$

Theorem 2. *Assume that data $\mathcal{I}$ and Γ, and weights Λ are given. Suppose that there exists a subset $J \subset \{1, 2, \dots, N\}$, $|J| = 2n + 1$, such that the interpolatory scheme of chords $\left\{ I(\theta_\ell, t_\ell) \right\}_{\ell \in J}$ is regular. Then there exists a unique harmonic polynomial $p \in \mathcal{H}_n$ for which the minimum in (3) is attained.*

Proof. Suppose p is a harmonic polynomial of degree at most n. Then p can be represented in the form

$$p = \sum_{k=0}^{2n} p_k h_k.$$

Since the Radon projection for a fixed line segment is a linear functional it follows that

$$\mathcal{R}_{\theta_i}(p; t_i) = \sum_{k=0}^{2n} p_k \mathcal{R}_{\theta_i}(h_k; t_i), \quad i = 1, \dots, N.$$

Hence, the problem (3) is equivalent to the problem

$$\Phi := \sum_{i=1}^{N} \lambda_i \left(\sum_{k=0}^{2n} p_k \mathcal{R}_{\theta_i}(h_k; t_i) - \gamma_i \right)^2 \to \min,$$

where Φ is a function of the coefficients $\{p_k\}_{k=0}^{2n}$.

Applying the necessary conditions for extrema

$$\frac{\partial \Phi}{\partial p_j} = 0, \qquad j = 0, \ldots, 2n,$$

we obtain the system of linear equations, for $j = 0, 1, \ldots, 2n$,

$$\sum_{k=0}^{2n} \left(\sum_{i=1}^{N} \lambda_i \mathcal{R}_{\theta_i}(h_k, t_i) \mathcal{R}_{\theta_i}(h_j, t_i) \right) p_k = \sum_{i=1}^{N} \lambda_i \gamma_i \mathcal{R}_{\theta_i}(h_j, t_i), \tag{4}$$

with respect to the coefficients $\{p_k\}_{k=0}^{2n}$.

In order to prove that (4) has a unique solution for arbitrary set Γ of Radon projections we consider the corresponding homogeneous system

$$\sum_{k=0}^{2n} \left(\sum_{i=1}^{N} \lambda_i \mathcal{R}_{\theta_i}(h_k, t_i) \mathcal{R}_{\theta_i}(h_j, t_i) \right) q_k = 0, \qquad j = 0, 1, \ldots, 2n. \tag{5}$$

Using the linearity of the functionals $\mathcal{R}_{\theta_i}(\cdot, t_i)$, we get

$$\sum_{i=1}^{N} \lambda_i \mathcal{R}_{\theta_i} \left(\sum_{k=0}^{2n} q_k h_k, t_i \right) \mathcal{R}_{\theta_i}(h_j, t_i) = 0, \qquad j = 0, 1, \ldots, 2n. \tag{6}$$

Denote

$$q := \sum_{k=0}^{2n} q_k h_k.$$

Let us note that q is a polynomial from $\mathcal{H}_n$. Then (5) may be rewritten as

$$\sum_{i=1}^{N} \lambda_i \mathcal{R}_{\theta_i}(q, t_i) \mathcal{R}_{\theta_i}(h_j, t_i) = 0, \qquad j = 0, 1, \ldots, 2n.$$

We now sum all the equations of (6) multiplied by the corresponding q_j and obtain

$$\sum_{i=1}^{N} \lambda_i \big(\mathcal{R}_{\theta_i}(q, t_i) \big)^2 = 0.$$

Hence, by the positivity of the weights λ_i, we have

$$\mathcal{R}_{\theta_i}(q, t_i) = 0, \qquad i = 1, \ldots, N.$$

Since there exists a subset $J \subset \{1, 2, \ldots, N\}$, $|J| = 2n + 1$, such that the interpolatory scheme of chords $\{I(\theta_\ell, t_\ell)\}_{\ell \in J}$ is regular, we conclude that $q \equiv 0$. Then

$$q_k = 0, \qquad k = 0, 1, \ldots, 2n,$$

i.e., the homogeneous system (5) has only the zero solution.

Therefore the linear system (4) has a unique solution, and the theorem is proved. $\qquad \square$

Remark 1. From the proof of Theorem 2, it can be seen that the coefficients $\{p_k\}_{k=0}^{2n}$ of the least-squares fitting polynomial $p = \sum_{k=0}^{2n} p_k h_k$ can be found as the solution of the following system of linear equations,

$$\sum_{k=0}^{2n} a_{jk} p_k = \sum_{i=1}^{N} b_{ji} \gamma_i, \qquad j = 0, \ldots, 2n.$$

In short, the vector $\underline{p}$ of coefficients is determined by

$$A\underline{p} = B\Gamma$$

with the symmetric and positive definite matrix $A = (a_{jk})_{j,k=0}^{2n}$ and the rectangular matrix $B = (b_{ji})_{j=0,i=1}^{2n,\,N}$ having entries

$$a_{jk} = \sum_{i=1}^{N} \lambda_i \mathcal{R}_{\theta_i}(h_k, t_i) \mathcal{R}_{\theta_i}(h_j, t_i), \qquad b_{ji} = \lambda_i \mathcal{R}_{\theta_i}(h_j, t_i).$$

These matrix entries can be computed using the formulas in Lemma 1 for the Radon projections of the harmonic basis functions.

3.2   Mixed Type of Data

Now, we shall consider a fitting problem for *mixed type of data* – both Radon projections and function values. Namely, let the data $\mathcal{I}$ and Γ, and the weights Λ be given as above in Section 3.1. Additionally we take values $U := \{u_j\}_{j=1}^{M}$ of the harmonic function u at arbitrary points $X := \{\mathbf{x}_j\}_{j=1}^{M}$ in the closed unit disk $\overline{D}$, i.e.,

$$u(\mathbf{x}_j) = u_j, \qquad j = 1, \ldots, M.$$

In particular, the points X can be chosen only on the unit circle ∂D. Let $\Omega := \{\omega_j\}_{j=1}^{M}$ be given weights corresponding to the function values.

The least squares fitting problem for mixed type of data is formulated as follows: given

- the data $\mathcal{I}$ and Γ, and corresponding weights Λ;
- function values U at points X and weights Ω;

find a harmonic polynomial $p \in \mathcal{H}_n$, $N > 2n + 1$, such that

$$\sum_{i=1}^{N} \lambda_i \big(\mathcal{R}_{\theta_i}(p, t_i) - \gamma_i\big)^2 + \sum_{j=1}^{M} \omega_j \big(p(\mathbf{x}_j) - u_j\big)^2 \to \min. \qquad (7)$$

Theorem 3. *Assume that mixed type of data $\mathcal{I}$, Γ, X, U, and weights Λ, Ω are given. Suppose that there exists a subset $J \subset \{1, 2, \ldots, N\}$, $|J| = 2n + 1$, such that the interpolatory scheme of chords $\{I(\theta_\ell, t_\ell)\}_{\ell \in J}$ is regular. Then there exists a unique harmonic polynomial $p \in \mathcal{H}_n$ for which the minimum in (7) is attained.*

Therefore including a regular interpolatory scheme from Section 2.2 into the set of chords $\mathcal{I}$ assures the uniqueness of the solution to the problem for mixed data. The proof of the theorem is similar to the proof of Theorem 2 and the coefficients of the least-squares minimizing polynomial can be computed by solving a linear system similar as in Remark 1.

4   Numerical Examples

4.1   Example 1

We approximate the harmonic function $u(x, y) = \exp(x) \cos(y)$ by a harmonic polynomial $p \in \mathcal{H}_n$ given $N = 2(2n + 1)$ values of its Radon projections: $2n + 1$ taken along the edges of a regular $(2n + 1)$-sided convex polygon (Figure 2, first picture), and $2n + 1$ along random chords. The weights are all set to 1. In Figure 3, we display the scheme of chords, the function u as well as the error $u - p$, where p is the least-squares fitting polynomial of degree $n = 7$ fitting information on 30 chords.

4.2   Example 2

We consider a similar problem as in Example 1, but in this case the weights are set to 1 for the chords forming a regular $(2n + 1)$-sided convex polygon, and to 100 for the remaining $N - (2n + 1)$ random chords.

In Figure 4, we plot the scheme of chords, and the error function $u - p$, where the degree of the least-squares fitting polynomial p is $n = 7$ and the number of chords is $N = 30$. No qualitative change in behavior from Example 1 is observed, and the error is of the same order of magnitude. This can be explained by the fact that in both examples, the polynomial p is already reasonably close to the best-approximating polynomial in the space $\mathcal{H}_7$. Indeed, adding additional data in the form of more chords does not significantly decrease the error in this example.

4.3   Example 3

We again approximate the harmonic function $u(x, y) = \exp(x) \cos(y)$, but consider the case of noisy data. We set up a sequence of chords $I_m = I(\theta_m, t_m)$ where the first 15 chords are chosen to form a regular convex polygon inscribed in the unit circle, and all following chords are chosen with $\theta_m \in [0, 2\pi)$ and $t_m \in (0, 1)$ sampled from a uniform random distribution over their respective intervals. Over each chord, we compute the exact Radon projection $\gamma_m = \int_{I_m} u \, d\mathbf{x}$ and then simulate noisy data $\tilde{\gamma}_m$ by adding to γ_m a random number drawn from a Gaussian distribution with mean 0 and standard deviation $\epsilon = 10^{-2}$.

Using this input data, we compare three reconstruction algorithms.

- First, the least squares method described in the present paper with degree $n = 7$ and a variable number of additional chords drawn from $\{I_m\}$. We also compare the case where the first $2n + 1$ chords are weighted with 10 and the remaining with 1 to the case where all chords are weighted uniformly.

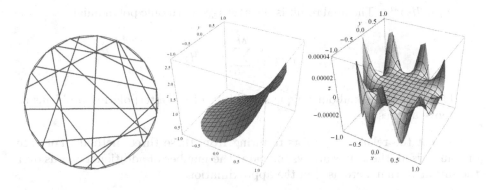

Fig. 3. Example 1, uniform weights, $n = 7, N = 30$: the scheme of chords, function u, error $u - p$

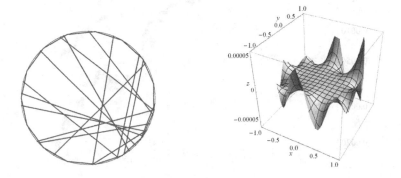

Fig. 4. Example 2, variable weights, $n = 7, N = 30$: the scheme of chords, error $u - p$

- Second, we use the basic harmonic interpolation approach described in Section 2.2. For this, we choose a number $M \in \mathbb{N}$, take the first $15M$ chords from $\{I_m\}$, and set up the interpolation problem (2) with interpolation degree $n = 7M$. Note that there is no theoretic justification for the interpolation problem to be uniquely solvable, but this is "almost always" the case, and we encounter no singular matrices in this example.
- Third, we set up an approach inspired by Monte Carlo methods. For this, we partition the input data into M sets of size 15 each, i.e., we consider the schemes of chords

$$(I_1, \ldots, I_{15}), \ (I_{16}, \ \ldots, I_{30}), \ \ldots, \ (I_{15M-14}, \ldots, I_{15M})$$

for some natural number M. On each of these sets, we solve an interpolation problem of degree $n = 7$ as in Section 2.2, yielding M interpolating polynomials

$(p_j \in \mathcal{H}_7)_{j=1}^{M}$. The final result is the averaged harmonic polynomial

$$p = \frac{1}{M} \sum_{j=1}^{M} p_j \in \mathcal{H}_7.$$

The same remark about unique solvability of the interpolation problems as above applies.

We plot the relative L_2-errors resulting from these three methods. For ease of comparison, the x axis always indicates the number of additional chords over the initial 15 that were used in the approximation.

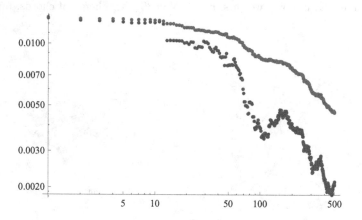

Fig. 5. Example 3: errors for the least-squares method with noisy data, degree $n = 7$. x-axis: number of additional chords. y-axis: relative L_2-error. The upper (purple) dots correspond to the case where the initial 15 chords are weighted 10 times stronger, while the lower (blue) dots correspond to the uniformly weighted case.

Figure 5 shows the errors for the least-squares approach for both uniform and variable weights. We observe that both choices of weights lead to a reduction of the error as more data is taken into account, however the uniform weights seem to produce lower errors overall.

Figure 6 shows the errors for the pure interpolation approach with increased degree of the interpolating harmonic polynomial. This method does not converge. This is due to the noise in the data being amplified by the large condition number of the interpolation matrix associated with randomly distributed chords. We point out that, in contrast, a uniformly bounded condition number was obtained in [8] for the case of equally spaced chords with constant distances to the origin.

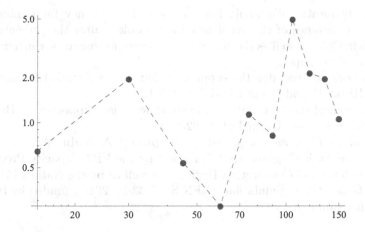

Fig. 6. Example 3: errors for the high-degree ($n = 7M$) interpolation with noisy data. x-axis: number of additional chords ($15(M - 1)$). y-axis: relative L_2-error

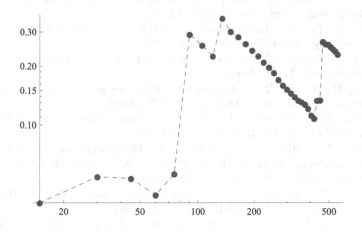

Fig. 7. Example 3: errors for the Monte Carlo-type method with noisy data, degree $n = 7$. x-axis: number of additional chords. y-axis: relative L_2-error

Figure 7 shows the errors for the Monte Carlo-type approach. Again no satisfactory convergence is observed. The reason for this is similar as for the pure interpolation method in that some of the small interpolation problems are ill-conditioned. Indeed, the jumps in the error plot can be traced back to problems with particularly large condition number of the corresponding interpolation matrix.

In summary, of the three methods, only the least-squares approach seems to yield satisfactory results for highly noisy data over large sets of randomly chosen chords.

Acknowledgments. We would like to thank Peter Binev for reminding us about the importance of the overdetermined problem after the presentation of our results in Oslo, as well as the anonymous referees for remarks which improved the quality of the paper.

The authors acknowledge the support by Bulgarian National Science Fund, Grant DMU 03/17 and Grant DFNI-T01/0001.

The research of the first and third authors was also supported by Bulgarian National Science Fund, Grant DDVU 0230/11.

The second author was supported by the project AComIn "Advanced Computing for Innovation", grant 316087, funded by the FP7 Capacity Programme Research Potential of Convergence Regions, as well as by the National Research Network "Geometry + Simulation" (NFN S117, 2012–2016), funded by the Austrian Science Fund (FWF).

References

1. Bojanov, B., Georgieva, I.: Interpolation by bivariate polynomials based on Radon projections. Studia Math. 162, 141–160 (2004)
2. Bojanov, B., Petrova, G.: Numerical integration over a disc. A new Gaussian cubature formula. Numer. Math. 80, 39–59 (1998)
3. Bojanov, B., Petrova, G.: Uniqueness of the Gaussian cubature for a ball. J. Approx. Theory 104, 21–44 (2000)
4. Bojanov, B., Xu, Y.: Reconstruction of a bivariate polynomial from its Radon projections. SIAM J. Math. Anal. 37, 238–250 (2005)
5. Cavaretta, A.S., Micchelli, C.A., Sharma, A.: Multivariate interpolation and the Radon transform, Part I. Math. Z. 174, 263–279 (1980)
6. Cavaretta, A.S., Micchelli, C.A., Sharma, A.: Multivariate interpolation and the Radon transform, Part II. In: DeVore, R., Scherer, K. (eds.) Quantitive Approximation, pp. 49–62. Acad. Press, New York (1980)
7. Davison, M.E., Grunbaum, F.A.: Tomographic reconstruction with arbitrary directions. Comm. Pure Appl. Math. 34, 77–120 (1981)
8. Georgieva, I., Hofreither, C.: Interpolation of harmonic functions based on Radon projections. Technical Report 2012-11, DK Computational Mathematics Linz Report Series (2012), https://www.dk-compmath.jku.at/publications/dk-reports/2012-10-17/view (submitted)
9. Georgieva, I., Hofreither, C., Koutschan, C., Pillwein, V., Thanatipanonda, T.: Harmonic interpolation based on Radon projections along the sides of regular polygons. Central European Journal of Mathematics (2012) (to appear); also available as Technical Report 2011-12 in the series of the DK Computational Mathematics Linz, https://www.dk-compmath.jku.at/publications/dk-reports/2011-10-20/view
10. Georgieva, I., Hofreither, C., Uluchev, R.: Interpolation of mixed type data by bivariate polynomials. In: Nikolov, G., Uluchev, R. (eds.) Constructive Theory of Functions, Sozopol (2010); In memory of Borislav Bojanov, pp. 93–107. Prof. Marin Drinov Academic Publishing House, Sofia (2012)
11. Georgieva, I., Ismail, S.: On recovering of a bivariate polynomial from its Radon projections. In: Bojanov, B. (ed.) Constructive Theory of Functions, Varna, pp. 127–134 (2005); Prof. Marin Drinov Academic Publishing House, Sofia (2006)

12. Georgieva, I., Uluchev, R.: Smoothing of Radon projections type of data by bivariate polynomials. J. Comput. Appl. Math. 215, 167–181 (2008)
13. Georgieva, I., Uluchev, R.: Surface reconstruction and lagrange basis polynomials. In: Lirkov, I., Margenov, S., Waśniewski, J. (eds.) LSSC 2007. LNCS, vol. 4818, pp. 670–678. Springer, Heidelberg (2008)
14. Hakopian, H.: Multivariate divided differences and multivariate interpolation of Lagrange and Hermite type. J. Approx. Theory 34, 286–305 (1982)
15. Hamaker, C., Solmon, D.C.: The angles between the null spaces of x-rays. J. Math. Anal. Appl. 62, 1–23 (1978)
16. Helgason, S.: The Radon Transform. In: Progress in Mathematics 5. Birkhauser, Boston (1980)
17. Jain, A.K.: Fundamentals of Digital Image Processing. Prentice Hall (1989)
18. John, F.: Abhängigkeiten zwischen den Flächenintegralen einer stetigen Funktion. Math. Anal. 111, 541–559 (1935)
19. Logan, B., Shepp, L.: Optimal reconstruction of a function from its projections. Duke Math. J. 42, 645–659 (1975)
20. Marr, R.: On the reconstruction of a function on a circular domain from a sampling of its line integrals. J. Math. Anal. Appl. 45, 357–374 (1974)
21. Natterer, F.: The Mathematics of Computerized Tomography. Classics in Applied Mathematics, vol. 32. SIAM (2001)
22. Nikolov, G.: Cubature formulae for the disk using Radon projections. East J. Approx. 14, 401–410 (2008)
23. Radon, J.: Über die Bestimmung von Funktionen durch ihre Integralwerte längs gewisser Mannigfaltigkeiten. Ber. Verch. Sächs. Akad. 69, 262–277 (1917)
24. Solmon, D.C.: The x-ray transform. J. Math. Anal. Appl. 56, 61–83 (1976)

Generating Functions for Uniform B-Splines

Ron Goldman

Department of Computer Science,
Rice University
Houston Texas 77005

Abstract. We derive a closed formula for the generating functions of the uniform B-splines. We begin by constructing a PDE for these generating functions starting from the de Boor recurrence. By solving this PDE, we find that we can express these generating functions explicitly as sums of polynomials times exponentials. Using these generating functions, we derive some known identities, including the Schoenberg identity, the two term formula for the derivatives in terms of B-splines of lower degree, and the partition of unity property. We also derive several new identities for uniform B-splines not previously available from classical methods, including formulas for sums and alternating sums, for moments and reciprocal moments, and for Laplace transforms and convolutions with monomials.

1 Introduction

Generating functions are a powerful tool for investigating the properties of discrete sequences. Explicit formulas and identities for elements of the sequence can often be readily derived once we have an explicit formula for their generating function [1].

The goal of this paper is to compute an explicit formula for the generating function of the uniform B-splines over arbitrary intervals. We shall then use these generating functions to derive several well known identities—including the Schoenberg identity, the two term formula for the derivatives in terms of B-splines of lower degree, and the partition of unity property—for the uniform B-splines. We will also derive several new identities for uniform B-splines not previously available from classical methods such as blossoming or the de Boor recurrence, including formulas for sums and alternating sums, for moments and reciprocal moments, and for Laplace transforms and convolutions with monomials.

This work is inspired by the papers of Y. Simsek [4–6], who computed explicit formulas for a novel collection of generating functions for the classical Bernstein bases

$$B_k^n(x) = \binom{n}{k} x^k (1-x)^{n-k} \qquad 0 \le k \le n, \ \ 0 \le n < \infty,$$

by summing over the degree n instead of over the index k. He then used these generating functions to derive many known and some new identities for the

M. Floater et al. (Eds.): MMCS 2012, LNCS 8177, pp. 172–188, 2014.
© Springer-Verlag Berlin Heidelberg 2014

Bernstein basis functions. We shall take a similar approach with the uniform B-splines $N_{k,n}(x)$, again summing over the degree n instead of the index k.

We proceed in the following fashion. In Section 2 we consider the simple special case of B-splines of degree n with knots at the integers $0, \ldots, n+1$ but restricted to the interval $[0, 1]$. Over this interval we find that the generating function has an especially simple form as an exponential, so we are encouraged to study the generating function over arbitrary intervals. In Section 3 we apply the de Boor recurrence to derive a PDE for the generating function, and in Section 4 we solve this PDE to find an explicit formula for the generating function over arbitrary intervals. This solution reveals a novel connection between uniform B-splines and exponentials. In Section 5 we show how to use this generating function to derive some classical identities for the uniform B-splines, including Schoenberg's identity, the formula for the derivatives of the B-splines in terms of B-splines of lower degree, and the fact that the B-splines form a the partition of unity. In Section 6 we apply the generating function to derive several new identities for the uniform B-splines not previously accessible from classical methods such as blossoming or the de Boor recurrence. These new identities for uniform B-splines include formulas for sums and alternating sums, for moments and reciprocal moments, and for Laplace transforms and convolutions with monomials. We close in Section 7 with a brief summary of our work along a short discussion of the limitations of our approach to deriving identities for the B-splines using generating functions. We also list a few natural problems involving generating functions and B-splines for future research.

2   A Simple Example: The Generating Function over the Interval $[0, 1]$

We shall begin by investigating the uniform B-splines with knots at the integers when restricted to the interval $[0, 1]$.

To fix our notation, let

$N_{k,n}(x) =$ the uniform B-spline of degree n with support $[k, k+n+1]$ and knots
at the integers $\{k, k+1, \ldots, k+n+1\}$.

We also introduce the generating functions

$$G_k(x, t) = \sum_{n=0}^{\infty} N_{k,n}(x) t^n.$$

Recall that for uniform B-splines, the functions $N_{k,n}(x)$ are just shifts of the functions $N_{0,n}(x)$—that is,

$$N_{k,n}(x) = N_{0,n}(x - k),$$

so

$$G_k(x, t) = G_0(x - k, t).$$

Thus to investigate the B-splines $N_{k,n}(x)$ and their generating functions $G_k(x,t)$, it is enough to study the B-splines $N_{0,n}(x)$ and their generating functions $G_0(x,t)$.

To investigate the B-splines $N_{0,n}(x)$, consider de Boor recurrence:

$$N_{0,n}(x) = \frac{x}{n} N_{0,n-1}(x) + \frac{n+1-x}{n} N_{1,n-1}(x).$$

For $x \leq 1$, we have $N_{1,n}(x) = 0$. Therefore

$$N_{0,n}(x) = \frac{x}{n} N_{0,n-1}(x) \qquad 0 \leq x \leq 1.$$

Hence in the interval $[0,1]$:

$$N_{0,0}(x) = 1, \ N_{0,1}(x) = x, \ N_{0,2}(x) = \frac{x^2}{2!}, \ \ldots, \ N_{0,n}(x) = \frac{x^n}{n!}.$$

Thus over the interval $[0,1]$, we have a remarkably simple explicit formula for the generating function $G_0(x,t)$ of the B-splines $N_{0,n}(x)$:

$$G_0(x,t) = \sum_{n=0}^{\infty} \frac{x^n t^n}{n!} = e^{xt} \qquad 0 \leq x \leq 1.$$

Our goal is to find explicit formulas for the generating function $G_0(x,t)$ over arbitrary intervals.

3   A PDE for the Generating Functions Built from the de Boor Recurrence

For a discrete sequence generated by a recurrence one classical way to derive an explicit formula for the generating function is first to use the recurrence to construct a functional equation for the generating function. One can then often solve this functional equation to find an explicit formula for the generating function. This technique works, for example, to derive an explicit formula for the generating function of the fibonacci numbers [1]. Here we shall apply this method to derive a PDE for the generating functions of the uniform B-splines. In the next section we will solve this PDE to find an explicit formula for the generating functions of the uniform B-splines over arbitrary intervals.

Theorem 1.

$$\frac{\partial G_0}{\partial t}(x,t) = xG_0(x,t) + (2-x)G_1(x,t) + t\frac{\partial G_1}{\partial t}(x,t). \tag{1}$$

Proof. We begin with the de Boor recurrence:

$$N_{0,n}(x) = \frac{x}{n} N_{0,n-1}(x) + \frac{n+1-x}{n} N_{1,n-1}(x).$$

Multiplying both sides by nt^{n-1} yields:

$$nN_{0,n}(x)t^{n-1} = xN_{0,n-1}(x)t^{n-1} + (n+1-x)N_{1,n-1}(x)t^{n-1}.$$

Now summing over n, we find that:

$$\sum_n nN_{0,n}(x)t^{n-1} = x\sum_n N_{0,n-1}(x)t^{n-1} + (2-x)\sum_n N_{1,n-1}(x)t^{n-1}$$
$$+ t\sum_n (n-1)N_{1,n-1}(x)t^{n-2}.$$

Therefore

$$\frac{\partial G_0}{\partial t}(x,t) = xG_0(x,t) + (2-x)G_1(x,t) + t\frac{\partial G_1}{\partial t}(x,t).$$

$\square$

4   Solving the PDE for the Generating Functions

We shall now derive an explicit formula for the generating function $G_0(x,t)$ by solving the PDE in Theorem 1. We begin with some special cases.

Over the interval $[0,1]$, we have $N_{1,n}(x) = 0$. so $G_1(x,t) = 0$. Hence the PDE in Equation (1) reduces to

$$\frac{\partial G_0}{\partial t}(x,t) = xG_0(x,t) \quad 0 \le x \le 1. \tag{2}$$

Therefore, as we observed in Section 2,

$$G_0(x,t) = e^{xt} \quad 0 \le x \le 1. \tag{3}$$

Over the interval $[1,2]$, we have

$$G_1(x,t) = G_0(x-1,t) = e^{(x-1)t} \quad 1 \le x \le 2. \tag{4}$$

Therefore the PDE in Equation (1) reduces to

$$\frac{\partial G_0}{\partial t}(x,t) = xG_0(x,t) + (2-x)e^{(x-1)t} + t(x-1)e^{(x-1)t} \quad 1 \le x \le 2. \tag{5}$$

One can now guess the solution must have terms with the exponentials e^{xt} and $e^{(x-1)t}$. By trial and error one soon finds that:

$$G_0(x,t) = e^{xt} - ((x-1)t+1)e^{(x-1)t} \quad 1 \le x \le 2, \tag{6}$$

which is easily verified by substituting Equation (6) into Equation (5) and seeing that the PDE is indeed satisfied. Proceeding in this manner, we find that we have the following general result.

Theorem 2. *For $x \in [p, p+1]$, the function*

$$G_0(x,t) = \sum_{j=0}^{p} (-1)^j \left(\frac{(x-j)^j t^j}{j!} + \frac{(x-j)^{j-1} t^{j-1}}{(j-1)!} \right) e^{(x-j)t} \qquad (7)$$

satisfies the PDE in Equation (1).

Proof. We proceed by induction on p. The cases $p = 0, 1$ have already been discussed. Suppose then that this result is true for $p-1$; then we must verify that this result is also valid for p. To simplify our notation, let

$$G_{0,p-1}(x,t) = \sum_{j=0}^{p-1} (-1)^j \left(\frac{(x-j)^j t^j}{j!} + \frac{(x-j)^{j-1} t^{j-1}}{(j-1)!} \right) e^{(x-j)t},$$

$$g_{0,p}(x,t) = \left(\frac{(x-p)^p t^p}{p!} + \frac{(x-p)^{p-1} t^{p-1}}{(p-1)!} \right) e^{(x-p)t}.$$

Then for $x \in [p, p+1]$

$$G_0(x,t) = G_{0,p-1}(x,t) + (-1)^p g_{0,p}(x,t),$$
$$G_1(x,t) = G_{0,p-1}(x-1,t) + (-1)^{p-1} g_{0,p-1}(x-1,t).$$

Moreover, by the inductive hypothesis

$$\frac{\partial G_{0,p-1}}{\partial t}(x,t) = x G_{0,p-1}(x,t) + (2-x) G_{1,p-1}(x,t) + t \frac{\partial G_{1,p-1}}{\partial t}(x,t).$$

Therefore it is enough to verify that

$$\frac{\partial g_{0,p}}{\partial t}(x,t) = x g_{0,p}(x,t) - (2-x) g_{0,p-1}(x-1,t) - t \frac{\partial g_{0,p-1}}{\partial t}(x-1,t).$$

But by direct computation:

$$\frac{\partial g_{0,p}}{\partial t}(x,t) = (x-p) \left(\frac{(x-p)^p t^p}{p!} + \frac{(x-p)^{p-1} t^{p-1}}{(p-1)!} \right) e^{(x-p)t}$$
$$+ \left(\frac{(x-p)^p t^{p-1}}{(p-1)!} + \frac{(x-p)^{p-1} t^{p-2}}{(p-2)!} \right) e^{(x-p)t}.$$

Thus

$$\frac{\partial g_{0,p}}{\partial t}(x,t) = x g_{0,p}(x,p) - \left(\frac{(x-p)^p t^p}{(p-1)!} + \frac{p(x-p)^{p-1} t^{p-1}}{(p-1)!} \right) e^{(x-p)t}$$
$$+ (x-p) \left(\frac{(x-p)^{p-1} t^{p-1}}{(p-1)!} + \frac{(x-p)^{p-2} t^{p-2}}{(p-2)!} \right) e^{(x-p)t},$$

or equivalently

$$\frac{\partial g_{0,p}}{\partial t}(x,t) = x g_{0,p}(x,p) - \left(\frac{(x-p)^p t^p}{(p-1)!} + \frac{p(x-p)^{p-1} t^{p-1}}{(p-1)!} \right) e^{(x-p)t}$$
$$+ ((x-2) + (2-p)) \left(\frac{(x-p)^{p-1} t^{p-1}}{(p-1)!} + \frac{(x-p)^{p-2} t^{p-2}}{(p-2)!} \right) e^{(x-p)t}.$$

Hence

$$\frac{\partial g_{0,p}}{\partial t}(x,t) = x g_{0,p}(x,t) - (2-x)g_{0,p-1}(x-1,t)$$
$$- \left(\frac{(x-p)^p t^p}{(p-1)!} + \frac{p(x-p)^{p-1}t^{p-1}}{(p-1)!}\right)e^{(x-p)t}$$
$$+ (2-p)\left(\frac{(x-p)^{p-1}t^{p-1}}{(p-1)!} + \frac{(x-p)^{p-2}t^{p-2}}{(p-2)!}\right)e^{(x-p)t},$$

so

$$\frac{\partial g_{0,p}}{\partial t} = x g_{0,p}(x,t) - (2-x)g_{0,p-1}(x-1,t) - \frac{(x-p)^p t^p}{(p-1)!}e^{(x-p)t}$$
$$+ (2-2p)\left(\frac{(x-p)^{p-1}t^{p-1}}{(p-1)!}\right)e^{(x-p)t} - \frac{(x-p)^{p-2}t^{p-2}}{(p-3)!}e^{(x-p)t}.$$

Therefore it is enough to verify that

$$t\frac{\partial g_{0,p-1}(x-1,t)}{\partial t} = \left(\frac{(x-p)^p t^p}{(p-1)!} + 2\frac{(x-p)^{p-1}t^{p-1}}{(p-2)!} + \frac{(x-p)^{p-2}t^{p-2}}{(p-3)!}\right)e^{(x-p)t}.$$

But by definition

$$g_{0,p-1}(x-1,t) = \left(\frac{(x-p)^{p-1}t^{p-1}}{(p-1)!} + \frac{(x-p)^{p-2}t^{p-2}}{(p-2)!}\right)e^{(x-p)t}.$$

Hence

$$t\frac{\partial g_{0,p-1}(x-1,t)}{\partial t} = \left(\frac{(x-p)^{p-1}t^{p-1}}{(p-2)!} + \frac{(x-p)^{p-2}t^{p-2}}{(p-3)!}\right)e^{(x-p)t}$$
$$+ (x-p)t\left(\frac{(x-p)^{p-1}t^{p-1}}{(p-1)!} + \frac{(x-p)^{p-2}t^{p-2}}{(p-2)!}\right)e^{(x-p)t},$$

so indeed

$$t\frac{\partial g_{0,p-1}(x-1,t)}{\partial t} = \left(\frac{(x-p)^p t^p}{(p-1)!} + 2\frac{(x-p)^{p-1}t^{p-1}}{(p-2)!} + \frac{(x-p)^{p-2}t^{p-2}}{(p-3)!}\right)e^{(x-p)t}.$$

$\square$

5   Deriving Identities for the Uniform B-Splines from their Generating Functions

With explicit formulas for the generating functions now in hand, we are finally ready to derive some identities for the uniform B-splines.

5.1   Schoenberg's Identity

Theorem 3. *(Schoenbergs Identity [3])*

$$N_{0,n}(x) = \frac{1}{n!} \sum_{j=0}^{p} (-1)^j \binom{n+1}{j} (x-j)^n \qquad p \leq x \leq p+1. \tag{8}$$

Proof. Schoenberg's identity for the B-splines follows immediately from the explicit formula for the generating functions. We simply compare coefficients of t^n on both sides of the generating function:

$$G_0(x,t) = \sum_{j=0}^{p} (-1)^j \left(\frac{(x-j)^j t^j}{j!} + \frac{(x-j)^{j-1} t^{j-1}}{(j-1)!} \right) e^{(x-j)t} \qquad p \leq x \leq p+1.$$

Expanding the exponential function on the right hand side, we find that

$$G_0(x,t) = \sum_{j=0}^{p} (-1)^j \left(\frac{(x-j)^j t^j}{j!} + \frac{(x-j)^{j-1} t^{j-1}}{(j-1)!} \right) \left(\sum_{k=0}^{\infty} \frac{(x-j)^k t^k}{k!} \right) \qquad p \leq x \leq p+1.$$

Now equating the terms with t^n on both side of this equation yields

$$N_{0,n}(x) t^n = \sum_{j=0}^{p} \left((-1)^j \frac{(x-j)^j t^j}{j!} \frac{(x-j)^{n-j} t^{n-j}}{(n-j)!} + \right.$$
$$\left. (-1)^j \frac{(x-j)^{j-1} t^{j-1}}{(j-1)!} \frac{(x-j)^{n-j+1} t^{n-j+1}}{(n-j+1)!} \right),$$

so

$$N_{0,n}(x) = \frac{1}{n!} \sum_{j=0}^{p} (-1)^j \left(\binom{n}{j} + \binom{n}{j-1} \right) (x-j)^n = \frac{1}{n!} \sum_{j=0}^{p} (-1)^j \binom{n+1}{j} (x-j)^n.$$

$\square$

5.2   The Derivative Formula

To derive a formula for the derivative of the uniform B-splines, we begin by deriving a functional equation for the derivative of their generating function.

Lemma 1.

$$\frac{\partial G_0}{\partial x}(x,t) = t G_0(x,t) - t G_1(x,t). \tag{9}$$

Proof. By Theorem 2:

$$G_0(x,t) = \sum_{j=0}^{p} (-1)^j \left(\frac{(x-j)^j t^j}{j!} + \frac{(x-j)^{j-1} t^{j-1}}{(j-1)!} \right) e^{(x-j)t}.$$

Therefore

$$\frac{\partial G_0}{\partial x}(x,t) = \sum_{j=0}^{p}(-1)^j\left(\frac{(x-j)^{j-1}t^j}{(j-1)!} + \frac{(x-j)^{j-2}t^{j-1}}{(j-2)!}\right)e^{(x-j)t}$$

$$+ t\sum_{j=0}^{p}(-1)^j\left(\frac{(x-j)^j t^j}{j!} + \frac{(x-j)^{j-1}t^{j-1}}{(j-1)!}\right)e^{(x-j)t},$$

or equivalently

$$\frac{\partial G_0}{\partial x}(x,t) = -t\sum_{j=0}^{p}(-1)^{j-1}\left(\frac{((x-1)-(j-1))^{j-1}t^{j-1}}{(j-1)!}\right.$$

$$+ \left.\frac{((x-1)-(j-1))^{j-2}t^{j-2}}{(j-2)!}\right)e^{((x-1)-(j-1))t}$$

$$+ t\sum_{j=0}^{p}(-1)^j\frac{(x-j)^j t^j}{j!} + \frac{(x-j)^{j-1}t^{j-1}}{(j-1)!}e^{(x-j)t}.$$

Hence

$$\frac{\partial G_0}{\partial x}(x,t) = tG_0(x,t) - tG_0(x-1,t) = tG_0(x,t) - tG_1(x,t).$$

Theorem 4. *(Derivative Formula)*

$$\frac{\partial N_{0,n}(x)}{\partial x} = N_{0,n-1}(x) - N_{1,n-1}(x). \tag{10}$$

Proof. From Lemma 1, we have the functional equation:

$$\frac{\partial G_0}{\partial x}(x,t) = tG_0(x,t) - tG_1(x,t).$$

Comparing the coefficients of t^n on both sides, we find that:

$$\frac{\partial N_{0,n}(x)}{\partial x}t^n = tN_{0,n-1}(x)t^{n-1} - tN_{1,n-1}(x)t^{n-1},$$

$$\frac{\partial N_{0,n}(x)}{\partial x} = N_{0,n-1}(x) - N_{1,n-1}(x).$$

$\square$

5.3   The de Boor Recurrence

Starting from the de Boor recurrence, we derived a PDE for the partial derivative of the generating function with respect to t. We can also go the other way: starting from this functional equation for the partial derivative of the generating function with respect to t, we can derive the de Boor recurrence. Thus this PDE is actually equivalent to the de Boor recurrence.

Theorem 5. *(de Boor Recurrence)*

$$N_{0,n}(x) = \frac{x}{n}N_{0,n-1}(x) + \frac{n+1-x}{n}N_{1,n-1}(x).$$

Proof. By Theorem 2, the generating function $G_0(x,t)$ satisfies the functional equation:

$$\frac{\partial G_0}{\partial t}(x,t) = xG_0(x,t) + (2-x)G_1(x,t) + t\frac{\partial G_1}{\partial t}(x,t).$$

Therefore

$$\sum_n nN_{0,n}(x)t^{n-1} = x\sum_n N_{0,n-1}(x)t^{n-1} + (2-x)\sum_n N_{1,n-1}(x)t^{n-1}$$
$$+ t\sum_n (n-1)N_{1,n-1}(x)t^{n-2}.$$

Comparing the coefficients of t^{n-1} on both sides yields:

$$nN_{0,n}(x) = xN_{0,n-1}(x) + (n+1-x)N_{1,n-1}(x).$$

Now dividing both sides by n, we conclude that:

$$N_{0,n}(x) = \frac{x}{n}N_{0,n-1}(x) + \frac{n+1-x}{n}N_{1,n-1}(x).$$

$\square$

5.4   Partition of Unity

Here we shall use the generating functions to show that the uniform B-splines form a partition of unity. We begin with some technical results.

Lemma 2.

$$\sum_{k=-d}^{0} G_k(x,t) = \sum_{k=-d}^{0} (-1)^{k+d}\left(\frac{(x-k)^{k+d}t^{k+d}}{(k+d)!}\right)e^{(x-k)t} \qquad 0 \le x \le 1. \qquad (11)$$

Proof. We proceed by induction on d. For $d=0$, this formula reduces to

$$G_0(x,t) = e^{xt},$$

which is just Equation (3). Now by the inductive hypothesis:

$$\sum_{k=-d}^{0} G_k(x,t) = \sum_{k=-d}^{0} (-1)^{k+d}\left(\frac{(x-k)^{k+d}t^{k+d}}{(k+d)!}\right)e^{(x-k)t}. \qquad (12)$$

Moreover,

$$G_{-(d+1)}(x,t) = \sum_n N_{-(d+1),n}(x)t^n$$

$$= \sum_{j=0}^{d+1} (-1)^j \left(\frac{(x+d+1-j)^j t^j}{j!} \right.$$

$$\left. + \frac{(x+d+1-j)^{j-1}t^{j-1}}{(j-1)!} \right) e^{(x+d+1-j)t}.$$

Reindexing by setting $i = j - 1$, we get

$$G_{-(d+1)}(x,t) = e^{(x+d+1)t} + \sum_{i=0}^{d} (-1)^{i+1} \left(\frac{(x+d-i)^{i+1}t^{i+1}}{(i+1)!} \right.$$

$$\left. + \frac{(x+d-i)^i t^i}{i!} \right) e^{(x+d-i)t}.$$

Now setting $k = i - d$, we arrive at

$$G_{-(d+1)}(x,t) = e^{(x+d+1)t} + \sum_{k=-d}^{0} (-1)^{k+d+1} \left(\frac{(x-k)^{k+d+1}t^{k+d+1}}{(k+d+1)!} \right.$$

$$\left. + \frac{(x-k)^{k+d}t^{k+d}}{(k+d)!} \right) e^{(x-k)t}.$$

Adding this last equation to (12) yields our result.   □

Lemma 3.

$$\sum_{-n \le k \le 0} N_{k,n}(x) = \sum_{k=0}^{n} (-1)^{n-k} \left(\frac{(x+k)^n}{k!(n-k)!} \right) \qquad 0 \le x \le 1. \qquad (13)$$

Proof. This result follows directly from Lemma 2 by setting $d = n$ and comparing the coefficients of t^n of both sides of Equation (11).   □

Lemma 4.

$$\sum_{k=0}^{n} (-1)^{n-k} \left(\frac{(x+k)^n}{k!(n-k)!} \right) = 1.$$

Proof. To establish this result, we shall use a divided difference argument. The following divided difference formula follows easily by induction on n:

$$f[0, 1, \ldots, n] = \frac{1}{n!} \sum_{k=0}^{n} (-1)^{n-k} \binom{n}{k} f(k).$$

Therefore

$$\sum_{k=0}^{n}(-1)^{n-k}\left(\frac{(x+k)^n}{k!(n-k)!}\right) = \frac{1}{n!}\sum_{k=0}^{n}(-1)^{n-k}\binom{n}{k}(x+k)^n = (x+\cdot)^n[0,1,\ldots,n].$$

But
$$f[0,1,\ldots,n] = \text{highest order coefficient of the polynomial interpolant,}$$
so
$$(x+\cdot)^n[0,1,\ldots,n] = 1.$$

$\square$

Proposition 1. *(Partition of Unity)*

$$\sum_{-n\le k\le 0} N_{k,n}(x) = 1.$$

Proof. By translation invariance, it is enough to prove this result for $0 \le x \le 1$. But for $0 \le x \le 1$, this result follows immediately from Lemmas 3 and 4. $\square$

6   New Identities for Uniform B-Splines

So far we have used our generating function to derive some well known identities for the uniform B-splines. In this section we shall derive some new identities for uniform B-splines using their generating functions.

6.1   New Identities from Specializing the Generating Functions

Here we derive new identities for the sums and alternating sums as well as for the moments and reciprocal moments of the uniform B-splines by considering special values of t in the generating functions $G_0(x,t)$. The reader may easily construct other identities for the B-splines by considering other specializations of their generating functions.

Theorem 6. *(Sums and Alternating Sums)*

$$\sum_{n=0}^{\infty} N_{0,n}(x) = e^x + \sum_{j=1}^{p}(-1)^j \frac{x(x-j)^{j-1}}{j!}e^{(x-j)} \qquad p \le x \le p+1. \tag{14}$$

$$\sum_{n=0}^{\infty}(-1)^n N_{0,n}(x) = e^{-x} + \sum_{j=1}^{p}\frac{(x-2j)(x-j)^{j-1}}{j!}e^{-(x-j)} \qquad p \le x \le p+1. \tag{15}$$

Proof. These results follow immediately by substituting $t = \pm 1$ on both sides of the generating function in Equation (7). $\square$

Theorem 7. *(Moments and Reciprocal Moments)*

$$\sum_{n=0}^{\infty} x^n N_{0,n}(x) = \sum_{j=0}^{p}(-1)^j \left(\frac{(x-j)^j x^j}{j!} + \frac{(x-j)^{j-1}x^{j-1}}{(j-1)!} \right) e^{(x-j)x}, \quad p \le x \le p+1.$$

(16)

$$\sum_{n=0}^{\infty} x^{-n} N_{0,n}(x) = \sum_{j=0}^{p}(-1)^j \left(\frac{(1-j/x)^j}{j!} + \frac{(1-j/x)^{j-1}}{(j-1)!} \right) e^{(1-j/x)}, \quad p \le x \le p+1.$$

(17)

Proof. These results follow immediately by substituting $t = x^{\pm 1}$ on both sides of the generating function in Equation (7). □

6.2   New Identities from Manipulating the Generating Functions

Here we derive new identities for the Laplace transform of the B-splines along with new convolution formulas for the B-splines with the monomials by manipulating the generating functions. Limited only by their imagination and ingenuity, readers may seek other identities for the B-splines by manipulating their generating functions.

Our explicit formula for the generating functions over the interval $[p, p+1]$ in Equation (7) is

$$\sum_{n} N_{0,n}(x)t^n = \sum_{j=0}^{p}(-1)^j \left(\frac{(x-j)^j t^j}{j!} + \frac{(x-j)^{j-1}t^{j-1}}{(j-1)!} \right) e^{(x-j)t}.$$

In this section we explore what happens when we move e^{xt} to the left hand side.

Theorem 8. *(Convolution Formulas)*

$$\sum_{j=0}^{n} \frac{(-1)^j}{j!} x^j N_{0,n-j}(x) = \sum_{j=0}^{p}(-1)^n \left(\frac{j^{n-j}(x-j)^j}{j!(n-j)!} - \frac{j^{n-j+1}(x-j)^{j-1}}{(j-1)!(n-j+1)!} \right)$$ (18)

for all n and $p \le x \le p+1$.

$$\sum_{k=0}^{n} \frac{(-1)^k}{k!}(x+\alpha)^k N_{0,n-k}(x) = (-1)^n \sum_{j=0}^{\min(p,n)} \left(\frac{(j+\alpha)^{n-j}}{j!(n-j+1)!}(x-j)^{j-1} \right.$$

$$\left. ((x-j)(n-j+1) - (j+\alpha)j) \right)$$

(19)

for all n, α and $p \le x \le p+1$.

Proof. To prove the first identity, start with the generating function in Equation (7):

$$\sum_n N_{0,n}(x)t^n = \sum_{j=0}^{p}(-1)^j\left(\frac{(x-j)^j t^j}{j!} + \frac{(x-j)^{j-1}t^{j-1}}{(j-1)!}\right)e^{(x-j)t}.$$

Now multiply both sides by e^{-xt}:

$$\sum_n N_{0,n}(x)t^n e^{-xt} = \sum_{j=0}^{p}(-1)^j\left(\frac{(x-j)^j t^j}{j!} + \frac{(x-j)^{j-1}t^{j-1}}{(j-1)!}\right)e^{-jt}.$$

Then expand the exponentials on both sides of this equation as power series and compare the coefficients of t^n. Thus

$$\sum_n\sum_{j=0}^{n}\frac{(-1)^j}{j!}x^j N_{0,n-j}(x)t^n = \sum_n\sum_{j=0}^{p}(-1)^n\left(\frac{j^{n-j}(x-j)^j}{j!(n-j)!} - \frac{j^{n-j+1}(x-j)^{j-1}}{(j-1)!(n-j+1)!}\right)t^n,$$

so comparing the coefficients of t^n it follows that for all n and $p \le x \le p+1$

$$\sum_{j=0}^{n}\frac{(-1)^j}{j!}x^j N_{0,n-j}(x) = \sum_{j=0}^{p}(-1)^n\left(\frac{j^{n-j}(x-j)^j}{j!(n-j)!} - \frac{j^{n-j+1}(x-j)^{j-1}}{(j-1)!(n-j+1)!}\right).$$

The second identity can be proved in a similar fashion by initially multiplying both sides of the generating function by $e^{-(x+\alpha)t}$ and proceeding as in the proof of the first identity. □

Next we shall investigate identities generated by taking the Laplace transform of the explicit formula for the generating functions. We begin by recalling a well-known result.

Lemma 5.

$$\int_0^\infty t^k e^{-at}dt = \frac{k!}{a^{(k+1)}}, \quad when \ a > 0. \tag{20}$$

Proof. Integrate by parts and apply induction on k. □

Theorem 9. *(Laplace Transforms)*

$$\sum_n \frac{n!N_{0,n}(x)}{(x+1)^{n+1}} = 1 + (x+1)\sum_{j=1}^{p}\frac{(-1)^j}{(j+1)^{j+1}}(x-j)^{j-1}. \tag{21}$$

$$\sum_n \frac{n!N_{0,n}(x)}{(x+\alpha)^{n+1}} = \frac{1}{\alpha} + (x+\alpha)\sum_{j=1}^{p}\frac{(-1)^j}{(j+\alpha)^{j+1}}(x-j)^{j-1} \tag{22}$$

for all $\alpha > 0$ and $p \le x \le p+1$.

Proof. To prove the first result, again we begin with the explicit formula for the generating function given in Equation (7):

$$\sum_n N_{0,n}(x)t^n = \sum_{j=0}^p (-1)^j \left(\frac{(x-j)^j t^j}{j!} + \frac{(x-j)^{j-1} t^{j-1}}{(j-1)!} \right) e^{(x-j)t} \quad p \le x \le p+1.$$

Now multiply both sides by $e^{-(x+1)t}$:

$$\sum_n N_{0,n}(x)t^n e^{-(x+1)t} = \sum_{j=0}^p (-1)^j \left(\frac{(x-j)^j t^j}{j!} + \frac{(x-j)^{j-1} t^{j-1}}{(j-1)!} \right) e^{-(j+1)t}$$

and integrate with respect to t:

$$\int_0^\infty \sum_n N_{0,n}(x)t^n e^{-(x+1)t} dt = \int_0^\infty \sum_{j=0}^p (-1)^j \left(\frac{(x-j)^j t^j}{j!} \right.$$
$$\left. + \frac{(x-j)^{j-1} t^{j-1}}{(j-1)!} \right) e^{-(j+1)t} dt.$$

Then

$$\sum_n N_{0,n}(x) \int_0^\infty t^n e^{-(x+1)t} dt = \sum_{j=0}^p (-1)^j \left(\frac{(x-j)^j}{j!} \int_0^\infty t^j e^{-(j+1)t} dt \right.$$
$$\left. + \frac{(x-j)^{j-1}}{(j-1)!} \int_0^\infty t^{j-1} e^{-(j+1)t} dt \right),$$

so by Lemma 5:

$$\sum_n \frac{n! N_{0,n}(x)}{(x+1)^{n+1}} = 1 + (x+1) \sum_{j=1}^p \frac{(-1)^j}{(j+1)^{j+1}} (x-j)^{j-1}.$$

The second identity can be proved in a similar fashion by initially multiplying both sides of the generating function by $e^{-(x+\alpha)t}$ and proceeding as in the proof of the first identity.   □

7   Summary, Conclusions, and Future Research

We derived a closed formula for the generating functions of the uniform B-splines, revealing a novel connection between uniform B-splines and exponential functions. Using this generating function, we established several classical identities for the uniform B-splines. These identities along with the corresponding functional equations for the generating functions are listed in Table 1. We also derived some new identities for uniform B-splines that cannot be derived by standard methods. These new identities are listed in Table 2.

Table 1. Some classical B-spline identities and the corresponding functional equations for their generating functions

B-Splines Identities	Generating Functions Functional Equations
$N_{0,n}(x) = \frac{x}{n} N_{0,n-1}(x)$ $+ \frac{n+1-x}{n} N_{1,n-1}(x)$	$\frac{\partial G_0}{\partial t}(x,t) = x G_0(x,t)$ $+ (2-x) G_1(x,t) + t \frac{\partial G_1}{\partial t}(x,t)$
$\frac{\partial N_{0,n}(x)}{\partial x} = N_{0,n-1}(x) - N_{1,n-1}(x)$	$\frac{\partial G_0}{\partial x}(x,t) = t G_0(x,t) - t G_1(x,t)$
$\sum_k N_{k,n}(x) \equiv 1$	$\sum_{k=-n}^{0} G_k(x,t) =$ $\sum_{k=-n}^{0}(-1)^{k+n}\left(\frac{(x-k)^{k+n}t^{k+n}}{(k+n)!}\right)e^{(x-k)t}$
$N_{0,n}(x) =$ $\frac{1}{n!}\sum_{j=0}^{p}(-1)^j \binom{n+1}{j}(x-j)^n$ $p \leq x \leq p+1$	$G_0(x,t) =$ $\sum_{j=0}^{p}(-1)^j\left(\frac{(x-j)^j t^j}{j!} + \frac{(x-j)^{j-1}t^{j-1}}{(j-1)!}\right)e^{(x-j)t}$ $p \leq x \leq p+1$

Sums and Alternating Sums

$$\sum_{n=0}^{\infty} N_{0,n}(x) = e^x + \sum_{j=1}^{p}(-1)^j \frac{x(x-j)^{j-1}}{j!}e^{(x-j)} \quad p \leq x \leq p+1$$

$$\sum_{n=0}^{\infty}(-1)^n N_{0,n}(x) = e^{-x} + \sum_{j=1}^{p}\frac{(x-2j)(x-j)^{j-1}}{j!}e^{-(x-j)} \quad p \leq x \leq p+1$$

Moments and Reciprocal Moments

$$\sum_{n=0}^{\infty} x^n N_{0,n}(x) = \sum_{j=0}^{p}(-1)^j\left(\frac{(x-j)^j x^j}{j!} + \frac{(x-j)^{j-1}x^{j-1}}{(j-1)!}\right)e^{(x-j)x} \quad p \leq x \leq p+1$$

$$\sum_{n=0}^{\infty} x^{-n} N_{0,n}(x) = \sum_{j=0}^{p}(-1)^j\left(\frac{(1-j/x)^j}{j!} + \frac{(1-j/x)^{j-1}}{(j-1)!}\right)e^{(1-j/x)} \quad p \leq x \leq p+1$$

Convolution Formulas

$$\sum_{j=0}^{n} \frac{(-1)^j}{j!} x^j N_{0,n-j}(x) = \sum_{j=0}^{p} (-1)^n \left(\frac{j^{n-j}(x-j)^j}{j!(n-j)!} - \frac{j^{n-j+1}(x-j)^{j-1}}{(j-1)!(n-j+1)!} \right)$$

for all n and $p \leq x \leq p+1$

$$\sum_{k=0}^{n} \frac{(-1)^k}{k!}(x+\alpha)^k N_{0,n-k}(x) = (-1)^n \sum_{j=0}^{\min(p,n)} \left(\frac{(j+\alpha)^{n-j}}{j!(n-j+1)!}(x-j)^{j-1} \right.$$

$$\left. \big((x-j)(n-j+1) - (j+\alpha)j\big) \right)$$

for all n, α and $p \leq x \leq p+1$

Laplace Transforms

$$\sum_{n} \frac{n! N_{0,n}(x)}{(x+1)^{n+1}} = 1 + (x+1) \sum_{j=1}^{p} \frac{(-1)^j}{(j+1)^{j+1}}(x-j)^{j-1}$$

$$\sum_{n} \frac{n! N_{0,n}(x)}{(x+\alpha)^{n+1}} = \frac{1}{\alpha} + (x+\alpha) \sum_{j=1}^{p} \frac{(-1)^j}{(j+\alpha)^{j+1}}(x-j)^{j-1}$$

for all $\alpha > 0$ and $p \leq x \leq p+1$

Table 2. Some new identities for the B-splines derived from their generating functions

Yet despite these successes, generating functions are not a panacea for deriving identities for uniform B-splines. The following two well known identities—the Marsden identity and the refinement equation—are not readily established using generating functions:

$$(x-t)^n = \sum_{k} (k+1-t) \cdots (k+n-t) N_{k,n}(x) \qquad (\textit{Marsden Identity})$$

$$N_{0,n}(x) = \sum_{k} \frac{\binom{n+1}{k}}{2^n} N_{0,n}(2x-k) \qquad (\textit{Refinement Equation})$$

We can, however, derive these identities directly or indirectly from the de Boor recurrence, which we have seen is equivalent to the PDE for the generating functions (see Table 1).

Currently our generating functions are restricted to B-splines with uniformly spaced knots— that is, knots in arithmetic progression

$$t_{k+1} = t_k + h \quad \text{(arithmetic progression)}.$$

In the future we hope to extend our generating functions to B-splines with knots in geometric or affine progression—that is, to B-splines with knot sequences where

$$t_{k+1} = qt_k \qquad \text{(geometric progression)}$$
$$t_{k+1} = qt_k + h \quad \text{(affine progression)}.$$

B-splines with knots in affine progression would also include B-splines with knots at the q-integers [2].

Finally we would also like to extend our generating functions to multivariate splines such as box splines, where simple recurrences for the basis functions are also available.

References

1. Graham, R., Knuth, D., Patashnik, O.: Concrete Mathematics: A Foundation for Computer Science. Addison Wesley, Reading (1992)
2. Phillips, G.: Interpolation and Approximation by Polynomials. Canadian Mathematical Society, Springer, New York (2003)
3. Schoenberg, I.J.: Contributions to the problem of approximation of equidistant data by analytic functions: Part A—On the problem of smoothing or graduation. A first class of analytic approximation formulae. Quarterly of Applied Mathematics 4, 45–99 (1964)
4. Simsek, Y.: Interpolation function of generalized q-bernstein-type basis polynomials and applications. In: Conference on Mathematical Methods for Curves and Surfaces, Avignon, France (2010)
5. Simsek, Y.: Generating functions for the Bernstein polynomials: A unified approach to deriving identities for the Bernstein basis functions, arXiv:1012.5538v1 (math.CA)
6. Simsek, Y.: Functional equations from generating functions: A novel approach to deriving identities for the Bernstein basis functions (2011) (preprint)

Planar Parametrization in Isogeometric Analysis

Jens Gravesen[1], Anton Evgrafov[1],
Dang-Manh Nguyen[2], and Peter Nørtoft[3],[*]

[1] Dept. of Appl. Math. & Comp. Sci., Technical University of Denmark, Denmark
{jgra,aaev}@dtu.dk
[2] Institute of Applied Geometry, Johannes Kepler University, Austria
Manh.Dang_Nguyen@jku.at
[3] Applied Mathematics, SINTEF ICT, Norway
peter@noertoft.net

Abstract. Before isogeometric analysis can be applied to solving a partial differential equation posed over some physical domain, one needs to construct a valid parametrization of the geometry. The accuracy of the analysis is affected by the quality of the parametrization. The challenge of computing and maintaining a valid geometry parametrization is particularly relevant in applications of isogemetric analysis to shape optimization, where the geometry varies from one optimization iteration to another. We propose a general framework for handling the geometry parametrization in isogeometric analysis and shape optimization. It utilizes an expensive non-linear method for constructing/updating a high quality reference parametrization, and an inexpensive linear method for maintaining the parametrization in the vicinity of the reference one. We describe several linear and non-linear parametrization methods, which are suitable for our framework. The non-linear methods we consider are based on solving a constrained optimization problem numerically, and are divided into two classes, geometry-oriented methods and analysis-oriented methods. Their performance is illustrated through a few numerical examples.

Keywords: Isogeometric analysis, shape optimization, parametrization.

1 Introduction

Isogeometric analysis is a modern computational method for solving partial differential equations (PDEs), which is based on a successful symbiosis between the variational techniques utilized in isoparametric finite element analysis with the geometric modelling tools from computer aided design [14,4]. A key ingredient of isogeometric analysis is the parametrization of the physical domain over which the PDE is posed, in many ways analogous to mesh generation in standard finite element analysis. Just as mesh quality affects the accuracy of a finite

---

[*] Presently at Dept. of Appl. Math. & Comp. Sci., Technical University of Denmark, Denmark.

M. Floater et al. (Eds.): MMCS 2012, LNCS 8177, pp. 189–212, 2014.

element approximation, the quality of the parametrization affects the accuracy of isogeometric analysis, see [21,2,34,35].

The question of computing and maintaining a valid geometry parametrization is particularly relevant in applications of isogemetic analysis to shape optimization problems, see e.g. [11,22,23,26]. Every time the geometry changes, that is, at every shape optimization iteration, one needs to update the parametrization in order to maintain the accuracy of the numerical approximation to the PDEs, governing the underlying physical model of the system. The algorithm for parametrization updates should therefore be (a) computationally inexpensive, as it is executed often; and (b) differentiable with respect to the variables determining the shape of the domain, which allows one to advantageously utilize gradient-based optimization algorithms thus reducing the total number of optimization iterations when compared with non-smooth or zero-order methods. One may again draw a parallel with the shape optimization based on the regular finite element analysis, which involves updating the mesh in between the shape optimization iterations.

The approach based on the discrete Coons patch [6] is a popular way of generating candidate parametrizations. This method is explicit and as a result it is very computationally inexpensive. Unfortunately, the resulting map needs not to be injective, and it is often necessary to invest further work in order to obtain even a valid, that is, a bijective parametrization. Even more work may be required to improve the quality of such a parametrization. Another approach to the same problem, which we have often utilized, is based on the spring model, cf. Section 3.1, in which the edges in the control net are modelled as elastic springs. In order to find a candidate parametrization one is required to solve a system of linear algebraic equations, thus rendering the method slightly more expensive than the discrete Coons patch. In our experience, however, the quality of the parametrizations obtained with this approach is slightly better.

If a good parametrization of a domain with a similar shape and patch layout is known, e.g., by using one of the methods in Section 4, one may employ one of the many methods developed for mesh generation [9,10,29] in order to compute a domain parametrization. We will in particular consider mean value coordinates [8,13], cf. Section 3.2. A new linear method of the same type is a quasi-conformal deformation method, cf. Section 3.3, which is inspired by conformal maps. Finally, any non-linear method may be linearized in the vicinity of a reference parametrization thereby resulting in a linear method.

We believe that no single linear method is capable of producing a high quality parametrization in all geometric configurations, and therefore we mainly investigate some non-linear methods. Many existing methods rely on the theory of harmonic functions on the physical domain. The method in [20] works on a triangulated volume and starts by constructing a parametrization of the boundary, i.e., the outer surface, using two harmonic functions with near orthogonal gradients. Then using harmonic functions in 3D the parametrization is propagated inwards to fill the entire volume. In [24] the inverse of the parametrization is constructed in a coordinate by coordinate fashion, using harmonic functions on the level set of the previously constructed coordinate functions. Finally the parametrization is defined as a tensor product spline approximation of the inverse map. The method in [25] demands that the inverse of the parametrization of a planar domain is a pair of harmonic functions and then proceeds to solving a uniquely solvable non-linear equation. This is mathematically equivalent to the last method in Section 4.1, where the Winslow functional is minimized. There is a unique minimizer whose inverse is the same pair of harmonic functions. The Winslow functional can also be interpreted as a condition number for the Jacobian and it is in that role that it is used in [12]. One may of course devise other methods based on the idea of finding extrema of geometric functionals, quantitatively assessing the quality of the parametrization, such as the area orthogonality functional and the Liao functional, cf. Section 4.1.

The final class of methods is based on estimating the approximation error and generating a parametrization that makes the estimate as small as possible. As test cases one can take problems with known analytical solutions and try to find the parametrization that minimizes the discrepancy between the exact and the numerical solutions, see [21,34] for a 1D eigenvalue problem and a 2D heat conduction problem, respectively. In practice one of course does not know the exact solution so instead a suitable error estimator is utilized. In Section 4.2 we try three different error estimators, where the first one is similar to the one used in [35].

The outline of the rest of this paper is as follows. In Section 2 we introduce the parametrization problem studied in this work, including the partial differential equation to be solved, namely Poisson's equation. In Section 3, we introduce three linear parametrization methods, and in Section 4, we describe a family of nonlinear, optimization-based parametrization methods based on two classes of quality measures, namely purely geometric and analysis-oriented measures. In Section 5, numerical results are presented, and in Section 6 we discuss extensions of the methods to shape optimization and to multiple patches. Finally, the current findings and some future challenges are summarized in Section 7.

2   Parametrization for Partial Differential Equations

In the following, we introduce the context in which the parametrization problem occurs, we formulate the parametrization problem, and we state a condition for the validity of a B-spline parametrization.

2.1   The Setting: Poisson's Problem

We consider a mixed boundary value problem for Poission's equation in two dimensions in a regular domain $\Omega \subset \mathbb{R}^2$ with piecewise-smooth boundary $\partial\Omega$. The boundary $\partial\Omega$ is represented as a closure of the union of two open disjoint subsets $\Gamma_D \neq \emptyset$ and Γ_N, on which we impose Dirichlet and Neumann boundary conditions. That is, we are interested in finding a function $u : \mathbb{R}^2 \to \mathbb{R}$, such that

$$-\Delta u = f \qquad \text{in } \Omega, \tag{1a}$$
$$u = g \qquad \text{on } \Gamma_D, \tag{1b}$$
$$\nabla u \cdot \mathbf{n} = h \qquad \text{on } \Gamma_N, \tag{1c}$$

where $f, g, h : \mathbb{R}^2 \to \mathbb{R}$ are given, and $\mathbf{n}$ is the outwards facing boundary normal. In the weak form, the boundary value problem (1) reads: Find $u \in \{ w \in H^1(\Omega) : w|_{\Gamma_D} = g \}$ such that

$$\int_\Omega \nabla u \cdot \nabla v \, dA = \int_\Omega f \, v \, dA + \int_{\Gamma_N} h \, v \, ds. \tag{2}$$

for all $v \in \{ w \in H^1(\Omega) : w|_{\Gamma_D} = 0 \}$.

2.2 The Challenge: Parametrize the Interior

In order to utilize isogeometric analysis for solving the boundary value problem (1) numerically, a suitable geometry parametrization $\mathbf{X}$ of the domain Ω is required. Constructing such a parametrization is akin to the mesh generation step required for the standard finite element analysis. The parametrization impacts the accuracy of the numerical solution to the problem [34]. Expectedly, a higher quality parametrization allows for numerical solution with higher accuracy, all other things being equal.

Assuming that the domain $\Omega \subset \mathbb{R}^2$ may be parametrized using a single patch, the challenge in two dimensions reads: given a parametrization $\mathbf{Y} : \partial[0,1]^2 \to \mathbb{R}^2$ of the boundary $\partial\Omega$, construct a parametrization of the interior $\mathbf{X} : [0,1]^2 \to \mathbb{R}^2$, such that $\mathbf{X}|_{\partial[0,1]^2} = \mathbf{Y}$.

In B-spline-based isogeometric analysis, the maps $\mathbf{Y}$ and $\mathbf{X}$ are splines, e.g.,

$$\mathbf{X}(\xi, \eta) = \begin{pmatrix} x(\xi, \eta) \\ y(\xi, \eta) \end{pmatrix} = \sum_{i,j} \mathbf{X}_{i,j} \, M_i(\xi) \, N_j(\eta), \tag{3}$$

where M_i and N_j are B-splines defined by polynomial degrees and knots vectors and $\mathbf{X}_{i,j}$ are the control points. The equivalent challenge is now to specify the interior control points given the boundary control points [34,2,11]. This problem is sketched in Fig. 1.

2.3 The Jacobian

As we assume the boundary map $\mathbf{Y} = \mathbf{X}|_{\partial[0,1]^2}$ is a parametrization, in particular a homeomorphism, the map $\mathbf{X} : [0,1]^2 \to \Omega$ is a diffeomorphism if and only if the Jacobian

$$\mathbf{J} = (\mathbf{X}_\xi \, \mathbf{X}_\eta) = \begin{pmatrix} \frac{\partial \mathbf{X}}{\partial \xi} & \frac{\partial \mathbf{X}}{\partial \eta} \end{pmatrix} = \begin{pmatrix} \frac{\partial x}{\partial \xi} & \frac{\partial x}{\partial \eta} \\ \frac{\partial y}{\partial \xi} & \frac{\partial y}{\partial \eta} \end{pmatrix} = \begin{pmatrix} x_\xi & x_\eta \\ y_\xi & y_\eta \end{pmatrix}. \tag{4}$$

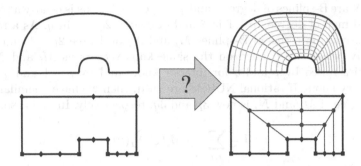

Fig. 1. Challenge: How do we go from a parametrization of the boundary of a domain to a parametrization that includes the interior of the domain?

is regular at every point. Therefore, in order to guarantee the validity of the parametrization, it is necessary that the determinant of the Jacobian does not vanish on Ω. At the four corners of the parameter domain $[0,1]^2$, both partial derivatives of $\mathbf{X}$ are determined by the boundary parametrization $\mathbf{Y}$. As a consequence of this, there are domains which are impossible to parametrize. Indeed, consider the V-shaped domain in Fig. 2. If the boundary parametrization is

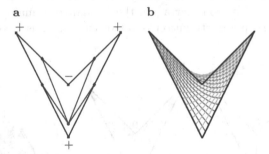

Fig. 2. An impossible domain. a: Control points and sign of the Jacobian determinant in the corners. b: The best quadratic parametrization when the edges are parametrized affinely.

regular, then the Jacobian has a positive determinant in the three convex corners, and a negative determinant in the concave corner (or vice versa if the orientation is reversed). So if the parametrization is C^1 on the closed parameter domain $[0,1]^2$, the determinant of the Jacobian attains both positive and negative values, and it is impossible to have a valid parametrization for this domain.

If we use B-splines to define the parametrization as in Equation (3), then the determinant can be written as

$$\det \mathbf{J} = \sum_{i,j,k,\ell} \det \begin{pmatrix} x_{i,j} & x_{k,\ell} \\ y_{i,j} & y_{k,\ell} \end{pmatrix} M_i'(\xi)\, N_j(\eta)\, M_k(\xi)\, N_\ell'(\eta)\,. \tag{5}$$

If M and N are B-splines of degree p and q, respectively, this is clearly a piecewise polynomial map of degree $2p - 1$ in ξ and of degree $2q - 1$ in η. As a result, it can be expressed in terms of B-splines $\widetilde{M}_i$ and $\widetilde{N}_j$ of degree $2p - 1$ and $2q - 1$, respectively, which are defined on the same knot vectors as M_i and N_j with multiplicities raised by p and q for interior knots and by $p - 1$ and $q - 1$ for the boundary knots. If rational NURBS are used, then we have a similar result, but the degree of $\widetilde{M}_i$ and $\widetilde{N}_j$ is now $3p$ and $3q$, respectively. In any case, we can write

$$\det \mathbf{J} = \sum_{i,j} d_{i,j}\, \widetilde{M}_i(\xi)\, \widetilde{N}_j(\eta)\,. \tag{6}$$

As the B-splines $\widetilde{M}_i$ and $\widetilde{N}_j$ are non negative, we immediately obtain

Theorem 1. *If the coefficients $d_{i,j}$ of the B-spline expansion (6) of the determinant of the Jacobian are positive then the parametrization is valid.*

Observe that this is a *sufficient condition* and not a *necessary* one. However, if we perform knot insertion, then more and more coefficients will become positive. Indeed, if $\det \mathbf{J} > 0$ on all of $[0, 1]^2$, then $d_{i,j} > 0$ for all i, j, after sufficiently many knot insertions. On the other hand, if the boundary parametrization has a zero derivative at some point, then the B-spline expansion (6) may have a negative coefficient no matter how many knot insertions we perform.

To demonstrate this, consider again the V-shaped domain, but now assume that the boundary parametrization is quadratic and has a zero derivative at

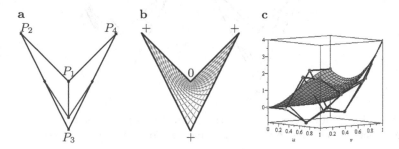

Fig. 3. The V-shaped domain with a singular boundary parametrization. a: Three control points are placed at the concave corner. b: The parametrization and the sign of the Jacobian determinant at the four corners. c: The Jacobian determinant.

the concave corner P_1, see Fig. 3. That is, the two edges meeting at P_1 are parametrized as

$$(1 - \xi^2)\, P_1 + \xi^2\, P_2 \quad \text{and} \quad (1 - \eta^2)\, P_1 + \eta^2\, P_4\,, \tag{7}$$

respectively. By letting the single inner control point be $\frac{1}{4}P_1 + \frac{3}{4}P_3$ we obtain a valid parametrization in the form of a bi-quadratic tensor product Bézier patch.

We may assume that $P_1 = 0$ and then

$$X(\xi, \eta) = P_2\, B_2^2(\xi)\, B_0^2(\eta) + \frac{3}{4} P_3\, B_1^2(\xi)\, B_1^2(\eta) + P_4\, B_0^2(\xi)\, B_2^2(\eta)$$
$$+ \frac{P_2 + P_3}{2}\, B_2^2(\xi)\, B_1^2(\eta) + \frac{P_3 + P_4}{2}\, B_1^2(\xi)\, B_2^2(\eta) + P_3\, B_2^2(\xi)\, B_2^2(\eta)\,. \quad (8)$$

The determinant of the Jacobian is a bi-cubic tensor product Bézier patch

$$\det \mathbf{J} = \sum_{i,j=0}^{3} d_{i,j}\, B_i^3(\xi)\, B_j^3(\eta)\,. \quad (9)$$

We see that $d_{0,0} = d_{1,0} = d_{0,1} = 0$, and $d_{1,1} = \det(P_2, P_4) < 0$ but it is not hard to see that $\det \mathbf{J} > 0$ on $]0, 1[^2$, see Fig. 3. This is still the case after any refinement of the knot vectors.

The fact that a change of the boundary parametrization of the V-shaped domain can make a parametrization of the interior possible was also noted in [32].

3   Linear Parametrization Methods

In this section, we present three linear methods for computing geometry parametrizations. The first of these, the spring model, operates without the need for any information apart from the boundary parametrization; this method may therefore be utilized for generating initial parametrizations for other linear or non-linear methods. The last two, the mean value coordinates and the quasi-conformal methods, rely on the knowledge of a reference parametrization of the interior. One may of course generate more linear methods by linearizing nonlinear ones around reference parametrizations, as discussed in Section 3.4.

3.1   The Spring Model

This method mimics a mechanical model, in which all edges in the control mesh are replaced with linear elastic springs. The mechanical equilibrium, which arises when the positions of the boundary control points are given, defines the position of the inner control points within this model. In this configuration, all inner control points are the averages of their four neighbours. That is, we have a set of simple linear equations:

$$4\mathbf{X}_{i,j} = \mathbf{X}_{i+1,j} + \mathbf{X}_{i-1,j} + \mathbf{X}_{i,j+1} + \mathbf{X}_{i,j-1}\,, \quad (10)$$

which is easily solved. By assigning different "spring constants" to different edges one obtains variations of the method.

3.2   Mean Value Coordinates

In recent years there has been a lot of work on parametrization of polygonal meshes [9,10,29]. If we use the control point formulation of our spline parametrization problem, some of these methods can be applied to our problem. A popular and appealing method is based on the mean value coordinates [8,13]. Here, points in the plane are given as a particular affine combination of the vertices of a closed polygon. The closed polygon is in our case the boundary of the control net.

Suppose we are given a *reference parametrization*, with inner control points $\widehat{\mathbf{X}}_{k,\ell}$, and a set of boundary control points $\widehat{\mathbf{X}}_i$, $i = 1, \ldots, n$, arranged in a counter clockwise fashion. Any point $\mathbf{x} \in \mathbb{R}^2$ can now be written as an affine combination of the boundary control points:

$$\mathbf{x} = \sum_{i=1}^{n} \lambda_i(\mathbf{x})\,\widehat{\mathbf{X}}_i\,, \quad \text{where} \quad \lambda_i(\mathbf{x}) = \frac{w_i(\mathbf{x})}{\sum_{i=1}^{n} w_i(\mathbf{x})}\,. \tag{11}$$

The weights $w_i(\mathbf{x})$ are defined by

$$
\begin{aligned}
w_i(\mathbf{x}) &= 2\frac{\tan(\alpha_{i-1}/2) + \tan(\alpha_i/2)}{\|\widehat{\mathbf{v}}_i\|} = \frac{2}{\|\widehat{\mathbf{v}}_i\|}\left(\frac{\sin\alpha_{i-1}}{1 + \cos\alpha_{i-1}} + \frac{\sin\alpha_i}{1 + \cos\alpha_i}\right) \\
&= \frac{2}{\|\widehat{\mathbf{v}}_i\|}\left(\frac{[\widehat{\mathbf{v}}_{i-1}, \widehat{\mathbf{v}}_i]}{\|\widehat{\mathbf{v}}_{i-1}\|\|\widehat{\mathbf{v}}_i\| + \langle\widehat{\mathbf{v}}_{i-1}, \widehat{\mathbf{v}}_i\rangle} + \frac{[\widehat{\mathbf{v}}_i, \widehat{\mathbf{v}}_{i+1}]}{\|\widehat{\mathbf{v}}_i\|\|\widehat{\mathbf{v}}_{i+1}\| + \langle\widehat{\mathbf{v}}_i, \widehat{\mathbf{v}}_{i+1}\rangle}\right),
\end{aligned} \tag{12}
$$

where

$$\langle\widehat{\mathbf{v}}, \widehat{\mathbf{w}}\rangle = v_1 w_1 + v_2 w_2 \quad \text{and} \quad [\widehat{\mathbf{v}}, \widehat{\mathbf{w}}] = v_1 w_2 - v_2 w_1 \tag{13}$$

are the inner product and the determinant of a pair of vectors $\widehat{\mathbf{v}}$ and $\widehat{\mathbf{w}}$, respectively. The angles and vectors are defined in Fig. 4a. If we have a parametrization

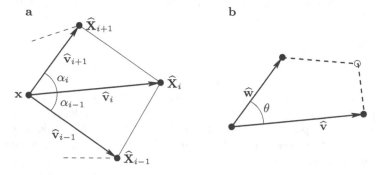

Fig. 4. a) Ingredients of mean value coordinates. b) Ingredients of the quasi conformal deformation.

of the boundary of another domain with new boundary control points $\mathbf{X}_i$, then we simply define the new inner control points as

$$\mathbf{X}_{k,\ell} = \sum_{i=1}^{n} \lambda_i(\widehat{\mathbf{X}}_{k,\ell}) \, \mathbf{X}_i \,, \tag{14}$$

i.e., we use the same normalized weights as in the reference control net.

3.3 Quasi Conformal Deformation

Once again we assume we have a reference parametrization $\widehat{\mathbf{X}}$. The idea is that we would like any other parametrization to have a control net that locally looks like a conformal deformation of the reference control net.

Consider a quadrilateral and two neighbouring edges in the reference control net, cf. Fig. 4b. We think of these edges as vectors $\widehat{\mathbf{v}}$ and $\widehat{\mathbf{w}}$ emanating from their common vertex. The method is based on a simple geometric identity $\|\widehat{\mathbf{v}}\|\widehat{\mathbf{w}} = \|\widehat{\mathbf{w}}\|\mathbf{R}(\theta)\,\widehat{\mathbf{v}}$, where $\mathbf{R}(\theta)$ is a rotation through the angle θ, that is:

$$\|\widehat{\mathbf{v}}\| \, \widehat{\mathbf{w}} = \|\widehat{\mathbf{w}}\| \, \mathbf{R}(\theta) \, \widehat{\mathbf{v}} = \frac{1}{\|\widehat{\mathbf{v}}\|} \begin{pmatrix} \langle \widehat{\mathbf{v}}, \widehat{\mathbf{w}} \rangle & -[\widehat{\mathbf{v}}, \widehat{\mathbf{w}}] \\ [\widehat{\mathbf{v}}, \widehat{\mathbf{w}}] & \langle \widehat{\mathbf{v}}, \widehat{\mathbf{w}} \rangle \end{pmatrix} \widehat{\mathbf{v}} \,. \tag{15}$$

If $\mathbf{v}$ and $\mathbf{w}$ are the corresponding edges in the new control net, then we can require that

$$\|\widehat{\mathbf{v}}\| \, \mathbf{w} = \frac{1}{\|\widehat{\mathbf{v}}\|} \begin{pmatrix} \langle \widehat{\mathbf{v}}, \widehat{\mathbf{w}} \rangle & -[\widehat{\mathbf{v}}, \widehat{\mathbf{w}}] \\ [\widehat{\mathbf{v}}, \widehat{\mathbf{w}}] & \langle \widehat{\mathbf{v}}, \widehat{\mathbf{w}} \rangle \end{pmatrix} \mathbf{v} \,, \tag{16}$$

for each such pair of edges. For each inner control point we have four linear algebraic equations of the type (16), and for every boundary control point, apart from the corners, we have two equations. This results in $4(MN - M - N)$ equations in $(M - 2)(N - 2)$ unknown inner control points. The resulting overdetermined system is then solved in the least squares sense.

One could also look after a conformal deformation of the reference parametrization by replacing the vectors $\widehat{v}$ and $\widehat{w}$ with the partial derivatives $\widehat{\mathbf{X}}_\xi$ and $\widehat{\mathbf{X}}_\eta$. That is, the new parametrization $\mathbf{X}$ should satisfy the equation

$$\|\widehat{\mathbf{X}}_\xi\| \, \mathbf{X}_\eta = \frac{1}{\|\widehat{\mathbf{X}}_\xi\|} \begin{pmatrix} \langle \widehat{\mathbf{X}}_\xi, \widehat{\mathbf{X}}_\eta \rangle & -[\widehat{\mathbf{X}}_\xi, \widehat{\mathbf{X}}_\eta] \\ [\widehat{\mathbf{X}}_\xi, \widehat{\mathbf{X}}_\eta] & \langle \widehat{\mathbf{X}}_\xi, \widehat{\mathbf{X}}_\eta \rangle \end{pmatrix} \mathbf{X}_\xi \,, \quad \text{in all of } [0,1]^2. \tag{17}$$

Similarly to the previous case, this family of equations could be solved in the least square sense.

3.4 Linearized Methods

In the following section we will introduce several non-linear methods that work by minimizing a certain quality measure c, and by a linearization of these, we may obtain new linear methods. One way of formalizing this is by considering a second order Taylor expansion of the quality measure in the vicinity of a reference parametrization $\widehat{\mathbf{X}}$. If we let $\mathbf{X}_1$ denote the known control points, typically the boundary control points, and let $\mathbf{X}_2$ denote unknown control points, typically the inner control points, then we can write

$$
\begin{aligned}
c(\mathbf{X}) \approx c(\widehat{\mathbf{X}}) &+ \left(\mathbf{G}_1(\widehat{\mathbf{X}})\ \mathbf{G}_2(\widehat{\mathbf{X}}) \right) \begin{pmatrix} \mathbf{X}_1 - \widehat{\mathbf{X}}_1 \\ \mathbf{X}_2 - \widehat{\mathbf{X}}_2 \end{pmatrix} \\
&+ \frac{1}{2} \left(\mathbf{X}_1^T - \widehat{\mathbf{X}}_1^T\ \ \mathbf{X}_2^T - \widehat{\mathbf{X}}_2^T \right) \begin{pmatrix} \mathbf{H}_{11}(\widehat{\mathbf{X}})\ \mathbf{H}_{12}(\widehat{\mathbf{X}}) \\ \mathbf{H}_{21}(\widehat{\mathbf{X}})\ \mathbf{H}_{22}(\widehat{\mathbf{X}}) \end{pmatrix} \begin{pmatrix} \mathbf{X}_1 - \widehat{\mathbf{X}}_1 \\ \mathbf{X}_2 - \widehat{\mathbf{X}}_2 \end{pmatrix}, \quad (18)
\end{aligned}
$$

where $\mathbf{G}_i$ and $\mathbf{H}_{ij}$ gives the gradient and the Hessian of c with respect to the control points of the parametrization. Assuming that the Hessian is positive definite, the right hand side is minimized when

$$
\mathbf{H}_{22}(\widehat{\mathbf{X}})\,\mathbf{X}_2 = \mathbf{H}_{22}(\widehat{\mathbf{X}})\,\widehat{\mathbf{X}}_2 - 2\mathbf{H}_{21}(\widehat{\mathbf{X}})\,(\mathbf{X}_1 - \widehat{\mathbf{X}}_1) - \mathbf{G}_2(\widehat{\mathbf{X}}), \qquad (19)
$$

which is a linear equation in the unknown control points $\mathbf{X}_2$.

4   Nonlinear Parametrization Methods

We proceed to presenting a family of nonlinear parametrization methods based on optimization, following the approach taken in, e.g. [34,35]. Thus, the interior parametrization is constructed by numerically maximizing quantitative measures of the parametrization quality. We divide these measure into two groups: the geometry-oriented and the analysis-oriented. Throughout, we assume that we are given a regular parametrization of the boundary with positive determinant of the Jacobian in the corners.

In order to have a valid parametrization, the Jacobian needs to have a non-vanishing determinant everywhere. Owing to our assumption about the sign in the corners, we will demand that the determinant is positive everywhere inside the domain, and we can then formulate the following max min problem

$$
\begin{aligned}
&\underset{\mathbf{X}}{\text{maximize}} && Z\,, && (20\text{a}) \\
&\text{such that} && \det \mathbf{J} \geq Z\,, \quad \text{in } [0,1]^2\,, && (20\text{b}) \\
&\text{where} && \mathbf{X}|_{\partial[0,1]^2} = \mathbf{Y}\,, && (20\text{c})
\end{aligned}
$$

In practice, we replace the condition (20b) with

$$
d_{i,j} \geq Z\,, \quad \text{for all } i,j\,, \qquad (21)
$$

where $d_{i,j}$ are the coefficients of the determinant of the Jacobian, cf. (6). In case an optimization algorithm terminates with a configuration, for which we have $Z > 0$, the resulting parametrization is necessarily valid. However, its quality does not have to be very high, cf. Fig. 5. Despite this drawback, the approach provides a simple way of generating valid initial parametrizations for other methods, which require such initialization.

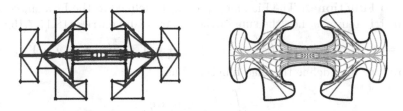

Fig. 5. Maximizing the smallest coefficient in the B-spline expansion of det $\mathbf{J}$

4.1   Geometric Measures

The first class of quality measures are geometric in nature, and thereby depend only on the parametrization itself. The methods in this class amount to solving an optimization problem, which can be formulated as

$$\underset{\mathbf{X}}{\text{minimize}} \qquad c(\mathbf{X}) , \tag{22a}$$

$$\text{such that} \qquad \det \mathbf{J} \geq \delta Z , \quad \text{in } [0,1]^2 , \tag{22b}$$

$$\text{where} \qquad \mathbf{X}|_{\partial[0,1]^2} = \mathbf{Y} , \tag{22c}$$

In the lower bound (22b) for det $\mathbf{J}$, the number $\delta \in [0,1]$ is an algorithmic parameter and the number Z is the result of the optimization (20). We have often successfully used $\delta = 0$.

When defining geometric quality measures for a parametrization, the Jacobian $\mathbf{J}$ and the first fundamental form g are important quantities:

$$g = \mathbf{J}^T \mathbf{J} = \begin{bmatrix} x_\xi^2 + y_\xi^2 & x_\xi x_\eta + y_\xi y_\eta \\ x_\xi x_\eta + y_\xi y_\eta & x_\eta^2 + y_\eta^2 \end{bmatrix} . \tag{23}$$

With these in mind, we proceed to define the area-orthogonality, the Liao, and the Winslow functionals, which are all well-known quantities for mesh generations, see e.g. [17,7].

The Area-Orthogonality Functional. The area-orthogonality measure m_{AO} is defined as the product of the diagonal entries of the metric tensor g, [7,15]

$$m_{\text{AO}} = g_{11} g_{22} = \left(x_\xi^2 + y_\xi^2 \right) \left(x_\eta^2 + y_\eta^2 \right) . \tag{24}$$

Based on this, we may define the area-orthogonality functional c_{AO} as the integral of the area-orthogonality measure m_{AO} over the parameter domain:

$$c_{\text{AO}} = \int_0^1 \int_0^1 m_{\text{AO}} \, \mathrm{d}\xi \, \mathrm{d}\eta . \tag{25}$$

The Liao Functional. The Liao measure m_L is defined as the Frobenius norm of the metric tensor g, i.e., the sum of the square of its entries [18,17,7,15]:

$$m_L = g_{11}^2 + g_{22}^2 + 2g_{12}^2 = \left(x_\xi^2 + y_\xi^2\right)^2 + \left(x_\eta^2 + y_\eta^2\right)^2 + 2\left(x_\xi x_\eta + y_\xi y_\eta\right)^2. \quad (26)$$

As above, we may define the Liao functional as

$$c_L = \int_0^1 \int_0^1 m_L \, d\xi \, d\eta. \quad (27)$$

The Winslow Functional. In this approach, the goal is to construct a parametrization as conformal as possible [33,11,22].

The parametrization $\mathbf{X}$ is conformal if and only if the Jacobian $\mathbf{J}$ is the product of a scaling and a rotation, or, equivalently, if the first fundamental form g is diagonal with identical diagonal elements. If we let λ_1 and λ_2 denote the eigenvalues of g, we need $\lambda_1 = \lambda_2$ to have conformality. We easily find that

$$\frac{\left(\sqrt{\lambda_1} - \sqrt{\lambda_2}\right)^2}{\sqrt{\lambda_1 \lambda_2}} = \frac{\lambda_1 + \lambda_2 - 2\sqrt{\lambda_1 \lambda_2}}{\sqrt{\lambda_1 \lambda_2}} = \frac{\lambda_1 + \lambda_2}{\sqrt{\lambda_1 \lambda_2}} - 2.$$

From this, we may define the Winslow measure m_W:

$$m_W = \frac{\lambda_1 + \lambda_2}{\sqrt{\lambda_1 \lambda_2}} = \frac{\text{tr}(g)}{\sqrt{\det(g)}} = \frac{x_\xi^2 + x_\eta^2 + y_\xi^2 + y_\eta^2}{x_\xi y_\eta - y_\xi x_\eta}, \quad (28)$$

where $\sqrt{\det(g)} = \det(\mathbf{J})$. As such, m_W is a pointwise measure of conformality. Using the Winslow function m_W we define the Winslow functional as:

$$c_W = \int_0^1 \int_0^1 m_W \, d\xi \, d\eta, \quad (29)$$

and use this as a global measure of conformality.

The Winslow functional has particularly nice mathematical properties. Indeed, if we switch the integration in (29) from the parameter domain $[0,1]^2$ to the physical domain Ω, then we obtain

$$c_W = \int_\Omega \left(\left(\frac{\partial \xi}{\partial x}\right)^2 + \left(\frac{\partial \xi}{\partial y}\right)^2 + \left(\frac{\partial \eta}{\partial x}\right)^2 + \left(\frac{\partial \eta}{\partial y}\right)^2 \right) dA. \quad (30)$$

This is the well known Dirichlet energy, and the unique minimizer is a pair of harmonic functions $\Omega \to [0,1]^2$ whose restriction to the boundary is the inverse $\mathbf{Y}^{-1}$ of the given boundary parametrization $\mathbf{Y} : \partial[0,1]^2 \to \partial\Omega$. As the target $[0,1]^2$ is convex, the Radó–Kneser–Choquet theorem [1,5,16,28] ensures that this pair of harmonic functions is a diffeomorphism on the interior. This means that our optimization problem (22), with the cost function (29), also has a unique minimum which is a diffeomorphism whose inverse is a pair of harmonic functions. This is not in conflict with the impossible domain shown in Fig. 2: the diffeomorphism is defined on the interior, and the maps may be non-differentiable at the boundary. In Fig. 6 we show the parametrization ensured by the theorem. Notice that the y coordinate is not differentiable in the concave corner, so the Jacobian is not defined in that corner.

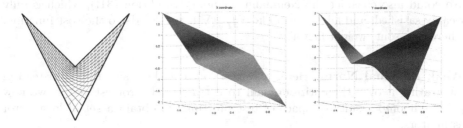

Fig. 6. Parametrization of the V-shape and the graphs of the x and y coordinates

4.2   Analysis-Oriented Measures

In the other class of non-linear variational methods for constructing parametrizations, we put analysis-oriented methods. Here, the explicit goal is to construct as accurate analysis of a given partial differential equation as possible. This accuracy needs to be estimated, which can be done by comparing the solutions from several analyses or by evaluating the residual. In any case, when using methods in this class we aim at *analysis-aware* parametrizations [34,2]. The quality measure for these methods depends not only on the parametrization, but also on the solution to the PDE at hand (the Poisson problem (1) in our case). The resulting optimization problems can be formulated as follows:

$$\underset{\mathbf{X}}{\text{minimize}} \qquad c(\mathbf{X}, u) , \tag{31a}$$

$$\text{such that} \qquad \det \mathbf{J} \geq \delta\, Z , \qquad \text{in } [0,1]^2 , \tag{31b}$$

$$\text{where} \qquad \mathbf{X}|_{\partial[0,1]^2} = \mathbf{Y} , \tag{31c}$$

$$-\Delta u = f , \qquad \text{in } \Omega , \tag{31d}$$

$$u = g , \qquad \text{on } \Gamma_D , \tag{31e}$$

$$\nabla u \cdot \mathbf{n} = h , \qquad \text{on } \Gamma_N . \tag{31f}$$

As before, in the lower bound (31b) for $\det \mathbf{J}$, the number $\delta \in [0,1]$ is an algorithmic parameter and the number Z is the result of the optimization (20). It goes without saying that if the Poisson problem is replaced by another problem, only the equations (31d)–(31f) are changed.

Strong Residual Norm. In this approach, we use the residual of the problem we are trying to solve as an error estimator. Hence, from Equation (31d) we set

$$m_{\mathrm{SR}} = (\Delta u + f)^2 . \tag{32}$$

We emphasize that at least a quadratic B-spline approximation of the field u must be employed. The exact expression for m_{SR} depends of course on the problem considered. As a result, we obtain the quality measure

$$c_{\mathrm{SR}} = \int_{\Omega} m_{\mathrm{SR}} \, \mathrm{d}A . \tag{33}$$

We could also consider the Neumann boundary condition (31f), which is only weakly satisfied, and add a term like $\alpha \int_{\Gamma_N} (\nabla u \cdot \mathbf{n} - h)^2 \, \mathrm{d}s$ to the cost function, where α is some weight factor.

Weak Residual Norm. Here, we again consider the residual, but instead of integrating it over the entire domain to get a global error estimator, we now project it onto a suitable space of test functions to obtain a set of local error estimators.

When the variational form of the PDE (2) is considered over a given space $\mathcal{S}_1$, the residual will belong to the orthogonal complement of this space owing to Galerkin's orthogonality. Therefore, we project the residual onto a larger space $\mathcal{S}_2 \supsetneq \mathcal{S}_1$ to obtain a meaningful, non-zero error estimator:

$$m_{\mathrm{WR},k} = \int_\Omega \nabla u \cdot \nabla R_k \, \mathrm{d}A - \int_{\Gamma_N} g \, R_k \, \mathrm{d}s - \int_\Omega f \, R_k \, \mathrm{d}A. \tag{34}$$

Here, the functions R_k are the basis functions for $\mathcal{S}_2$ stemming from tensor product B-splines on the parameter domain $[0,1]^2$. There are many possibilities in choosing $\mathcal{S}_2$. One obvious choice is by halving all knot segments (h-refinement), and another is degree elevation (p-refinement). As the integration is performed knot segment by knot segment the latter yields cheaper integration, so this is the one we have tested. Again, the exact expression for m_{WR} depends on the problem considered.

In this method, we consider the quality measure

$$c_{\mathrm{WR}} = \sum_k m_{\mathrm{WR},k}^2. \tag{35}$$

Of course, we could also introduce weights α_k on $m_{\mathrm{WR},k}$, e.g. the area of the support of the basis function R_k.

Enrichment Error Norm. As in the previous subsection we consider two different spline spaces $\mathcal{S}_1 \subsetneq \mathcal{S}_2$, but now we seek two approximate solutions $u_1 \in \mathcal{S}_1$ and $u_2 \in \mathcal{S}_2$ and regard their difference as an error estimator:

$$u_1 - u_2 = \sum_k m_{\mathrm{EE},k} R_k, \tag{36}$$

where the R_k as above is the basis for $\mathcal{S}_2$. Therefore, the quality measure is

$$c_{\mathrm{EE}} = \sum_k m_{\mathrm{EE},k}^2, \tag{37}$$

and again, we could introduce weights α_k on $m_{\mathrm{EE},k}$. Note that we have to solve the equation twice in this approach, so it is a rather expensive method.

5   Numerical Examples

In this section, we study two numerical examples of the parametrization problem outlined in Section 2, and we compare the resulting parametrizations based on the nonlinear methods described in Section 4. The methods are implemented in MATLAB® [19] and Octave [27]. The optimization is done using IPOPT, a non-linear optimization package based on an interior point method [30]. In both examples, the geometries are represented by quadratic splines, while the scalar field u is approximated using cubic splines. The equations are discretized using a Galerkin method as described in [4] and the knots are in all cases uniformly spaced. The weak residual and enrichment error methods are based on a degree elevation of the analysis spline by one, i.e., the spline spaces $\mathcal{S}_1$ and $\mathcal{S}_2$ consists of cubic and quartic C^2 splines, respectively.

5.1   Poisson's Equation on a Wedge-Shaped Domain

We consider the parametrization problem for a boundary value problem (BVP) with a known analytical solution. The example is taken from [34]. The domain under consideration is $\Omega = \{(x,y) \mid -1 \leq y \leq x^2, 0 \leq x \leq 1\}$, and we impose homogeneous boundary conditions $u = 0$ on the entire boundary $\partial\Omega$, as depicted in Fig. 7a. The field $u^* = \sin(\pi(y - x^2))\sin(\pi x)\sin(\pi y)$ obviously fulfills the

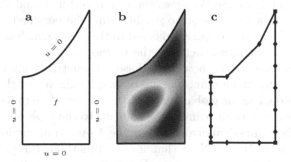

Fig. 7. Wedge-shaped domain. a: Domain and boundary conditions. b: Analytical solution of the boundary value problem. c: Boundary control points.

boundary conditions, and therefore is the unique solution to the BVP corresponding to $f = -\Delta u^*$. This solution is shown in Fig. 7b, and the control points of the boundary are depicted in Fig. 7c.

We solve the parametrization problem for this BVP by optimizing the location of the 12 interior control points, yielding a total of 24 design variables for the optimization. We initialize all methods using the spring model in Section 3.1. Fig. 8 depicts, for each of the six parametrization methods, the optimal control net, the corresponding parametrization, and the numerical error, computed as the difference $|u_h - u^*|$ between the computed solutions u_h and the analytical solution u^*. The depicted error is based on a discretization of the state variable u with $\sim 10^4$

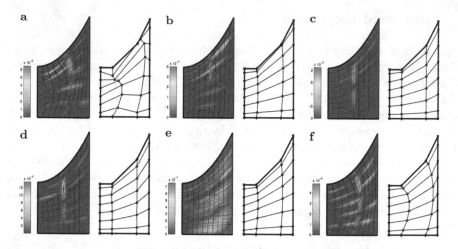

Fig. 8. Wedge-shaped domain: Isoparametric lines and numerical error (left) and control net (right) for the parametrization based on area-orthogonality (a), Liao (b), Winslow (c), strong residual (d), weak residual (e), and enrichment error (f)

degrees-of-freedom, while the optimization for the analysis-oriented methods are performed on a coarser discretization of u with $\sim 10^3$ degrees-of-freedom. We note that the optimal control net and the corresponding parametrizations are quite similar for the Liao, the Winslow, the strong residual, and the weak residual methods, whereas the area-orthogonality, and the enriched error methods differ somewhat. This is also clearly reflected in the error, which is found to vary by several orders of magnitude between the methods.

An interesting question is, how well these parametrizations reproduce the analytical solution when we refine the analysis. The answer to this is shown in Fig. 9. The figure depicts the global numerical error ϵ as a function of the number of basis functions used to approximate the solution to the PDE for each of the six methods. As global numerical error, we use the L_2-norm of the local numerical error: $\epsilon^2 = \int_\Omega |u_h - u^*|^2 \, dA$. Note that for each method, the parametrization is kept fixed during these experiments. For not too coarse discretizations, we see that the error varies by several orders of magnitude between the methods, clearly emphasizing the importance of the way the domain is parametrized. The smallest error is found for the weak residual method, while the highest error is found for the area-orthogonality method. Additionally, for this example the error for the weak residual method converges faster than for the other methods, which have practically identical convergence orders.

We conclude this example by emphasizing that the computational expenses vary significantly between the two classes of methods. The geometrically based methods (area-orthogonality, Liao, and Winslow) converged within ~ 30 optimization iterations, whereas the analysis-oriented methods (strong residual, weak residual, and enrichment error) converged after ~ 300 iterations. Even more importantly, the analysis-oriented methods require solving the PDE in each optimization step, unlike the geometrical methods.

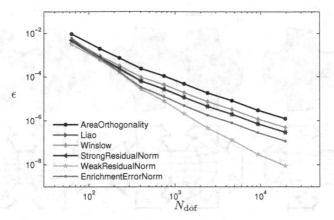

Fig. 9. Wedge-shaped domain: error as a function of number of degrees-of-freedom for the different parametrization methods

5.2 Poisson's Equation on a Jigsaw puzzle

We consider the Poisson problem (1) posed over the jigsaw puzzle piece shown in Fig. 10a. We use the field

$$u_G^* = \sum_{i=1}^{2} \exp\left(-\frac{(x - \tilde{x}_i)^2}{a_i^2} - \frac{(y - \tilde{y}_i)^2}{b_i^2}\right) \tag{38}$$

as boundary condition on $\partial\Omega$ with given parameters $\tilde{\mathbf{x}}, \tilde{\mathbf{y}}, \mathbf{a}, \mathbf{b} \in \mathbb{R}^2$, and with $f = -\Delta u_G^*$, u_G^* is the unique solution to the BVP. The field is depicted in Fig. 10b. The boundary conditions are enforced strongly through the least square fit of the traces in the trial space to the field (38). The boundary control points are shown in Fig. 10c.

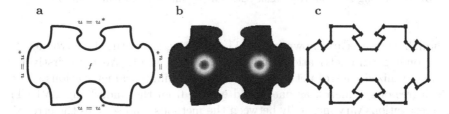

Fig. 10. Jigsaw puzzle. a: Domain and boundary conditions. b: Analytical solution of the boundary value problem. c: Boundary control points.

We solve the parametrization problem using all six nonlinear methods by optimizing the position of the 64 interior control points, giving us a total of 128 design variables. In this example, we initialize the geometric methods from the spring model in Section 3.1, and the analysis-oriented methods from the Winslow

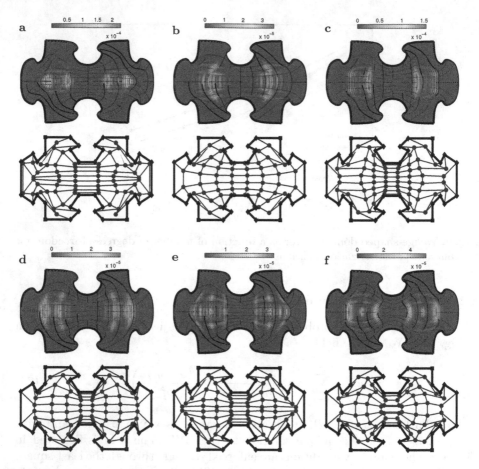

Fig. 11. Jigsaw puzzle piece: Isoparametric lines and numerical error (top) and control net (bottom) for the parametrization based on area-orthogonality (a), Liao (b), Winslow (c), strong residual (d), weak residual (e), and enrichment error (f)

method. The results are shown in Fig. 11, depicting the optimal control net, the corresponding parametrization, and the numerical error. We note firstly that all the optimized control net and their corresponding parametrizations show a high degree of symmetry, as one would expect from the underlying BVP. The parametrizations vary markedly between the methods, and so does the error size and distribution. To examine the numerical error more closely, we compare again the methods in terms of the L_2-norm of the error when the analysis is refined. This is shown in Fig. 12, displaying the global numerical error ϵ as a function of the number of degrees-of-freedom for the analysis for each of the six methods. We note that for sufficiently fine discretizations, the global error convergence order is the same for all methods. The superconvergence of the weak residual method observed in the previous example in Fig. 8 is no longer seen. The difference in the global error varies by approximately one order of magnitude between the

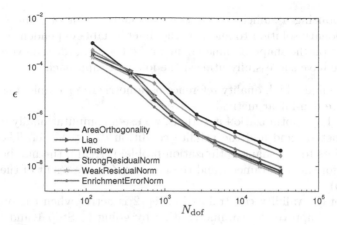

Fig. 12. Jigsaw puzzle: error as a function of number of degrees-of-freedom for the different parametrization methods

methods. Again, the weak residual method yields the lowest error, while the area-orthogonality gives the highest.

In terms of computational expenses, the geometry-oriented methods converged again significantly faster than the analysis-oriented methods. And as each geometric iteration is significantly cheaper than a corresponding analysis-oriented iteration, the computational time is orders of magnitude smaller for the geometric methods than the analysis-oriented ones.

6   Discussion

The solution to the parametrization problem is particularly important in the context of shape optimization, where a parametrization needs to be recomputed repeatedly as the shape of the physical domain is updated by the shape optimization algorithm, cf. [31]. In addition, most realistic industrial problems can only be realized based on multiple patches, and the problems are most often three-dimensional and not planar. In the present section, we further discuss these challenges.

6.1   Shape Optimization

The authors are especially interested in using IGA for shape optimization, which imposes further requirements on the parametrization method. In addition to producing a valid parametrization of high quality, they have to be computationally inexpensive and robust. Last but not the least, they should produce parametrizations, which depend in a differentiable way on the parameters, determining the shape of the domain. For this purpose, the non-linear reparametrization methods are often too expensive and too slow in practice. Furthermore, should the numerical algorithm for solving the optimization problems (22) or (31) stop without producing a sufficiently precise stationary point or "jump" from one locally stationary solution to another, the differentiable dependence of the parametrization on the shape parameters might be lost. In order to overcome these problems we have successfully utilized the following approach:

1. First, we find a high quality reference parametrization, employing a possibly expensive non-linear method.
2. During shape optimization iterations, we use a computationally inexpensive linear method and add the validity condition $d_{i,j} \geq \delta Z$, cf. Theorem 1 as constraints to the shape optimization problem. Again, the number $\delta \in [0, 1]$ is an algorithmic parameter and the number Z is the result of the optimization (20).
3. If any of the validity constraints in Step 2 is active when the optimization stops, we improve the parametrization by going to Step 1 and restart the optimization.

In the papers [22,23,26] this method has been successfully applied to 2D shape optimization problems. The Winslow functional is minimized in Step 1 and the linearized Winslow functional is used in Step 2, except for [22] where quasi conformal deformation was used.

6.2 Multiple Patches

So far we have only considered a single patch, but extending the non-linear methods and their linearizations to several patches is straightforward. We simply let the control points for the inner boundary be variables in the optimization formulations such as (22) and (31). It is interesting to observe how the Winslow functional distributes the angles between patches meeting a common corner, cf. Fig. 13.

 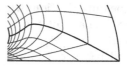

Fig. 13. An inner boundary of a multi-patch configuration. To the left the initial parametrization, to the right the parametrization obtained by minimizing the Winslow functional

6.3 Higher Dimensions

Due to the Radó–Kneser–Choquet theorem the method of minimizing the Winslow functional has a sound mathematical underpinning in dimension two. Unfortunately, there is no version of this theorem in higher dimensions, and there is no unique way to generalize the Winslow functional to higher dimensions either. The analysis-oriented methods, on the other hand, generalize verbatim to higher dimensions, as does the Liao Functional.

7 Conclusion and Outlook

The construction of geometry parametrizations in isogeometric analysis is of vital importance for obtaining reliable and accurate numerical results. In applications of isogeometric analysis to shape optimization, the requirements to computational algorithms for constructing geometry parametrizations increase further, owing to the repeated updates to the geometry made by the shape optimization process. In the present work, we have proposed several methods, both linear and non-linear, for constructing a parametrization, which meet these requirements.

The linear methods are computationally inexpensive, but do not guarantee that the resulting parametrization is injective. We have outlined the spring model, the mean values coordinates, and the quasi conformal deformation methods. Some of these can be used as an initial guess for other methods, and some work well in the vicinity of a known valid parametrization. The injectivity of the parametrization can be guaranteed by controlling the determinant of the Jacobian, which in turn can be controlled by its coefficients in a B-spline expansion.

Two classes of non-linear parametrization methods have been considered, which are based on maximizing a quantitative measure of the quality of the parametrization. One class is based on the geometric quality measures, and uses some of the methods known from mesh generation. Specifically, we have investigated the area-orthogonality, the Liao, and the Winslow functionals. The other class of quality measures is analysis-oriented, and rely on error estimates. Among many estimators available for adaptive meshing, we have tested three, namely the strong residual, the weak residual, and the enrichment error norm. The non-linear methods require more computational effort than the linear ones, in particular the analysis-oriented methods. At the same time they produce valid parametrizations, typically of higher quality.

We ensure the validity of the parametrization by adding the positivity of the determinant of the Jacobian as constraints to the optimization-based parametrization methods. In our computational experience, these constraints are not active at the end of the optimization, i.e., the functional we minimize has a local minimum in the set of valid parametrizations. This is guaranteed in the case of the Winslow functional which has a unique minimum. To safeguard from numerical errors, we keep the positivity of the determinant of the Jacobian as constraints even in this case.

The analysis-oriented methods strive to make the numerical solution of the PDE at hand as accurate as possible with respect to a given error estimator. For the few examples of elliptic boundary value problems we have considered, they seem to work well. However, a word of caution is required. Conceivably, instead of making the approximation error smaller we may expose flaws in the error estimator and end up with a useless parametrization after all, which nevertheless results in a small estimated error.

We are particularly interested in using isogeometric analysis for shape optimization, and that puts conflicting demands on the parametrization algorithm. It has to be fast, differentiable, robust, and reliable. We have solved the problem by using a cheap and fast linear method most of the time, and only use an expensive non-linear method when it is required. We have considered a range of 2D shape optimization problems and we have successfully used the Winslow functional as the non-linear method and the linearized Winslow functional for the linear method

The future work in the field of parametrizations in isogeometric analysis has both practical/experimental and theoretical aspects. First of all, more tests of the proposed optimization methods are needed on more geometries and other equations, including non-elliptic problems, both in 2D and 3D.

For example, isogeometric analysis is known to perform very well for numerically approximating the eigenvalues, providing the small error even for the optical/high frequency part of the spectrum, apart from a few highest frequency modes [3]. In a simple 1D example with a known spectrum, the error can be made small for *all* eigenvalues by adjusting the parametrization of the geometry. As eigenvalue approximation errors have far reaching implications for the numerical accuracy of other problems with the same operator, it would be very interesting to know whether such parametrization adjustments generalize to problems in higher dimensions and can be achieved without knowing the exact spectrum.

Another fundamental issue directly related to geometry parametrization is that of generating the patch layout in case several patches are needed. This can be done "by hand," but automated methods are of course highly desirable.

It would be very interesting to characterize the minima of the analysis-oriented parametrization methods. For example, for which BVPs/error estimators can we guarantee the validity of the resulting parametrization without explicitly enforcing it?

We believe that no universal linear method for generating a geometry parametrization exists. We formulate it as a conjecture, and the proof of this fact would of course be very interesting:

Conjecture 1. Let $F : C^1(\partial I^2, \mathbb{R}^2) \to C^1(I^2, \mathbb{R}^2)$ be an affine map such that $F(\mathbf{Y})|_{\partial I^2} = \mathbf{Y}$ for all $\mathbf{Y} \in C^1(\partial I^2, \mathbb{R}^2)$. Then there is a regular map $\mathbf{Y} \in C^1(\partial I^2, \mathbb{R}^2)$ with a positive Jacobian determinant in the corners such that $F(\mathbf{Y})$ has a negative Jacobian determinant at least at one point.

Another conjecture is related to minimizing the Winslow functional over the finite-dimensional spaces of splines:

Conjecture 2. Let $\mathbf{Y} \in C^1(\partial I^2, \partial \Omega)$ be a valid spline parametrization of the boundary of a domain $\Omega \in \mathbb{R}^2$ with a positive Jacobian determinant in the corners. Then there exists a finite-dimensional spline space $\mathcal{S} \subset C^1(\partial I^2, \mathbb{R}^2)$ and a minimizer $\mathbf{X} \in \mathcal{S}$ of the Winslow functional, such that $\mathbf{X}$ is a valid parametrization of Ω with $\mathbf{X}|_{\partial I^2} = \mathbf{Y}$.

Acknowledgement. The authors would like to thank Thomas A. Hogan, The Boeing Company, USA, for valuable discussions and for suggesting the enrichment error norm as a quality measure.

References

1. Choquet, G.: Sur un type de transformation analytique généralisant la représentation conforme et défini au moyen de fonctions harmoniques. Bull. Sci. Math. 69, 156–165 (1945)
2. Cohen, E., Martin, T., Kirby, R.M., Lyche, T., Riesenfeld, R.F.: Analysis-aware modeling: understanding quality considerations in modeling for isogeometric analysis. Comput. Meth. Appl. Mech. Engrg. 199, 334–356 (2010)
3. Cottrell, J.A., Reali, A., Bazilevs, Y., Hughes, T.J.R.: Isogeometric analysis of structural vibrations. Comput. Meth. Appl. Mech. Engrg. 195, 5257–5296 (2006)
4. Cottrell, J.A., Hughes, T.J.R., Bazilevs, Y.: Isogeometric Analysis: Toward Integration of CAD and FEA. John Wiley & Sons, Chichester (2009)
5. Duren, P., Hengartner, W.: Harmonic Mappings of Multiply Connected Domains. Pac. J. Math. 180, 201–220 (1997)
6. Farin, G., Hansford, D.: Discrete Coons Patches. Comput. Aided Geom. Des. 16, 691–700 (1999)
7. Farrashkhalvat, M., Miles, J.P.: Basic Structured Grid Generation: With an introduction to unstructured grid generation. Butterworth-Heinemann, Burlington (2003)
8. Floater, M.S.: Mean Value Coordinates. Comput. Aided Geom. Des. 20, 19–27 (2003)
9. Floater, M.S., Hormann, K.: Parameterization of Triangulations and Unorganized Points. In: Iske, A., Quak, E., Floater, M.S. (eds.) Tutorials on Multiresolution in Geometric Modelling, pp. 287–315. Springer, Heidelberg (2002)
10. Floater, M.S., Hormann, K.: Surface Parameterization: a Tutorial and Survey. In: Dodgson, N.A., Floater, M.S., Sabin, M.A. (eds.) Advances in Multiresolution for Geometric Modelling, pp. 157–186. Springer, Heidelberg (2005)
11. Gravesen, J., Evgrafov, A., Gersborg, A.R., Nguyen, D.M., Nielsen, P.N.: Isogeometric Analysis and Shape Optimisation. In: Eriksson, A., Tibert, G. (eds.) Proc. of the 23rd Nordic Seminar on Computational Mechanics, pp. 14–17 (2010)
12. Hormann, K., Greiner, G.: MIPS: An efficient global parametrization method. In: Laurent, P.-J., Sablonnire, P., Schumaker, L.L. (eds.) Curve and Surface Design: Saint-Malo 1999. Innovations in Applied Mathematics, pp. 153–162. Vanderbilt University Press, Nashville (2000)
13. Hormann, K., Floater, M.S.: Mean Value Coordinates for Arbitrary Planar Polygons. ACM Transactions on Graphics 25, 1424–1441 (2006)
14. Hughes, T.J.R., Cottrell, J.A., Bazilevs, Y.: Isogeometric Analysis: CAD, Finite Elements, NURBS, Exact Geometry and Mesh Refinement. Comput. Meth. Appl. Mech. Engrg. 194, 4135–4195 (2005)
15. Khattri, S.K.: Grid Generation and Adaptation by Functionals. Comput. Appl. Math. 26, 235–249 (2007)
16. Kneser, H.: Lösung der Aufgabe 41, Jahresber. Deutsch. Math.-Verein. 35, 123–124 (1926)
17. Knupp, P.M.: Algebraic Mesh Quality Metrics. SIAM J. Sci. Comput. 23, 193–218 (2001)
18. Liao, G.: Variational Approach to Grid Generation. Num. Meth. Part. Diff. Eq. 8, 143–147 (1992)
19. MATLAB. Version 7.14.0.739 (R2012a) The MathWorks Inc. Natick, Massachusetts (2012)

20. Martin, T., Cohen, E., Kirby, R.M.: Volumetric Parameterization and Trivariate B-Spline Fitting Using Harmonic Functions. Comput. Aided Geom. Des. 26, 648–664 (2009)
21. Nguyen, D.M., Nielsen, P.N., Evgrafov, A., Gersborg, A.R., Gravesen, J.: Parametrisation in Iso Geometric Analysis: A first report, DTU (2009), http://orbit.dtu.dk/services/downloadRegister/4040813/first-report.pdf
22. Nguyen, D.M., Evgrafov, A., Gersborg, A.R., Gravesen, J.: Isogeometric Shape Optimization of Vibrating Membranes. Comput. Meth. Appl. Mech. Engrg. 200, 1343–1353 (2011)
23. Nguyen, D.M., Evgrafov, A., Gravesen, J.: Isogeometric Shape Optimization for Electromagnetic scattering problems. Prog. in Electromagn. Res. B 45, 117–146 (2012)
24. Nguyen, T., Jüttler, B.: Parameterization of Contractible Domains Using Sequences of Harmonic Maps. In: Boissonnat, J.-D., Chenin, P., Cohen, A., Gout, C., Lyche, T., Mazure, M.-L., Schumaker, L. (eds.) Curves and Surfaces 2011. LNCS, vol. 6920, pp. 501–514. Springer, Heidelberg (2012)
25. Nguyen, T., Mourain, B., Galigo, A., Xu, G.: A Construction of Injective Parameterizations of Domains for Isogeometric Applications. In: Proc. of the 2011 International Workshop on Symbolic-Numeric Computation, pp. 149–150. ACM, New York (2012)
26. Nørtoft, P., Gravesen, J.: Isogeometric Shape Optimization in Fluid Mechanics, Struct. Multidiscip. Opt., doi:10.1007/s00158-013-0931-8
27. Octave community: GNU/Octave (2012), http://www.gnu.org/software/octave
28. Radó, T.: Aufgabe 41. Jahresber. Deutsch. Math.-Verein. 35, 49 (1926)
29. Sheffer, A., Praun, E., Rose, K.: Mesh Parameterization Methods and their Applications. Foundations and Trends in Computer Graphics and Vision 2, 105–171 (2006)
30. Wächter, A., Biegler, L.T.: On the Implementation of a Primal-Dual Interior Point Filter Line Search Algorithm for Large-Scale Nonlinear Programming. Math. Program. 106, 25–57 (2006)
31. Wall, W.A., Frenzel, M.A., Cyron, C.: Isogeometric Structural Shape Optimization. Comput. Meth. Appl. Mech. Engrg. 197, 2976–2988 (2008)
32. Weber, O., Ben-Chen, M., Gotsman, C., Hormann, K.: A complex view of barycentric mappings. Computer Graphics Forum 30, 1533–1542 (2011); Proc. of SGP
33. Winslow, A.: Numerical Solution of the Quasilinear Poisson Equation in a Nonuniform Triangle Mesh. J. Comput. Phys. 2, 149–172 (1967)
34. Xu, G., Mourrain, B., Duvigneau, R., Galligo, A.: Optimal Analysis-Aware Parameterization of Computational Domain in Isogeometric Analysis. In: Mourrain, B., Schaefer, S., Xu, G. (eds.) GMP 2010. LNCS, vol. 6130, pp. 236–254. Springer, Heidelberg (2010)
35. Xu, G., Mourrain, B., Duvigneau, R., Galligo, A.: Parameterization of Computational Domain in Isogeometric Analysis: Methods and Comparison. Comput. Meth. Appl. Mech. Engrg. 200, 2021–2031 (2011)

Realistic Plant Modeling from Images Based on Analysis-by-Synthesis

Jérôme Guénard[1], Géraldine Morin[1],
Frédéric Boudon[2], and Vincent Charvillat[1]

[1] IRIT - VORTEX - University of Toulouse
[2] INRIA - Cirad - Montpellier

Abstract. Plants are essential elements of virtual worlds to get pleasant and realistic 3D environments. Even if mature computer vision techniques allow the reconstruction of challenging 3D objects from images, due to high complexity of plant topology, dedicated methods for generating 3D plant models must be devised. We propose an analysis-by-synthesis method which generates 3D models of a plant from both images and a priori knowledge of the plant species.

Our method is based on a skeletonisation algorithm which allows to generate a possible skeleton from a foliage segmentation. Then, we build a 3D generative model, based on a parametric model of branching systems that takes into account botanical knowledge. This method extends previous works by constraining the resulting skeleton to follow a natural branching structure. A first instance of a 3D model is generated. A reprojection of this model is compared with the original image. Then, we show that selecting the model from multiple proposals for the main branching structure of the plant and for the foliage improves the quality of the generated 3D model. Varying parameter values of the generative model, we produce a series of candidate models. A criterion based on comparing 3D virtual plant reprojection with the original image selects the best model. Finally, results on different species of plants illustrate the performance of the proposed method.

1 Introduction

Procedural methods to generate plant models can build a complex plant architecture from few simple rules [1]. In his pioneering work [2], Lindenmayer proposes the formalism of L-systems as a general framework. By carefully parameterising these rules, it is possible to achieve a large variety of realistic plant shapes [3,4]. However, a strict recursive application of rules leads to self-similar structures and thus, to enhance realism, irregularities may be generated through probabilistic approaches [5,1]. Adjusting stochastic parameters to achieve realistic models requires intensive botanical knowledge [6]. Another approach consists in modeling plant irregularities as a result of the competition for space between the different organs of the plants [7]. In this case, the volume of a plant is specified by the user and a generative process grows a branching structure with branches competing between each other. Competition can be biased to favor certain types of

M. Floater et al. (Eds.): MMCS 2012, LNCS 8177, pp. 213–229, 2014.

structures. However, automatic control of competition parameters to achieve a given shape is still complicated.

All these first works are derived from computer graphics community. Other approaches use information provided by images to increase the degree of realism. A couple of research directions should be investigated. Clearly, a plant should follow the biological property of its species and also ressemble a picture of an existing instance. That is typically the subject of our work. Our idea is not to exactly reconstruct the plant from an image, including its hidden parts (which seems impracticable) but rather to drive the instantiation of the plant 3D model by minimising the difference between its reprojection and the original plant in the image.

Fig. 1. On the left, an original image of a vine plant before and after a metric rectification. In the middle, a possible architecture of the branching extracted with our skeletonisation method. At the right, a corresponding 3D model of the plant.

Unlike existing methods detailed in section 2, ours must be able to get a 3D model of a plant without any human interaction from images with possibly no visible branches. By integrating biological knowledge of the plant species, we propose a simple fully-automatic process to extract the structure of a plant from the shape of its foliage. The picture can be taken in arbitrary conditions and may be of poor quality (for example, in the vine case, the image is degraded after a metric rectification due to the assumption that all the principal branches are in a plane as shown at the left on the Fig. 1). We start by presenting a new skeletonisation algorithm in section 3 in the vine case (as we can see in the middle of the Fig. 1) and we explain a possible extension to our skeletonisation method for other kinds of plants with 3D branching architecture in section 3.5. Then a 3D model is generated thanks to our 3D generative model (section 4). Finally, an analysis-by-synthesis scheme allows to improve this reconstruction insuring that the foliage model reprojection matches closely the original foliage like explained in section 5 (right on the Fig. 1). The last section shows results and validation comparing with data provided by experts.

2   State of the Art: Generating Plants from Images

Realistic plants are challenging objects to model and recent advances in automatic modeling can be explained by the convergence of computer graphics and computer vision [8]. We start this state of the art with the first method of plant modeling from images. Then, we continue with the ones starting by reconstructing clouds of 3D points. After, we talk about other methods using several images to finish with approaches using a single image as ours.

A pioneering work on the reconstruction of trees from images was made by Shlyakhter et al. [9] who reconstruct the visual hull of the tree from silhouettes deduced from the images. A skeleton is computed from the hull using a Medial Axis Transform (MAT) and is used as main branches. Branchlets and leaves are then generated with an L-system. The skeleton determined from the MAT does not necessarily look like a realistic branching system. Also, the density of the original tree is not taken into account.

Quan et al. [10,11] and Tan et al. [12] also use multiple images to reconstruct a 3D model of trees or plants. In order to avoid searching features correspondances in different images, they use views close to each other (more than 20 images for any plants). Thus, they obtain a quasi-dense cloud of points by structure from motion. For simple plants, a parametric model is first fitted on each set of points representing a leaf. They then generate branches based on information given by the user. For trees, they start by reconstructing visible branches to create branch pattern that they combine in a fractal way until reaching leaves. Reche-Martinez et al. [13] propose another reconstruction from multiple images, based on billboards. Neubert et al. [14] construct a volume encompassing the plant in the form of voxels using image processing techniques and fill it with particles. Particles paths toward the ground and a user given general skeleton define the branching system.

Wang et al. [15] model different species of trees using images of tree samples from the real world which are analysed to extract similar elements. A stochastic model to assemble these elements is also derived and parameterised from the image. The resulting model can generate many similar trees. The goal in this case is not necessarily to reconstruct a specific tree instance corresponding to an image. Similarly, Li et al. [16] propose a probabilistic approach to reconstruct a tree parameterized from videos. For these methods, the only source of information is the given images leading to template branching patterns. If the set of patterns is rich enough, it will produce aesthetically pleasing results, but without guarantee to be representative of its species. Additionally, a user must specify a draft of the structure on the image to avoid segmentation. Talton et al., in [17], propose to fit a grammar-based procedural methods using Markov Chain Monte Carlo technique to model objects from a 2D or 3D binary shape. Their results are aesthetically very convincing but their optimisation routine requires long computation time.

Other approaches explore the use of a single image [18,19]. In [18], from a manual segmentation of the plant in the image, they extract a skeleton. A 3D representation of the branches is derived from visible parts, then the leaves are added. Here, an user sketching step is required. In [19], a graph topology is first extracted from a single image of a branching system (a tree without foliage). Then the 3D tree model is reconstructed by rotating the branches.

In general, methods of the literature, such as [12] and [18] require visible branches to learn about the structure of the skeleton. In our case, branches are directly derived from the foliage structure. Fig. 16 (top row) shows branching structures devised manually by experts from image: we see that the branches are deduced on one hand from the knowledge of a space filled by a branch and its attached leaves and on the other hand from the silhouette of the foliage. We propose a generalised recursive skeletonisation algorithm together with an analysis-by-synthesis mechanism to determine the branches and their attached foliage that is the 3D model. Our approach is fully-automatic, that is, does not require any user interaction.

3   Skeletonisation

3.1   General Field Skeletonisation Method

Skeletonisation is a classical topic in image processing. We have followed and completed the analysis of different approaches as proposed in [20] and as illustrated in Tab. 1. This table summarises the different properties that skeletons respect like the thinness or the robustness and compares them to our requirements shown in the left column. All these properties are detailed in [20].

'Homotopic' and 'connected' are topology preserving properties. As the foliage may have holes, we prefer not to be attached to its topology. For the same reason, reconstruction is not very relevant. On the opposite robust is an important property, since the foliage may not be stable (for example, the foliage may change in case of wind). Being reliable means that any point in the foliage should be visible from a point on the skeleton. For our setting, it is not necessary to request a branch in each bump of silhouette. The transformation invariance property (in particular affine invariance) is not important in our setting as we do depend on orientation. For example, vine branches grow vertically. In our case, the smooth property is important to get realistic branches but we do not need a centred skeleton. Finally, we shall see that for plants more complex than vines, the hierarchic nature of the skeleton is used (section 3.5). For these reasons, we choose to adapt the general field method, and in particular the work of Cornea [21], since their approach fits the best our needs.

Cornea et $al.$ original method [21] consists in computing the skeleton (Fig. 2 (c)) from a vector field (Fig. 2 (b)). For each interior pixel $\mathbf{p_i}$ of the binary shape $\mathcal{B}$, a force vector $\vec{\mathbf{f_i}}$ is computed as a weighted average of unit vectors to the boundary pixels:

$$\vec{\mathbf{f_i}} = \sum_{\mathbf{m_j} \in \Omega} \frac{1}{||\overrightarrow{\mathbf{m_j p_i}}||^2} \frac{\overrightarrow{\mathbf{m_j p_i}}}{||\overrightarrow{\mathbf{m_j p_i}}||}$$

where Ω contains the contour pixels $\mathbf{m_j}$ of $\mathcal{B}$ (Fig. 2 (a)). Then, points where the magnitude of the force vector vanishes, so-called $critical$ $points$ (Fig. 2 (b)), are connected by following the force direction pixel by pixel. The results of this method can be seen in Fig. 2. The problem here is that this method is not robust

Table 1. Summary of properties achievable by different skeletonisation methods. In green, the characteristics compliant with our needs.

	Medial Axis	Thinning		Distance Field	Geometric		General Field		Our needs
G: general		G	[22]	G	G	[23]	G	[21]	
Homotopic	Y	Y	Y		Y	Y	N	N	N
Connected	Y	Y	Y				Y	N	Y
Reconstruction	Y	N	N		N	N	N	N	N
Robust	N	N	N	N	N	Y	Y	Y	Y
Reliable	Y		Y				N		N	N
Transf. Invariance	Y		N	Y		Y	Y	Y	N
Centred	Y	Y			Y		Y		N
Smooth	N		N			N	Y	Y	Y
Hierarchic	Y	N	N			Y	Y	Y	Y

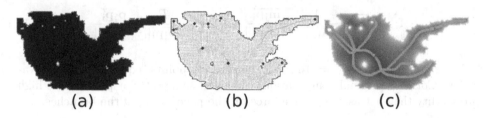

Fig. 2. Cornea *et al.* original method. (a) shape $\mathcal{B}$ with contours pixels $\in \Omega$ represented in red. (b) vector field with critical points in blue. (c) extracted skeleton in green.

to the holes in the binary shape. Furthermore, sometimes only one branch grows when two or more are required.

3.2 A New Computation of the Vector Field

By redefining the set of contour points, we manage to use Cornea's vector field method to get a realistic skeleton in 2D.

Based on botanical expertise, we assume that different branches of relatively similar size coexist and share the space of the plant crown. A large convex silhouette is usually explained by more than one branch. For the skeleton to reflect this hierarchy of branches, we propose a strategy to partition the foliage surface into subsets by positioning artificial *contour points* in the shape. Fig. 3 shows that by adding *contour points* within the shape, we define an appropriate branch set.

We compute a probability map $\mathcal{P}$ on $\mathcal{B}$ containing, for each interior point $\mathbf{p_i}$, the probability to be considered as a *contour point*. The new force vector $\overrightarrow{\mathbf{f_i}}$

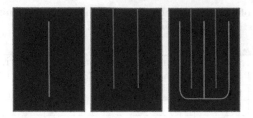

Fig. 3. At the left, the skeleton (in green) extracted with Cornea's original method. A spatial partition can be generated (red lines) to constraint the skeleton to have more branches. At the right, the skeleton computed by Cornea's method when all red points are considered as contour points.

does not depend only on the points $\mathbf{m_j} \in \Omega$ but on all the points of $\mathcal{B}$. The new formula to compute the vector field is:

$$\vec{\mathbf{f}_i} = \sum_{\mathbf{m_j} \in \Omega} \frac{1}{||\overrightarrow{\mathbf{m_j p_i}}||^2} \frac{\overrightarrow{\mathbf{m_j p_i}}}{||\overrightarrow{\mathbf{m_j p_i}}||} + \sum_{\substack{\mathbf{p_j} \in \mathcal{B} \setminus \Omega \\ j \neq i}} \frac{\mathcal{P}_j}{||\overrightarrow{\mathbf{p_j p_i}}||^2} \frac{\overrightarrow{\mathbf{p_j p_i}}}{||\overrightarrow{\mathbf{p_j p_i}}||} \tag{1}$$

We can see that if $\mathcal{P}_j = 0$ for all the interior points, equation 1 is equivalent to Cornea's original computation whereas if some interior points have high probability they act as a repulsive force on the positioning of the branches.

3.3   Definition of the Probability Map

We assume here that n the number of branches is given. We compute the *probability map* $\mathcal{P}$ with an iterative algorithm. The first step is the choice of cuts in $\mathcal{B}$. The cuts are segments with one *starting point* and one *ending point* and represent the possible positions of the separations between the n branches in the shape. Assuming that the shoots grow vertically from the cane, we propose to place trivially the *ending points* $\mathbf{e_i}, i = 1..n - 1$ of the cuts uniformly in the bottom of $\mathcal{B}$ (Fig. 4). Then, the *starting points* are computed one by one. To do that, we compute the DCE (*Discrete Curve Evolution*) of Ω as in [24]. It provides a simplified polygonal boundary composed of N vertices $(\mathbf{s_l})_{,l=1..N}$ like shown in Fig. 5. Usually, we choose $N = 2n$. An angle α_l can be associated with each vertex, representing clockwise angle between the 2 segments around the vertex. A set of points $(\mathbf{c_k})_{,k=1..K}$ uniformly discretises the polygon.

Then a new probability ρ_k to be a *starting point* is computed for each point $\mathbf{c_k}$ taking into account two values:

− the proximity to an inward angle:

$$\rho_k^1 \sim \sum_{l=1}^{N} \frac{1}{d(\mathbf{c_k}, \mathbf{s_l})} \left(1 - \frac{\alpha_l}{2\pi}\right)$$

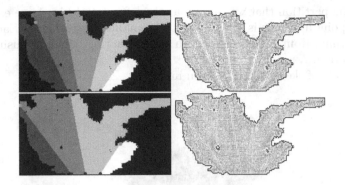

Fig. 4. At the left, examples of cuts with $n = 4$ and $n = 5$ branches. Ending points are represented in blue and starting points in green. At the right new vector fields.

Fig. 5. DCE algorithm examples. At the left, the original image with the contour points around the foliage. The two other images show the DCE algorithm with a 24-point polygon in the middle and a 8-point polygon on the right. The inward angles are represented in blue.

– the distance along a boundary to the set $\mathcal{H}$ of already chosen *starting points*:

$$\rho_k^2 \sim \min_{\mathbf{c} \in \mathcal{H}} d(\mathbf{c_k}, \mathbf{c}).$$

The mix probability ρ_k is proportional to $\phi(\rho_k^1, 1, \sigma) + \phi(\rho_k^2, 1, \sigma)$ where $\phi(., 1, \sigma)$ represents the gaussian function with a mean equals to 1 and a standard deviation equals to σ (here, $\sigma = 0.4$).

A *starting point* $\{\mathbf{c_k}\}$ is selected according to the probability ρ_k. Then it is associated with an *ending point* $\mathbf{e_i}$ and accepted if:

$$\left| \frac{\#pixels \in \mathcal{B} \ on \ the \ left \ of \ (\mathbf{e_i c_k})}{\#pixels \in \mathcal{B}} - \frac{i}{n} \right| \leq \tau \qquad (2)$$

where τ is a parameter allowing the created partitions of the binary shape to have the same size or not. In the vines case, $\tau = 15\%$.

Finally, when all the cuts have been accepted, the new vector field is computed like shown at the right part of Fig. 4.

3.4   Adjusting First Order Branches

We now have a vector field coherent with the n branches assumption. We want to extract branches from this vector field. For each row i of the image and each area p of the partition that we can see at the left part of Fig. 4, we extract the attracting point $\mathbf{a}_i^P$ which is the one with the smallest vector norm. Each branch $\mathbf{b_p}$ is a Catmull-Rom curve adjusted on the attracting points $\mathbf{a}_i^P$, using a least square criterion.

An example of skeleton can be seen in Fig. 6.

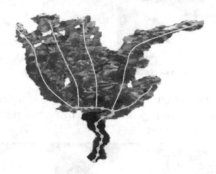

Fig. 6. A final skeleton with 5 branches

The following algorithm summarises the proposed approach:

ALGORITHM:
input: binary shape $\mathcal{B}$ and number of branches n.
output: a skeleton model of the plant consisting in n branches.

---

* Compute the DCE of $\mathcal{B}$ with $N = 2n$ vertices and extract candidate $\{\mathbf{c_k}\}$ uniformly along the polygon.
* Initialise $\mathcal{H} = \emptyset$.
* Place the *ending points* $\{\mathbf{e_i}, \mathbf{i} = \mathbf{1..n-1}\}$ in the bottom of $\mathcal{B}$.
* For i from 1 to $n - 1$
  · Compute ρ_k for each $(\mathbf{c_k})$.
  · while (!equation (2))
    ○ Randomly select a *starting point* $\mathbf{c}$ among $\{\mathbf{c_k}\}\backslash\mathcal{H}$ in function of ρ_k and an *ending point* among $\{\mathbf{e_i}\}$.
  · $\mathcal{H} = \mathcal{H} \cup \mathbf{c}$.
* Compute the probability map $\mathcal{P}$.
* Compute the vector field using equation (1).
* Extract the points $\mathbf{a}_i^p$ for each row i of each area p of the partition.
* Adjust a branch in each area p of the partition.
* For each node of each branch, extract a foliage width.

3.5 Higher Order Branches and Depth Information

In the case of the vines, all the main branches are assumed in a same plane and there are only first order branches. We adapt this planar setting to a more recursive structure and to a 3D setting where rotational symmetry is assumed, like in typical monopodial plants (i.e. plants organised around a main trunk).

Iterative Skeletonisation Algorithm
An automatic colour based segmentation computes a 2D binary shape from the foliage. Then, the branches are extracted using a modified version of our algorithm presented above. The *ending points* are placed on the vertical line passing through the trunk of the *monopodial tree* and the condition (2) is replaced by a condition checking that the angle between the cut and the trunk is coherent (i.e. around $\frac{\pi}{2}$ in the bottom of the tree, $\frac{\pi}{6}$ in the top and with an angle computed linearly between these two values for an intermediate cut). This algorithm is applied recursively to get second order branches for each partition. We can see an example of cuts with a *Liquidambar tree* in Fig. 7.

Fig. 7. An example of skeleton for the Liquidambar tree

Depth of Branches
To generate 3D information, we drew inspiration from Zeng *et al.* [19] and Okabe *et al.* [25]. The goal is to deduce depth information for the branches in the 2D skeleton to make a realistic plant from other views, preserving the appearance from the original viewpoint as it is shown in Fig. 8. First, we compute the convex hull of our 2D skeleton. Then, revolving this convex hull around the line passing through the trunk, we obtain a encompassing volume of the plant. Considering an orthographic projection onto the ground, for each branch which does not touch the 2D convex hull, we change depth information for that the end of this branch touches the boundary of the bounding volume. We have two possibilities, at the front or at the back. We choose the one which maximises the angles between the projections of all the branches to the ground, adding the branch one by one.

4   3D Generative Model

Now we have a possible structure of the plant, the next step is to generate a 3D model of this plant. To do that, we need to build a 3D generative model thanks to all the *a priori* knowledge of the plant.

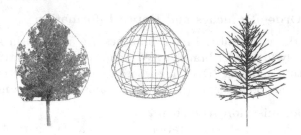

Fig. 8. At the left, we can see the 2D convex hull of the foliage, in the middle, the-bounding volume and at the right, the final 3D skeleton of a Liquidambar tree

In our work, we combine procedural methods to generate a plant and image based approaches. Procedural methods makes it possible to take into account botanical constraints such as possible regular arrangements of organs (for instance leaves). The procedural model uses stochastic parameters in the positionning of the branchlets and the leaves.

We choose to generate the constrained model with L-systems, using the L-Py modeller [26]. An L-system [1] is a formal grammar, most commonly used to model the growth processes of plant development. The main idea of L-systems is to rewrite a string of modules representing the structure of the plant. Rewriting rules express the creation and change of state of the various modules of the plant over time. Our model include a deterministic part which is controlled by the understanding of plant images and a stochastic part to allow a more realistic result. The model is deduced by both learning from a large number of plants and also knowledge given by specialists.

We choose to generate our model in two stages: the branching system model and the foliage.

Branching System Model

Each branch is set by a number of 3D nodes which are the control nodes of the branch. From each of these nodes one or more lateral branches of the same nature may grow. Then, to model the 3D structure, each branch is a generalised cylinder along a curve passing through all nodes. A B-Spline curve is built with a local interpolation scheme of degree 3 [27]. A radius is assigned to each of these nodes to determine the radius of the generalised cylinder in these nodes. This radius is linearly interpolated between two nodes. Textures taken from real images are then applied to branches.

Foliage Model

We begin by extracting leaf textures from real images. At each node defining the branches structure is also assigned a value R which is the radius of the cylinder encompassing the leaves. Thus, to model the foliage, *branchlets* are generated randomly along the branches. Stems are placed along main branches and *branchlets*. Their density and their length is a random variable distributed normally with mean R and standard deviation $\frac{R}{4}$. On each of these stems a leaf modeled by a Bezier surface is placed on which a randomly chosen texture is mapped (Fig. 9b.).

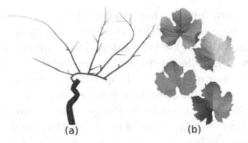

Fig. 9. Example with the case of vine. (a) Branch structure modelisation. (b) Textures of vine leaves.

5   Reprojection Criterion

The last step goals to evaluate the quality of the 3D reconstructed 3D model. We reproject the 3D model in an image with the same viewpoint of the original image. We obtain a binary shape $\mathcal{I}_i$ (1 if foliage, 0 elsewhere). In the same way, the original segmented foliage forms a second binary image $\mathcal{B}$. The error reprojection is computed as:

$$error_i = \frac{\#((\mathcal{I}_i - \mathcal{B})^2 == 1)}{\#pixels(\mathcal{B})} \tag{3}$$

Fig. 10 illustrates the comparison between the projections and the original image.

Fig. 10. At the left, the original image. At the right, the reprojected model. In the middle, the projection errors map. White pixels correspond to pixels where the original image and the reprojected one are superposed and red pixels are *wrong* pixels.

The idea now is to improve the proposed 3D model using an analysis-by-synthesis strategy which allows to merge information from the *a priori* botanical knowledge and from the image. An increasing number of authors propose to use *external* knowledge for easing reconstruction from images. Indeed providing knowledge about the scene to be reconstructed simplifies the image processing steps. In [28], Tu *et al.* define generative models for faces, text, and generic regions which are activated by bottom-up proposals learnt using probabilistic methods. These proposals are then accepted or rejected using a stochastic criterion. Yuille *et al.* [29] claim that this approach, which allows to deal with the complexity of natural images, has intriguing similarities to the brain. They present a method where low level features are used to make bottom-up proposals, finally validated by high-level models. In a similar analysis-by-synthesis method where *a priori* knowledge consists in geometric and mechanical properties, Gupta *et al.* [30] iteratively make proposals for interpreting parts (blocks) of the image. We use a similar approach, but not iterative.

In our case, we give more freedom to the generative model. For example, we do not impose the number of branches of the plant that we do not *a priori* know. Thanks to all the knowledge of the plant, we can model each parameter (like the position of the cuts, the leaves densities or the number of branches) by a random variable, and thus generate numerous models.

We select the best candidate proposed by the generative model using the following formula:

$$\mathcal{M}_{i_0} = \operatorname*{argmax}_{\mathcal{M}_i} p(\mathcal{M}_i | \mathcal{I}_i) = \operatorname*{argmax}_{\mathcal{M}_i} p(\mathcal{M}_i) p(\mathcal{I}_i | \mathcal{M}_i) \tag{4}$$

We choose $p(\mathcal{I}_i | \mathcal{M}_i) = 1 - error_i = 1 - \frac{\#((\mathcal{I}_i - \mathcal{B})^2 == 1)}{\#pixels(\mathcal{B})}$ and $p(\mathcal{M}_i)$ is a product of terms which are probabilities function of all the knowledge of the plant. For the vine case example, one of the term of $p(\mathcal{M}_i)$ is a gaussian representing the probability of the number of shoots. Fig. 11 shows different error maps. White pixels correspond to pixels where the reprojected model and the original image are superposed.

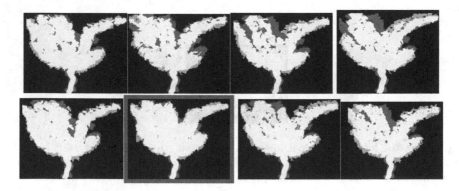

Fig. 11. Different errors maps with different numbers of branches, different distributions of leaves and different densities. The map outlined in red is the error map of the selected 3D model because there is the smallest percentage of gray pixels.

6   Results and Validation

Our method has been tested on a large number of images and videos. Some results are shown in Fig. 15, 14 and 12. To validate our model, we use the error criterion of equation (3). The average error for the case of vines is 6.9%, 7.2% for

Fig. 12. Vine plants modelisation. At the top the original images. At the bottom, rendering of automatically generated vine models using our approach.

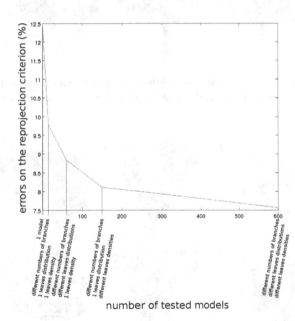

Fig. 13. Reprojection errors according to the number of tested models. Different variations are tested as the number of branches or the leaves density. For example, if we test different 3D models with only one leaves distribution and 1 leaves density, the error is almost 10%. Then, this error decreases when we add different leaves distributions or different leaves density.

Fig. 14. Walnut. At the left, the original image. At the right, the 3D model with the same viewpoint.

Fig. 15. Liquidambar. At the left, the original image. In the middle, the 3D model with the same viewpoint. At the right, the 3D model with an other viewpoint.

Fig. 16. Expert skeletons. At the top, a viticulture expert has drawn skeletons on vine images (in red). At the bottom, the projections of the skeletons of our method (in yellow).

Table 2. The first row represents the number of the vine image. The second and the third rows represent the numbers of shoots drawn by the experts. The last row represent the number of shoots of the 3D models generated with our method from these images.

Image	1	2	3	4	5	6	7	8	9
First expert estimation	3	6	5	5	6	5	3	7	7
Second expert estimation	2	6	4	5	4	4	2	5	4
Our method	2	5	4	5	6	6	3	5	6

the Walnut and 8.5% for the Liquidambar. In Fig. 13, we show the reprojection error according to the number of tested models for an example of vine (the third example of Fig. 12). It is interesting to note that the greater is the number of tested models, the lower is the reprojection error. The curve decreases very quickly between 1 and 15, fairly quickly until 100. This proves the effectiveness of our skeletonisation method which restricts significantly the search space. This curve illustrates the importance to test several models but of course, the final selected model is not necessarily the last one.

A second validation is to compare our solution to the one provided by viticulture experts (Fig. 16). It seems difficult to find a significant measure by comparing the ground truth to our skeletons. Furthermore, for us, the most important is the final appearance with leaves. Indeed, for the first example in Fig. 16, our algorithm found a very similar skeleton to the expert one at the left but it is not the one which has been validated by our method at the right. So, we used our algorithm on the drawn ground truth skeletons with different leaves distributions and different leaves densities to find the best 3D model. The improvement of the reprojection criterion in comparison to our automatically generated skeleton is only 0.2% in average. This small difference proves the performance of our method which does not require human intervention.

In Tab. 2, we can see the number of shoots drawn by two differents experts from vines images. The last row shows the number of shoots of the 3D models generated with our method from the same images. We almost reconstruct a 3D model with the same number of shoots drawn by one of the two experts.

For the 3D case, we have shown that our method can reconstruct realistic 3D trees from a single image. However, the branching system is mostly based on branches.

7 Conclusion

Combining analysis and synthesis, we have proposed a new fully-automatic method of plant modeling from a low resolution image without any branching pattern unlike [12,19].

The leading contribution of this paper is a new skeletonisation algorithm able to extract the structure of a plant from an image of its foliage. Then, we built 3D parametric generative models for different plants using the knowledge about the species. A final analysis-by-synthesis step improves the quality of the final 3D model by comparing the original image with a large number of 3D models generating varying different parameters of the 3D generative model. The skeletons are used to make proposals to the 3D generative model. The reprojection criterion insures the similarity between the proposed 3D model and the original image. We further validated our proposed model by comparing it to ground truth given by experts in the case of vine.

In future work, we are first interested in extending our setting to non monopodial plant. The current algorithm can be applied to non monopodial plants, like the Walnut, but the result is still not very satisfying. Moreover, we could investigate the automatisation of the use of *a priori*. Indeed, the construction of the generative model could be done by learning from a large data set avoiding the necessity prior knowledge on the branching structure of the plant species. The process would be evolving in loop, where the analysis could give feedback to the generative models.

References

1. Prusinkiewicz, P., Lindenmayer, A.: The algorithmic beauty of plants. Springer (1990)
2. Lindenmayer, A.: Mathematical models for cellular interaction in development: Parts i and ii. Journal of Theoretical Biology 18 (1968)
3. Weber, J., Penn, J.: Creation and rendering of realistic trees. In: Proceedings of the 22nd Annual Conference on Computer Graphics and Interactive Techniques, SIGGRAPH 1995, pp. 119–128. ACM, New York (1995)
4. Deussen, O., Lintermann, B.: Digital Design of Nature: Computer Generated Plants and Organics. Springer (2005)
5. de Reffye, P., Edelin, C., Françon, J., Jaeger, M., Puech, C.: Plant models faithful to botanical structure and development. In: Proceedings of the 15th Annual Conference on Computer Graphics and Interactive Techniques, SIGGRAPH 1988, pp. 151–158. ACM, New York (1988)
6. Markov and semi-markov switching linear mixed models used to identify forest tree growth components. Biometrics (2009)
7. Palubicki, W., Horel, K., Longay, S., Runions, A., Lane, B., Měch, R., Prusinkiewicz, P.: Self-organizing tree models for image synthesis. SIGGRAPH, 1–10 (2009)
8. Quan, L.: Image-based Plant Modeling. Springer (2010)
9. Shlyakhter, I., Rozenoer, M., Dorsey, J., Teller, S.: Reconstructing 3d tree models from instrumented photographs. IEEE Comput. Graph. Appl., 53–61 (2001)
10. Quan, L., Tan, P., Zeng, G., Yuan, L., Wang, J., Kang, S.B.: Image-based plant modeling. ACM TOG, 599–604 (2006)
11. Quan, L., Wang, J., Tan, P., Yuan, L.: Image-based modeling by joint segmentation. IJCV, 135–150 (2007)
12. Tan, P., Zeng, G., Wang, J., Kang, S.B., Quan, L.: Image-based tree modeling. ACM TOG 87 (2007)
13. Reche-Martinez, A., Martin, I., Drettakis, G.: Volumetric reconstruction and interactive rendering of trees from photographs. ACM TOG, 720–727 (2004)
14. Neubert, B., Franken, T., Deussen, O.: Approximate image-based tree-modeling using particle flows. ACM TOG (Proc. of SIGGRAPH) (2007)

15. Wang, R., Hua, W., Dong, Z., Peng, Q., Bao, H.: Synthesizing trees by plantons. Vis. Comput. 22(4), 238–248 (2006)
16. Li, C., Deussen, O., Song, Y.Z., Willis, P., Hall, P.: Modeling and generating moving trees from video. ACM Trans. Graph. 30(6), 127:1–127:12 (2011)
17. Talton, J.O., Lou, Y., Lesser, S., Duke, J., Měch, R., Koltun, V.: Metropolis procedural modeling. ACM Trans. Graph. 30(2), 11:1–11:14 (2011)
18. Tan, P., Fang, T., Xiao, J., Zhao, P., Quan, L.: Single image tree modeling. ACM SIGGRAPH, 1–7 (2008)
19. Zeng, J., Zhang, Y., Zhan, S.: 3d tree models reconstruction from a single image. In: ISDA, pp. 445–450 (2006)
20. Cornea, N.D., Silver, D., Min, P.: Curve-skeleton properties, applications, and algorithms. TVCG, 530–548 (May 2007)
21. Cornea, N.D., Silver, D., Yuan, X., Balasubramanian, R.: Computing hierarchical curve-skeletons of 3d objects. The Visual Computer (2005)
22. Yang, X., Bai, X., Yang, X., Liu, W.: An efficient quick thinning algorithm. In: Proceedings of the 2008 Congress on Image and Signal Processing, CISP, vol. 3, pp. 475–478. IEEE Computer Society, Washington, DC (2008)
23. Bai, X., Latecki, L.J., Society, I.C., Yu Liu, W.: Skeleton pruning by contour partitioning with discrete curve evolution. IEEE Trans. Pattern Anal. Mach. Intell. 29, 449–462 (2007)
24. Latecki, L.J., Lakämper, R.: Shape similarity measure based on correspondence of visual parts. PAMI, 1185–1190 (2000)
25. Okabe, M., Owada, S., Igarashi, T.: Interactive design of botanical trees using freehand sketches and example-based editing. Computer Graphics Forum 24(3), 487–496 (2005)
26. Boudon, F., Pradal, C., Cokelaer, T., Prusinkiewicz, P., Godin, C.: L-py: an l-system simulation framework for modeling plant development based on a dynamic language. Frontiers in Plant Science 3(76) (2012)
27. Piegl, L., Tiller, W.: The NURBS book, 2nd edn. Springer-Verlag New York, Inc. (1997)
28. Tu, Z., Chen, X., Yuille, A.L., Zhu, S.C.: Image parsing: Unifying segmentation, detection, and recognition. In: Ponce, J., Hebert, M., Schmid, C., Zisserman, A. (eds.) Toward Category-Level Object Recognition. LNCS, vol. 4170, pp. 545–576. Springer, Heidelberg (2006)
29. Yuille, A., Kersten, D.: Vision as bayesian inference: Analysis by synthesis? introduction: Perception as inference. Psychology, 1–15 (2006)
30. Gupta, A., Efros, A.A., Hebert, M.: Blocks world revisited: Image understanding using qualitative geometry and mechanics. In: Daniilidis, K., Maragos, P., Paragios, N. (eds.) ECCV 2010, Part IV. LNCS, vol. 6314, pp. 482–496. Springer, Heidelberg (2010)

Hermite Interpolation with Rational Splines with Free Weights

Carsten Hamm[1], Tomas Sauer[2], and Florian Zimmermann[1,2]

[1] Siemens AG,
Industry Sector - Drive Technologies Division,
D-91056 Erlangen, Germany
[2] University of Passau, Innstr. 43,
D-94032 Passau, Germany

Abstract. In this text we present an approach for Hermite interpolation with rational splines without predefined weight factors. We rearrange the equation of the derivative of the rational spline function into a homogeneous linear system of equations in homogeneous space. We use this linear system to formulate different interpolation problems, with the weight factors as well as the control points as a solution. In the first approach, we solve the linear system directly by adding only one inhomogeneous equation to normalise the weights. This approach has some significant constraints. The second approach uses the linear system as a secondary condition for maximizing the minimum weight. This way allows us to obtain method more open regarding the number of interpolation points. In the third approach, we reduce the number of interpolation points to approximate the values of the function between the interpolation points.

1   Introduction

Importing bevel gear tooth flanks into a CAD program immediately leads to the problem of finding a good approximation for the gear geometry by means of standard geometric primitives. Since, for principal reasons, the geometry of a tooth flank is part of a sphere, smooth piecewise *rational* surfaces or NURBs are the primitives of choice. Moreover, the gear contact and its quality is tied to properties of the tangent; hence, a Hermite interpolation or approximation has to be applied.

The naive approach to address this task with NURBS is first to select the weight factors and then simply solve a linear system like in case of polynomial splines, see [2,3]. Although the results for different *a priori* weights vary significantly, the choice of the weight factors often is purely heuristic. Farin, for example, addresses this topic in his book as follows [2, P.240]: *"We have not yet addressed the problem of how to choose the weights [..] for the data points [..]. No known algorithms exist for this problem. It seems reasonable to assign high weights in regions where the interpolant is expected to curve sharply."*. An approach presented in [4] uses the weights as additional degrees of freedom, transforms the rational spline into homogeneous coordinates and solves a *homogeneous linear system* for Lagrange interpolation. This idea is extended by

M. Floater et al. (Eds.): MMCS 2012, LNCS 8177, pp. 230–237, 2014.
© Springer-Verlag Berlin Heidelberg 2014

building the derivative in the homogeneous space and setting up a linear system. However, this approach only interpolates the derivative in the *homogeneous* space which is not equal to the derivative of the curve. In this paper, we extend the Lagrange approach from [4] to a Hermite interpolation and approximation method based on derivatives in euclidean space.

2 Definitions

A *knot sequence* $T_{m,n} := \{t_1, \ldots, t_{n+m+1}\} \subset \mathbb{R}$ of order $m \in \mathbb{N}$ is defined to be an ascending, finite sequence with at most $m + 1$ identical elements:

$$t_1 \leq \ldots \leq t_{n+m+1}, \qquad t_j < t_{j+m+1}, \qquad j = 1, \ldots, n.$$

The polynomial B-splines b_i^m of degree m with respect to the knot sequence T are given by the usual recurrence relation

$$b_j^k(\cdot|T) := \frac{\cdot - t_j}{t_{j+k} - t_j} b_j^{k-1}(\cdot|T) + \frac{t_{j+k+1} - \cdot}{t_{j+k+1} - t_{j+1}} b_{j+1}^{k-1}(\cdot|T),$$
$$b_j^0(\cdot|T) := \chi_{[t_j, t_{j+1})}$$

and we write $\mathbf{b} := (b_1, \ldots, b_n)^t$ for the vector of all B–splines of degree m with respect to the knot sequence T.

The weights $\mathbf{w}$ for the rational B–splines are the nomalized nonnegative vector

$$\mathbf{w} = (w_1, \ldots, w_n)^t \in \mathbb{R}_+^n, \qquad \mathbf{w}^t \mathbf{1} = n.$$

We find it convenient to arrange the *control points* $\mathbf{d}_i \in \mathbb{R}^d$, $i = 1, \ldots, n$ into the matrix $\mathbf{D} = (\mathbf{d}_1 \ldots \mathbf{d}_n) \in \mathbb{R}^{d \times n}$. Analogously, we define the i-th *homogeneous control point* as $\begin{pmatrix} w_i \\ \overline{\mathbf{d}}_i \end{pmatrix} = \begin{pmatrix} w_i \\ w_i \mathbf{d}_i \end{pmatrix} \in \mathbb{R}^{d+1}$ and arrange them into the matrix

$$\begin{pmatrix} w_1 \ldots w_n \\ \overline{\mathbf{d}}_1 \ldots \overline{\mathbf{d}}_n \end{pmatrix} =: \begin{pmatrix} \mathbf{w}^t \\ \overline{\mathbf{d}}_{x_1}^t \\ \vdots \\ \overline{\mathbf{d}}_{x_d}^t \end{pmatrix} \in \mathbb{R}^{(d+1) \times n}.$$

We reshape the control matrix one more time, and call

$$\overline{d} = \left(w^t \; \overline{\mathbf{d}}_{x_1}^t \; \ldots \; \overline{\mathbf{d}}_{x_d}^t \right)^t \in \mathbb{R}^{(d+1)n}$$

the vector of all homogeneous control points.

In this notation, a rational spline curve $\mathbf{f} : [t_m, t_n] \longrightarrow \mathbb{R}^d$ is given as:

$$\mathbf{f}(t) := \frac{\mathbf{D}\mathbf{W}\mathbf{b}(t)}{\mathbf{w}^t \mathbf{b}(t)}, \qquad \mathbf{W} := \operatorname{diag} \mathbf{w} = \begin{pmatrix} w_1 & & \\ & \ddots & \\ & & w_n \end{pmatrix} \in \mathbb{R}^{n \times n}. \tag{1}$$

We call the spline curve polynomial if $\mathbf{w} = \mathbf{1}$. In this case, equation (1) simplifies and becomes the well–known

$$\mathbf{f}(t) := \mathbf{D}\mathbf{b}(t).$$

3   Derivatives and Conditions for Interpolation

The r-th derivative of a polynomial spline curve of degree m is a polynomial spline curve of degree $m - r$ which can be calculated using a simple matrix multiplication, see [1], as $\frac{d^r}{dt^r}\mathbf{f}(t) = \mathbf{D}\,\mathbf{G}_r\,\mathbf{b}^{m-r}(t)$ with $\mathbf{G}_r := \mathbf{G}(m, T) \cdots \mathbf{G}(m - r + 1, T)$ where

$$\mathbf{G}(m, T) := \begin{pmatrix} \frac{m}{\Delta_m t_1} & -\frac{m}{\Delta_m t_2} & & \\ & \ddots & \ddots & \\ & & \frac{m}{\Delta_m t_n} & -\frac{m}{\Delta_m t_{n+1}} \end{pmatrix} \in \mathbb{R}^{n \times n+1}$$

uses the difference operator $\Delta_m t_j := t_{j+m} - t_j$, $j = 1, \ldots, n + 1$, with the convention that $\frac{1}{\Delta_m t_j} := 0$ if $\Delta_m t_j = 0$.

With the matrix $\mathbf{G}$ we can also recursively define the derivative of a rational spline curve:

$$\frac{d^r}{dt^r}\mathbf{f}(t) = -\sum_{s=1}^{r} \binom{r}{s} \frac{\mathbf{w}^t \mathbf{G}_s \mathbf{b}(t)}{\mathbf{w}^t \mathbf{b}(t)} \frac{d^{r-s}}{dt^{r-s}}\mathbf{f}(t) + \frac{\mathbf{D}\mathbf{W}\mathbf{G}_r \mathbf{b}(t)}{\mathbf{w}^t \mathbf{b}(t)} \tag{2}$$

Equation (2) can be deduced by deriving $\mathbf{w}^t \mathbf{b}(t)\,\mathbf{f}(t)$ with Leibniz's rule, as described in [3, p.125].

Let $\mathbf{p}_1^r, \ldots, \mathbf{p}_k^r \in \mathbb{R}^d$ be given *interpolation values* and $\tau_1, \ldots, \tau_k \in \mathbb{R}$ be the associated *interpolation nodes* or *interpolation sites*, $r = 0, \ldots, \widetilde{r}$. A function $\mathbf{f}$ is said to be a solution to the Hermite interpolation problem if

$$\frac{d^r}{dt^r}\mathbf{f}(\tau_j) = \mathbf{p}_j^r, \qquad j = 1, \ldots, k, \ r = 0, \ldots, \widetilde{r}.$$

We substitute $\mathbf{p}_j^r$ and τ_j into equation (2) and obtain

$$\mathbf{0} = -\mathbf{p}_j^r\,\mathbf{w}^t\mathbf{b}(\tau_j) - \sum_{s=1}^{r} \binom{r}{s} \mathbf{p}_j^{r-s}\,\mathbf{w}^t\mathbf{G}_s\mathbf{b}(\tau_j) + \begin{pmatrix} \overline{\mathbf{d}}_{x_1}^t \\ \vdots \\ \overline{\mathbf{d}}_{x_d}^t \end{pmatrix} \mathbf{G}_r\mathbf{b}(\tau_j)$$

$$= -\sum_{s=0}^{r} \binom{r}{s} \mathbf{p}_j^{r-s}\,\mathbf{b}^t(\tau_j)\mathbf{G}_s^t\mathbf{w} + \begin{pmatrix} \mathbf{b}^t(\tau_j)\mathbf{G}_r^t\overline{\mathbf{d}}_{x_1} \\ \vdots \\ \mathbf{b}^t(\tau_j)\mathbf{G}_r^t\overline{\mathbf{d}}_{x_d} \end{pmatrix}$$

$$= \left(\mathbf{P}_r^t(\tau_j)\ \mathbf{B}_r^t(\tau_j)\right) \begin{pmatrix} \mathbf{w} \\ \mathbf{d} \end{pmatrix}, \tag{3}$$

with

$$\mathbf{P}_r^t(\tau_j) = -\sum_{s=0}^{r} \binom{r}{s} \mathbf{p}_j^{r-s}\, \mathbf{b}^t(\tau_j)\, \mathbf{G}_s^t \in \mathbb{R}^{d\times n},$$

$$\mathbf{B}_r^t(\tau_j) = \begin{pmatrix} \mathbf{b}^t(\tau_j)\mathbf{G}_r^t & & \\ & \ddots & \\ & & \mathbf{b}^t(\tau_j)\mathbf{G}_r^t \end{pmatrix} \in \mathbb{R}^{d\times nd}.$$

4   Solving the Linear System

By putting together all the Hermite interpolation conditions (3) as well as a normalization condition for the weights, we end up with the (inhomogeneous) linear system

$$\begin{pmatrix} \mathbf{P}^t(\tau_1)\ \mathbf{B}^t(\tau_1) \\ \vdots \quad\ \vdots \\ \mathbf{P}^t(\tau_k)\ \mathbf{B}^t(\tau_k) \\ 1 \qquad 0 \end{pmatrix} \begin{pmatrix} \mathbf{w} \\ \overline{\mathbf{d}} \end{pmatrix} = \begin{pmatrix} \mathbf{0} \\ n \end{pmatrix} \tag{4}$$

where

$$\mathbf{P}^t(\tau_j) = \begin{pmatrix} \mathbf{P}_0^t(\tau_j) \\ \vdots \\ \mathbf{P}_{\widetilde r}^t(\tau_j) \end{pmatrix}, \qquad \mathbf{B}^t(\tau_j) = \begin{pmatrix} \mathbf{B}_0^t(\tau_j) \\ \vdots \\ \mathbf{B}_0^t(\tau_j) \end{pmatrix}.$$

This linear system is of dimension $1 + dk(\widetilde r + 1) \times n(d+1)$. For the existence of a *unique* solution, the linear system must be square which means that the dimension of the spline space $n \in \mathbb{N}$ has to be $n = \frac{1+dk(\widetilde r+1)}{d+1}$. Unfortunately, it is not always possible to find an integer for n, as we can directly see in the case of Hermite interpolation $\widetilde r = 1$ in $\mathbb{R}^3$, where $n = \frac{1}{4} + k \cdot \frac{3}{2} \notin \mathbb{N}$.

To find a solution for non–square systems, we reformulate the interpolation problem as an optimization problem, where the minimal weight u is maximized. This approach guarantees that the the solution is as polynomial as possible and, in particular, it gives the polynomial solution of the problem whenever such a solution exists. Hence, we consider

$$\min_{\mathbf{w},\overline{\mathbf{d}}} (0\ 0\ -1) \begin{pmatrix} \mathbf{w} \\ \overline{\mathbf{d}} \\ u \end{pmatrix}, \text{ subject to } \begin{pmatrix} \mathbf{I}\ 0\ -1 \\ 0\ 0\ \ 1 \end{pmatrix} \begin{pmatrix} \mathbf{w} \\ \overline{\mathbf{d}} \\ u \end{pmatrix} \geq \begin{pmatrix} \mathbf{0} \\ 0 \end{pmatrix} \tag{5}$$

and the linear equality constraints

$$\begin{pmatrix} \mathbf{P}^t(\tau_1)\ \mathbf{B}^t(\tau_1)\ 0 \\ \vdots \quad\ \vdots \quad\ \vdots \\ \mathbf{P}^t(\tau_k)\ \mathbf{B}^t(\tau_k)\ 0 \\ 1 \qquad 0 \qquad 0 \end{pmatrix} \begin{pmatrix} \mathbf{w} \\ \overline{\mathbf{d}} \\ u \end{pmatrix} = \begin{pmatrix} 0 \\ \vdots \\ 0 \\ n \end{pmatrix}.$$

To avoid singularities, we added one inequality constraint, requiering u and therefor all weights to be nonnegative.

From Figure 1 we see that the NURBS curve resulting from this optimization converges to the polynomial spline as n grows. This behaviour is obvious, since as soon as the side conditions can be satisfied, the uniform weight vector of the polynomial spline is the unique solution of the optimization problem. If we want to use additional degrees of freedom to smooth the spline further, we need to add more data points. If the points $\mathbf{p}_j$ are sampling points of a function $\mathbf{p}$ at the values τ_j, we can extend the approach by also approximating $\mathbf{p}$ in between the interpolation points $\mathbf{p}(\tau_j)$. To that end, we minimize the integral over the squared residual of the linear system (3),

$$
\int_{t_1}^{t_n} \left\| (\mathbf{P}_0^t(\tau)\, \mathbf{B}_0^t(\tau)) \begin{pmatrix} \mathbf{w} \\ \mathbf{d} \end{pmatrix} \right\|^2 d\tau
$$
$$
= \left(\mathbf{w}^t\ \overline{\mathbf{d}}^t \right) \int_{t_1}^{t_n} \begin{pmatrix} \mathbf{P}_0(\tau)\,\mathbf{P}_0^t(\tau) & \mathbf{P}_0(\tau)\,\mathbf{B}_0^t(\tau) \\ \mathbf{B}_0(\tau)\,\mathbf{P}_0^t(\tau) & \mathbf{B}_0(\tau)\,\mathbf{B}_0^t(\tau) \end{pmatrix} d\tau \begin{pmatrix} \mathbf{w} \\ \mathbf{d} \end{pmatrix}
$$

This leads to the quadratic optimization problem

$$
\min_{\mathbf{w},\overline{\mathbf{d}}} \left(\mathbf{w}^t\ \overline{\mathbf{d}}^t \right) \begin{pmatrix} \int \mathbf{P}_0 \mathbf{P}_0^t & \int \mathbf{P}_0 \mathbf{B}_0^t \\ \int \mathbf{B}_0 \mathbf{P}_0^t & \int \mathbf{B}_0^t \mathbf{B}_0 \end{pmatrix} \begin{pmatrix} \mathbf{w} \\ \mathbf{d} \end{pmatrix}, \tag{6}
$$

with equality constraints that encode the interpolation conditions:

$$
\begin{pmatrix} \mathbf{P}^t(\tau_1)\ \mathbf{B}^t(\tau_1) \\ \vdots \qquad \vdots \\ \mathbf{P}^t(\tau_k)\ \mathbf{B}^t(\tau_k) \\ \mathbf{1}^t \qquad 0 \end{pmatrix} \begin{pmatrix} \mathbf{w} \\ \overline{\mathbf{d}} \end{pmatrix} = \begin{pmatrix} \mathbf{0} \\ \vdots \\ \mathbf{0} \\ n \end{pmatrix}.
$$

Since the quadratic form in (6) is positive definite, it is equivalent to the following linear system involving the Lagrange multipliers $\lambda_1, \ldots, \lambda_k \in \mathbb{R}^d$ and $\lambda_{k+1} \in \mathbb{R}$:

$$
\begin{pmatrix} \int \mathbf{P}_0 \mathbf{P}_0^t & \int \mathbf{P}_0 \mathbf{B}_0^t & \mathbf{P}(\tau_1) & \ldots & \mathbf{P}(\tau_k) & 1 \\ \int \mathbf{B}_0 \mathbf{P}_0^t & \int \mathbf{B}_0 \mathbf{B}_0^t & \mathbf{B}(\tau_1) & \ldots & \mathbf{B}(\tau_k) & 0 \\ \mathbf{P}^t(\tau_1) & \mathbf{B}^t(\tau_1) & 0 & \ldots & 0 & 0 \\ \vdots & \vdots & \vdots & \ddots & \vdots & \vdots \\ \mathbf{P}^t(\tau_k) & \mathbf{N}^t(\tau_k) & 0 & \ldots & 0 & 0 \\ 1 & 0 & 0 & \ldots & 0 & 0 \end{pmatrix} \begin{pmatrix} \mathbf{w} \\ \overline{\mathbf{d}} \\ \lambda_1 \\ \vdots \\ \lambda_k \\ \lambda_{k+1} \end{pmatrix} = \begin{pmatrix} \mathbf{0} \\ \mathbf{0} \\ \mathbf{0} \\ \vdots \\ \mathbf{0} \\ n \end{pmatrix}.
$$

In order to satisfy the interpolation constraints, the spline space has to be of dimension $n = k(\widetilde{r}+1) \in \mathbb{N}$. The converse depends on a more intricate interplay between the weights, the interpolation sites and the position of the knots.

5   Examples

Figures 1 - 3 illustrate the Hermite interpolation methods for an equidistant sampling of the cosine function on the interval $[0, \pi]$.

$$\mathbf{p}(t) := \cos(t) \in \mathbb{R}, \qquad \mathbf{p}_k^{(0/1)} := \mathbf{p}^{(0/1)}(t_k),$$

$$t_k := \frac{k-1}{5}\pi, \qquad k = 1, \ldots, 6,$$

$$m := 3, \qquad n := 7, 9, 12,$$

$$T := \left[0, 0, 0, 0, \frac{1}{m-n} \cdot \pi, \ldots, \frac{m-n-1}{m-n}\pi, \pi, \pi, \pi\right].$$

In addition, Figure 4 compares the approximation of an original gear profile by an interpolating polynomial spline with that of a rational spline computed by the method from equation (6). It can be seen that even based on a smaller

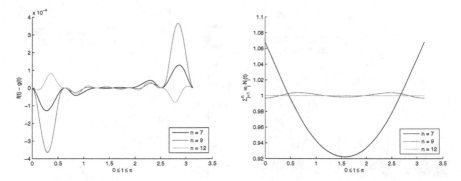

Fig. 1. Method from equation 5 with different dimensions of the spline space n
(a): The interpolation error   (b): The weight function

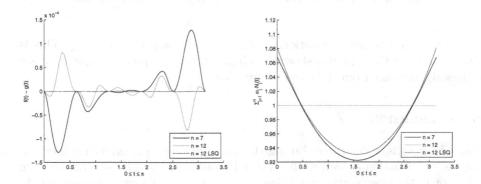

Fig. 2. Comparison between method from equation 5 and LSQ method from equation 6
(a): The interpolation error   (b): The weight function

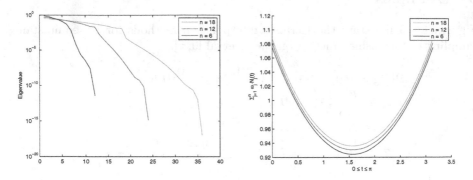

Fig. 3. Method from equation 6 with different dimensions of the spline space n
(a): The eigenvalues of the LSQ matrix   (b): The weight function

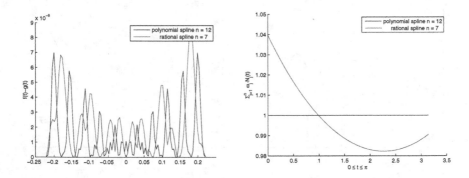

Fig. 4. Comparison between a polynomial and a rational spline approximation of the gear profile
                (a): The interpolation error   (b): The weight function

number of data points, the rational spline provides superior accuracy. This is due to the fact that, as also shown in Figure 4, the computed weight function differs significantly from the constant function.

6   Conclusion

The best results were achieved using the Least-Square-Approximation method, for the test function as well as the original problem. This is no surprise since in addition to the interpolation conditions an overall approximation is requested. As a result of the better fit of the weights, the error of the function was reduced by an order of magnitude of 10^2 compared to the polynomial case, hence automatically adopting the weight vector is definitely worth the effort. The costs of the better approximation, however, are longer computing time and less stable results: for higher values of n the linear system has a severe numerical rank defect, as can be seen in the eigenvalue profile in Figure 3. This leads to ambiguities in the weight vector as well as in the coefficients of the approximating function.

To overcome this problem, we rely on the fact that the best weight function is tied to the approximated function and therefore should be mostly independent of the number of interpolation points. Hence, to obtain a stable solution, we first solve the problem for a relatively small number of interpolation points and use this result to approximate the weight function. Then we add more interpolation points with a fixed weight function to obtain a better approximation. A nice side effect of this approach is that in the second phase we have to solve only relatively simple linear problems.

References

1. de Boor, C.: A Practical Guide to Splines. Springer, New York (2001)
2. Farin, G.: Curves and Surfaces for CAGD. Academic Press, San Diego (2002)
3. Piegl, L., Tiller, W.: The NURBS Book. Springer, Berlin (1997)
4. Schneider, F.-J.: Interpolation, Approximation und Konvertierung mit rationalen B-Splines. Dissertation TH Darmstadt (1993)

Direct Pixel-Accurate Rendering
of Smooth Surfaces

Jon Hjelmervik

Sintef ICT and University of Oslo
Jon.M.Hjelmervik@sintef.no

Abstract. High-quality rendering of B-spline surfaces is important for a range of applications. Providing interactive rendering with guaranteed quality gives the user not only visually pleasing images, but also trustworthy information about the model. In this paper we present a view-dependent error estimate for parametric surfaces. This estimate forms the basis of our surface rendering algorithm, which makes use of the hardware tessellator functionality of GPUs.

We use the screen space distance between the tessellated surface and the corresponding surface point as an error metric. This makes the algorithm particularly useful when visualizing additional attributes attached to the surface. An example of this is isogeometric analysis, in which simulation results are visualized along with the surface.

1 Introduction

Smooth surfaces are used in settings ranging from the entertainment industry to CAD applications. In the entertainment industry, the model's sole purpose is to create visually pleasing images. The CAD-related industry, on the other hand, use visualization both to get an overview of models and to investigate their geometric qualities, for example the smoothness of a car's body. Most CAD models are at some point also used in an analysis setting, e.g., stress analysis, to investigate the model's physical properties. It is therefore a great demand for fast rendering methods that provide accurate, and visually pleasing results that includes associated data such as textures and simulation results.

One way of computing a correct rendering of a surface is to find the first intersection between the surface and rays originating from a virtual camera. This is called ray-casting, which is computationally expensive. GPUs are designed to rasterize triangles and rendering performance is often dominated by the number of triangles in a scene. Therefore, the main challenge is to determine which triangles to draw, and how to invoke their rendering the main challenge.

Single-pass rendering is the most common way to render triangles, because each object is sent only once through the rendering pipeline. *Multi-pass* methods, on the other hand, write partial results to a framebuffer (usually an off-screen buffer) for use in a succeeding rendering pass. A GPU delivers its best performs when it can render a large set of triangles in parallel. Multi-pass algorithms therefore impose a performance penalty when a rendering pass must wait for the

M. Floater et al. (Eds.): MMCS 2012, LNCS 8177, pp. 238–247, 2014.

previous the completion of a previous pass. A CAD model can be composed of a large set of parts, each represented by a number of boundary surfaces. In such settings, with hundreds or thousands of surfaces it may be difficult to implement multi-pass algorithms without adding a large overhead due to pipeline stalls.

We propose a single-pass rendering method driven by the guarantee of pixel accurate rendering. Obviously, this indicates a view-dependent tessellation, i.e., the surface tessellation depends on the location and orientation of the scenes virtual camera. Our focus is an algorithm suitable for isogeometric models.

1.1   Related Work

Traditionally, algorithms could either provide error guarantees *or* be interactive. Filip et al. [1] proposed to use bounds on the second derivatives to create a semi-uniform tessellation of a C^2 continuous surface. The main idea is to split the surface into a set of patches, and find the tessellation levels independently for each patch. In order to create a watertight tessellation without cracks, each patch boundary has a separate tessellation level matching abutting patch edges. Their implementation is CPU based, but their approach fits very well with the OpenGL tessellator, and will be the basis for our work.

Cook et al. [2] propose to recursively split a surface until each triangle is less than a pixel. Their tessellations will for most surfaces be much denser than required. A fast CPU based algorithm for generating these tessellations was presented by Fisher et al. [3]. Since the triangles is recomputed based on the view position, they must be transmitted through the PCI-express bus each frame, limiting the rendering speed.

If we scarify guaranteed accuracy, there are numerous methods for view-dependent tessellation of smooth surfaces. Guthe et al. [4] proposed to use the CPU for deciding the required tessellation level of a semi-uniform tessellation, and use first generation shader technology to evaluate the surface. Hjelmervik and Hagen [5] proposed a two-pass algorithm using the GPU both for determining tessellation level and the evaluation itself. It is based on early GPUs, and is therefore not optimized for graphics hardware of today.

Lutterkort [6] developed an algorithm for computing piecewise linear enclosure of polygonal surfaces, called slefes. Yeo et al. [7] developed a interactive rendering algorithm with guaranteed pixel accuracy based on these slefes for DX11 compatible hardware. Their algorithm use separate rendering passes for generation of slefe boxes (estimate of the surface's deviation from linear), determining tessellation level, and the tessellation itself.

1.2   Hardware Tessellator

Hardware tessellators as exposed by OpenGL 4 and DirectX 11 are used to create semi-uniform tessellations. It lets us concentrate on error estimates, tessellation levels and surface evaluation, without concern for generating the triangles and managing their connectivity.

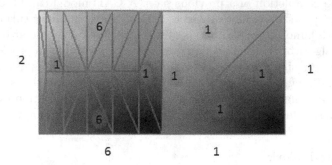

Fig. 1. Example tessellation with interior and boundary tessellation levels. Note that it generates triangulation with consistent connectivity.

The hardware tessellator allows for implementations in which the entire tessellation is performed at the GPU, and the triangles are directly rasterized without being transferred to off-chip memory. The tessellator acts on individual surface patches, which in our case will be the same as the Bézier patches. Each patch is tessellated individually, as illustrated in Figure 1.

The hardware tessellator adds two programmable shader stages to the existing OpenGL graphics pipeline that all triangles undergo. The first stage, the *tessellation control shader*, controls the tessellation process by specifying the parameters for the sampling density for each patch. A set of triangles complying with these criteria is then automatically created, and the *tessellation evaluation shader* has the responsibility to compute the position of each new vertex, based on its parameter value. To ensure that the triangles form a valid triangulation without holes, the the control shader specifies the sampling density along each boundary edge in addition to the parameter directions.

Note that all sampling parameters are set in the control shader, before any triangle is created. Traditional algorithms based on iteratively refining the triangles do therefore not fit with this setup. To fully take advantage of the hardware tessellator, the triangles should never leave the chip, meaning the tessellation will have to be redone each frame.

1.3 Contribution

Our research was performed independently of work by Yeo et al. [7] based on slefes, but contains many of the same features. Both methods use the same error metric to determine the required tessellation level, but estimate the error in different ways. Where Yeo et al. use slefes, where the theory is only fully developed for polynomial surfaces, our algorithm can be used for any C^2 continuous surface with bounds on the second order derivatives. This also allows our approach to be extended such that the lowest possible tessellation is less than one triangle per Bézier patch.

CAD-models often consist of a large number of objects, each described by their boundary surfaces. Special surfaces such as cylinders and swept surfaces, which are of different polynomial order in the parameter directions, play important roles in CAD. However, when used in a simulation it is common to use the same polynomial order in both parameter directions. Our algorithm allows for more dense sampling in the parameter direction with the highest second order derivative. Furthermore, in contrast to slefe based tessellation, all computations are performed after camera transformation, and thus, the tessellation is therefore ignorant to which space the object is modeled in.

As a starting point, we took the algorithm for semi-uniform tessellation described in Filip et al. [1], and analyzed how the error estimate is affected by projecting the surface to the screen. Their implementation, guarantees that the tessellation is within the error tolerance at the uniformly tessellated interior of each patch, and at the boundary curves separating the patches. However, the ring of triangles connecting the patch boundary with its interior may not. Some of our test cases included patches where one boundary curve was linear, and hence sampled only at its endpoints. This leads to visual artifacts, which is remedied in our implementation.

2   Algorithm

To use the tessellator we need a predicate which defines the tessellation level without the need of actually sampling the surface. One of the simplest predicates is to measure the size of the surface's bounding box when it is projected to the screen. Such a predicate would lead to overtessellation of flat surfaces or undertessellation of the more complex parts. We therefore need a view-dependent predicate taking into account the shape and parametrization of the surface.

The rasterization solves the ray-casting problem of finding the intersection between a ray through each pixel center and the triangle. The texture coordinates (or parameter value) at a pixel is therefore associated to its center. Since we can use this parameter value to evaluate the surface and any additional data (textures of simulation data), it will be pixel accurate if the surface evaluated at the given parameter value belongs within the same pixel. We will use this property to define our error metric as follows:

$$e(u, v) = \|\texttt{proj}(S(u, v) - T(u, v))\|_\infty. \tag{1}$$

Here S is the original surface, T is the tessellated version and $\texttt{proj}$ is the projection from eye space to the screen. In this section we will ignore the projection and focus on the approximation error of a linear interpolation of a C^2 continuous surface.

A well known upper bound for linear interpolation of a C^2 continuous curve g is

$$|I_2 g - g|_\infty \le \left(\frac{\Delta^2}{8}\right) \max |g''|, \tag{2}$$

  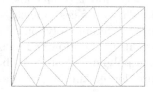

(a) Original tessellation   (b) Interior samples moved   (c) Final tessellation
closer to the left boundary

Fig. 2. Figure (a) shows a tessellation where the left boundary has only two sample points. Figure (b) shows how to move the interior sample points closer to the left boundary. Finally, in (c) an extra column of triangles is inserted.

in which $I_2 g$ is a linear interpolation of g with sampling distance Δ. Thus, we can choose the sampling distance to meet any given tolerance.

The interior of each patch is tessellated by triangles in which two of the edges follow the parameter directions of the surface. Therefore, we we can estimate the approximation error by

$$|I_2 f - f|_\infty \leq \frac{\Delta_u^2}{8} \max |f_{uu}| + \frac{\Delta_u \Delta_v}{4} \max |f_{uv}| + \frac{\Delta_v^2}{8} \max |f_{vv}|, \qquad (3)$$

where Δ_u and Δ_v are sampling distance in parameter direction u and v respectively. Again, if we can find upper bounds of the second order derivatives we can adjust the sampling distances to meet any given error tolerance. However, since there are two unknowns and only one requirement, the solution is not unique. The optimal solution is the one that has the lowest triangle count, whilst still fulfilling the error tolerance. Filip et al. chose different sampling densities in each parameter direction based on the estimates of the second order derivatives, while You et al. decided to use the same sampling densities. We use a greedy iterative process to determine the sampling distances.

In contrast to Filip et al., we experienced a breach of pixel accuracy for triangles adjacent to the patch boundaries. Any adjustment to the sampling density or position along the patch boundaries would create cracks in the tessellation, because patch boundaries are shared by the neighboring patch. However, in the interior of the patch, we may apply an transformation to the parameter values of each sample point before evaluating the surface. We chose to define an affine transformation that will narrow the boundary band as illustrated in Figure 2. Each patch is sampled uniformly in the inner, resulting in a uniform tessellation. The shapes of the boundary triangles are implementation specific, making it impossible to make a hard guarantee on the error in this area. We therefore assume that each triangle has one edge along the boundary curve and one edge parallel to the other parameter direction when computing the width of the boundary band.

Moving the interior samples closer to an edge will increase the size of the interior triangles, which may violate the pixel correctness. It may therefore be necessary to increase the number of interior triangles. Remember that the tessellation levels and the affine transformation is computed based on information of the second derivative, without the need of actually generating the triangles to check their feasibility. Therefore, all decisions are made in the control shader. The control shader therefore performs the following steps:

1. compute the sampling distance for each edge
2. compute the height for each boundary band
3. compute the interior sampling distance
4. use the computed sampling distances to set the tessellation levels and affine transformation if necessary.

The hardware tessellator will create the required triangles and call the evaluation shader for each generated vertex, which applies the affine transformation to all interior points before evaluating the surface. Evaluation of B-spline surfaces in a shader is straight-forward and explained in Guthe et al. [4].

3   Pixel Accurate Tessellation

In Section 2, we described a tessellation algorithm that generates a tessellation with a guaranteed maximal distance from the original C^2 continuous surface. However, the error metric did not take into account the viewing distance or view direction, leading to a static tessellation. In this section, we will study the same error metric evaluated in screen space instead of model space, i.e., the error in terms of pixels on the screen.

3.1   Projected Error

As a first step, we study what a perturbation of a point in eye space leads to as a perturbation in screen space. Vertices in a 3D scene are represented by four dimensional homogeneous coordinates, and the x, y and z components are divided by the w component as a part of the fixed function perspective division in GPUs. The homogeneous coordinates allows perspective projection to be formulated as a matrix-vector multiplication. Here, we use we use the OpenGL projection matrix, which can be written as

$$\begin{bmatrix} A & 0 & B & 0 \\ 0 & C & D & 0 \\ 0 & 0 & E & F \\ 0 & 0 & -1 & 0 \end{bmatrix} \begin{bmatrix} x_{es} \\ y_{es} \\ z_{es} \\ w_{es} \end{bmatrix} = \begin{bmatrix} x_{cs} \\ y_{cs} \\ z_{cs} \\ w_{cs} \end{bmatrix}, \tag{4}$$

where $\mathbf{x_{es}}$ and $\mathbf{x_{cs}}$ are vectors in eye space and clip space respectively, see Shreiner et al. [8] for details. Let

$$\mathbf{x} = \begin{bmatrix} x \\ y \\ z \\ w \end{bmatrix}, \text{ and, } \hat{\mathbf{x}} = \begin{bmatrix} \hat{x} \\ \hat{y} \\ \hat{z} \\ \hat{y} \end{bmatrix} \tag{5}$$

be a point in eye space and its projection to the screen respectively. Then

$$\hat{y} = -\frac{Cy + Dz}{z}, \text{ and its inverse, } y = -\frac{\hat{y}z + Dz}{C} \tag{6}$$

describes the relationship between the y component in eye space and its projected counterpart. Let ϵ be the perturbation of $\mathbf{x}$ and $\hat{\epsilon}$ be the corresponding perturbation of of $\hat{\mathbf{x}}$. Then their relation can be written

$$\frac{y + \epsilon_y}{z + \epsilon_z} = -\frac{\hat{y} + \hat{\epsilon}_y + D}{C}, \tag{7}$$

which leads to

$$\hat{\epsilon}_y = C\frac{\epsilon_z y - \epsilon_y z}{z(\epsilon_z + z)}, \tag{8}$$

which describes the projected perturbation given by position and perturbation before projection. Assuming ϵ_z is neglectable in the term $z(\epsilon_z + z)$ we arrive at

$$|\hat{\epsilon}_y| \lesssim C\left(\left|\frac{\epsilon_z y}{z^2}\right| + \left|\frac{\epsilon_y}{z}\right|\right). \tag{9}$$

What remains is to use the estimates from Section 2 to express the error projected error by the bounding box of the surface and its second order derivatives.

Inserting (2) into (9) and solving for Δ, we arrive at

$$\Delta \lesssim \sqrt{\frac{8\epsilon^y}{C} \bigg/ \left(\frac{\lceil g_z'' \rceil \lceil y \rceil}{\lfloor z^2 \rfloor} + \frac{\lceil g_y'' \rceil}{\lfloor z \rfloor}\right)} \tag{10}$$

as the expression for the maximal sample distance of a curve, given tolerances ϵ_x and ϵ_y. Here, $\lfloor \cdot \rfloor$ and $\lceil \cdot \rceil$ denotes the minimal and maximal absolute value respectively. To restrict the error to be less than one pixel, ϵ_y and ϵ_x are set to $0.5/window\_width$ and $0.5/window\_height$ respectively.

For the curve case, we were able to derive an explicit formula for the tessellation levels. For the surface case we must first choose a strategy for balancing the tessellation levels in the two parameter directions. Clearly, the tessellation level in the interior of the Bézier patch must be at least as dense as the sampling of the boundary curves. We propose to use the most dense boundary tessellation in each parameter direction as an initial guess, and iteratively refine the parameter direction that reduces the approximation error the most. The maximal error given the tessellation level for each parameter direction can be estimated by inserting (9) into (3),

$$\frac{8\epsilon^x}{A} \geq \Delta_u^2 \left(\frac{\lceil z_{uu} \rceil \lceil x \rceil}{\lfloor z^2 \rfloor} + \frac{\lceil x_{uu} \rceil}{\lfloor z \rfloor}\right)$$

$$+ \Delta_v^2 \left(\frac{\lceil z_{vv} \rceil \lceil x \rceil}{\lfloor z^2 \rfloor} + \frac{\lceil x_{vv} \rceil}{\lfloor z \rfloor}\right)$$

$$+ 2\Delta_u\Delta_v \left(\frac{\lceil z_{uv} \rceil \lceil x \rceil}{\lfloor z^2 \rfloor} + \frac{\lceil x_{uv} \rceil}{\lfloor z \rfloor}\right).$$

For most cases the iteration terminates quickly and does not represent a bottleneck in the control shader.

Fig. 3. Surface color illustrates how close the approximation error is to the given tolerance. Gray is used when the error is less than 10% of the tolerance, and blue indicates that the error larger. The error does not exceed the given tolerance.

4   Results

The tessellations generated by this algorithm are expected to be without any visual artifacts, as it provides pixel accurate rendering. What remains to be seen is whether the objects get excessively tessellated, meaning that more triangles than necessary are generated. It is also interesting to see which areas will have the highest triangle density.

In our experience, the tolerance is only violated when the calculated tessellation level is higher than the hardware limit of the GPU. To measure the quality of the tessellation we study the maximal error at each patch. If the maximal error is much less than the tolerance it indicates that the model is excessively tessellated. Figure 3 is colored based on the error relative to the given tolerance. Note that the error is largest along the boundary band, which is expected since these triangles may be larger than the interior triangles. Due to the shape of the boundary triangles the error estimate does not apply there, but experiments show good results here as well.

As expected, the result is pixel accurate and any tessellation algorithm with this property will produce indistinguishable images. The main objective of our work is to provide fast, high-quality, reliable rendering, which may also be achieved by relaxed the error tolerance beyond one pixel. The increased tolerance will improve the rendering speed, while keeping the assurance that the result will be within the given tolerance of the real model. Coarse tessellations near silhouette edges are easily detected. Several algorithms, including Dyken et al. [9] target this problem directly, and refines near silhouette triangles. As shown in Figure 4 no such special treatment is required here, as Bézier patches near silhouette edges generate smaller triangles compared to other areas. The triangles are also concentrated in high-curvature areas. Using a GeForce GTX 580 we are able to tessellate 7 million cubic Bézier patches per second, which is less than Yeo et al. [7]. This is both due to our algorithm being more compute intensive and due to lack of optimization of our code.

Fig. 4. Color encoding of triangle sizes. Gray triangles have maximal edge length of more than 5 pixels, yellow have 5-2.5, and green triangles less than 2.5 pixels.

5   Conclusion and Future Work

This work has been performed in parallel to the pixel accurate algorithm by Yeo et al. [7] with an almost identical goal. Their work was published at the time of writing this paper, making it natural to discuss the main differences which come from different strategies for estimating the rendering error. We require a one-pass algorithm to facilitate easy integration with large number of simple surfaces, and our focus is on CAD models.

Single-pass algorithms such as ours avoid latency introduced by multi-pass rendering, which is an advantage when rendering small surfaces. However, the control shader must recompute the tessellation levels based on surface coefficients each frame making it potentially slower for large models due to repeated computations. Also, patches sharing an edge will always choose compatible tessellation levels, removing the need to store adjacency information for the model to reach watertight tessellations.

The resulting tessellations from both approaches appear to be of similar quality and seems to have approximately the same triangle count for most examples, but we have not yet performed a head-to-head comparison. CAD surfaces such as swept surfaces and cylinder parts will take advantage of our approach where the parameter directions does not use the same tessellation levels.

Bézier patches fully outside the view frustum are efficiently detected and the tessellation level is set to zero in both algorithms. Small patches on the other hand, will result in a minimum of two triangles, even if they are much smaller than a pixel. For some applications it may therefore be better to treat more than one Bézier patch in each tessellation patch. Our error estimate can be used in this setting, since does not require a polynomial surface.

References

1. Filip, D., Magedson, R., Markot, R.: Surface algorithms using bounds on derivatives. Comput. Aided Geom. Des. 3(4), 295–311 (1987)
2. Cook, R.L., Carpenter, L., Catmull, E.: The reyes image rendering architecture. In: Proceedings of the 14th Annual Conference on Computer Graphics and Interactive Techniques, SIGGRAPH 1987, pp. 95–102. ACM, New York (1987)

3. Fisher, M., Fatahalian, K., Boulos, S., Akeley, K., Mark, W.R., Hanrahan, P.: Diagsplit: parallel, crack-free, adaptive tessellation for micropolygon rendering. ACM Trans. Graph. 28(5), 150:1–150:10 (2009)
4. Guthe, M., Balázs, A., Klein, R.: Gpu-based trimming and tessellation of nurbs and t-spline surfaces. ACM Trans. Graph. 24(3), 1016–1023 (2005)
5. Hjelmervik, J., Hagen, T.: GPU-based screen space tessellation. In: Dæhlen, M., Mørken, K., Schumaker, L.L. (eds.) Mathematical Methods for Curves and Surfaces: Tromsø 2004, pp. 213–221. Nashboro Press (2005)
6. Lutterkort, D.C.: Envelopes of nonlinear geometry. PhD thesis, Purdue University, West Lafayette, IN, USA (2000); AAI3017831
7. Yeo, Y.I., Bin, L., Peters, J.: Efficient pixel-accurate rendering of curved surfaces. In: Garland, M., Wang, R., Spencer, S.N., Gopi, M., Yoon, S.E. (eds.) I3D, pp. 165–174. ACM (2012)
8. Shreiner, D., Woo, M., Neider, J., Davis, T.: OpenGL Programming Guide: The Official Guide to Learning OpenGL, Version 2. Addison-Wesley Longman Publishing Co., Inc., Boston (2005)
9. Dyken, C., Reimers, M., Seland, J.: Real-time GPU silhouette refinement using adaptively blended bézier patches. Computer Graphics Forum 27(1), 1–12 (2008)

Non-uniform Interpolatory Subdivision Based on Local Interpolants of Minimal Degree

Kęstutis Karčiauskas[1] and Jörg Peters[2]

[1] Vilnius University, Lithuania
[2] University of Florida, USA
jorg@cise.ufl.edu

Abstract. This paper presents new univariate linear non-uniform interpolatory subdivision constructions that yield high smoothness, C^3 and C^4, and are based on least-degree spline interpolants. This approach is motivated by evidence, partly presented here, that constructions based on high-degree local interpolants fail to yield satisfactory shape, especially for sparse, non-uniform samples. While this improves on earlier schemes, a broad consideration of alternatives yields two technically simpler constructions that result in comparable shape and smoothness: careful pre-processing of sparse, non-uniform samples and interlaced fitting with splines of increasing smoothness. We briefly compare these solutions to recent non-linear interpolatory subdivision schemes.

1 Introduction

For non-uniformly spaced samples, uniform linear interpolatory curve subdivision algorithms [DL02, Sab10] often results in dramatic overshoot and oscillation. Starting with [War95], non-uniform constructions have been proposed such that new knots are inserted at the midpoints of knot-intervals. Mid-point insertion yields locally uniform knot spacings that meet at the original data points. The data points thereby become isolated 'extraordinary points' where left and right knot intervals may differ; and extraordinary point neighborhoods become the focus of the analysis. Recent examples of such non-uniform constructions are the edge parameter subdivisions [BCR11b, BCR11a] and C^1, C^2, C^3 and C^4 interpolatory curves [KP13b].

However, for higher smoothness, even these new non-uniform constructions exhibit shape problems for non-uniform data such as shown in Fig. 2a, pointing to the classical trade-off between smoothness, convexity and interpolation (cf. Fig. 10(c)). For example Warren's C^2 6-point scheme [War95] as well as a C^3 6-point scheme of [KP13b] unexpectedly loose the convexity of the piecewise linear interpolant to the samples (see Fig. 1c (top)); and a C^4 10-point scheme visibly oscillates. By contrast the Catmull-Rom-inspired construction $\text{CR}^2_{3/256}$ (see Fig. 1 for the meaning of super-and subscripts) fares considerably better. We think these and many other examples indicate that large support, resulting from high-degree interpolants, causes problems and not just because of the increased complexity of the rules. To wit, Warren's scheme is based on local polynomial interpolants of degree 5, the C^3 6-point scheme uses interpolants of degree 5 and 3 and the 10-point scheme polynomials of degree 7 – whereas the $\text{CR}^2_{3/256}$ construction is based on local interpolants of degree 2.

M. Floater et al. (Eds.): MMCS 2012, LNCS 8177, pp. 248–264, 2014.

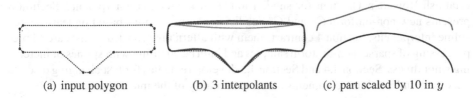

| (a) input polygon | (b) 3 interpolants | (c) part scaled by 10 in y |

Fig. 1. Thumb tag data. The constructions use centripetal knot spacing [Lee89] since, for non-uniform samples, centripetal is superior to chordal. (b) brown = 10-point scheme of [KP13b] with $w = 0.00098$, dipping down in the center; black = Catmull-Rom C^2 construction $\mathrm{CR}^2_{3/256}$ (*The notation of [KP13b] exposes, in the superscript, continuity and possibly the degrees of local interpolants and, in the subscript the setting of the free parameter w of the construction*); red = C^3 6-point scheme $\mathrm{A}^{3,5:3:3}_{0.0141}$ visually identical to $\mathrm{A}^{2,5:3}_{3/256}$, i.e. Warren's C^2 6-point scheme[War95]. (c) The roofs of the T-shaped polygon of $\mathrm{A}^{2,5:3}_{3/256}$ (top) and $\mathrm{CR}^2_{3/256}$ (bottom) are displayed with different offset for clarity and scaled by 10 in the y direction to emphasize the curvature oscillation of $\mathrm{A}^{2,5:3}_{3/256}$.

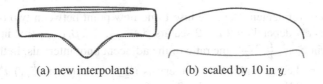

| (a) new interpolants | (b) scaled by 10 in y |

Fig. 2. Minimal degree d local interpolant constructions: blue = C^3 6-point, $d = 3$; cyan = almost C^4 8-point, $d = 4$; green = C^4 10-point, $d = 4$

While reproduction of polynomials of degree k is important for approximation, minimal degree of the interpolants seems consistently advantageous both for controlling shape and for simplicity of downstream use. Hence, in this paper, we construct new C^3 and C^4 non-uniform schemes using only the local interpolants of minimal degree $d = 3, 4$. Indeed, the construction using $d = 3$ clearly improves on earlier schemes. But $d = 4$ interpolants used in a new 8-point scheme of Hölder regularity > 3.96 as well as in a C^4 10-point scheme, provide only slight improvement and loose convexity for highly non-uniform data. By contrast, as illustrated in Fig. 2, the curve generated by the 6-point scheme with $d = 3$ preserves the expected convexity.

This partial failure led us to explore a broader set of alternatives: initial refinement of data with lower-order schemes followed by higher-order schemes to achieve the required smoothness; and, secondly, interlaced fitting with splines of increasing smoothness. We also consider, in Section 4.2, locally-determined knot spacings that reduce th enon-uniformity by spreading it out.

For ease of comparison, we illustrate all our experiments with curves derived from the 'thumb tag' data Fig. 1(a). Many other data sets were tested with like results, e.g. the 'bread loaf' data of Fig. 10(a).

Structure of the paper. Section 2 reviews the analysis of non-uniform interpolatory midpoint-insertion subdivision schemes of [KP13b] adding improved techniques to establish Rouché's Theorem for subdivision from low-degree interpolants. Section 3 presents new non-uniform C^3 and C^4 subdivision constructions based on least-degree spline interpolants. Section 4 contrasts them with alternative constructions: careful pre-processing of sparse, non-uniform samples and interlaced fitting with splines of increasing smoothness. Section 4.4 and Section 4.5 develop remedies for fast changing discrete curvature and Section 5 comments on the minimality of the interpolants.

2 Non-uniform Symmetric Interpolatory Midpoint Subdivision

Except for Section 2.1, this section closely follows the exposition of [KP13b]. Given a sequence of increasing scalars $\{t_i\}$, called knots, and a sequence of points $\{\mathbf{p}_i\}$ in $\mathbf{R}^d$, the $k+1$st point sequence is derived from the kth, starting with $\mathbf{p}_i^0 := \mathbf{p}_i$, by

$$\mathbf{p}_{2i}^{k+1} := \mathbf{p}_i^k, \qquad \mathbf{p}_{2i+1}^{k+1} := \sum_{j=1}^{2n} e_{ij}\mathbf{p}_{i-n+j}^k. \tag{1}$$

That is, in every refinement step, we insert one new point between two old ones. The $2n$ coefficients e_{ij} depend on $2n-2$ scalars $\beta_{i-n+2}, \ldots, \beta_{i+n-1}$ that in turn depend on the knots via $\beta_i := \frac{t_{i+1}-t_i}{t_i-t_{i-1}}$, the ratio of the adjacent knot intervals. In the following, new knots are picked as midpoints of intervals $t_{2i+1}^{k+1} := \frac{1}{2}(t_i^k + t_{i+1}^k), t_{2i}^{k+1} := t_i^k$, a choice that [DGS99] calls semi-regular. Therefore

$$\beta_{2i}^{k+1} := \beta_i^k, \qquad \beta_{2i+1}^{k+1} := 1. \tag{2}$$

All constructions will be invariant under the replacements
symmetry: $e_{ij} \to e_{i,2n+1-j}$ $\beta_{i-n+2+s} \to (\beta_{i+n-1-s})^{-1}$,
translation: $e_{ij} \to e_{i+s,j}$ $\beta_{i-n+2}, \ldots, \beta_{i+n-1} \to \beta_{i-n+2+s}, \ldots, \beta_{i+n-1+s}$.
As in [KP13b], we follow [War95] and first establish the smoothness in the uniform case $\beta_i = 1$, then focus on the extraordinary points corresponding to an isolated $\beta \neq 1$.

For uniform knots $\beta_i = 1$ for all i and we may abbreviate the coefficients to $\bar{e}_j$. Since $\bar{e}_{2n-i} = \bar{e}_i$ for the schemes in this paper, Table 1 displays only $\bar{e}_j, j = 1, \ldots, n$ of the relevant generalizations of the classical 4-point scheme. The uniform schemes are analyzed using z-transforms, see [Dyn92, DL02, DFH04].

Table 1. Uniform symmetric C^{n-1} $2n$-point interpolatory schemes with parameter w [Wei90, KLY07]

$2n$	$\bar{e}_j, j = 1, \ldots, n$	C^{n-1} range for w
6	$w, -3w - \frac{1}{16}, 2w + \frac{9}{16}$	$(0 \ldots 0.042]$
8	$-w, 5w + \frac{3}{256}, -9w - \frac{25}{256}, 5w + \frac{75}{128};$	$[0.0016 \ldots 0.0084]$
10	$w, -7w - \frac{5}{2048}, 20w + \frac{49}{2048}, -28w - \frac{245}{2048}, 14w + \frac{1225}{2048}$	$[0.0005 \ldots 0.0016]$

The now isolated non-uniform locations are analyzed by the following four steps of which especially the last benefits from symbolic computation.

1. Repeated knot insertion at the middle of intervals surrounds each knot where $\beta \neq 1$ by knots with $\beta = 1$. The β at this isolated extraordinary knot is denoted γ in the following.
2. Uniform subdivision applies where $\beta = 1$. Table 1 gives the w-ranges for C^m continuity.
3. The $(4n - 1) \times (4n - 1)$ subdivision matrix L for the isolated extraordinary point has the rows

$$L_1 := (E_0, \mathbf{0}^{2n-1}); \quad L_{4n-1} := (\mathbf{0}^{2n-1}, E_0);$$

$$L_{2i} := \left(\mathbf{0}^{n-1+i}, 1, \mathbf{0}^{3n-1-i}\right), \; i = 1, \ldots, 2n - 1, \tag{3}$$

$$L_{2i+1} := \left(\mathbf{0}^i, E_i, \mathbf{0}^{2n-1-i}\right), \; i = 1, \ldots, 2n - 2,$$

where $\mathbf{0}^s$ is a sequence of s zeros,

$$E(\beta_{i-n+2}, \ldots \beta_{i+n-1}) := (e_{i1}, \ldots, e_{i,2n}),$$

maps the $2n - 2$ ratios β to the $2n$ coefficients e_{ij}, and, with $\mathbf{1}^s$ a sequence of s ones, $E_k := E(\mathbf{1}^{2n-2-k}, \gamma, \mathbf{1}^{k-1})$, $k = 1, \ldots, 2n - 2$, $E_0 := E(\mathbf{1}^{2n-2})$. For an example see e.g. [War95, Sec.5].

Since the constructions are chosen to reproduce polynomials up to degree m, the matrix L has eigenvalues $1, \frac{1}{2}, \ldots, \frac{1}{2^m}$ whose eigenfunctions are the polynomials $1, t, \ldots, t^m$. For analysis, the characteristic polynomial $\chi(\lambda)$ of L is best factored into

$$\chi(\lambda) = const(\lambda - 1)(\lambda - \frac{1}{2}) \cdots (\lambda - \frac{1}{2^m})\ell(\lambda)r(\lambda), \tag{4}$$

where $\ell(\lambda)$ is of the form $(\lambda \pm w)^k$ that allows immediate checking whether its roots are strictly *dominated* in absolute value by $\frac{1}{2^m}$. To establish smoothness, it then suffices to show that the absolute values of the roots of $r(\lambda)$ are dominated, i.e. strictly less than $\frac{1}{2^m}$.

4. Note that the polynomial $r(\lambda)$ also depends on the extraordinary ratio γ and the parameter w. We pick a suitable candidate value w after numerical experiments. To prove that the roots of the polynomial $r(\lambda)$ are dominated by $\underline{\lambda} := \frac{1}{2^m}$, we use Rouché's Theorem [Lan85] in the following way.

   a. Let $\tilde{r}(\lambda)$ be the polynomial obtained by replacing $\gamma \to \frac{1}{\gamma}$. By checking that $\tilde{r}(\lambda)\gamma^{\tilde{m}} = r(\lambda)$ for some $\tilde{m}$, we may assume that $\gamma \in (0, 1]$.
   b. $r(\lambda) := \sum_{s=0}^{p} d_s(\gamma)\lambda^s$ has coefficients $d_s(\gamma)$ that are themselves polynomials (with Bézier coefficients d_i^s) of degree k over $[0, 1]$.
   c. We check, separately for each i, by symbolic computation that

$$\sum_{s=0}^{p-1} |d_i^s|\bar{\lambda}^s - d_i^p\bar{\lambda}^p < 0, \quad d_i^p > 0. \tag{5}$$

Let $g(z) := \sum_{s=0}^{p} d_s(\gamma)z^s$ and $h(z) := d_p(\gamma)z^p$ for z on a circle of radius $\bar{\lambda}$. Then (5) implies the strict inequality in

$$|g(z) - h(z)| = |\sum_{s=0}^{p-1} d_s z^s| \leq \sum_{s=0}^{p-1} |d_s||z^s| < |h(z)|.$$

Rouché's Theorem [Lan85] implies that g and h have the same number p of roots in the $\bar{\lambda}$-disk, i.e. by the degree of g all roots of g are confined to the $\bar{\lambda}$-disk and hence $r(\lambda)$ is dominated by $\bar{\lambda}$.

2.1   Details of Proving Root Domination

Compared to constructions based on higher-degree interpolants, our minimal degree constructions have a smaller, easily checked factor $\ell(\lambda)$ but a more complex factor $r(\lambda)$. Using Rouché's Theorem, we show that the roots of the polynomial $r(\lambda)$ are dominated by $\bar{\lambda} > \underline{\lambda} := 1/2^m$, where $\bar{\lambda} = 1/5$ for the 6-point C^3 scheme of Section 3.1, $\bar{\lambda} = 1/7$ for the 8-point almost C^4 scheme of Section 3.2 and $\bar{\lambda} = 1/8$ for the 10-point C^4 scheme of Section 3.3.

To show that the roots of $r(\lambda)$ are dominated by $\underline{\lambda}$, $r(z)$ is considered as a complex function over the annulus $\underline{\lambda} \le |z| \le \bar{\lambda}$, $z := x + \mathrm{i}y$. We define $F_1(x, y) := |r(z)|^2$ and $F_2(x, y) := F_1(-y, x)$ and parameterize the positive quarter-annulus by

$$\rho := (\underline{\lambda}(1 - u) + \bar{\lambda}u)\left(\frac{1 - v^2}{1 + v^2}, \frac{2v}{1 + v^2}\right), \quad (u, v) \in [0 \mathinner{\ldotp\ldotp} 1]^2.$$

We further define $f_i(u, v, \gamma) := F_i \circ \rho(u, v)$, $i = 1, 2$. After scaling the denominator by $(1 + v^2)^d$, f_i becomes a polynomial (of high degree) in the variables (u, v, γ). It is converted to trivariate Bézier form. By looking at the coefficients, we can verify (in Section 3.1 3.2, 3.3) that these f_i and hence the functions F_1, F_2 are strictly positive. The proof for the other two quadrants not covered by f_i then follows by substituting the complex conjugate $z \to \bar{z}$ and observing that $r(z)$ has real coefficients.

3   Highly Smooth Non-uniform Interpolatory Subdivision

This section presents three constructions that yield respectively C^3, almost C^4 and C^4 curves. Denoting by $\mathbf{f}_i^k$ the polynomial of degree k that interpolates, for $s = 0, \ldots, k$, the points $\mathbf{p}_{i-\kappa+s}$ at the values $t_{i-\kappa+s}$, $\kappa := \lfloor \frac{k}{2} \rfloor$, we define the localized interpolant to be

$$\check{\mathbf{f}}_{i,j}^k(u) := \mathbf{f}_j^k((1 - u)t_i + ut_{i+1}), \quad u \in [0 \mathinner{\ldotp\ldotp} 1]. \tag{6}$$

3.1 C^3 6-point Scheme from Cubic Interpolants

Construction of new points $\tilde{\mathbf{p}}_{2i+1}$
1. The interpolating curves $\check{\mathbf{f}}_{i,i-1}^3$, $\check{\mathbf{f}}_{i,i}^3$ and $\check{\mathbf{f}}_{i,i+1}^3$ of degree 3 are expressed in Bézier form of degree 5 with coefficients $\mathbf{b}_k^l$, $\mathbf{b}_k^m$, $\mathbf{b}_k^r$, $k = 0, \ldots, 5$.
2. The Bézier coefficients $\mathbf{b}_k$ of a degree 5 curve $\mathbf{g}$ are defined as

$$\mathbf{b}_k := \frac{\mathbf{b}_k^l + \mathbf{b}_k^m}{2}, k = 0, 1, 2; \quad \mathbf{b}_k := \frac{\mathbf{b}_k^r + \mathbf{b}_k^m}{2}, k = 3, 4, 5.$$

3. Set

$$\tilde{\mathbf{p}}_{2i+1} := \tilde{\omega}\mathbf{g}(\frac{1}{2}) + (1 - \tilde{\omega})(\frac{1}{12}(\mathbf{b}_0 + \mathbf{b}_5) + \frac{5}{12}(\mathbf{b}_2 + \mathbf{b}_3)),$$

$$\tilde{\omega} := 16 - 1152w, \quad w := \frac{5}{384} \approx 0.01302.$$

Analysis. The analysis via z-transforms confirms that the construction for uniform knots is C^3. Since $\ell(\lambda) = (\lambda - w)^2$, we only needed to analyze the degree 5 polynomial $r(\lambda)$ according to Section 2.1 to confirm C^3 continuity. All Bézier coefficients of f_2 are strictly positive. Proving strict positivity of f_1 is only possible after subdividing the domain $[0 \mathinner{\ldotp\ldotp} 1]^3$ in the u- and γ-directions as shown in Fig. 3: The restriction of f_1 to each of the subdomains has strictly positive Bézier coefficients.

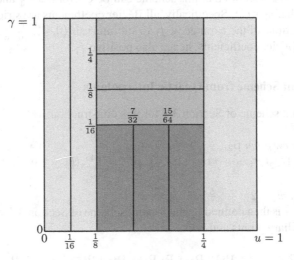

Fig. 3. Subdivision of the (u, γ) coordinates of f_1 to prove strict positivity and hence C^3 continuity of the new 6-point scheme

Comparison The global shape improvement over the C^3 schemes from [KP13b] can be observed in Fig. 2.

3.2   An almost C^4 8-point Scheme

Construction of new point $\tilde{\mathbf{p}}_{2i+1}$
1. We set $\mathbf{p}^l := \mathbf{f}_{i-1}^4(et_{i-1} + (1-e)t_i)$, $\mathbf{p}^r := \mathbf{f}_{i+2}^4((1-e)t_{i+1} + et_{i+2})$.
2. By $\mathbf{f}^l$ and $\mathbf{f}^r$ we denote degree 4 polynomials that interpolate respectively

$$\mathbf{f}^l(t_{i-1}) = \mathbf{p}_{i-1}, \ \mathbf{f}^l(et_{i-1} + (1-e)t_i) = \mathbf{p}^l, \ \mathbf{f}^l(t_i) = \mathbf{p}_i,$$

$$\mathbf{f}^l(t_{i+1}) = \mathbf{p}_{i+1}, \ \mathbf{f}^l(t_{i+2}) = \mathbf{p}_{i+2},$$

$$\mathbf{f}^r(t_{i-1}) = \mathbf{p}_{i-1}, \ \mathbf{f}^r(t_i) = \mathbf{p}_i, \ \mathbf{f}^r(t_{i+1}) = \mathbf{p}_{i+1},$$

$$\mathbf{f}^r((1-e)t_{i+1} + et_{i+2}) = \mathbf{p}^r, \ \mathbf{f}^r(t_{i+2}) = \mathbf{p}_{i+2}.$$

3. Set

$$\tilde{\mathbf{p}}_{2i+1} := \tilde{\omega}\frac{1}{2}(\mathbf{f}_i^4 + \mathbf{f}_{i+1}^4)(\frac{t_i + t_{i+1}}{2}) + (1 - \tilde{\omega})\frac{1}{2}(\mathbf{f}^l + \mathbf{f}^r)(\frac{t_i + t_{i+1}}{2}), \qquad (7)$$

$$\tilde{\omega} := \frac{1}{3(2-e)}(6 - 3e - 256we - 512w).$$

Analysis. For a uniform knot sequence, e cancels out. For the choice $w := 0.0038$, the analysis via z-transforms confirms that the construction for uniform knots is C^3. Careful numerical treatment shows that Hölder regularity exceeds 3.96 and that a nearby value of w yields an upper bound of 4.04 [Hor12]. That is, the analysis neither confirms C^4 continuity nor does it rule out C^4 continuity.

Analysis of the non-uniform case yields $\ell(\lambda) = (\lambda + w)^2$ and $r(\lambda)$ of degree 8. The analysis of Section 2.1 shows that this scheme can be C^4 for $e := \frac{1}{4}$ and $w := 0.0038$, if the uniform scheme is C^4. Specifically, all Bézier coefficients are strictly positive for f_2 and the restrictions of the u-range of f_1 to subintervals $(0, \frac{1}{32}, \frac{1}{16}, \frac{1}{8}, \frac{1}{4}, \frac{1}{2}, 1)$ yields strictly positive Bézier coefficients, hence also positive f_1.

3.3 C^4 10-point Scheme from Quartic Interpolants

We use the 8-point scheme of Section 3.2 for this construction.

Construction of new point $\tilde{\mathbf{p}}_{2i+1}$
1. We set $\mathbf{p}^l := \mathbf{f}^4_{i-2}(et_{i-3} + (1-e)t_{i-2})$, $\mathbf{p}^r := \mathbf{f}^4_{i+3}((1-e)t_{i+3} + et_{i+4})$, where $e := \frac{5-2048w}{5+6144w}$.
2. We set $\tilde{e} := \frac{1}{2}$, $\tilde{w} := 2w + \frac{5}{2048}$,
3. The point $\tilde{\mathbf{p}}_{2i+1}$ is then defined by the 8-point scheme of Section 3.2 with parameters $\tilde{e}$ and $\tilde{w}$ and auxiliary points and knots

$$\begin{array}{ccccccccc} \mathbf{p}^l & & \mathbf{p}_{i-2} \ \mathbf{p}_{i-1} \ \mathbf{p}_i \ \mathbf{p}_{i+1} \ \mathbf{p}_{i+2} \ \mathbf{p}_{i+3} & & \mathbf{p}^r \\ et_{i-3} + (1-e)t_{i-2} & t_{i-2} & t_{i-1} \ t_i \ t_{i+1} \ t_{i+2} \ t_{i+3} & (1-e)t_{i+3} + et_{i+4} \end{array}$$

Analysis The standard analysis of Section 2 yields $\ell(\lambda) = (\lambda - w)^2$ and $r(\lambda)$ of degree 12. For our choice of $w = 0.0014$, the analysis described in Section 2.1 yields that all Bézier coefficients are strictly positive and hence $f_i > 0$.

Comparison While the new 10-point construction improves the global shape compared to the 10-point scheme in Fig. 1, the improvement is not impressive and the global shape is worse than that of the almost C^4 8-point scheme that is also based on quartic interpolants but has smaller support.

4   Alternative Approaches to Improve Quality

Given the lack of decisive improvement for C^4 continuity, we explored a broader set of alternatives to deal with highly non-uniform data.

4.1 C^2 6-point Preparation

The simple C^2 6-point interpolatory scheme CR^2_w of [KP13b] consistently exhibits a good global shape with the choice $w = 3/256$ yielding a better curvature distribution than another natural choice $w = 1/192$, see Fig. 4b,c. Wide support and degree interpolants that exhibit poor global shape for highly non-uniform samples benefit from

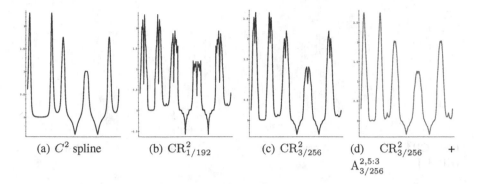

(a) C^2 spline   (b) $\text{CR}^2_{1/192}$ (c) $\text{CR}^2_{3/256}$ (d) $\text{CR}^2_{3/256}$ + $A^{2,5:3}_{3/256}$

Fig. 4. Curvature plots. Visually all curves are similar to $\text{CR}^2_{3/256}$ shown as black curve in Fig. 1b. (a) C^2 Catmull-Rom spline [KP13b] from which are derived (b) $\text{CR}^2_{1/192}$ and (c) $\text{CR}^2_{3/256}$. (c) has better curvature distribution and is used for preparation. (d) one step of $\text{CR}^2_{3/256}$ followed by $A^{2,5:3}_{3/256}$ thereafter.

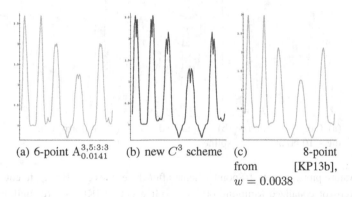

(a) 6-point $A^{3,5:3:3}_{0.0141}$ (b) new C^3 scheme   (c)   8-point
                                                                from   [KP13b],
                                                                $w = 0.0038$

Fig. 5. Curvature of C^3 subdivision curves after preparation with $\text{CR}^2_{3/256}$

applying a single initial step of $\text{CR}^2_{3/256}$. After this *6-point preparation*, the curves generated by the schemes improve both in global shape and in curvature distribution, to a degree usually observed only for uniformly distributed data. For example, 6-point preparation of the inferior 10-point scheme of [KP13b] in Fig. 6a demonstrates that a curvature distribution can be achieved, as good as for the new 10-point scheme Fig. 6b. The 6-point preparation also dramatically improves the red curve in Fig. 1c to a curve visually identical to $\text{CR}^2_{3/256}$ but with better curvature distribution. We note that repeated pre-processing leaves the global shape visually unchanged but appears to harm the curvature distribution (Fig. 6c,d).

4.2 Equalizing Knots Disappoint

By inserting the knots to make the spacing more uniform, we hoped to maintain curvature quality while switching from complicated non-uniform rules to simple uniform

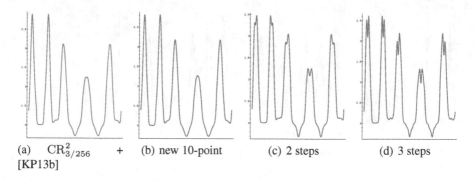

(a) $CR^2_{3/256}$ +   (b) new 10-point   (c) 2 steps   (d) 3 steps
[KP13b]

Fig. 6. Curvature plots of C^4 10-point subdivision curves after $CR^2_{3/256}$ preparation: (a) 10-point, $w = 0.00098$ from [KP13b] after one step of $CR^2_{3/256}$; (b) one (c) two (d) three steps followed by new 10-point scheme.

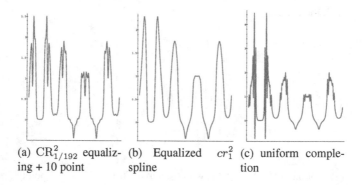

(a) $CR^2_{1/192}$ equaliz-   (b) Equalized cr^2_1 (c) uniform comple-
ing + 10 point   spline   tion

Fig. 7. Curvature plots. (a) new 10-point scheme after three steps of $CR^2_{1/192}$ to equalize knots; (b) Three steps of equalized sampling othe C^2 spline cr^2_1 of [KP13b]. (c) uniform 10-point scheme ($w = 0.0014$) after 4 steps of equalizing sampling of C^2 spline.

ones. We applied up to three steps of the adaptive $CR^2_{3/256}$ construction of [KP13b] to be able to define new points with equalizing knots according to [SD05]:

$$t^{k+1}_{2i+1} := (1 - \bar{t})t^k_i + \bar{t}t^k_{i+1}, \quad \bar{t} := \frac{\sqrt{t^k_{i+1} - t^k_{i-1}}}{\sqrt{t^k_{i+1} - t^k_{i-1}} + \sqrt{t^k_{i+2} - t^k_i}}. \tag{8}$$

However Fig. 7a illustrates that equalizing, followed by the new 10-point scheme, only harms the curvature distribution shown in Fig. 4c. (A referee has suggested that this follows from the lack of curvature continuity of the curves generated in [SD05], as recently shown in [FBCR1x]).

Several equalizing upsampling steps of a quartic C^2 spline (see Fig. 4a) shows no improvement in Fig. 7b despite increased effort. Subsequent uniform 10-point subdivision when the knot interval ratios are close to 2 clearly yields no progress; see Fig. 7c.

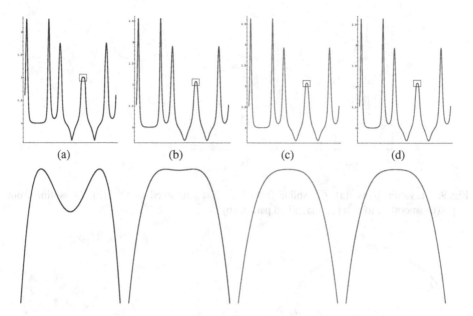

Fig. 8. (*top*) Curvature plots (visually the splines are hard to distinguish). (*bottom*) magnified vicinity of curvature distribution marked by the box in the respective top figure. Note, (a) is the same as Fig. 4(a). (b): C^2 spline (a) smoothed to C^3; (c): C^3 spline (b) smoothed to C^4; (d): C^4 spline (c) smoothed to C^5.

4.3 Subdivision Tracking Repeated-Smoothing-Interpolating Splines

[KP13b] introduced the idea of interlaced smoothing of interpolatory splines. Initially splines of low continuity and degree determine the global shape. Then the degree of these splines is raised and the additional degrees of freedom used to make the spline smoother, while still closely conforming to the initial shape. For example, we start with C^1 Catmull-Rom splines and express them as splines of degree 4. Then we smooth the spline, modifying the C^1 constraint and enforcing at the same time a new C^2 constraint to arrive at C^2 quartic splines. In a next step these C^2 quartic splines are raised to degree 6 followed by enforcing C^2 and C^3 constraints.

Interlacing degree-raising with smoothing is important for quality. Compared to immediately setting the degree and enforcing smoothness, interlacing yields better curvature distribution as demonstrated in [KP13b]; see Fig. 8, 9. New higher-order smoothing formulas for Fig. 8c,d are given in the Appendix. The progression of Fig. 8 typifies the beneficial effect of the smoothing process.

We can disguise the pure spline construction as a subdivision scheme by uniformly upsampling their Bézier segments as detailed in the Appendix Section 6.2, labelling each segment and inserting a new point at the middle of the interval. Spline construction plus upsampling amortizes over repeated subdivision steps so that after a few steps the approach is as efficient as simple uniform subdivision with higher continuities.

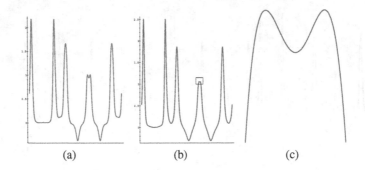

(a)   (b)   (c)

Fig. 9. Curvature plots. (a): C^2 spline from Fig. 8(a) smoothed to C^4; (b): C^3 spline from Fig. 8(b) smoothed to C^5; (c): magnified part of (b)

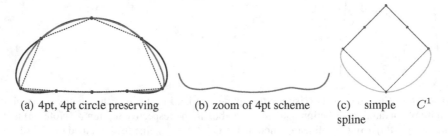

(a) 4pt, 4pt circle preserving   (b) zoom of 4pt scheme   (c) simple C^1
spline

Fig. 10. Interpolating C^1 curves. (a): red = 4-point scheme [DLG88], black = geometric (non-linear) circle preserving scheme from [DH12]; (b): vertical scaling of [DLG88] (also [DH12, SD05] visibly oscillates in (a)). (c) convex C^1 spline with collapsed control segment to deal with the classical trade-off between smoothness, convexity and interpolation, already present in the functional data $x_i = i, y_i = |i|, i \in -2, \ldots, 2$.

4.4   Relaxed Interpolation

One of the motivations of geometric (non-linear) subdivision is reproduction of basic shapes, such as the circle. While this is achieved in pieces, the transition between pieces of different shape often suffers (Fig. 10a, 11a).

An alternative is relaxed interpolation [ADS10]. Relaxed interpolation is akin to quasi-interpolation and generally, compared to strict interpolation, improves the shape for mildly changing data. However, as the oscillation in Fig. 11a illustrates, even relaxation does not cope well with rapidly changing discrete curvature. For denser samples, the reproduction property of geometric subdivision improves the shape but transitions remain a challenge (see Fig. 11d).

4.5   Curvature-Sensitive Interpolation

Not all data admit interpolation by smooth, convex curves, especially where local discrete curvature changes rapidly, see Fig. 10c. For the specific case, we can modify C^1 spline interpolation to preserve interpolation and convexity, the natural requirements for

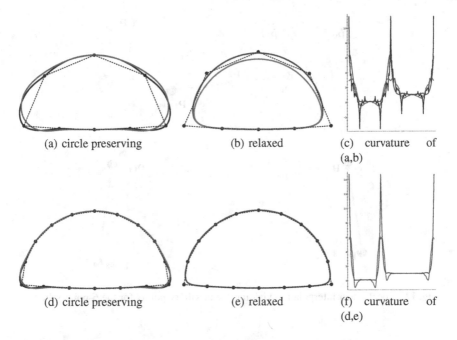

(a) circle preserving   (b) relaxed   (c) curvature of (a,b)

(d) circle preserving   (e) relaxed   (f) curvature of (d,e)

Fig. 11. Relaxed interpolation. (a,d) Non-linear subdivision curves: black = circle preserving [SD05], red = relaxed circle preserving [Sab10]. (b,e) blue = convex C^2 cubic, relaxed curvature-sensitive spline [KP13a], (b) green = cubic C^2 B-spline;

fair curves. We then have to give up on geometric smoothness, for example by collapsing a control segment. More generally, quasi-interpolation with splines does not fare well unless it is made 'curvature-sensitive' [KP13a] as shown in Fig. 11b. We think that, without taking into account discrete local curvature of the input data, any curve construction, whether subdivision or splines, can and will oscillate. Since curvature-sensitive splines switch depending on local discrete curvature, we do not present a subdivision analog. Such an analogue is surely difficult to analyze, especially since rigorous proofs of smoothness of simpler non-linear, geometric subdivision are still a challenge. (For splines we have at least smoothness by construction.) We note that curvature-sensitive splines can be modified to reproduce a circle [KP12] as illustrated in Fig. 11e and that while cubic C^2 B-splines are indeed of high quality, quasi-interpolating (relaxed) curvature-sensitive splines are closer to the input data, see Fig. 11b.

5   Least Degree

In our title and later on, we refer to interpolants of 'minimal degree' and give a rationale for seeking low degree. Indeed, our schemes use the natural midpoint insertion that generates piecewise uniformly-spaced knot subsequences and, according to [Dyn00], C^k uniform interpolating schemes have to reproduce all polynomials of degree k. Consequently we used interpolants of degree at least k for our C^k constructions.

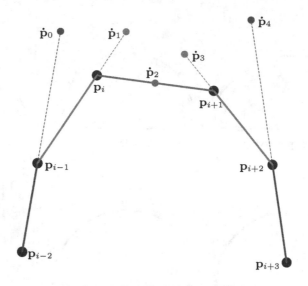

Fig. 12. Local linear interpolation yielding the auxiliary points for the 6-point scheme

A more pedantic but precise naming is *constructive* interpolants of minimal degree since, somewhat surprisingly, the interpolating schemes of high continuity can be obtained using only linear interpolants! Specifically we can build from linear interpolants Warren's C^2 6-point scheme (HERE originally local interpolating degree 5), our new C^3 6-point scheme (degree 3) and the almost C^4 8-point (degree 4). We conjecture that such formulas can also be found for the new 10-point scheme etc. But we rush to point out that, to find such linear-interpolation-based formulas, we first constructed a good scheme. To find good schemes from linear interpolants does not seem promising.

Formulas for symmetric $2n$-point schemes for linear interpolants (hence only half the formulas are needed) have the following construction (cf. Fig. 12). For a fixed i, we set

$$\dot{\mathbf{p}}_s := \mathbf{f}^1_{i-n+1+s}(\tfrac{1}{2}(t_i + t_{i+1})), \quad s = 0, \ldots, 2n - 2 \tag{9}$$

and denote $\dot{\beta}_s := \beta_{i-n+1+s}$, $s = 1, \ldots, 2n - 2$. Then the new point $\tilde{\mathbf{p}}_{2i+1}$ can be expressed via the following points $\dot{\mathbf{p}}_s$

$$\tilde{\mathbf{p}}_{2i+1} := \sum_{s=0}^{2n-2} \alpha_s \dot{\mathbf{p}}_s \text{ with symmetry } \dot{\beta}_s \to (\dot{\beta}_{2n-1-s})^{-1} \Rightarrow \alpha_r \to \alpha_{2n-2-r}. \tag{10}$$

where for the uniform
4-point * $\alpha_0 := 2w$.
6-point $\alpha_0 := -\tfrac{2}{3}w$, $\alpha_1 := \tfrac{8}{3}w + \tfrac{1}{8}$.
8-point $\alpha_0 := \tfrac{2}{5}w$, $\alpha_1 := -\tfrac{12}{5}w - \tfrac{1}{128}$, $\alpha_2 := 6w + \tfrac{5}{32}$.
10-point $\alpha_0 := -\tfrac{2}{7}w$, $\alpha_1 := \tfrac{16}{7}w + \tfrac{1}{1024}$, $\alpha_2 := -8w - \tfrac{7}{512}$, $\alpha_3 := 16w + \tfrac{175}{1024}$.
(*linear interpolating functions are consistent with C^1).

As an example of a non-uniform scheme, for CR_w^2, the formulas are

$$\alpha_4 := -\frac{8w}{(1+\dot{\beta}_3)(1+\dot{\beta}_4)(1+2\dot{\beta}_3)},$$

$$\alpha_3 := \frac{(1+\dot{\beta}_2)(1+2\dot{\beta}_3)(1+\dot{\beta}_4) + 32w(2+\dot{\beta}_2+2\dot{\beta}_3+\dot{\beta}_4+2\dot{\beta}_3\dot{\beta}_4)}{4(1+\dot{\beta}_2)(1+\dot{\beta}_3)(1+\dot{\beta}_4)(1+2\dot{\beta}_3)}.$$

6 Conclusions

While uniform schemes can be constructed algebraically and analyzed with z-transforms (see for example [KLY07]), it is difficult to see a similar calculus for non-uniform schemes. And with a rigorous prediction of shape and curvature not even available for linear subdivision, it is not surprising that the more complicated non-linear setting does not provide proofs. Hence we shared observations and corresponding recipes.

Starting from C^3, interpolatory non-uniform subdivision rules are not only quite complex, but the shape is unsatisfactory both for highly non-uniform samples and, as for all curve constructions, for strong change in discrete curvature. While careful minimal interpolant-based constructions yield some progress they are still not satisfactory. An initial step of the simple 6-point scheme $CR_{3/256}^2$, based on degree 2 interpolants, addresses non-uniform samples well. And strong change in discrete curvature can be handled by relaxing interpolatory requirements and possibly adding curvature-sensitive averaging.

Acknowledgments. The work was supported in part by NSF Grant CCF-1117695. We thank the referees. One referee pointed out an alternative construction for interpolants [BCR13] that appeared after our submission.

References

[ADS10] Ursula, H., Augsdörfer, N.A.: Dodgson, and Malcolm A. Sabin. Variations on the four-point subdivision scheme. Computer Aided Geometric Design 27(1), 78–95 (2010)

[BCR11a] Beccari, C., Casciola, G., Romani, L.: Polynomial-based non-uniform interpolatory subdivision with features control. Journal of Computational and Applied Mathematics 16(235), 4754–4769 (2011)

[BCR11b] Beccari, C.V., Casciola, G., Romani, L.: Non-uniform interpolatory curve subdivision with edge parameters built upon compactly supported fundamental splines. BIT Numerical Mathematics 51(4), 781–808 (2011)

[BCR13] Beccari, C.V., Casciola, G., Romani, L.: Construction and characterization of non-uniform local interpolating polynomial splines. J. Computational Applied Mathematics 240 (2013)

[DFH04] Dyn, N., Floater, M.S., Hormann, K.: A C^2 four-point subdivision scheme with fourth order accuracy and its extensions. In: Daehlen, M., Morken, K., Schumaker, L.L. (eds.) Mathematical Methods for Curves and Surfaces, Tromsoe, pp. 145–156 (2004)

[DGS99]   Daubechies, I., Guskov, I., Sweldens, W.: Regularity of irregular subdivision. Constr. Approx. 15(3), 381–426 (1999)

[DH12]   Dyn, N., Hormann, K.: Geometric conditions for tangent continuity of interpolatory planar subdivision curves. Computer Aided Geometric Design 29(6), 332–347 (2012)

[DL02]   Dyn, N., Levin, D.: Subdivision schemes in geometric modelling. Acta Numerica 11, 73–144 (2002)

[DLG88]   Dyn, N., Levin, D., Gregory, J.: A 4-point interpolatory subdivision scheme for curve design. Computer Aided Geometric Design 4(4), 257–268 (1988)

[Dyn92]   Dyn, N.: Subdivision schemes in computer-aided geometric design. In: Light, W. (ed.) Advances in Numerical Analysis II, pp. 36–104. Oxford University Press (1992)

[Dyn00]   Dyn, N.: Interpolatory subdivision schemes. In: Iske, A., Quak, E., Floater, M.S. (eds.) Tutorials on Multiresolurion in Geometric Modelling, pp. 25–50. Springer, Heidelberg (2000)

[FBCR1x]   Floater, M., Beccari, C.V., Cashman, T., Romani, L.: A smoothness criterion for monotonicity-preserving subdivision. Advances in Computational Mathematics 240 (2013)

[Hor12]   Hormann, K.: Private communication (October 2012)

[KLY07]   Ko, K.P., Lee, B.-G., Yoon, G.J.: A study on the mask of interpolatory symmetric subdivision schemes. Applied Mathematics and Computation 187(2), 609–621 (2007)

[KP12]   Karčiauskas, K., Peters, J.: Curvature-sensitive splines. Presentation at: 8th Itl. Conference on Mathematical Methods for Curves and Surfaces, Oslo, Norway (2012)

[KP13a]   Karčiauskas, K., Peters, J.: Curvature-sensitive splines and design with basic curves. Computer-Aided Design (45), 415–423 (2013)

[KP13b]   Karčiauskas, K., Peters, J.: Non-uniform interpolatory subdivision via splines. Journal of Computational and Applied Mathematics, MATA 2012 Issue 240, 31–41 (2013)

[Lan85]   Lang, S.: Complex Analysis, 2nd edn. Springer, New York (1985)

[Lee89]   Lee, E.: Choosing nodes in parametric curve interpolation. Computer Aided Design 21(6) (1989); Presented at the SIAM Applied Geometry Meeting, Albany, N.Y. (1987)

[Sab10]   Sabin, M.: Analysis and Design of Univariate Subdivision Schemes. Geometry and Computing, vol. 6. Springer, New York (2010)

[SD05]   Sabin, M.A., Dodgson, N.A.: A circle-preserving variant of the four-point subdivision scheme. In: Mathematical Methods for Curves and Surfaces: Tromsø 2004, Modern Methods in Mathematics (2005)

[War95]   Warren, J.: Binary subdivision schemes for functions over irregular knot sequences. In: Dæhlen, M., Lyche, T., Schumaker, L.L. (eds.) Proceedings of the First Conference on Mathematical Methods for Curves and Surfaces (MMCS 1994), Nashville, USA, June 16-21, pp. 543–562. Vanderbilt University Press (1995)

[Wei90]   Weissman, A.: A 6-point interpolatory subdivision scheme for curve design. PhD thesis, Tel Aviv University (1990)

Appendix

6.1   Interlaced Spline Smoothing

Since formulas for C^1, C^2 and C^3 constructions appeared in [KP13b, Section 2], we list here only formulas for smoothness higher than C^3. Denote the Bézier control points of two consecutive curve segments of degree m, connected with geometric continuity parameter β, by $\tilde{\mathbf{b}}_0, \ldots, \tilde{\mathbf{b}}_m$, respectively $\mathbf{b}_0, \ldots, \mathbf{b}_m$, $\tilde{\mathbf{b}}_m = \mathbf{b}_0$. Each step modifies the C^{k-1} constraint and enforces a new C^k constraint.

Smoothing $C^2 \to C^4$ The curves are assumed C^2 connected. We set (redefine)

$$\mathbf{b}_3 := a_0 \tilde{\mathbf{b}}_{m-4} + a_1 \tilde{\mathbf{b}}_{m-2} + a_2 \mathbf{b}_0 + a_3 \mathbf{b}_2 + a_4 \mathbf{b}_4 ,$$

$$\tilde{\mathbf{b}}_{m-3} := \tilde{a}_0 \tilde{\mathbf{b}}_{m-4} + \tilde{a}_1 \tilde{\mathbf{b}}_{m-2} + \tilde{a}_2 \mathbf{b}_0 + \tilde{a}_3 \mathbf{b}_2 + \tilde{a}_4 \mathbf{b}_4 , \tag{11}$$

$$a_0 := -\frac{\beta^4}{4(1+\beta)} , \quad a_1 := \frac{1}{2}\beta^2(1+\beta) , \quad a_2 := -\frac{1}{4}(1+\beta)^3 ,$$

$$a_3 := 1 + \beta , \quad a_4 := \frac{1}{4(1+\beta)} ;$$

and $\tilde{a}_k(\beta) := a_{4-k}(\frac{1}{\beta})$, $k = 0, \ldots, 4$.

Smoothing $C^3 \to C^5$ The curves are assumed C^3 connected. We set (redefine)

$$\mathbf{b}_4 := a_0 \tilde{\mathbf{b}}_{m-5} + a_1 \tilde{\mathbf{b}}_{m-3} + a_2 \tilde{\mathbf{b}}_{m-2} + a_3 \mathbf{b}_2 + a_4 \mathbf{b}_3 + a_5 \mathbf{b}_5 ,$$

$$\tilde{\mathbf{b}}_{m-4} := \tilde{a}_0 \tilde{\mathbf{b}}_{m-5} + \tilde{a}_1 \tilde{\mathbf{b}}_{m-3} + \tilde{a}_2 \tilde{\mathbf{b}}_{m-2} + \tilde{a}_3 \mathbf{b}_2 + \tilde{a}_4 \mathbf{b}_3 + \tilde{a}_5 \mathbf{b}_5 , \tag{12}$$

$$a_0 := \frac{\beta^5}{5(1+\beta)} , \quad a_1 := -\frac{3}{5}\beta^3(1+\beta) , \quad a_2 := \frac{2}{5}\beta^2(1+\beta)^2 ,$$

$$a_3 := -\frac{3}{5}(1+\beta)^2 , \quad a_4 := \frac{7}{5}(1+\beta) , \quad a_5 := \frac{1}{5(1+\beta)} ;$$

and $\tilde{a}_k(\beta) := a_{5-k}(\frac{1}{\beta})$, $k = 0, \ldots, 5$.

6.2 Interpolatory Subdivision Replicating an Underlying Spline

Let $\mathbf{f}_i$ be pieces of the spline, of any continuity ≥ 0, in Bézier form of degree m, defined over $[0 .. 1]$.

– The spline is sampled at the points $\mathbf{f}_r(\frac{j}{m})$, $j = 0, \ldots, m$, $j = 0, \ldots, m - 1$, $(\mathbf{f}_{r-1}(\frac{m}{m}) = \mathbf{f}_r(\frac{0}{m}))$. To the segment $\mathbf{f}_r(\frac{j}{m})$, $\mathbf{f}_r(\frac{j+1}{m})$ the label j is assigned. Sampled points are denoted by $\mathbf{p}_i$ and the knot spacing t_i is uniform.
– The new point with label s, corresponding to the segment $(\mathbf{p}_i, \mathbf{p}_{i+1})$, is the value at $\frac{t_i + t_{i+1}}{2}$ of the polynomial interpolant of degree m to the points $\mathbf{p}_{i-s+j}$, $j = 0, \ldots, m$. The interpolant coincides with the initial $\mathbf{f}_r$ of this segment defined over interval $[t_{i-s} .. t_{i-s+m}]$. Hence insertion rules indeed depend only on label s and are easily pre-calculated.
– New subsegments are labeled s.

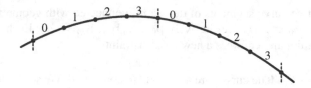

Fig. 13. Labelling the upsampled spline (case $m = 4$)

Pre-calculation of insertion rules We take Lagrange polynomial of degree m interpolating at i the points $\mathbf{p}_i$, $i = 0, \ldots, m$, and evaluate it at $s + \frac{1}{2}$. The coefficients for $\mathbf{p}_i$ form a mask of a new point corresponding to label s. Due to symmetry, only half the entries, $s = 0, \ldots, \frac{m}{2} - 1$, are displayed and the entries must be divided by 2^D

$m := 4$: $D = 7$
$\quad s := 0$ $(35, 140, -70, 28, -5)$;
$\quad s := 1$ $(-5, 60, 90, -20, 3)$.
$m := 6$: $D = 10$
$\quad s := 0$ $(231, 1386, -1155, -924, -495, 154, -21)$;
$\quad s := 1$ $(-21, 378, 945, -420, 189, -54, 7)$;
$\quad s := 2$ $(7, -70, 525, 700, -175, 42, -5)$.
$m := 8$: $D = 15$
$\quad s := 0$ $(6435, 51480, -60060, 72072, -64350, 40040, -16380, 3960, -429)$;
$\quad s := 1$ $(-429, 10296, 36036, -24024, 18018, -10296, 4004, -936, 99)$;
$\quad s := 2$ $(99, -1320, 13860, 27720, -11550, 5544, -1980, 440, -45)$;
$\quad s := 3$ $(-45, 504, -2940, 17640, 22050, -5880, 1764, -360, 35)$.

Representation of Motion Spaces Using Spline Functions and Fourier Series

Thomas Kronfeld, Jens Fankhänel, and Guido Brunnett

Technische Universität Chemnitz, Computer Graphics Lab
{tkro,guido.brunnett}@cs.tu-chemnitz.de

Abstract. Natural looking human motion are difficult to create and to manipulate because of the high dimensionality of motion data. In the last years, large collections of motion capture data are used to increase the realism in character animation. In order to simplify the generation of motion, we present a mathematical method to create variations in motion data. Given a few samples of motion data of a particular activity, our framework generates a high dimensional continuous motion space. Therewith our motion synthesis framework is able to synthesize motion by varying boundary conditions. Furthermore, we investigate the different properties of spline functions and Fourier series and their suitability for the description of complex human motion. We have derived an optimization heuristic, which is used to automatically generate the initial motion space. We have evaluated our system by comparison against ground-truth motion data and alternative methods.

1 Introduction

Digital human models for modelling and simulating human-like motion are used in a wide variety of different applications, including humanoid robotics, biomechanics, virtual prototyping, and character animation. Beside these classical applications digital humans are often used for product and process design within a digital production setting. Here the motion of the human model is used to analyse load handling, field of view and reachability tasks. The results of there ergonomic analyses can be used to optimize the motion sequence and the workspace in order to minimize exhausting and hazardous situations. Common digital human models used within the digital factory are RAMSIS and HumanBuilder, for example. A survey of digital humans and their application can be found in [15]. They are either integrated modules for use within CAx systems or standalone products which expose several import/export interfaces to communicate with CAx systems. In order to generate motion sequences, they use forward or inverse kinematics. This means these systems basically rely on keyframing methods. Keyframing has the advantage, that it provides the user with absolute control over the positioning process. However, it is a notoriously time consuming task to synthesize complex movement, such as pick up a box or hammering a nail. The main reason is that a typical articulated human model usually has at least 90 degrees of freedom. The animator must then painstakingly animate each of

M. Floater et al. (Eds.): MMCS 2012, LNCS 8177, pp. 265–282, 2014.

these degrees of freedom, one at a time. In order to speed up the motion generation, the user can define macros to store frequently repeated motion sequences within a motion database. There are several studies that examine the practical usability of these human models. Many of these studies, such as Spanner-Ulmer et al. [16], show that there is an increased demand for better and faster motion adjustment. In order to generate reliable motion sequences, most of the systems mentioned above are able to use motion capture data as input motion. Here the motion of a real life person is captured using a tracking device. However these methods are limited to playback of existing motion sequences.

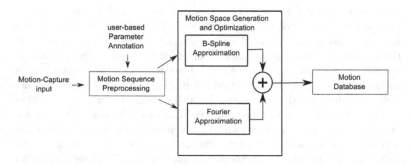

Fig. 1. Overview of our motion-processing framework

The goal of our framework is to provide a motion synthesis database, which generates biomechanical and physically correct movement. Therefore, we have to fulfil two competing goals: on one hand, we would like to have an adaptable motion which covers a whole range of possible activities. On the other hand, the used storage space should be as small as possible, in order to handle large system which is able to generate all kinds of motions. Figure 1 gives an overview of the proposed framework.

In our current experiments we rely completely on motion capture data as input to our system. However, it is also possible to use sequences of interpolated keyframes as starting point.

In the reminder of this paper, Section 2 places our work in the context of the recent literature. Section 3 gives an overview of our framework. Section 4 describes the mathematical definition of motion spaces as well as the developed algorithms. Section 5 presents the experimental results, while conclusions as well as further research directions are outlined in section 6.

2   Related Work

In computer animation, data-driven motion synthesis is an important technique to generate realistic motions from recorded motion capture data. Due to the frequent use of motion capture data, efficient editing tools of such data become more and more important. There have been an abundance of research results addressing the problem of developing motion editing tools to modify motion data [7].

The framework developed in this paper is similar in spirit to the work of Rose et al. [18]. The authors also separate motion data into activities, that they call verbs. A verb describes the base motion of the activity. In order to create certain styles, i.e. for the verb "walking" one wants to create the style "sneaky walking", they use adverbs to encode this. The adverbs are b-splines which describe the difference between the base motion ("walking") and the style ("sneaky"). In order to create a series of motion styles, they need to interpolate between the different styles and therefore need to apply b-spline interpolation.

In contrast to their work, we use splines and Fourier series to approximate the given sets of motion data. The motion data is annotated with certain parameters by the user, similar to the adverbs of Rose et al., but there is no need to identify a base motion of a certain activity.

Splines are extensively studied in computer sciences and engineering. They are also used in statistics to approximate scattered data [5]. However, most of these methods use fixed knot spline fitting and thus their use is very limited. If the knots are considered free, the approximation can be improved significantly. The free knot spline approximations are non-linear optimization algorithms where the spline coefficients and the knot positions must be determined. There are basically two different approaches to solve the problem. The first class of algorithms tries to solve the whole problem in one step, while the second class separates the problem in a linear least square optimization with fixed knots and a non-linear optimization to determine the knot position.

A member of the first class is the method developed by Holt and Fletcher [8]. They developed a special algorithm for least square problems with specially structured constraints and applied it to the problem of spline approximation with free knots. A serious problem of these approaches is the existence of a potentially high number of local extrema in the solution space of the approximation function. Jupp [11] calls this problem the lethargy-problem. With a logarithmic change of variables he transforms the constraint problem into an unconstrained one. He shows that this transformation in a certain sense reduces the difficulties related with the original formulation. When solving the resulting unconstrained non-linear least squares problem, he makes special use of its structure and separates the linear and non-linear aspects. Dierckx [6] developed an algorithm, where the constraints for the knot positions are treated via a certain barrier term in the objective function. The unconstrained problem obtained in this way is solved by the Fletcher/Reeves conjugate gradient method. Recently, Suchomski [22] proposed a new method of optimal variable-knot spline interpolation in the discrete L_2-norm. He explicitly enforces the Schoenberg-Whitney regularity condition, equation (14), which leads to additional constraints for the knots. With a logarithmic change of variables similar to that of Jupp, he transforms the constrained non-linear least squares problem into an unconstrained one.

De Boor and Rice [4] also consider the case of L_2-approximation by cubic splines with variable knots. Thereby they do not allow the knots to coalesce and

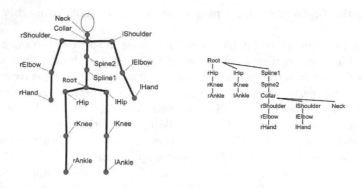

Fig. 2. The structure of a virtual human skeleton (left) and the corresponding hierarchical tree structure (right)

they use a certain "coordinate relaxation" where each knot is varied cyclically so as to minimize the L_2-error as a function of this knot alone. In the proposed framework we adopt this method and describe a possible generalization to multivariate spline approximation. Furthermore, we simplify the termination criterion described by de Boor and Rice [4].

Fourier series are used in a wide range of mathematical and physical applications. They are also used to process motion data. Bruderlin and Williams [1] applied a number of different signal processing techniques to motion data to allow editing. Unuma et al. [23] used Fourier principle to interpolate and extrapolate motion data in the frequency domain. The method described in this paper uses them to approximate motion data.

3   Motion Spaces

The configuration of a virtual human model is specified by its joint angles and the position of the root segment. The position of the root is given by a 3-dimensional vector and the joint rotations are given by Euler angles. An example of a skeleton structure is shown in figure 2. Motion data consists of a set of motion signals. Those signals are sampled at a sequence of discrete time steps with a uniform interval to form a motion sequence. The sampled values of the signals determine the configuration of an articulated figure at a time-step, which is called a pose.

A motion sequence of length m is given as a $n \times m$-matrix

$$\mathbf{M} = \begin{pmatrix} p_{1,1} & \cdots & p_{1,m} \\ \vdots & \ddots & \vdots \\ p_{n,1} & \cdots & p_{n,m} \end{pmatrix}. \tag{1}$$

Each column $\mathbf{p}_j = (p_{1,j}, \ldots, p_{n,j})^T$ (for $1 \leq j \leq m$) of $\mathbf{M}$ describes a pose of the virtual human, where n are the degrees of freedom of the skeleton. In the following, we consider each degree of freedom, i.e. the rows of $\mathbf{M}$, individually.

A row is denoted by an m-dimensional vector $\varphi_i = (p_{i,1}, \ldots, p_{i,m})$. A motion sequence is thus given by $\mathbf{M} = (\varphi_1, \ldots, \varphi_n)^T$.

In our case we have k motion sequences of a certain activity, for example motion sequences of pick up a box at different heights. In the first step, the user has to define the height parameter $\{h_1, \ldots, h_k\} = D_h$ (where $|D_h| = k$ and $h_1 < \ldots < h_k$) for each motion sequence.

$$\mathbf{M}_{h_j} = \begin{pmatrix} p_{1,1}^{h_j} & \cdots & p_{1,m}^{h_j} \\ \vdots & \ddots & \vdots \\ p_{n,1}^{h_j} & \cdots & p_{n,m}^{h_j} \end{pmatrix} = \begin{pmatrix} \varphi_1^{h_j} \\ \vdots \\ \varphi_n^{h_j} \end{pmatrix}, \quad \begin{array}{c} 1 \le j \le |D_h|. \\ h_i \in D_h \end{array} \tag{2}$$

We assume, that each motion sequence has the same length m, that is the same number of poses.

In order to describe motion sequences of an arbitrary activity we associate r parameters:

$$\{w_1^1, \ldots, w_{n_1}^1\} = D_{w^1} \text{ where } |D_{w^1}| = n_1 \text{ and } w_1^1 < \ldots < w_{n_1}^1 \\ \vdots \tag{3} \\ \{w_1^r, \ldots, w_{n_r}^r\} = D_{w^r} \text{ where } |D_{w^r}| = n_r \text{ and } w_1^r < \ldots < w_{n_r}^r$$

with each motion sequence:

$$\mathbf{M}_{w_{i_1}^1, \ldots, w_{i_r}^r} = \begin{pmatrix} \varphi_1^{w_{i_1}^1, \ldots, w_{i_r}^r} \\ \vdots \\ \varphi_n^{w_{i_1}^1, \ldots, w_{i_r}^r} \end{pmatrix}, \quad \begin{array}{c} w_{i_1}^1 \in D_{w^1} \\ \vdots \\ w_{i_r}^r \in D_{w^r}. \end{array} \tag{4}$$

The discrete parameters are either defined by the user or given as annotation of the captured motion. Each of these motion sequences of a certain activity is an element of a discrete motion space. The discrete motion space is defined as a structure, where the motion signals of each degree of freedom are arranged in a separate set:

$$\mathbf{\Phi}^{D_{w^1}, \ldots, D_{w^r}} = \left\{ \Phi_1^{D_{w^1}, \ldots, D_{w^r}}, \ldots, \Phi_n^{D_{w^1}, \ldots, D_{w^r}} \right\} \tag{5}$$

where

$$\Phi_i^{D_{w^1}, \ldots, D_{w^r}} = \left\{ \varphi_i^{w_{i_1}^1, \ldots, w_{i_r}^r} \mid w_{i_1}^1 \in D_{w^1} \wedge \ldots \wedge w_{i_r}^r \in D_{w^r} \right\} \tag{6}$$

for $i = 1, \ldots, n$.

4   Calculation of Motion Spaces

What we intend to do is to calculate a continuous motion space, where the user can select any parameter within the ranges of the discrete motion space. To achive this, we approximate the given discrete motion space using regression analysis.

The motion signal of an arbitrary activity within a continuous motion space is given as a $r+1$ dimensional function:

$$\varphi_i(v_1,\ldots,v_r,t) \tag{7}$$

where $1 \leq i \leq n$ and $1 \leq t \leq m$ is the time step. The parameters v_j (for $1 \leq j \leq r$) of the function $\varphi_i(\cdot)$ correspond to the parameters $w^j \in D_{w^j}$ defined within the discrete motion space.

In order to calculate a pose of the virtual human at a discrete time step t one has to evaluate all motion signals of the continuous motion space:

$$\Theta(v_1,\ldots,v_r,t) = \begin{pmatrix} \varphi_1(v_1,\ldots,v_r,t) \\ \vdots \\ \varphi_n(v_1,\ldots,v_r,t) \end{pmatrix}, \quad \text{where } v_j \in [w_1^j, w_{n_i}^j] \subset \mathbb{R}. \tag{8}$$

4.1 Regression

Regression analysis is a statistical technique for modelling and analysing the relationship between several variables. The goal is to model the dependence of a response variable y on one or more predictor variables $x_1,\ldots,x_{l-1}$ with given realizations $N = (x_1^k,\ldots,x_{l-1}^k, y_k)_{k=1}^r$ where $(x_1^k,\ldots,x_{l-1}^k, y_k) \in \mathbb{R}^l$ for all $k \in 1,\ldots,r$. The point set N represents the unknown function $f : \mathbb{R}^{l-1} \to \mathbb{R}$ where

$$y_k = f(x_1^k,\ldots,x_{l-1}^k) + \varepsilon_k \quad \forall k = 1,\ldots,n.$$

The additive term ε_i characterizes the statistical errors, which are independent and identically distributed with zero mean and constant variance. The aim of regression analysis is to use the data to construct a function $\hat{f}(x_1,\ldots,x_n)$ that can serve as a reasonable approximation to f. Here reasonable means accurate in the least squares sense, since the function $\hat{f}$ is used to approximate f at points which are not part of the initial point set.

The basic least squares method consists of finding the function $\hat{f}$ which minimizes the sum of squared errors

$$\varrho = \sum_{k=1}^{r} \left(y_k - \hat{f}(x_1^k,\ldots,x_{l-1}^k) \right)^2 \tag{9}$$

With regard to motion spaces, the focus lies on the relationship between several independent variables, the parameters of the motion sequences and one dependent variable, the corresponding joint angle. The input motion space is specified in equation (5) as a set of sets. The elements are the motion signals of a certain degree of freedom of the skeleton, defined in equation (6).

The optimization is performed separately for each degree of freedom, by utilizing the least square criterion:

$$\varrho_i = \sum_{t=1}^{m} \sum_{w^1 \in D_{w^1}} \cdots \sum_{w^r \in D_{w^r}} (p_{i,t}^{w^1,\ldots,w^r} - \varphi_i(w^1,\ldots,w^r,t))^2 \tag{10}$$

where $1 \leq i \leq n$ denotes the degree of freedom.

4.2   B-Spline Approximation

This section describes the approximation method using b-splines. In the first part we consider a single motion signal and approximate them by cubic b-spline with variable number of knots and knot positions. To simplify the mathematical model, a motion signal of the ith degree of freedom of the underlying skeleton is given by $\varphi_i = (p_{i,1}, \ldots, p_{i,m})$, all other parameters are omitted in the following illustration. The goal of the optimization process is to calculate a function $\hat{f}$ which minimizes:

$$
\begin{aligned}
\varrho_i &= \sum_{j=1}^{m} (p_{i,j} - \hat{f}(j))^2 \\
&= \sum_{j=1}^{m} (p_{i,j} - \sum_{l=-d}^{k} (c_l \cdot B_l^{(d)}(j)))^2
\end{aligned}
\tag{11}
$$

where $d = 3$ is the degree of the basis function defined on the vector of knots $(\lambda_{-d}, \ldots, \lambda_{k+d+1})$ with coincident boundary knots $\lambda_{-d} = \ldots = \lambda_0 = 1$ and $\lambda_{k+1} = \ldots = \lambda_{k+d+1} = m$. Equation (11) can be rewritten in matrix notation:

$$
\varrho_i = (\varphi_i^T - \mathbf{B}\mathbf{c})^T (\varphi_i^T - \mathbf{B}\mathbf{c}) = \|\varphi_i^T - \mathbf{B}\mathbf{c}\|^2
\tag{12}
$$

where $\| \cdot \|$ denotes the euclidean vector norm and

$$
\mathbf{B} = \begin{pmatrix} B_{-d}^{(d)}(1) & \ldots & B_k^{(d)}(1) \\ \vdots & \ddots & \vdots \\ B_{-d}^{(d)}(m) & \ldots & B_k^{(d)}(m) \end{pmatrix}, \qquad \mathbf{c} = \begin{pmatrix} c_{-d} \\ \vdots \\ c_k \end{pmatrix}
\tag{13}
$$

are a $m \times (k+d+1)$ dimensional observation matrix containing the basis functions and a $(k + d + 1)$ dimensional vector of coefficients, respectively. The goal is to determine the unknown coefficients c_i for $i \in \{-d, \ldots, k\}$. Therefore one has to solve an overdetermined linear system

$$
\mathbf{B} \cdot \mathbf{c} = \varphi_i^T.
$$

In order to solve this, the matrix $\mathbf{B}$ has to have full rank $k + d + 1$. According to Cox [2] this will be the case if and only if the knot vector satisfies the Schoenberg-Whitney condition, i.e. if there exists $\{u_{-d}, \ldots, u_k\} \subset \{1, \ldots, m\}$ with $u_j < u_{j+1}$ such that

$$
\lambda_j < u_j < \lambda_{j+d+1}, \qquad j = -d, \ldots, k.
\tag{14}
$$

The optimization algorithm used herein is a simplified version of the algorithm described by de Boor and Rice [4]. The algorithm consists of two nested loops. The inner loop relocates the knots in a greedy fashion, while the outer loop inserts knots in intervals with high L_2-error.

The method starts with the initial knot-vector Λ_0 which has only one interior knot $\lambda_1^0 = \frac{m+1}{2}$ at the center of the time interval. After each calculation of the least square optimization the L_2-error for each interior knot is calculated. For the relocation of the knots, the intervals $[\lambda_{i-1}, \lambda_i, \lambda_{i+1}]$ for $1 \leq i \leq k$ are considered and processed in order of decreasing L_2-error. For each of the three nodes of the considered interval, the L_2-error $e(\lambda_{i-1})$, $e(\lambda_i)$ and $e(\lambda_{i+1})$ is calculated. These three values are used to determine the minimum of the parabola $p(t)$ satisfying $p(\lambda_{i-1}) = e(\lambda_{i-1})$, $p(\lambda_i) = e(\lambda_i)$ and $p(\lambda_{i+1}) = e(\lambda_{i+1})$, respectively. It is assumed, that the minimum is located within the interval $[p(\lambda_{i-1}), p(\lambda_{i+1})]$, otherwise the interval is skipped. Therefore only intervals where $e(\lambda_{i-1}), e(\lambda_{i+1}) \geq e(\lambda_i)$ are considered. During the relocation process the inner knots are moved within the "safe interval" defined by de Boor and Rice [3]:

$$[\lambda_{i-1} + 0.0625(\lambda_{i+1} - \lambda_{i-1}), \lambda_{i+1} - 0.0625(\lambda_{i+1} - \lambda_{i-1})] . \tag{15}$$

so that the Schoenberg-Whitney condition remains satisfied. After relocation, the involved knots are labelled as processed and are no longer considered for the current iteration. Intervals that contain a processed knot are skipped. After processing the hole knot vector, the method continues with the least-square optimization and starts the next iteration. The inner loop stops when either the knots begin to oscillate or the L_2-error is not reduced significantly. Therefore, during the relocation step the position of the knots as well as the current L_2-error are recorded. De Boor and Rice [4] suggest to perform at least four iterations of the inner loop. After these four loops the reduction of the error of each loop is calculated. The assumption is, that the method is converging linearly and the error is reduced at each cycle. The inner loop terminates if the reduction falls below a user defined threshold $\varepsilon_{reduction}$ i.e. $\left| \frac{\varrho_{i,j-1} - \varrho_{i,j}}{\varrho_{i,j}} \right| \leq \varepsilon_{reduction}$, where $\varrho_{i,j}$ denotes the L_2-error of the ith motion signal after the jth iteration.

The outer loop checks if the L_2-error of the current approximation spline is below a user defined threshold. If this is not the case, then a knot is inserted near the location of the maximum error. In order not to violate the Schoenberg-Whitney condition, the inserted knot has to lie within the safe interval defined in equation (15).

The other dimensions are relatively sparse compared to the uniformly sampled time dimension. Therefore there are cases where a large region contains a few data points or the number of spline coefficients is close to the number of data points. This may causes a rank deficiency observation matrix, thus the least square solution is no longer unique. The points of the given set are, however, located on a rectangular grid, making the Schoenberg-Whitney condition remain valid. In the following, the two-dimensional case is discussed. The extension to higher dimensions is straightforward and outlined at the end of this section.

Considering a set of motion signals with one additional parameter $w_{i_1}^1 \in D_{w^1}$ for $1 \leq i_1 \leq |D_{w^1}| = n_1$ and the point set $N_i = \{(w_{i_1}^1, t, p_{i,t}^{w_{i_1}^1}) | 1 \leq t \leq m \wedge w_{i_1}^1 \in D_{w^1} \wedge 1 \leq i_1 \leq |D_{w^1}| = n_1\}$. Therewith one has a rectangular domain $[1, m] \times [w_1^1, w_{n_1}^1]$. The problem here is that the distribution of the points in N_i is relative sparse, which leads to the fact, that one can not solve the linear equation system. For this reason, the motion signals are interpolated linearly. The interpolated point set is denoted by $\tilde{N}_i$ and the associated parameters make up the set $\tilde{D}_{w^1}$. The interpolated values only provide a rough approximation to the real values. In order to reduce their influence, a weighted least square optimization is used in the following.

Here the weights of the data-points in N_i are set to $\alpha_{i,t}^{w^1} = 1$. For the interpolated data points $\tilde{N}_i$, we suggest to use weights of $\alpha_{i,t}^{w^1} \leq 0.3$. The least square optimization is then calculated for the point-set $\hat{N}_i = N_i \cup \tilde{N}_i$ and the parameter set $\hat{D}_{w^1} = D_{w^1} \cup \tilde{D}_{w^1}$.

$$\varrho_i = \sum_{t=1}^{m} \sum_{w \in \hat{D}_{w^1}} (\alpha_{i,t}^{w} \, p_{i,t}^{w} - \alpha_{i,t}^{w} \, \varphi_i(w,t))^2 \tag{16}$$

where $\alpha_{i,t}^{w}$ are the individual weights and

$$\varphi_i(w,t) = \sum_{l_1=-d}^{k_1} \sum_{l_2=-d}^{k_2} c_{l_1,l_2} B_{l_1}^{(d)}(t) B_{l_2}^{(d)}(w) \tag{17}$$

is a tensor product spline of degree d. As in the univariate case, the unknown spline coefficients c_{l_1,l_2} can be determined as the least-square solution of the overdetermined linear system:

$$\alpha_{i,t}^{w} \sum_{l_1=-d}^{k_1} \sum_{l_2=-d}^{k_2} c_{l_1,l_2} B_{l_1}^{(d)}(t) B_{l_2}^{(d)}(w) = \alpha_{i,t}^{w} \, p_{i,t}^{w}. \tag{18}$$

This can also transformed in matrix notation:

$$\mathbf{B} \, \mathbf{vec}(\mathbf{c}) = \mathbf{vec}(\mathbf{z}) \tag{19}$$

where $\mathbf{B}$ is a $(k_1+d+1)(k_2+d+1) \times m$ matrix with $\alpha_{i,t}^{w} B_{l_1}^{(d)}(t) B_{l_2}^{(d)}(w)$ as elements. By $\mathbf{vec}(A)$ we denote the column vector obtained by putting the columns of the matrix A underneath each other in their natural order. The matrices are therefore:

$$\mathbf{c} = \begin{pmatrix} c_{-d,-d} & \cdots & c_{k_1,-d} \\ \vdots & \ddots & \vdots \\ c_{-d,k_2} & \cdots & c_{k_1,k_2} \end{pmatrix} \tag{20}$$

and

$$\mathbf{z} = \begin{pmatrix} \alpha_{i,1}^{w_1} p_{i,1}^{w_1} & \cdots & \alpha_{i,m}^{w_1} p_{i,m}^{w_1} \\ \vdots & \ddots & \vdots \\ \alpha_{i,1}^{w_{\hat{n}_1}} p_{i,1}^{w_{\hat{n}_1}} & \cdots & \alpha_{i,m}^{w_{\hat{n}_1}} p_{i,m}^{w_{\hat{n}_1}} \end{pmatrix} \tag{21}$$

where $w_i \in \hat{D}_w$ for $1 \leq i \leq |\hat{D}_w| = \hat{n}_1$. Furthermore it is assumed, that the w_i is sorted in ascending order.

In the two-dimensional case the number of knots and their positions are optimized similarly to the one-dimensional case. In order to create an initial knot vector, the knots of the time dimension, calculated with the one dimensional least square optimization can be used. Therefore, one can choose the approximation

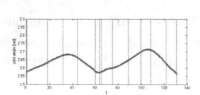
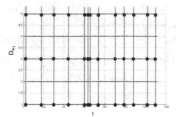

Fig. 3. The one dimensional spline approximation with free knots is shown in the left figure. Initialization of the two dimensional spline approximation is shown in the right figure. Here the knots are marked as black dots.

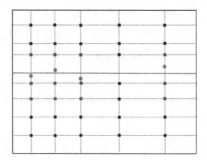

Fig. 4. Example of the knot relocation step. The knots of the intervals considered during the relocation step, are outlined in red and the calculated minimum of the corresponding parabola is shown as a red dot. The knots are then repositioned along the red line.

of an arbitrary motion signal, since these knots are relocated in the following optimization progress. For the other dimension, it is useful to start with $n_1/2$ equally spaced inner knots. An example of the initialization is shown in figure 3.

During the optimization the knots need to be relocated. The relocation of the knots is performed separately in every dimension. Therefore one can extend the method proposed by de Boor and Rice [4] (see figure 4). At first all intervals where $e(\lambda_{i,j-1}), e(\lambda_{i,j+1}) \geq e(\lambda_{i,j})$ (for $1 \leq i \leq k_1$ and $1 \leq j \leq k_2$) holds are marked. After that for each i the number of marked and unmarked intervals is compared. If the majority of the intervals is marked, then all knots are relocated to the weighted average of the calculated minima. This is performed iteratively for every dimension. As termination criterion, the criteria defined for the one dimensional case are used.

The knot insertion is realized analogue to the one dimensional case. At first one has to determine the location of maximal L_2-error. The new knots are inserted along the line passing through that maximum. This process is performed separately for every dimension (see figure 5). The insertion of the knots in the

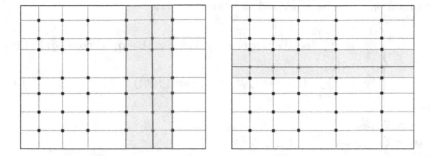

Fig. 5. Example of the knot insertion. To maintain the lattice structure of the knots, one must insert an entire row/column of knots.

parameter dimension is not trivial, since there are regions where the point set N_i is sparse. Therefore, prior to every least square optimization, one has to check if the Schoenberg-Whitney condition is satisfied. If there are intervals where this is not the case, one has to augment the point set $\tilde{N}_i$ and the parameter set $\tilde{D}_{w_1}$, respectively.

In case of higher dimensional motion spaces, we apply an $r+1$-dimensional tensor product spline regression. The optimization begins with the approximation of the one-dimensional motion signal. After that a $r+1$-dimensional lattice is generated, and the knot optimization is performed analogue to the two dimensional case.

4.3   Fourier Approximation

As a second approach we approximate the motion signals using a trigonometric polynomial. Considering a single motion signal $\varphi = (p_1, \ldots, p_m)$ and the corresponding point set $N = \{(t_j, p_j)\}$ for $1 \leq j \leq m$ and $t_j = j$. We use the following approach for approximating function

$$\varphi(t) = \hat{p}_{l-1}(t) = \sum_{j=0}^{l-1} \xi_j \cdot \exp\left(\mathtt{i} jqt\right) \tag{22}$$

where $\mathtt{i}$ is the imaginary unit, q is a constant scaling factor and $l \leq m$. Using that approach, we have to solve the following minimization problem:

$$\varrho(l) = \min_{\xi \in \mathbb{C}^l} \sum_{j=1}^{m} \left(p_j - \hat{p}_{l-1}\left(t_j\right)\right)^2. \tag{23}$$

If the independent variables t_j are equidistant, then we could use the discrete Fourier transform to solve the minimization problem (23). In this case the polynomial $\hat{p}_{l-1}$ can be expressed as follows:

$$\hat{p}_{l-1}(x) = \begin{cases} \frac{\alpha_0}{2} + \sum\limits_{j=1}^{l/2-1} (\alpha_j \cdot \cos{(jqx)} + \beta_j \cdot \sin{(jqx)}) + \alpha_{l/2} \cdot \cos{\left(\frac{l}{2}qx\right)}, & l \text{ even} \\ \frac{\alpha_0}{2} + \sum\limits_{j=1}^{(l-1)/2} (\alpha_j \cdot \cos{(jqx)} + \beta_j \cdot \sin{(jqx)}), & l \text{ odd} \end{cases}$$

where

$$\begin{aligned} \alpha_0 &= 2 \cdot \beta_0; \\ \alpha_j &= \xi_j + \xi_{l-j}, & 1 \leq j < \tfrac{l}{2}; \\ \alpha_{l/2} &= \xi_{l/2}, & l \text{ even}; \end{aligned}$$

and

$$\mathrm{i} \cdot \beta_j = \xi_j - \xi_{l-j}, \qquad 1 \leq j < \tfrac{l}{2}.$$

$$(24)$$

The unknown coefficients α_i and β_j can be arranged in the vector γ as follows:

$$\gamma = \begin{bmatrix} \beta_s \\ \beta_{s-1} \\ \vdots \\ \beta_0 \\ \alpha_t \\ \alpha_{t-1} \\ \vdots \\ \alpha_1 \end{bmatrix} \in \mathbb{R}^{l-1} \quad \text{where} \quad t := \left\lfloor \frac{l}{2} \right\rfloor \quad \text{and} \quad s := \left\lfloor \frac{l-1}{2} \right\rfloor. \tag{25}$$

Then one has:

$$\begin{aligned} r(l) &= \min_{\xi \in \mathbb{C}^l} \sum_{j=1}^{n} (y_j - \hat{p}_{l-1}(x_j))^2 \\ &= \min_{\gamma \in \mathbb{R}^{l-1}, \alpha_0 \in \mathbb{R}} \sum_{j=1}^{n} (y_j - \hat{p}_{l-1}(x_j))^2. \end{aligned} \tag{26}$$

If one sets $l = m$ then there is always a trigonometric polynomial $\hat{p}_{l-1}$ with $r(l) = (y_j - \hat{p}_{l-1}(x_j))^2 = 0$. The basic idea is to calculate the Fourier-transform using $l = m$ in a first step. If we consider the magnitude of values stored in γ, then the values with higher magnitude contribute more to the final shape of the function $\varphi(t)$. Therefore we select a value $k \leq m$ and select the k largest values of γ:

$$\varphi(t) = \frac{\alpha_0}{2} + \sum_{j=1}^{u} \alpha_{l_j} \cdot \cos{(l_j qt)} + \sum_{j=1}^{v} \beta_{l_j} \cdot \sin{(l_j qt)} \tag{27}$$

with $\quad u + v = k$

where $\quad \alpha_{l_1}, ..., \alpha_{l_u}, \beta_{l_1}, ..., \beta_{l_v} \quad$ are the k largest absolute values of the vector $\gamma \in \mathbb{R}^{l-1}$. The idea is based on the fact that the functions $\left(\frac{\cos x}{\sqrt{\pi}}, \frac{\sin x}{\sqrt{\pi}}, \frac{\cos 2x}{\sqrt{\pi}}, \right.$ $\frac{\sin 2x}{\sqrt{\pi}}, \frac{\cos 3x}{\sqrt{\pi}}, \frac{\sin 3x}{\sqrt{\pi}}, \left. ... \right)$ span an orthonormal basis in the space $L_2(0, 2\pi)$, which are the squared trigonometric function on the interval $[0, 2\pi]$.

In practice we have a global error threshold defined by the user. Therewith we sort the vector γ in ascending order and remove the values of smallest magnitude until the resulting error exceeds the given threshold.

Unfortunately, this method cannot be generalized as easily as the b-spline regression outlined in the last paragraph. In order to calculate the two-dimensional Fourier transformation, we need equidistant sampling points in both dimensions. Considering the discrete motion space with 1 parameter, we have a uniform sampling of each motion signal itself. However we cannot assume, that the user defined parameter $w \in D_{w_1}$ is chosen equidistant. Here the solution is to calculate a one-dimensional Fourier-transform first and obtain the coefficients:

$$\alpha_{0,l_2}, \alpha_{1,l_2}, ..., \alpha_{t_1,l_2}, \beta_{1,l_2}, \beta_{2,l_2}, ..., \beta_{s_1,l_2} \qquad \forall l_2 \in \{0,1,...,n_1 = |D_{w_1}|\}$$
$$\text{where} \qquad t_1 := \left\lfloor \frac{m}{2} \right\rfloor \qquad \text{and} \qquad s_1 := \left\lfloor \frac{m-1}{2} \right\rfloor .$$

In the second step the following equations must be solved:

$$H \cdot \begin{bmatrix} a_{l_1,0} \\ a_{l_1,1} \\ \vdots \\ a_{l_1,t_2} \\ c_{l_1,1} \\ c_{l_1,2} \\ \vdots \\ c_{l_1,s_2} \end{bmatrix} = \begin{bmatrix} \alpha_{l_1,0} \\ \alpha_{l_1,1} \\ \vdots \\ \alpha_{l_1,n_1} \end{bmatrix} ; \qquad l_1 = 0, 1, ..., t_1;$$

$$\text{and} \qquad H \cdot \begin{bmatrix} b_{l_1,0} \\ b_{l_1,1} \\ \vdots \\ b_{l_1,t_2} \\ d_{l_1,1} \\ d_{l_1,2} \\ \vdots \\ d_{l_1,s_2} \end{bmatrix} = \begin{bmatrix} \beta_{l_1,0} \\ \beta_{l_1,1} \\ \vdots \\ \beta_{l_1,n_1} \end{bmatrix} ; \qquad l_1 = 1, 2, ..., s_1. \tag{28}$$

where

$$t_2 := \left\lfloor \frac{n_1}{2} \right\rfloor ;$$
$$s_2 := \left\lfloor \frac{n_1-1}{2} \right\rfloor ;$$
$$h_{kl} := \begin{cases} \cos\left((l-1) \cdot q_2 \cdot (w_{k-1} - w_0)\right), & 1 \le l \le t_2+1 \\ \sin\left((l-t_2-1) \cdot q_2 \cdot (w_{k-1} - w_0)\right), & t_2+1 < l \le n_1+1 \end{cases} \tag{29}$$
$$k = 1, 2, ..., n_1$$
$$\text{and} \qquad H := [h_{kl}]_{k,l=1}^{n_1} .$$

The two step approximation method can directly be generalized to arbitrary dimension.

4.4   Combining the Approximation Results

As a last preprocessing step we combine the approximated motion signals to form a continuous motion space. By doing that, we attempt to compress the data. For the spline function one has to store the values of the coefficients and the knot vector. In order to evaluate the Fourier function we have to store the selected coefficients values and their indices. For each degree of freedom we select the approximated motion function, where fewer values need to be stored. Here we compare the number of Fourier coefficients and b-Spline coefficients used. The smaller number is used to form the motion space.

5   Experimental Results

To evaluate our framework we used four different activities, which are common within a production setting. These are pick up box, hammering a nail, manual

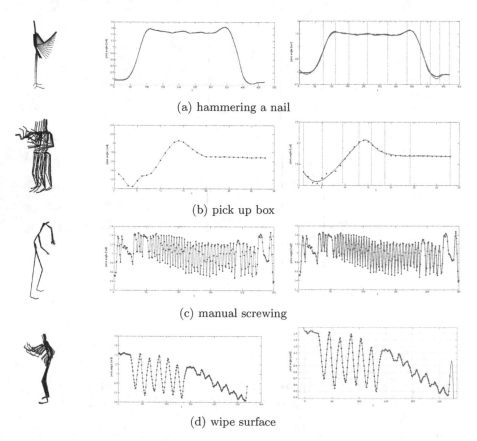

(a) hammering a nail

(b) pick up box

(c) manual screwing

(d) wipe surface

Fig. 6. Examples of the approximated activities. The left column shows the motion of the skeleton. The corresponding motion signal of the right elbow joint is shown in the middle column. The approximation results are presented on the right.

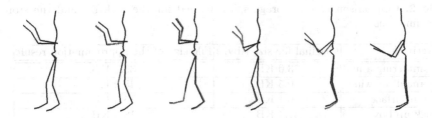

Fig. 7. The figure shows the approximation results for a single motion. The original motion capture data is shown in black. The approximated motion is drawn in red.

Table 1. Number of coefficients needed to approximate the left elbow joint, with a mean error of $\varepsilon = 0.5$ degree. ε_{max} specifies the maximum angular error in degrees.

Activity	length of the motion signal	# of Fourier coefficients	# of B-spline coefficients
hammering a nail	460	30 ($\varepsilon_{max} = 6.5$)	10 ($\varepsilon_{max} = 1.2$)
manual screwing	250	50 ($\varepsilon_{max} = 2.3$)	90 ($\varepsilon_{max} = 15.01$)
wipe surface	270	34 ($\varepsilon_{max} = 0.5$)	96 ($\varepsilon_{max} = 4.3$)
pick up box	33	24 ($\varepsilon_{max} = 2.2$)	8 ($\varepsilon_{max} = 1.8$)

screwing and wiping of a surface. Examples of these activities are shown in figure 6.

In order to verify the quality of the approximation methods we approximated a single motion. The results are shown in figure 7. The individual poses were placed over each other. The black poses correspond to the original motion capture data. The approximated motion is placed underneath in red. For approximating an average error of 0.5 degrees was used.

An example of the interpolation results of the framework shown in figure 8. Here the calculated continuous movement space of the box lift motion was used. The black poses correspond to the recorded motion data. The blue poses are generated by evaluation of the motion spaces.

An example of the motion generated by evaluating the continuous motion spaces is shown in figure 9. In table 1 the number of coefficients needed as well as the maximal approximation error is shown.

In order to measure the compression archived with the proposed approximation methods, we compared the file size of the original bvh file with the file size of the motion spaces. Both files are encoded in ASCII format. The bvh file consists of two parts. The first part specifies the skeleton hierarchy and the second part lists the Euler angles for all poses. The motion space file is specified in a similar fashion. Here the first part also specifies the skeleton hierarchy, while the second part specifies the b-spline and Fourier functions. In order to reconstruct the b-spline functions, one has to store the coefficients and the knot positions. In case of Fourier function the coefficients and their indices need to be stored. In table 2 the corresponding file sizes are listed.

280   T. Kronfeld, J. Fankhänel, and G. Brunnett

Table 2. Comparison of the storage space needed for the .bvh file and the stored approximations

Activity	Original file size (.bvh)	File size of the approximation result
hammering a nail	73.6 KB	39.5 KB
manual screwing	80.5 KB	45.2 KB
wipe surface	40.1 KB	25.3 KB
pick up box	37.4 KB	18.8 KB

Fig. 8. A two handed pick action sampled across the hight parameter. The black poses are motion capture data and the blue poses are generated by our approach. The height parameter increases from top to bottom (0.8; 0.9; 1.0; 1.1; 1.2 meter).

Fig. 9. The example shows the activity pick up a box, generated by evaluating the continuous motion space. The heights of the boxes are: 0cm, 40cm, 80cm, 100cm and 120cm (left to right).

6 Conclusion and Further Work

We have shown, that our approximation method enhances the capabilities of the captured human motion while we are also able to reduce the amount of storage space needed. We also plan to apply this method to long connected motion sequences, in order to archive a good compression rate. It may also be possible to apply additional compression methods, in order to further reduce the storage space needed.

Also, it may be useful to examining further spline models and their suitability for the approximation and compression of motion data.

Acknowledgements. The authors would like to thank imk automotive GmbH and the Institut of Mechatronics for providing the motion capture data. The work reported in this publication has been financially supported by the European Union and the federal state of Saxony, in the junior research group "The Smart Virtual Worker".

References

1. Bruderlin, A., Williams, L.: Motion signal processing. In: Proceedings of the 22nd Annual Conference on Computer Graphics ans Interactive Techniques, pp. 97–104 (1995)
2. Cox, M.G.: Practical spline approximation. In: Topics in Numerical Analysis. Lecture Notes in Mathematics, vol. 965, pp. 79–112 (1981)
3. de Boor, C., Rice, J.R.: Least square cubic spline approximation I - fixed knots. Computer Sciences Technical Reports 20 (1986)
4. de Boor, C., Rice, J.R.: Least square cubic spline approximation, II - variable knots. Computer Science Technical Reports, 149 (1986)
5. de Boor, C.: A practical guide to splines. Springer, Heidelberg (1978)
6. Dierckx, P.: Net aanpassen van krommen en oppervlakken aan meetpunten met behulp van spline funkties. PhD thesis, Katholieke Universiteit Leuven (1979)
7. Gao, Y., Ma, L., Wu, X., Chen, Z.: From keyframing to motion capture. In: Human Interaction with Machines, pp. 35–42 (2006)

8. Holt, J.N., Fletcher, R.: An algorithm for constrained non-linear least squares. J. Inst. Math. Appl. 23, 449–463 (1979)
9. Hu, Y.: An algorithm for data reduction using splines with free knots. IMA J. Numer. Anal. 13, 365–381 (1993)
10. Jupp, D.L.B.: The Lethargy theorem - a property of approximation by γ-polynomials. Journal of Approximation Theory 14, 204–217 (1975)
11. Jupp, D.L.B.: Approximation to data by splines with free knots. SIAM J. Numer. Anal. 15(2), 328–343 (1978)
12. Lee, J., Chai, J., Reitsma, P.S.A., Hodgins, J.K., Pollard, N.S.: Interactive control of avatars animated with human motion data. ACM Trans. Graph. 21(3), 491–500 (2002)
13. Lee, K.H., Park, J.P., Lee, J.: Movement classes from human motion data. Transactions on Edutainment VI, 122–131 (2011)
14. Lyche, T., Mørken, K.: A data reduction strategy for splines with application to the approximation of function and data. IMA J. Numer. Anal. 8, 185–208 (1988)
15. Mühlstedt, J., Kaußler, H., Spanner-Ulmer, B.: Programme in Menschengestalt: Digitale Menschmodelle für CAx- und PLM-Systeme. Zeitschrift für Arbeitswissenschaft 02/2008
16. Spanner-Ulmer, B., Mühlstedt, J.: Digitale Menschmodelle als Werkzeuge virtueller Ergonomie. Industrie Management 26, 69–72 (2010)
17. Park, J.P., Lee, K.H., Lee, J.: Finding Syntactic Structures from Human Motion Data. Computer Graphics Forum 30(8), 2183–2193 (2011)
18. Rose, C., Cohen, M.F., Bodenheimer, B.: Verbs and adverbs: multidimensional motion interpolation. IEEE Computer Graphics and Applications 18(5), 32–40 (1998)
19. Schoenberg, I.J., Whitney, A.: On Polya frequency functions III. Trans. AMS 74, 246–259 (1953)
20. Schütze, T.: Diskrete Quadratmittenapproximation durch Splines mit freien Knoten. Ph.D. thesis, Dresden (1998)
21. Schwetlick, H., Schütze, T.: Least squares approximation by splines with free knots. BIT 35(3), 361–384 (1995)
22. Suchomski, P.: Method for optimal variable-knot spline interpolation in the L_2 discrete norm. Internat. J. Systems Sci. 22, 2263–2274 (1991)
23. Unuma, M., Anjyo, K., Takeuchi, R.: Fourier Principles for Emotion-based Human Figure Animation. In: Proceedings of the 22nd Annual Conference on Computer Graphics and Interactive Techniques, pp. 91–96 (1995)
24. van Welbergen, H., van Basten, B.J.H., Egges, A., Ruttkay, Z.M., Overmars, M.H.: Real Time Animation of Virtual Humans: A Trade-off Between Naturalness and Control. Computer Graphics Forum 29(8), 2530–2554 (2010)

Bilinear Clifford-Bézier Patches on Isotropic Cyclides

Rimvydas Krasauskas, Severinas Zubė, and Salvatore Cacciola

Vilnius University, Faculty of Mathematics and Informatics,
Naugarduko g. 24, 03225 Vilnius, Lithuania

Abstract. We study Bézier-like formulas with weights in geometric algebra for parametrizing a special class of rational surfaces in isotropic 3-space. These formulas are useful for constructing isotropic-Möbius invariant surfaces that are dual to rational offset surfaces in euclidean 3-space. Our focus is on bilinear Clifford-Bézier patches. We derive their implicitization formula and characterize them as patches on special quartic surfaces called *isotropic cyclides*. Finally we present one modeling application with rational surfaces admitting rational offsets.

Keywords: isotropic cyclides, Clifford algebra, geometric algebra, rational offset surfaces.

1 Introduction

In this paper we develop the parallel theory to the recently introduced bilinear quaternionic-Bézier patches on Darboux cyclides (see [7] and [12]).

The main motivation for these studies is in the following theorem due to Pottmann and Peternell [9,11]: *There is a 1–1 correspondence via duality between non-developable rational offset surfaces (PN-surfaces) in euclidean space and rational surfaces in isotropic space* (see details in Section 5.1).

For example, pipe surfaces with quadratic spines, or non-singular quadrics (and their offsets) in $\mathbb{R}^3$ are duals of certain isotropic cyclides.

We hope that these results can be the starting point of a new approach that will help to simplify the general theory and also will provide new computational and modeling tools for rational offset surfaces.

Isotropic space and its geometric algebra are introduced in Section 2. Bilinear Clifford–Bézier (CB) patches are defined and their properties are described in Section 3. Section 4 is devoted to isotropic cyclides: their geometry is studied and relations with CB-patches are established. In Section 5 modeling applications with Pythagorean normal surfaces (PN-surfaces) are illustrated. Technical results characterizing both versions (Darboux and isotropic) of cyclides are postponed to Section 6. Conclusions and further research directions are drawn in Section 7.

2 Isotropic Space and Its Geometric Algebra

2.1 Isotropic Geometry

Define an *isotropic* space $\mathbb{R}^3_{++0}$ as a vector space $\mathbb{R}^3$ with a scalar (interior) product having signature $(+, +, 0)$, i.e. with an orthonormal basis $\{e_1, e_2, e_3\}$, such that

$$e_1 \cdot e_1 = e_2 \cdot e_2 = 1, \quad e_3 \cdot e_3 = 0, \quad e_i \cdot e_j = 0, \ i \neq j.$$

M. Floater et al. (Eds.): MMCS 2012, LNCS 8177, pp. 283–303, 2014.

284   R. Krasauskas, S. Zubė, and S. Cacciola

Therefore, $x \cdot x = x_1^2 + x_2^2$ and distances in *isotropic geometry* are measured as Euclidean distances in the projection to the first two coordinates $(x_1, x_2, x_3) \mapsto (x_1, x_2)$, which is called a *top view*. There are distinguished vertical lines and vertical planes that are called *isotropic lines* and *isotropic planes*.

Definition 1. *An isotropic sphere (or i-sphere) is a quadric $S \subset \mathbb{R}^3_{++0}$ defined by the equation*

$$a(x_1^2 + x_2^2) + L(x_1, x_2) + dx_3 + b = 0, \tag{1}$$

for some $a, b, d \in \mathbb{R}, a \neq 0, L$ a linear form.

The motivation for this definition is simple: i-spheres are isotropic inversions of planes (see Lemma 2). Similar to planes, i-spheres can be of two different types (depending on the parameter d):

- i-spheres of *parabolic type* – paraboloids of revolution with vertical axis ($d = -1$);
- i-spheres of *cylindric type* – cylinders with top view circles ($d = 0$).

An *isotropic circle* (or i-circle) is the intersection between an i-sphere of parabolic type and a plane. This definition implies that an i-circle is either an ellipse with a circle as top view or a parabola with vertical axis.

Thus in $\mathbb{R}^3_{++0}$, we can consider the counterpart to Möbius geometry, where the main objects are *i-M-spheres* that are either i-spheres of parabolic type or non-isotropic planes. Their intersections are called *i-M-circles* that can be i-circles of both types or non-isotropic lines. The space of *isotropic Möbius* geometry is defined as an extension $\mathbb{R}^3_{++0} \cup \mathbb{R}$, where the ideal line $\mathbb{R}$ of infinite points is attached. Every i-M sphere $x_3 = a(x_1^2 + x_2^2) + L(x_1, x_2) + b$ has a unique ideal point $2a \in \mathbb{R}$ on that line (see details in Section 4.1).

2.2 Isotropic Möbius Transformations

The group of isotropic Möbius (i-M) transformations is generated by inversions with respect to i-spheres (see details in [10]):

(i) inversions with respect to i-spheres of cylindrical type are most similar to classical inversions, since they are defined by a central point p and radius r:

$$\mathrm{inv}_{p,r}(x) = r^2 \mathrm{inv}(x - p) + p, \quad \mathrm{inv}(x) = \frac{x}{x \cdot x}, \tag{2}$$

i.e. $\mathrm{inv}(x)$ is the special case of $\mathrm{inv}_{p,r}(x)$ when p is the origin and $r = 1$;

(ii) inversions with respect to i-spheres of parabolic type $x_3 = f(x_1, x_2), f(x_1, x_2) = a(x_1^2 + x_2^2) + L(x_1, x_2) + b$,

$$\mathrm{inv}^f(x) = (x_1, x_2, 2f(x_1, x_2) - x_3). \tag{3}$$

Lemma 1. *All i-M transformations are generated by*

- *uniform scalings $S_r(x) = rx$;*
- *translations $T_p(x) = x + p$;*

- *inversion* $\mathrm{inv}(x) = x/(x \cdot x)$;
- *vertical reflections* $\mathrm{inv}^0(x) = (x_1, x_2, -x_3)$.

Proof. Any inversion of type (i) can be obtained using the first three generators, since

$$\mathrm{inv}_{p,r} = (T_p \circ S_r) \circ \mathrm{inv} \circ (T_p \circ S_r)^{-1}.$$

Also if p is a vertex of the paraboloid $x_3 = f(x_1, x_2)$ then

$$\mathrm{inv}^f = T_p \circ \mathrm{inv} \circ \mathrm{inv}^a \circ \mathrm{inv} \circ T_p^{-1}, \quad \mathrm{inv}^a = T_{(0,0,a)} \circ \mathrm{inv}^0 \circ T_{(0,0,a)}^{-1}.$$

Hence by adding the vertical reflection inv^0 one can generate all inversions of type (ii).

The following lemma shows that, in analogy to the classical Möbius geometry, i-M circles and i-M spheres are preserved by i-M transformations.

Lemma 2. *If S is an i-M sphere and C is an i-M circle, then*

(1) $\mathrm{inv}_{p,r}(S)$ *is an i-M sphere;*
(2) $\mathrm{inv}_{p,r}(S)$ *is a plane if and only if S passes through p;*
(3) $\mathrm{inv}_{p,r}(C)$ *is an i-M circle;*
(4) $\mathrm{inv}_{p,r}(C)$ *is a line if and only if C passes through p.*

Proof. First we note that items (1) and (2) in the particular case when $\mathrm{inv}_{p,r} = \mathrm{inv}$ (i.e. p is the origin and $r = 1$) follow from Lemma 7 in Section 6. Indeed, inv coincides with g-inversion when $g = x^2 + y^2$ and g-Möbius spheres are exactly i-spheres or planes (the class of i-M spheres is a bit smaller but it is invariant with respect to inversions).

The general case can be reduced to this particular case, since according to Lemma 1 $\mathrm{inv}_{p,r}$ is a composition of translations, uniform scalings and the map inv, and for every i-M sphere S, $p \in S$, if and only if $T_p^{-1}S$ passes through the origin.

Items (3) and (4) now can be proved by representing any i-M circle as the intersection of two i-M spheres.   □

2.3   The Geometric Algebra $\mathcal{G}(\mathbb{R}^3_{++0})$

In the geometric algebra (Clifford algebra) generated by $\mathbb{R}^3_{++0}$ the *geometric* product of vectors is defined to be associative and distributive with respect to addition, with the additional relation: $x \cdot x = x^2 \in \mathbb{R}$. This algebra will be denoted by $\mathcal{G} = \mathcal{G}(\mathbb{R}^3_{++0})$. For vectors $v, u \in \mathbb{R}^3_{++0}$ the geometric product is a sum of interior and exterior products

$$vu = v \cdot u + v \wedge u. \tag{4}$$

The algebra $\mathcal{G}(\mathbb{R}^3_{++0})$ has the same underlying vector space as the usual exterior algebra $\bigwedge(\mathbb{R}^3)$, namely it is a vector space of dimension 8, that can be decomposed as a direct sum $E_0 \oplus E_1 \oplus E_2 \oplus E_3$ of subspaces with the following bases

$$\{1\}, \quad \{e_1, e_2, e_3\}, \quad \{e_{12}, e_{23}, e_{13}\}, \quad \{e_{123}\}, \tag{5}$$

where $e_{ij} = e_i e_j$ and $e_{123} = e_1 e_2 e_3$. The vector spaces $E_0, \ldots, E_3$ are scalars, vectors, bi-vectors, and tri-vectors respectively. For any $x \in \mathcal{G}(\mathbb{R}^3_{++0})$, its k-grade component $\langle x \rangle_k$ is the projection to the subspace E_k of grade k.

A *reversion* operation in the algebra $\mathcal{G}$ will also be used (see [3,4,8] for details). If x is a product of vectors $x = v_1 v_2 \cdots v_{n-1} v_n$, then its reversion is $\tilde{x} = v_n v_{n-1} \cdots v_2 v_1$. If all v_i are non-zero, then $x\tilde{x} = (v_n \cdot v_n) \cdots (v_2 \cdot v_2)(v_1 \cdot v_1) \in \mathbb{R}$. Hence it is easy to calculate the inverse element $x^{-1} = \tilde{x}/(x\tilde{x})$. Reversion is similar to conjugation in quaternions. For example, if x is an element of even grade, then

$$x = a + be_{12} + ce_{23} + de_{13}, \quad \tilde{x} = a - be_{12} - ce_{23} - de_{13}. \tag{6}$$

In terms of the algebra $\mathcal{G}(\mathbb{R}^3_{++0})$ we get a simple formula for inversion in $\mathbb{R}^3_{++0}$

$$\mathrm{inv}(x) = x^{-1}, \quad \mathrm{inv}_{p,r}(x) = r^2(x - p)^{-1} + p. \tag{7}$$

Similar to the representation of a euclidean circular arc by a quaternionic–Bézier (QB) curve of degree 1 in [7,13], one can derive the following Clifford–Bézier (CB) formula

$$C(t) = (p_0 w_0(1 - t) + p_1 w_1 t)(w_0(1 - t) + w_1 t)^{-1} \in \mathbb{R}^3_{++0}, \tag{8}$$

where p_i, $i = 0, 1$, are two endpoints of an i-circular arc C in $\mathbb{R}^3_{++0}$, and w_i, $i = 0, 1$, are weights defined using some interior point q of the arc

$$w_0 = (q - p_0)^{-1}, \quad w_1 = (p_1 - q)^{-1}. \tag{9}$$

Here points and weights are vectors in $\mathbb{R}^3_{++0}$, and all algebraic operations are defined in the algebra $\mathcal{G}(\mathbb{R}^3_{++0})$. In order to avoid degenerate cases we suppose that all three points have different top views. If the points p_0, q, p_1 are collinear, then $C(t)$ is a line segment. So in general $C(t)$ is the arc of an i-M circle (see Remark 1 below).

Obviously one can divide by w_0 and get the equivalent couple of weights $w'_0 = 1$ and $w'_1 = w_1 w_0^{-1} = (p_1 - q)^{-1}(q - p_0)$. Now the weights are not vectors but still by the product formula (4) they are elements of the even subalgebra $\mathcal{G}_{\text{even}} = E_0 \oplus E_2 \subset \mathcal{G}$. The bi-vector component of w'_1 has a simple geometric meaning: it defines a vector plane generated by $p_1 - q$ and $q - p_0$, in particular it contains the vector $p_1 - p_0$. The latter condition is equivalent to the equation

$$(p_1 - p_0) \wedge \langle w_1 w_0^{-1} \rangle_2 = 0, \tag{10}$$

which is a necessary and sufficient condition for the weights in (8) to define an i-M circle.

If just one weight is multiplied by a real number $\lambda > 0$, then the arc is reparametrized.

Remark 1. The formula (8) for a circular arc C is invariant with respect to i-M transformations. Indeed, it is enough to check this invariance for all generators of the i-Möbius group (see Lemma 1). For example,

$$\mathrm{inv}(C(t)) = C(t)^{-1} = (w_0(1 - t) + w_1 t)(p_0 w_0(1 - t) + p_1 w_1 t)^{-1}.$$

Therefore, $\mathrm{inv}(C(t)) = C'(t)$ is defined by the same formula (8) with $w'_i = p_i w_i$ and $p'_i = w_i(w'_i)^{-1} = w_i w_i^{-1} p_i^{-1} = \mathrm{inv}(p_i)$, $i = 0, 1$.

Lemma 3. *$C(t)$ can be represented in classical Bézier form with the following control points and real weights:*

$$P_0 = p_0, \quad P_1 = (p_0 w_0 \tilde{w}_1 + p_1 w_1 \tilde{w}_0)/(w_0 \tilde{w}_1 + w_1 \tilde{w}_0), \quad P_2 = p_1, \quad (11)$$

$$W_0 = w_0 \tilde{w}_0, \quad W_1 = (w_0 \tilde{w}_1 + w_1 \tilde{w}_0)/2, \quad W_2 = w_1 \tilde{w}_1. \quad (12)$$

Proof. Denoting numerator and denominator in (8) by $F(t)$ and $W(t)$ respectively and using the reversion operation in $\mathcal{G}(\mathbb{R}^3_{++0})$, one can express $C(t)$ as a fraction with real denominator:

$$C(t) = F(t)W(t)^{-1} = F(t)\widetilde{W}(t)(W(t)\widetilde{W}(t))^{-1}. \quad (13)$$

Then calculating separately $F(t)\widetilde{W}(t)$ and $W(t)\widetilde{W}(t)$

$$F(t)\widetilde{W}(t) = p_0 w_0 \tilde{w}_0 (1-t)^2 + (p_0 w_0 \tilde{w}_1 + p_1 w_1 \tilde{w}_0)(1-t)t + p_1 w_1 \tilde{w}_1 t^2$$

$$W(t)\widetilde{W}(t) = w_0 \tilde{w}_0 (1-t)^2 + (w_0 \tilde{w}_1 + w_1 \tilde{w}_0)(1-t)t + w_1 \tilde{w}_1 t^2$$

we detect Bézier control points (11) and weights (12). □

Corollary 1. *An arc C defined by (8) with vectors p_i and $w_i \neq 0$, $i = 0, 1$, is contained in $\mathbb{R}^3_{++0}$ if and only if $p_0 w_0 \tilde{w}_1 + p_1 w_1 \tilde{w}_0$ is a vector, i.e. $\langle p_0 w_0 \tilde{w}_1 + p_1 w_1 \tilde{w}_0 \rangle_3 = 0$.*

3 Clifford–Bézier Surface Patches

We call a curve or a surface *i-spherical* if it is contained in an i-M sphere (so this includes the planar case). For example, two i-M circles are *i-cospherical* if they are on the same i-M sphere.

An i-spherical triangle with corner points p_0, p_1, p_2 bounded by three i-circular arcs (such that these three i-circles intersect in a point q) has the parametrization formula with weights: $w_i = (p_i - q)^{-1}$, $i = 0, 1, 2$,

$$P(s,t) = (p_0 w_0 (1-s-t) + p_1 w_1 s + p_2 w_2 t)(w_0 (1-s-t) + w_1 s + w_2 t)^{-1} \quad (14)$$

with the triangular parameter domain: $s, t \geq 0$, $s+t \leq 1$. It is natural to call this surface patch a *linear triangular CB-patch*. Any formula (14) with arbitrary weights, such that the boundary curves are i-M circles, define a patch on an i-M sphere.

Any i-M-sphere is i-M equivalent to the plane $x_3 = 0$ in $\mathbb{R}^3_{++0}$ which can be identified with the euclidean plane $\mathbb{R}^2$. Therefore, all propositions about spherical triangular and quadrangular QB-patches of any degree (see [7]) are valid also for i-spherical CB-patches of parabolic type.

Let us switch to the simplest non-i-spherical case. Consider a bilinear CB-patch defined by the formula (here the fraction $\frac{a}{b}$ means ab^{-1}):

$$P(s,t) = \frac{p_0 w_0 (1-s)(1-t) + p_1 w_1 s(1-t) + p_2 w_2 (1-s)t + p_3 w_3 st}{w_0 (1-s)(1-t) + w_1 s(1-t) + w_2 (1-s)t + w_3 st}, \quad (15)$$

with the square parameter domain $0 \leq s, t \leq 1$. We consider only the case when the image is contained in the vector space $\mathbb{R}^3_{++0}$. Formula (15) is i-M invariant similar to the i-circle case (see Remark 1).

Lemma 4. *If two adjacent boundary circles of a bilinear CB-patch P are i-cospherical, then the bilinear CB-patch is either i-spherical or it is a patch of a double ruled quadric (including its i-M transformations).*

Proof. Denote by C_{ij} a boundary circle joining vertices p_i and p_j. Assuming that the circles C_{01} and C_{02} are i-cospherical, there are two cases: the circles intersect in two points p_0 and $q \neq p_0$; the circles are tangent in p_0 (a double point).

In the first case we apply inversion with a center q and use the same notation for the transformed patch. Now C_{01} and C_{02} are line segments, and one can assume (after a reparametrization) $w_0 = w_1 = w_2 = 1$. If $w_3 \in \mathbb{R}$, then the CB-patch P is on a doubly ruled quadric (or plane). Otherwise $\langle w_3 \rangle_2 \neq 0$, and according to (10) the vectors $p_3 - p_1$ and $p_3 - p_2$ lie on the plane generated by the bi-vector $\langle w_3 \rangle_2$. Hence the two boundary circles C_{13} and C_{23} lie on the same plane Π going through the three points p_1, p_2, p_3. All weights along these circles have the same bi-vector direction, since they are linear averages between w_3 and 1. Similarly it follows that all other circles on the CB-patch are on the same plane Π, and P is planar, i.e. i-spherical.

In the second case after the same inversion (with a center p_0) we get the base point at $(s, t) = (0, 0)$. Therefore we consider a subpatch P' which is the restriction of P on a rectangular domain $(s, t) \in [0, 1] \times [\epsilon, 1]$, with a small $\epsilon > 0$. The same argument as earlier allows us to get weights $w_0 = w_1 = 1$, and w_2 very close to 1. Then the subpatch P' is as close as we wish to the plane Π going through the three points p_1, p_2, p_3. So the whole P is planar. $\square$

Lemma 5. *Let four circles $C_{01}, C_{02}, C_{13}, C_{23}$ in $\mathbb{R}^3_{++0}$ be defined by four control points and weights $\{(p_i, w_i)\}_{i=0,\dots,3}$, and suppose that any two adjacent i-circles are not i-cospherical. Then there is a unique non-zero number*

$$\lambda = -\frac{\langle p_1 w_1 \widetilde{w}_2 + p_2 w_2 \widetilde{w}_1 \rangle_3}{\langle p_0 w_0 \widetilde{w}_3 + p_3 w_3 \widetilde{w}_0 \rangle_3} \in \mathbb{R}, \tag{16}$$

such that the same control points with weights $w_0, w_1, w_2, \lambda w_3$ define a bilinear CB-patch.

Proof. Let $P = FW^{-1}$ be a fraction as in formula (15) with control points p_i, $i = 0, \dots, 3$, and weights $w_0, w_1, w_2, \lambda w_3$. Following the proof of Lemma 3 we modify the expression $FW^{-1} = F\widetilde{W}(W\widetilde{W})^{-1}$ and expand its numerator $F\widetilde{W}$ in the biquadratic Bernstein basis with control points q_{ij} (multiplied by their weights) of the corresponding rational biquadratic Bézier surface. Boundary control points are in $\mathbb{R}^3_{++0}$, since they represent i-M circular arcs in $\mathbb{R}^3_{++0}$. The middle control point multiplied by its weight has the following form:

$$q_{11} = (p_1 w_1 \widetilde{w}_2 + p_2 w_2 \widetilde{w}_1) + \lambda(p_0 w_0 \widetilde{w}_3 + p_3 w_3 \widetilde{w}_0),$$

where both expressions in brackets have non-zero 3-grades (otherwise, adjacent boundary circles would be i-cospherical, cf. Corollary 1). Solving the equation $\langle q_{11} \rangle_3 = 0$ for λ we get exactly (16). $\square$

Theorem 1. *Let C_0, C_{01}, C_1 be i-circles in $\mathbb{R}^3_{++0}$, and let $p_0 = C_0 \cap C_{01}$ and $p_1 = C_{01} \cap C_1$ be the unique points of their transversal intersections. Then for almost any other point $p_2 \in C_0$, $p_2 \neq p_0$, there exists a unique bilinear CB-patch (up to trivial reparametrization) with control points p_i, $i = 0, 1, 2$, and $p_3 \in C_1$ with three boundary arcs lying on the given three i-circles.*

Proof. We choose any point $q \in C_1$, $q \neq p_1$, and apply inversion with center in q. Using the same notation, we see that C_1 is a line and, up to multiplication by a real number, we can find unique weights $w_1 = 1$, w_0 and w_2, that correspond to the given i-circles. The point p_2 and the weight w_2 determine a plane Π where the fourth boundary i-circle C_{23} can lie (see the condition (10)). So we can find a point p_3 as an intersection $\Pi \cap C_1$ with a weight $w_3 = 1$. The exceptional case when Π is parallel to the line C_1 can happen only when the initial point q (before the inversion) has been chosen as p_3. Finally we use Lemma 5 in order to fill the closed contour of i-circles.   □

Our next theorem allows us to find the implicit equation of the patch $P(s, t)$ defined by (15) as an algebraic surface in $\mathbb{R}^3_{++0}$. Without loss of generality, we can assume that the weights of $P(s, t)$ are in the even subalgebra $\mathcal{G}_{\text{even}} = E_0 \oplus E_2$ (e.g dividing by w_0, see Section 2.3). We denote by $F(s, t)$ the numerator and by $W(s, t)$ the denominator of the fraction $P(s, t)$ in (15). Then formally $P(s, t)$ is in the vector subspace $E_1 \oplus E_3 \subset \mathcal{G}(\mathbb{R}^3_{++0})$ generated by the basis $\{e_1, e_2, e_3, e_{123}\}$. Let us introduce a formal element $X = xe_1 + ye_2 + ze_3 + ue_{123}$, multiply both sides of the equation $X = F(s, t)W(s, t)^{-1}$ by $W(s, t)$, and move all terms to the left side

$$XW(s, t) - F(s, t) = 0.$$

We treat this equation as a system of 4 real linear equations with 4 unknowns

$$(1 - s)(1 - t), \quad s(1 - t), \quad (1 - s)t, \quad st.$$

The 4×4 matrix M of this system has 4 columns filled with coordinates of vectors $(X - p_i)w_i \in E_1 \oplus E_3$, $i = 0, \ldots, 3$, in the same basis. Hence the entries of the matrix M are linear forms in x, y, z, u, and the polynomial

$$F(x, y, z, u) = \det M \tag{17}$$

must vanish on every point X of the patch $P(s, t)$. Therefore, $F(x, y, z, u) = 0$ defines at most a quartic equation in the variables x, y, z, u.

Theorem 2. *The implicit equation of a bilinear CB-patch $P(s, t)$ in $\mathbb{R}^3_{++0}$ is a factor of the polynomial $F(x, y, z, 0)$, where $F(x, y, z, u) = \det M$ has degree at most 4.*   □

The following example illustrates this theorem.

Example 1. Consider the CB-patch (15) with control points and weights

$$p_0 = 2e_2, \quad p_1 = e_2, \quad p_2 = -2e_2, \quad p_3 = -e_2,$$
$$w_0 = 1, \quad w_1 = e_{12} - e_{23}, \quad w_2 = 2e_{12} + 3e_{23}, \quad w_3 = -3(2 + e_{13}).$$

We compute 4 columns of the matrix M as the coordinates of elements $(X - p_i)w_i$, $i = 0, \ldots, 3$, in the basis $\{e_1, e_2, e_3, e_{123}\}$:

$$M = \begin{pmatrix} x & 1 - y & -2(2 + y) & -6x \\ -2 + y & x & 2x & -6(1 + y) \\ z & (1 - y - u) & (y - 2u + 2) & -3(x + 2z) \\ u & (-x + z) & (x + 2z) & 3(1 + y - 2u) \end{pmatrix}.$$

Finally the implicit equation of the surface is obtained by substituting $u = 0$ in the determinant $\det M = F(x, y, z, u)$ and multiplying by $-1/9$ for simplicity

$$(x^2 + y^2)^2 - 8x^2 - 5y^2 + 12z^2 + 4 = 0. \tag{18}$$

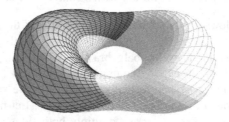

Fig. 1. The CB-patch on the full algebraic surface defined by (18)

4   Isotropic Cyclides

4.1   Model of Isotropic Möbius Geometry on the Blaschke Cylinder

In analogy with the standard conformal model of Möbius geometry in $\mathbb{R}^3$, we introduce the model of isotropic Möbius geometry defined on the Blaschke cylinder $\mathcal{B}$: $x_1^2 + x_2^2 + x_4^2 = x_5^2$ in $\mathbb{P}^4$ (see Section 5). Here it will be convenient to use slightly different coordinates in $\mathbb{P}^4$: instead of the standard basis $\{e_1, \ldots, e_5\}$, we will use the basis $\{e_0, e_1, e_2, e_3, e_\infty\}$ with $e_\infty = e_4 + e_5$, $e_0 = (-e_4 + e_5)/2$. Then $x_\infty = (x_4 + x_5)/2$, $x_0 = -x_4 + x_5$, and the equation of $\mathcal{B}$ is $x_1^2 + x_2^2 - 2x_\infty x_0 = 0$. Actually $\mathcal{B}$ is a cone in $\mathbb{P}^4$ with a vertex $v = [e_3]$ (i.e. $v = [0, 0, 0, 1, 0]$ in the current basis).

We treat the isotropic space $\mathbb{R}^3_{++0}$ as an affine part $x_0 = 1$ in $\mathbb{P}^3$. Define a *stereographic projection* from $\mathcal{B}$ to $\mathbb{R}^3_{++0}$ as a restriction of the linear projection from the point $c = [e_\infty] = [0, 0, 0, 0, 1]$:

$$\pi : \mathbb{P}^4 \to \mathbb{P}^3, \quad (x_0, x_1, x_2, x_3, x_\infty) \mapsto (x_0, x_1, x_2, x_3).$$

The inverse of this stereographic projection is

$$\sigma : \mathbb{R}^3_{++0} \to \mathcal{B}, \quad (x_1, x_2, x_3) \mapsto [1, x_1, x_2, x_3, \tfrac{1}{2}(x_1^2 + x_2^2)].$$

Theorem 3. *The inverse stereographic projection defines a 1–1 correspondence between i-M spheres (respectively i-M circles) and hyperplanes (respectively 2-planes) in* $\mathbb{P}^4$ *that do not contain the vertex* v *of the Blaschke cylinder:*

- *i-M spheres* S *go to hyperplane sections* $\sigma(S) = H \cap \mathcal{B}$;
- *i-M circles* C *go to 2-plane sections* $\sigma(C) = \Pi \cap \mathcal{B}$;

Moreover, isotropic lines go to lines on $\mathcal{B}$ *and the ideal line corresponds to a line* l_∞ *going through two points* $c = [e_\infty]$ *and* $v = [e_3]$.

Proof. Consider a hyperplane $H \subset \mathbb{P}^4$ given by the equation

$$ax_\infty + \frac{1}{2}L(x_0, \dots, x_3) = 0, \tag{19}$$

where L a linear form in the variables $x_0, \dots, x_3$. The equation of the projected surface $\pi(H \cap \mathcal{B})$ is obtained by eliminating the variable x_∞ from the system of equations $x_1^2 + x_2^2 - 2x_0 x_\infty = 0$ and (19). From the first equation we have $x_\infty = (x_1^2 + x_2^2)/(2x_0)$. Therefore, $\pi(\mathcal{A} \cap \mathcal{B})$ is defined by

$$a(x_1^2 + x_2^2) + x_0 L(x_0, \dots, x_3) = 0. \tag{20}$$

The hyperplane H does not contain $v = [e_3]$ if and only if $L(x_0, \dots, x_3)$ depends on x_3, i.e. the latter equation (20) defines an i-M sphere. Actually hyperplanes containing v correspond to i-spheres of cylindric type or isotropic planes.

Any i-M circle C can be represented as the intersection of two i-M spheres $C = S_1 \cap S_2$. Then $\sigma(C) = \sigma(S_1) \cap \sigma(S_2) = (H_1 \cap H_2) \cap \mathcal{B}$ for some hyperplanes H_1 and H_2 that do not contain v. So $\Pi = H_1 \cap H_2$ is a 2-plane, $v \notin \Pi$, and $\sigma(C) = \Pi \cap \mathcal{B}$.

Any isotropic line in $\mathbb{R}^3_{++0}$ can be parametrized by $t \mapsto (x_1, x_2, t)$ for some fixed x_1, x_2. Then $\sigma(x_1, x_2, t) = [1, x_1, x_2, t, (x_1^2 + x_2^2)/2]$, $t \in \mathbb{R}$, defines a line in the Blaschke cylinder $\mathcal{B}$. In this way just one line in $\mathcal{B}$ will be missed: the line l_∞ going through c and v. Let us parametrize its affine part by $\mathbb{R}$, $t \mapsto [te_3 + e_\infty]$. Then the i-M sphere $S : x_3 = a(x_1^2 + x_2^2) + L(1, x_1, x_2)$ corresponds to the hyperplane $ax_\infty + \frac{1}{2}(L(x_0, x_1, x_2) - x_3) = 0$ (see (19)), that intersects l_∞ at the point $t = 2a \in \mathbb{R}$. This is exactly the ideal point of S. $\qquad\square$

4.2 Basics on Isotropic Cyclides

Definition 2. *An* isotropic cyclide *is a surface in* $\mathbb{R}^3_{++0}$ *defined by an equation of the form*

$$a(x_1^2 + x_2^2)^2 + L(x_1, x_2, x_3)(x_1^2 + x_2^2) + Q(x_1, x_2, x_3) = 0, \tag{21}$$

where $a \in \mathbb{R}$ *and* $L(x_1, x_2, x_3), Q(x_1, x_2, x_3)$ *are polynomials,* $\deg L \le 1, \deg Q \le 2$.

Theorem 4. *The intersection of any quadric hypersurface* $\mathcal{A} \subset \mathbb{P}^4$ *with the Blaschke cylinder* $\mathcal{A} \cap \mathcal{B}$ *is mapped by the stereographic projection to an isotropic cyclide in* $\mathbb{R}^3_{++0}$. *Moreover every isotropic cyclide can be defined in this way.*

Proof. Consider a quadric $\mathcal{A} \subset \mathbb{P}^4$ given by the equation

$$ax_\infty^2 + \frac{1}{2}x_\infty L(x_0, \ldots, x_3) + \frac{1}{4}Q(x_0, \ldots, x_3) = 0, \tag{22}$$

where L and Q are a linear and a quadratic form respectively in the variables $x_0, \ldots, x_3$. The equation of the projected surface $\pi(\mathcal{A} \cap \mathcal{B})$ is obtained by eliminating the variable x_∞ as in the proof of Theorem 3

$$a(x_1^2 + x_2^2)^2 + x_0(x_1^2 + x_2^2)L(x_0, \ldots, x_3) + x_0^2 Q(x_0, \ldots, x_3) = 0.$$

In the affine part $\{x_0 = 1\}$ of $\mathbb{P}^3$ this equation represents an isotropic cyclide (21). $\square$

Let A and B be symmetric real 5x5 matrices defining, respectively, a quadric hypersurface $\mathcal{A}$ and the Blaschke cylinder $\mathcal{B}$, such that $\sigma(D) = \mathcal{A} \cap \mathcal{B}$ for an isotropic cyclide D. For example, in the standard basis, $B = \operatorname{diag}(1, 1, 0, 1, -1)$. Then $\sigma(D)$ is the carrier of the pencil of quadrics $\mathcal{P}_D : A - tB$, where t is a real parameter. Moreover, as in [12], all families of i-circles on D can be found by looking at singular quadrics (cones) in the pencil $\mathcal{P}_D$ defined by the roots of the polynomial $\det(A - tB)$.

More precisely, suppose that the isotropic cyclide D contains an i-circle C. Let Π be the 2-plane in $\mathbb{P}^4$ containing $C' = \sigma(C) = \Pi \cap \mathcal{B}$ (see Theorem 3). Choose a point $p \in \Pi \setminus C'$, not contained in $D' = \sigma(D)$. There is a unique quadric Q_C in the pencil $\mathcal{P}_D$ such that $p \in Q_C$. Then $Q_C \cap \Pi \supset C' \cup \{p\}$, which implies that $\Pi \subset Q_C$. Since Q_C is a quadric containing a 2-plane, Q_C is a quadratic cone. We say that the cone Q_C is *associated* with the i-circle C.

Actually any cone Q in a pencil $\mathcal{P}_D$ can be associated with many i-circles. Indeed, a general hyperplane section $H \cap Q$ is a real quadric surface of three possible types:

- a quadric with no lines, e.g. an ellipsoid;
- a cone with one family of lines (rank $Q = 3$);
- a double ruled quadric with two families of lines (rank $Q = 4$).

These 0, 1, or 2 families of lines define families of 2-planes in cones that correspond to families of i-circles on the isotropic cyclide D.

Definition 3. *Families of i-circles on a given isotropic cyclide D are called* paired *if their associated cones in the pencil $\mathcal{P}_D$ coincide. Otherwise they are called* unpaired. *A family of i-circles associated with a cone of rank 3 is called* single.

In the following theorem we collect several properties of i-circle families on isotropic cyclides, that are similar to the case of Darboux cyclides (see [12]) .

Theorem 5. *Any two i-circles from different paired families (or from one single family) are i-cospherical. Two i-circles from non-paired families are not i-cospherical, and they intersect exactly at 1 point (might be on the ideal line).*

Proof. If the i-circles C_1 and C_2 belong to different paired families or to the same single family, then in both cases their associated cone Q is the same. Let Π_1 and Π_2 be their 2-planes, i.e. $\sigma(C_i) = \Pi_i \cap \mathcal{B}$, $i = 1, 2$. For a general hyperplane H, the quadric surface $Q \cap H$ contains the lines $l_i = \Pi_i \cap H$, $i = 1, 2$. There are two cases: the quadric $Q \cap H$ is double ruled (then l_1 and l_2 are from different families of lines) or it is a cone. In both cases these two lines intersect in a certain point p. Hence, $\Pi_1 \cap \Pi_2$ contains a line going through the point p and the vertex of Q. Then Π_1 and Π_2 are in some hyperplane H', and both i-circles C_1 and C_2 are in the i-sphere $\pi(H' \cap B)$. Hence they are i-cospherical.

If two i-circles C_1 and C_2 belong to non-paired families, then their associated cones Q_1 and Q_2 are different. Suppose C_1 and C_2 are i-cospherical. Then the corresponding 2-planes Π_1 and Π_2 are contained in one hyperplane H. This implies that their intersection contains a line, say l, and $l \subset Q_1 \cap Q_2$. If $l \subset B$, then $\Pi_1 \cap B \supset \sigma(C_1) \cup l$. This is a contradiction because B is a quadric, so intersecting B with any 2-plane we cannot get a curve of degree higher than 2. Hence $l \not\subset B$. In particular there exists a point $p \notin B$ such that $p \in Q_1 \cap Q_2$. This is again a contradiction because there exists a unique quadric in the pencil $\mathcal{P}_D$ containing the point p. Therefore C_1 and C_2 are not i-cospherical.

Now let us show that $C_1 \cap C_2$ is one point. Note that $\sigma(C_1)$ and $\sigma(C_2)$ have to intersect because their 2-planes Π_1 and Π_2 have a non-empty intersection in $\mathbb{P}^4$, and again they cannot have a common point outside B. Of course this intersection point might appear on the ideal line, e.g. when both i-circles are of parabolic type.   □

4.3 Degenerate Isotropic Cyclides

We say that an isotropic cyclide D is *degenerate* if it is defined by a pencil of quadrics $\mathcal{P}_D : A - tB$ such that $\det(A - tB) = 0$ for all $t \in \mathbb{R}$. In this case all quadrics in the pencil $\mathcal{P}_D$ are cones, so that a degenerate isotropic cyclide carries infinitely many families of i-circles.

Note that in the classical conformal model this case cannot happen because the sphere $S^3 \subset \mathbb{P}^4$ is not a degenerate quadric.

Suppose now that D is a degenerate isotropic cyclide, such that for general $t \in \mathbb{R}$ the matrix $A - tB$ has rank 4, so that the general quadric in the pencil has only one point as a vertex.

The theorem below (originally proved by C. Segre) is a combination of Theorem 21.2 and Corollary 21.3 in [2] for the case of $\mathbb{P}^4$.

Theorem 6 (Segre). *For a pencil of quadrics $\mathcal{P}$ in $\mathbb{P}^4$ let*

$$V = \bigcup_{Q \in \mathcal{P}, \mathrm{rank} Q = 4} \mathrm{Vert}(Q).$$

Then exactly the following three cases may occur:

(1) V is a point, and all quadrics in the pencil have the same vertex;

(2) the Zariski closure of V is a line l, and the vertices of all quadrics in the pencil (with rank $= 4$) are contained in this line; moreover all such cones are tangent along this line.

(3) the Zariski closure of V is a conic contained in a plane Π, and all quadrics in the pencil contain that plane.

Theorem 7. *Degenerate isotropic cyclides can be of two kinds:*

(1) union of isotropic lines, i.e. cylinders over bi-circular quartic curves in the horizontal plane $\mathbb{R}^2$

$$a(x_1^2 + x_2^2)^2 + L(x_1, x_2)(x_1^2 + x_2^2) + Q(x_1, x_2) = 0 \qquad (23)$$

(2) i-M equivalent to paraboloids $x_3 = ax_1^2 + bx_2^2$, $a \neq b$.

Proof. We apply Theorem 6 to the pencil of quadrics $\mathcal{P} = \mathcal{P}_D$.

In case (1) of Theorem 6 the quadric $\mathcal{A}$ must be a cone with a vertex v, the same as the vertex of $\mathcal{B}$. Hence $\mathcal{A}$ is a union of lines in $\mathcal{B}$, and projects to a union of isotropic lines in $\mathbb{R}^3_{++0}$ (see Theorem 3). Then the equation of $\mathcal{A}$ does not depend on x_3, i.e. it is of the form (23). Hence $\mathcal{A}$ is a cylinder over a 2-circular curve in the plane $x_3 = 0$. Note that this cylinder is not a rational surface when this curve is not rational.

In case (2) of Theorem 6 the vertices of all quadrics in the pencil are contained in a line l. Moreover all such cones are tangent along this line. The projection of this line to the isotropic space is an isotropic line. Therefore one can move the line l by inversion to the ideal line. Then it corresponds to the line on the Blaschke cylinder which goes through its vertex $v = [e_3]$ and the center of the stereographic projection $c = [e_\infty]$, i.e. the line l is defined by three equations $l : \{x_0 = x_1 = x_2 = 0\}$. Then an arbitrary quadric $\mathcal{A}$ which contains l has the following equation

$$x_0 L_0 + x_1 L_1 + x_2 L_2 = 0,$$

where L_0, L_1, L_2 are linear forms in the variables $x_0, \dots, x_\infty$. Moreover, the tangent hyperplane to the Blaschke cylinder $T_c(\mathcal{B}) = \{x_0 = 0\}$ must coincide with $T_p(\mathcal{A})$ for any point $p = [se_3 + te_\infty] \in l$. This means that

$$\frac{\partial \mathcal{A}}{\partial x_1}(p) = L_1(p) = 0, \qquad \frac{\partial \mathcal{A}}{\partial x_2}(p) = L_2(p) = 0,$$

Therefore L_1 and L_2 depend only on x_0, x_1, x_2, and we see that the projected surface $\pi(\mathcal{A} \cap \mathcal{B})$ has equation

$$L_0(x_0^2, x_0x_1, x_0x_2, x_0x_3, x_1^2 + x_2^2) + x_1 L_1(x_0, x_1, x_2) + x_2 L_2(x_0, x_1, x_2) = 0.$$

In the affine space $x_0 = 1$ this is a paraboloid $x_3 = f(x_1, x_2)$ defined by a quadratic polynomial f. Using translation and rotation in the (x_1, x_2)-plane, one can transform this paraboloid into the standard one $x_3 = ax_1^2 + bx_2^2$. Note that $a \neq b$, since otherwise this paraboloid would be an i-sphere of parabolic type having just one common point with the line l.

Case (3) of Theorem 6 is very special, since all quadrics in the pencil contain the plane Π, which implies that D also contain Π as one of its irreducible components. Hence, in particular, D is not irreducible. $\qquad\qquad\square$

4.4   Isotropic Cyclides and CB-patches

Theorem 8. *Let* $T \subset \mathbb{R}^3_{++0}$ *be an algebraic surface of degree* ≤ 4. *If* $\mathrm{inv}(T)$ *has a component* V *of degree* ≤ 4, *then* V *is an isotropic cyclide. In particular* T *is an isotropic cyclide or it is the union of two isotropic cyclides.*

Proof. Apply Theorem 11 with $g = x^2 + y^2$. $\qquad\qquad\qquad\qquad\qquad\qquad\qquad\square$

Corollary 2. *Any non-i-spherical bilinear CB-patch is an isotropic cyclide patch.*

Proof. Let S be a non-i-spherical bilinear CB-patch. Then by Theorem 2 $\deg S \leq 4$, whence S is contained in a surface $T \subseteq \mathbb{R}^3$ of degree at most 4. Then $S' := \mathrm{inv}(S)$ is again a bilinear CB-patch, so that $\deg S' \leq 4$. This implies that $\mathrm{inv}(T)$ has a component V of degree at most 4 such that $S' \subseteq V$. Hence, by Theorem 8, V is an isotropic cyclide. Therefore S' is an isotropic cyclide patch, so that the same holds for S. $\qquad\square$

Any bilinear CB-patch generates 2 families of i-circles as its isoparametric curves. The following theorem characterizes such patches on the given isotropic cyclide.

Theorem 9. *Any two non-paired families of i-circles from different pairs on a given isotropic cyclide D are generated by a bilinear CB-patch. Two families of i-circles from the same pair can be generated only by a CB-patch defined by rulings of a double ruled quadrics (or its i-M equivalent).*

Proof. Take two i-cirlces C_0 and C_1 from the family $\mathcal{F}_1$ and take an i-circle C_{01} from the other family $\mathcal{F}_2$. Then by Theorem 5 the pairs of i-circles C_0, C_{01} and C_1, C_{01} intersect in the unique points p_0 and p_1. By Theorem 1 there exists a unique CB-patch S bounded by these three circles. But by Corollary 2 we know that every CB-patch is contained in an isotropic cyclide. Hence if we show that D is the only isotropic cyclide going through the three circles, then $S \subseteq D$, so that D generates the families $\mathcal{F}_1$ and $\mathcal{F}_2$.

Consider the cone Q associated with the family $\mathcal{F}_1$ in the pencil of quadrics $\mathcal{P}_D$. Then the projections of the conics $\sigma(C_0)$, $\sigma(C_1)$, $\sigma(C_{01})$ to a general hyperplane H in $\mathbb{P}^4$ from the vertex of Q are two lines l_0, l_1 and a conic C, such that the pairs l_0, C and l_1, C intersect exactly in one point.

Now if D' is a different isotropic cyclide that goes through C_0, C_1 and C_{01}, then a cone Q' associated with C_0 and C_1 has the same vertex as Q. Similarly projecting the corresponding conics to the hyperplane H, we get the same configuration of lines l_0, l_1 and the conic C. If the lines l_0, l_1 are skew, then by Lemma 6 there is a unique quadric containing this configuration. This implies that $Q' = Q$, so that $D' = D$ and the theorem is proved. If the lines l_0, l_1 intersect, then the family $\mathcal{F}_1$ is single and we need to prove the uniqueness of a quadratic cone going through l_0, l_1 and C in H, which is easy. $\qquad\qquad\qquad\qquad\qquad\qquad\qquad\qquad\qquad\qquad\qquad\qquad\qquad\square$

Lemma 6. *Given two skew lines (or segments) l_1, l_2 and a conic C in $\mathbb{R}^3$ intersecting them, there exists a unique quadric surface that contains l_1, l_2 and C.*

Proof. Assume there are two quadrics Q_1 and Q_2 containing l_1, l_2 and C. Then $Q_1 \cap Q_2 = C \cup l_1 \cup l_2$. Choose a point $p \notin C$ but on the same plane H with C. Then there is a quadric Q in the pencil generated by Q_1 and Q_2, such that $p \in Q$. Since Q contains p and C, it contains also the plane H. Therefore, Q is a union of two planes and both lines l_1, l_2 should lie on the other plane, but they are skew. This is a contradiction. $\qquad\square$

Two bilinear CB-patches have the same *type* if they generate the same families of i-circles, i.e. they define rational parametrizations that differ by an automorphism of $\mathbb{P}^1 \times \mathbb{P}^1$.

Corollary 3. *There are exactly 12 different types of bilinear CB-patches on an isotropic cyclide with 6 real families of i-circles that are grouped into 3 pairs.*

Proof. The existence of such an isotropic cyclide is proved in Example 2. According to Theorem 9 we choose 2 pairs of families from 3 available, and then we choose from every family a pair of circles. Thus we get $3 \cdot 2 \cdot 2 = 12$ possibilities. $\square$

Example 2. The isotropic cyclide from Example 1 given by the equation

$$(x^2 + y^2)^2 - 8x^2 - 5y^2 + 12z^2 + 4 = 0,$$

has 6 families if i-circles. In Figure 2 one can see 6 i-circles from different families going through one point. Paired families are represented by pairs if i-circles which intersect in two points.

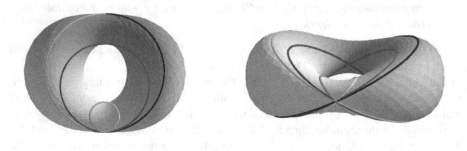

Fig. 2. An isotropic cyclide with 6 i-circles from different families

Corollary 4. *A patch on the degenerate isotropic cyclide $D : x_3 = ax_1^2 + bx_2^2, a \neq b$, can be parametrized as a bilinear CB-patch if and only if its top view is a parallelogram.*

Proof. All i-M circles on D are either i-circles of parabolic type or non-isotropic lines. According to Section 4.3 their associated cones in $\mathbb{P}^4$ have vertices on the exceptional line l_∞ that correspond to the ideal line. Therefore two i-circles on D are associated to the same cone if their ideal points coincide. Also different families of i-circles are mapped by the top view projection to different families of parallel lines in the plane $\mathbb{R}^2$. Hence the top view of a bilinear CB-patch will be a parallelogram.

Let us consider two adjacent boundary lines of a given parallelogram $P \subset \mathbb{R}^2$, which are top views of two i-circles C and C' on D. The latter i-circles can be either from unpaired or from paired families. In both cases by Theorem 9 we can find a CB-patch with a top view P. $\square$

5   Applications to Rational Surfaces with Rational Offsets

5.1   PN-surfaces and the Blaschke Model of Laguerre Geometry

Pythagorean-normal (PN) surfaces are rational surfaces $F(s, t)$ in the euclidean space $\mathbb{R}^3$ together with a field of rational unit normals $n(s, t)$. PN-surfaces are important in geometric modeling applications, since they are rational surfaces with rational offsets. Their d-offsets for any $d \in \mathbb{R}$ can be easily parametrized rationally:

$$F_d(s,t) = F(s,t) + dn(s,t), \quad |n(s,t)| = 1.$$

Following [9,11] (see also survey in [5]), we use a dual approach to PN-surfaces. We treat such a surface as the set of its oriented tangent planes

$$T^{\text{or}} : n_1 x + n_2 y + n_3 z + h = 0, \quad n_1^2 + n_2^2 + n_3^2 = 1,$$

and represent them as points

$$[n_1, n_2, h, n_3, -1] \tag{24}$$

on the Blaschke cylinder $\mathcal{B} : x_1^2 + x_2^2 + x_4^2 - x_5^2 = 0$ in $\mathbb{P}^4$.

The following theorem is due to Pottmann and Peternell [9,11].

Theorem 10. *This duality defines a 1–1 correspondence between non-developable (respectively developable) PN-surfaces in the euclidean space $\mathbb{R}^3$ and rational surfaces (respectively curves) in the Blaschke cylinder $\mathcal{B} \subset \mathbb{P}^4$.*

Oriented tangent planes and oriented spheres in the standard Laguerre geometry correspond to points and hyperplane sections respectively in the Blaschke cylinder. Laguerre transformations on $\mathcal{B}$ are all projective transformations of the ambient space $\mathbb{P}^4$ that preserve $\mathcal{B}$. So one can try to model curves and surfaces in $\mathcal{B}$ and then by duality go back to euclidean space. Unfortunately to work directly in this 3-dimensional cylinder in $\mathbb{P}^4$ is not so easy.

Therefore we project the Blaschke cylinder to the isotropic space $\mathbb{R}^3_{++0}$ using stereographic projection. First we define a point (24) (which is dual to T^{or}) in slightly different coordinates $x_0, \ldots, x_\infty$

$$\delta(T^{\text{or}}) = [n_3 + 1, n_1, n_2, h, \frac{1}{2}(n_3 - 1)] \in \mathbb{P}^4,$$

and then we apply the stereographic projection π

$$\phi(T^{\text{or}}) = \pi(\delta(T^{\text{or}})) = \frac{-1}{n_3 + 1}(n_1, n_2, h) \in \mathbb{R}^3_{++0}. \tag{25}$$

Note that for oriented planes $-z + h = 0$ with normal vector $n = (0, 0, -1)$, this formula is not defined. In this case we assign the image as a point (actually a number) $h \in \mathbb{R}$ on the ideal line in the extended isotropic space $\mathbb{R}^3_{++0} \cup \mathbb{R}$.

Now oriented tangent planes and oriented spheres can be recognized as points and i-M spheres respectively in the isotropic space (see Theorem 3). Intersections of i-M spheres are i-M circles. Therefore i-M circles correspond to oriented circular cones (or cylinders), since the latter are envelopes of all common oriented tangent planes for pairs of oriented spheres.

In the following section we will illustrate how bilinear CB-patches can be used for non-trivial PN-surface construction.

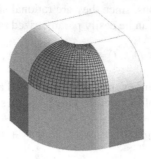

Fig. 3. A torus patch blending the corner

5.2 Blending Example

The goal is to construct a PN-surface blend of a corner bounded by three orthogonally intersecting planes in $\mathbb{R}^3$, where three edges are rounded with circular cylinders of different radii $r_1 < r_2 < r_3$.

We are going to blend these three cylinders and the top horizontal plane using the special patch of a PN-surface that is dual to a particular bilinear CB-patch.

First we start from the simpler case $r_1 = r_2 < r_3$ (see Fig. 3) where one can blend the corner using the obvious patch of a torus with the smaller radius $r_1 = r_2$ and the bigger radius $R = r_3 - r_1$. Later on in this section (see too Fig. 4) we will change radius r_1.

Choose the radii $r_1 = r_2 = 1$, $r_3 = 3/2$ and three planes $\Pi_1 : x - 3/2 = 0$, $\Pi_2 : y - 3/2 = 0$, $\Pi_3 : z - 1 = 0$. In the isotropic space they correspond to the control points of a CB-patch (see (25)):

$$p_0 = \phi(\Pi_1) = -e_1 + \frac{3}{2}e_3, \quad p_1 = \phi(\Pi_2) = -e_2 + \frac{3}{2}e_3, \quad p_2 = p_3 = \phi(\Pi_3) = \frac{1}{2}e_3.$$

Then we calculate the intermediate points q_{ij} on arcs joining p_i and p_j as the ϕ images of corresponding tangent planes on cylinders (we choose positions with rational coordinates avoiding square roots):

$$q_{01} = -\frac{3}{5}e_1 - \frac{4}{5}e_2 + \frac{3}{2}e_3, \quad q_{02} = -\frac{1}{3}e_1 + \frac{13}{18}e_3, \quad q_{03} = -\frac{1}{3}e_2 + \frac{13}{18}e_3.$$

Now we correct the point $q_{02} := q_{02} + \frac{1}{10}e_3$ in order to change the radius of the cylinder between tangent planes Π_1 and Π_3. Then we calculate weights (see Section 2)

$$w_0 = 1,$$
$$w_1 = (p_1 - q_{01})^{-1}(q_{01} - p_0) = 1 - e_{12},$$
$$w_2 = (p_2 - q_{02})^{-1}(q_{02} - p_0) = 2 - \frac{1}{10}e_{13},$$
$$w_3 = \lambda(p_3 - q_{13})^{-1}(q_{13} - p_1)w_1 = \lambda(2 - e_{13} - e_{23} - 2e_{12}),$$

where the constant $\lambda = \frac{19}{10}$ is determined using Lemma 5.

In Fig. 4 (left) we see the CB-patch with the calculated control points and weights and its top view in wire-frame. It is a degenerate quad patch with one edge collapsed into a point.

The resulting blending surface in Fig. 4 (right) is generated by applying the inverse of the stereographic projection and calculating a dual surface as the envelope of its tangent planes. This is a PN-surface patch that can be converted into a tensor product Bézier surface of bidegree $(3, 4)$.

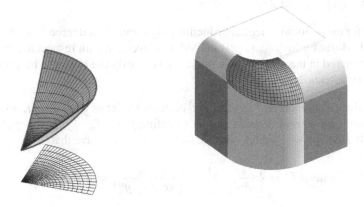

Fig. 4. A CB-patch and its dual PN-surface patch

6 g-Möbius Geometry

In classical Möbius geometry in $\mathbb{R}^3$ the main objects considered are Möbius spheres: these can be spheres or planes. In other words we can define a Möbius sphere as a surface in $\mathbb{R}^3$ defined by a polynomial of degree 2, whose homogeneous part of highest degree has the form $a(x^2 + y^2 + z^2)$, for some $a \in \mathbb{R}$ (possibly zero).

In this section we collect several technical results about g-Möbius geometry, where the polynomial $x^2 + y^2 + z^2$ is replaced with any homogeneous polynomial $g(x, y, z)$ of degree 2, irreducible over $\mathbb{R}$. Actually $g = x^2 + y^2$ corresponds to isotropic Möbius geometry.

Moreover we will give an algebraic definition of a g-cyclide as a generalization of a Darboux cyclide in Möbius geometry. The main result is Theorem 11, which gives a sufficient condition for a surface to be a g-cyclide.

6.1 g-spheres

Definition 4. *A g-sphere $S \subseteq \mathbb{R}^3$ is an algebraic surface whose equation is of the form*

$$ag(x, y, z) + L(x, y, z) + b = 0,$$

where $a \neq 0, b \in \mathbb{R}$ and L is an homogeneous polynomial of degree 1 (or it is zero). A g-Möbius sphere is either a g-sphere or a plane.

Definition 5. *Let* $G = \{g = 0\} \subset \mathbb{R}^3$. *A* g-*inversion (with respect to the origin) is the map* $\mathrm{inv}_g : \mathbb{R}^3 \setminus G \to \mathbb{R}^3 \setminus G$, *defined by*

$$\mathrm{inv}_g(x, y, z) = \left(\frac{x}{g(x, y, z)}, \frac{y}{g(x, y, z)}, \frac{z}{g(x, y, z)} \right).$$

Given an algebraic subset $T \subset \mathbb{R}^3$, *such that no irreducible components of* T *are contained in* G, *we define* $\mathrm{inv}_g(T) := \overline{\mathrm{inv}_g(T \setminus G)}$, *where the closure is with respect to the euclidean topology in* $\mathbb{R}^3$.

Note that for every homogeneous irreducible polynomial g of degree 2, we have that $(\mathrm{inv}_g)^2 = \mathrm{id}$. Moreover $\mathrm{inv}_{x^2+y^2+z^2}$ is the Möbius inversion with respect to the sphere of radius 1 centered in the origin, while $\mathrm{inv}_{x^2+y^2}$ is exactly the isotropic inversion inv that appears in Section 2.2.

Remark 2. Given a polynomial $f \in \mathbb{R}[x, y, z]$ of degree n, let us write $f = f_n + \cdots + f_0$, where f_i is homogeneous of degree i and let us define $f_g := f_n + f_{n-1}g + f_{n-2}g^2 + \cdots + f_0 g^n$. If $T := \{f = 0\}$, then, because $(\mathrm{inv}_g)^2 = \mathrm{id}$, we have that

$$\mathrm{inv}_g(T) = \left\{ \frac{f_g}{g^n} = 0 \right\} = \left\{ \frac{f_g}{\gcd(f_g, g^n)} = 0 \right\}.$$

Lemma 7. *S is a g-Möbius sphere if and only if* $\mathrm{inv}_g(S)$ *is a g-Möbius sphere. Moreover S passes through the origin if and only if* $\mathrm{inv}_g(S)$ *is a plane.*

Proof. A g-Möbius sphere S has the equation $ag(x, y, z) + L(x, y, z) + b = 0$, for some $a, b \in \mathbb{R}$. Then $\mathrm{inv}_g(S) = \{bg(x, y, z) + L(x, y, z) + a = 0\}$ is again a g-Möbius sphere. Moreover S passes through the origin if and only if $b = 0$, if and only if $\mathrm{inv}_g(S)$ is a plane.   □

6.2 g-cyclides

Definition 6. We say that an algebraic surface $S \subset \mathbb{R}^3$ is a g-*cyclide* if it is the zero-set of the polynomial

$$f(x, y, z) = \lambda g(x, y, z)^2 + g(x, y, z)L(x, y, z) + Q(x, y, z),$$

where $\lambda \in \mathbb{R}$, L is an homogeneous polynomial of degree 1 (or it is zero) and Q is a polynomial of degree at most 2.

Note that a $(x^2 + y^2 + z^2)$-cyclide is a Darboux cyclide and a $(x^2 + y^2)$-cyclide is an isotropic cyclide.

Lemma 8. *X is a g-cyclide if and only if* $\mathrm{inv}_g(X)$ *is a g-cyclide.*

Proof. Let $X := \{\lambda g^2 + gL + Q = 0\} = \{\lambda g^2 + gL + Q_2 + Q_1 + Q_0 = 0\}$, where $\deg Q_i = i$. Then $\mathrm{inv}_g(X)$ has the equation $\lambda + L + Q_2 + Q_1 g + Q_0 g^2$, so that $\mathrm{inv}_g(X)$ is a g-cyclide.   □

Theorem 11. *Let $T \subseteq \mathbb{R}^3$ be a surface of degree ≤ 4. If $\mathrm{inv}_g(T)$ has a component V of degree ≤ 4, then V is a g-cyclide. In particular either T is a g-cyclide or T is the union of two g-cyclides.*

Proof. Let $T = \{f = 0\} \subset \mathbb{R}^3$, where $f \in \mathbb{R}[x, y, z]$ and $\deg f \leq 4$. Let us suppose $\deg f = 4$ (the other cases are similar). Let $f_g = f_4 + f_3 g + f_2 g^2 + f_1 g^3 + f_0 g^4$, as in Remark 2. Note that in general $\deg f_g \leq 8$ and $\deg f_g = 8$ if and only if $f_0 \neq 0$. Let us suppose that in fact $f_0 \neq 0$ (the other cases are similar).

Denote by T' the variety obtained by performing the inversion inv_g of T. Consider a polynomial h such that $f_g = hg^k$, for some $0 \leq k \leq 4$, and $\gcd(g, h) = 1$. Then, by Remark 2, $T' = \{h = 0\}$, so that, by hypothesis, h has a factor a of degree ≤ 4, defining a component V of T'. Let us suppose for simplicity that $\deg a = 4$. Then $f_g = ab$ is the product of two polynomials of degree 4. By lemma 9 it follows that V is a g-cyclide.

The last part of the theorem follows because we have proved that $T' = \mathrm{inv}_g(T)$ is a union of g-cyclides, so that the same holds for T. $\square$

Remark 3. If a surface T satisfies the conditions of Theorem 11, then it can be a union of two g-cyclides but it might happen that it is not a g-cyclide itself. For example, T might be the union of two quadrics and $\mathrm{inv}_g(T)$ might be a union of two quartic cyclides.

Lemma 9. *Let f and f_g be as in the proof of Theorem 11. If $f_g = ab$ is the product of two polynomials of degree 4, then a and b define two g-cyclides.*

Proof. Remember that $f_g := f_4 + f_3 g + f_2 g^2 + f_1 g^3 + f_0 g^4$, and for all i, the polynomials $f_i g^{4-i}$ are homogeneous of degree $8 - i$. Let us write $a := a_4 + a_3 + a_2 + a_1 + a_0$ and $b := b_4 + b_3 + b_2 + b_1 + b_0$, where, for every i, a_i and b_i are homogeneous of degree i. To prove the lemma, we need to prove that $g^2 | a_4$, $g^2 | b_4$, $g | a_3$ and $g | b_3$.

By considering the part of degree 8 of the equation $f_g = ab$, we get that

$$a_4 b_4 = f_0 g^4$$

(remember that we are assuming $f_0 \neq 0$). This implies that $a_4 = \gamma_a g^2$, $b_4 = \gamma_b g^2$, where $\gamma_a, \gamma_b \in \mathbb{R} \setminus \{0\}$.

Now, the part of degree 6 of the equation $f_g = a \cdot b$ gives that

$$f_2 g^2 = a_4 b_2 + a_3 b_3 + a_2 b_4 = g^2(\gamma_a b_2 + \gamma_b a_2) + a_3 b_3.$$

Hence $a_3 b_3 = g^2 \cdot (f_2 - \gamma_a b_2 + \gamma_b a_2)$. If $f_2 - \gamma_a b_2 + \gamma_b a_2 \neq 0$, then $g^2 | a_3 b_3$, which implies that $g | a_3$ and $g | b_3$ and we are done. If $f_2 - \gamma_a b_2 + \gamma_b a_2 = 0$, then $a_3 b_3 = 0$. Without loss of generality we can then assume then $b_3 = 0$. In this case we also consider the part of degree 7, which gives that

$$a_3 b_4 + b_3 a_4 = f_1 g^3.$$

This implies that $a_3 \gamma_b g^2 = f_1 g^3$, so that $g | a_3$, and we are done. $\square$

7 Conclusions

A Bézier-like rational surface construction was introduced with weights in the geometric (Clifford) algebra generated by isotropic space. It is shown that these Clifford-Bézier (CB) patches are isotropic Möbius invariant. The bilinear case is studied in most detail: their implicitization formula is derived, they are characterized as patches on isotropic cyclides and the uniqueness of patches with three given boundary isotropic circles is proved.

The developed theory allows us to model Pythagorean-normal (PN) surfaces employing duality between the standard model of Laguerre geometry and the isotropic one. One example is presented where a PN-surface (dual to a CB-patch) is used to blend three cylinders of different radii and one plane.

Further research directions include: detailed classification of isotropic cyclides, generalization of bilinear CB-patches to higher degrees, more applications to PN-surface modeling, e.g. one can expect that CB-patches of bidegree $(1, 2)$ can reproduce the branching blend in [6].

Acknowledgements. This paper has been partially financed by the Marie-Curie Initial Training Network SAGA (ShApes, Geometry, Algebra) FP7-PEOPLE contract PITN-GA-2008-214584.

The Maple Package for Clifford Algebra [1] was used extensively for symbolic computations in this research.

References

1. Ablamowicz, R., Fauser, B.: A Maple 10 Package for Clifford Algebra Computations, Version 10 (2007), http://math.tntech.edu/rafal/cliff10
2. Ciliberto, C., van der Geer, G.: Andreotti–Mayer loci and the Schottky problem. Documenta Mathematica 13, 453–504 (2008)
3. Dorst, L., Fontijne, D., Mann, S.: Geometric Algebra to Computer Science. Morgan-Kaufmann (2007)
4. Goldman, R.: A Homogeneous Model for Three-Dimensional Computer Graphics Based on the Clifford Algebra for $\mathbb{R}^3$. In: Dorst, L., Lasenby, J. (eds.) Guide to Geometric Algebra in Practice, pp. 329–352 (2011)
5. Krasauskas, R., Peternell, M.: Rational offset surfaces and their modeling applications. In: Emiris, I.Z., Sottile, F., Theobald, T. (eds.) IMA Volume 151: Nonlinear Computational Geometry, pp. 109–135 (2010)
6. Krasauskas, R.: Branching blend of natural quadrics based on surfaces with rational offsets. Computer Aided Geometric Design 25, 332–341 (2008)
7. Krasauskas, R., Zube, S.: Bezier-like parametrizations of spheres and cyclides using geometric algebra. In: Guerlebeck, K. (ed.) Proceedings of 9th International Conference on Clifford Algebras and their Applications in Mathematical Physics, Weimar, Germany (2011)
8. Perwass, C.: Geometric Algebra with Applications in Engineering. Series: Geometry and Computing, vol. 4. Springer (2009)
9. Peternell, M., Pottmann, H.: A Laguerre geometric approach to rational offsets. Computer Aided Geometric Design 15, 223–249 (1998)

10. Pottmann, H., Grohs, P., Mitra, N.J.: Laguerre Minimal Surfaces, Isotropic Geometry and Linear Elasticity. Advances in Computational Mathematics 31, 391–419 (2009)

11. Pottmann, H., Peternell, M.: Applications of Laguerre geometry in CAGD. Computer Aided Geometric Design 15, 165–186 (1998)

12. Pottmann, H., Shi, L., Skopenkov, M.: Darboux cyclides and webs from circles. Computer Aided Geometric Design 29, 77–97 (2012)

13. Zube, S.: A circle represenatation using complex and quaternion numbers. Lithuanian Journal of Mathematics 46, 298–310 (2006)

Algorithms and Data Structures for Truncated Hierarchical B–splines

Gábor Kiss[1], Carlotta Giannelli[2], and Bert Jüttler[2]

[1] Doctoral Program "Computational Mathematics"
[2] Institute of Applied Geometry
Johannes Kepler University Linz, Altenberger Str. 69, 4040 Linz, Austria
gabor.kiss@dk-compmath.jku.at,
{carlotta.giannelli,bert.juettler}@jku.at

Abstract. Tensor–product B–spline surfaces are commonly used as standard modeling tool in Computer Aided Geometric Design and for numerical simulation in Isogeometric Analysis. However, when considering tensor–product grids, there is no possibility of a localized mesh refinement without propagation of the refinement outside the region of interest. The recently introduced truncated hierarchical B–splines (THB–splines) [5] provide the possibility of a local and adaptive refinement procedure, while simultaneously preserving the partition of unity property. We present an effective implementation of the fundamental algorithms needed for the manipulation of THB–spline representations based on standard data structures. By combining a quadtree data structure — which is used to represent the nested sequence of subdomains — with a suitable data structure for sparse matrices, we obtain an efficient technique for the construction and evaluation of THB–splines.

Keywords: hierarchical tensor–product B–splines, truncated basis, THB–splines, isogeometric analysis, local refinement.

1 Introduction

The *de facto* standard in computer aided geometric design is the tensor–product B–spline model together with its non–uniform rational extension (NURBS). Among other fundamental properties, like minimum support, efficient refinement and degree–elevation algorithms, B–splines are nonnegative and form a partition of unity. This implies that a B–spline curve/surface is completely contained in the *convex hull* of a certain set of points, usually referred to as control net. The shape of the control net directly influences the shape of the B–spline representation, so that the designer can use it to manipulate the corresponding parametric representation in a fairly intuitive way. Unfortunately, an unavoidable drawback of the tensor–product structure is a global nature of the mesh refinement which excludes the possibility of a local refinement scheme as illustrated in Figure 1(a–c).

M. Floater et al. (Eds.): MMCS 2012, LNCS 8177, pp. 304–323, 2014.
© Springer-Verlag Berlin Heidelberg 2014

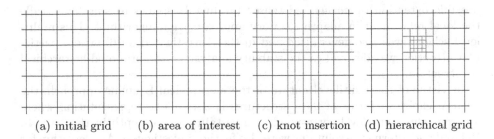

(a) initial grid   (b) area of interest   (c) knot insertion   (d) hierarchical grid

Fig. 1. Adaptive refinement of an initial tensor–product grid (a) with respect to a localized region (b) may be achieved by avoiding a propagation of the refinement due to the tensor–product structure (c) through a hierarchical approach (d).

Despite an increasing interest in the identification of adaptive spline spaces and related applications, see e.g., [7,15,18], local mesh refinement remains a nontrivial and computationally expensive issue. A suitable trade–off between the quality of the geometric representation (in terms of degrees of freedom needed to obtain a certain accuracy) and the complexity of the mesh refinement algorithm has necessarily to be taken into account. Different approaches have been proposed which all extend the standard tensor–product model by allowing T–junctions between axis aligned mesh segments. Among others, this led to the introduction of hierarchical B–splines (HB–splines) [4,11,12], T–splines [16], polynomial splines over T–meshes [2] and – more recently – truncated hierarchical B–splines (THB–splines) [5] and locally refined B–splines [3].

The idea of performing surface modeling by manipulating the parametric representation at different levels of details was originally proposed by Forsey and Bartels [4]. In order to localize the editing of detailed features, the refinement is iteratively adapted on restricted patches of the surface in terms of a sequence of *overlays* with nested knot vectors. Subsequently, Kraft [11,12] showed that the hierarchical structure enforced on the mesh refinement procedure can be complemented by a simple and automatic identification of basis functions which naturally generalizes some of the fundamental properties of tensor–product B–splines — such as nonnegativity and linear independence — to the case of HB–splines.

The multilevel approach allows to break the rigidity of a tensor–product configuration by simultaneously preserving a highly organized structure as shown in Figure 1(d). An example of hierarchical refinements over rectangular–shape regions is presented in Figure 2.

The hierarchical B–spline model found applications in data interpolation and approximation [10,11,13], as well as in finite element and isogeometric analysis [1,14,18]. Alternative spline hierarchies were also considered in the literature, see e.g., [9,19].

Kraft's basis for HB–splines does not possess the partition of unity property without additional scaling and it possesses only limited stability properties. Truncated hierarchical B–splines [5] have the potential to overcome these limitations and provide improved sparsity properties. They were introduced as a possible extension of *normalized* tensor–product B–splines to suitably handle the local refinement in adaptive surface approximation algorithms. This multi-level scheme was also generalized and further investigated in [6], where partic-ular attention was devoted to the stability analysis of the proposed hierarchical construction.

In virtue of the multilevel nature of the hierarchical B–spline approach, the natural choice in terms of data structures is a tree–like representation where a given refinement level correspond to a certain level of depth in the tree [4]. Related and alternative solutions were also further investigated. An algorithm for scattered data interpolation and approximation by multilevel bicubic B–splines based on a hierarchy of control lattices was described in [13]. An implementation of hierarchical B–splines in terms of a tree data structure whose nodes represent the B–splines from different levels was recently presented in [1]. Another solution consists of storing in each node of the tree the data related to a knot span of a certain level, in particular the significant basis functions acting on it [14].

The goal of the present paper is to introduce an effective implementation of data structures and algorithms for the newly introduced THB–splines. To represent the subdomain hierarchy we use a *quadtree* data structure in com-bination with *sparse matrices*. The *quadtree* provides an efficient and dynamic data structure for representing the subdomains. It also facilitates the needed update which may be caused by an iterative refinement process. One key moti-vation for this choice is to reduce the memory overhead in need for storing the subdomain hierarchy as much as possible. The selection of (possibly truncated) basis functions proceeds as described in [5] by means of certain queries which use the quadtree. The result is encoded by a sequence of *sparse matrices*. The quadtree and the related sparse matrices are initially created and subsequently updated during the refinement procedure. For the hierarchical spline evaluation algorithm, however, only the access to the sparse matrices is required. This leads to a reasonable trade–off with respect to memory and time consumption during both the *construction* of THB–splines from an underlying subdomain hierarchy and their *evaluation* for given parameter values.

The paper is organized as follows. In Section 2 we describe the hierarchical ap-proach to adaptive mesh refinement together with the definition and evaluation of the THB–spline basis. Section 3 introduces the data structures and algorithms used for the representation of the subdomain hierarchy, while Section 4 explains the construction of the matrices needed during the THB–spline evaluation in more detail. Some numerical results are then presented in Section 5 to illustrate the performance of our approach, while the extension of the proposed approach to more general knot configurations and refinements is discussed in Section 6. Finally, Section 7 concludes the paper.

2   THB–splines

We define an adaptive extension of the classical tensor–product B–spline con-struction in terms of a certain number N of hierarchical *levels* that correspond

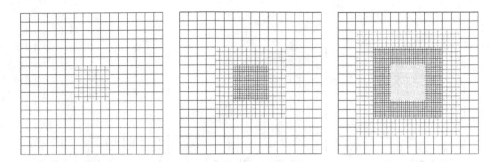

Fig. 2. An example of hierarchical refinement over rectangular–shape regions where the central area of the mesh is always refined up to the maximum level of detail: two levels (left), three levels (middle) and four levels (right).

to an increasing level of detail. At each refinement step we select a specific tensor–product grid associated with the current level. Provided that the sequence of tensor–product grids corresponds to a nested sequence of spline spaces $V^0, \ldots, V^{N-1}$ which satisfies

$$V^{\ell-1} \subset V^{\ell},$$

for $\ell = 1, \ldots, N-1$, the hierarchical framework allows to consider different types of grid refinement. Following the notation of [17], the superscript ℓ denotes the hierarchical level ℓ throughout the paper.

The present paper focuses on the *bivariate* tensor–product case. However, the framework can easily be adapted to the multivariate setting and even to more general spline spaces [5,6]. Nevertheless, even if the representation model we are going to introduce may in principle be used to handle non–uniform mesh refinement and even spaces generated by degree elevation, we will consider only the dyadic uniform case throughout this paper. The modifications which are required to extend the proposed approach to more general knot configurations are discussed in Section 6.

More precisely, we assume that the coarsest spline space V^0 is spanned by bivariate tensor–product B–splines with respect to two bi–infinite uniform knot sequences. The finer spaces V^ℓ are obtained by iteratively applying dyadic subdivision, i.e., each cell of the original tensor-product grid is split uniformly into four cells.

Let Ω^0 be a rectangular planar domain whose edges are aligned with the tensor-product grid of V^0, and let $\{\Omega^\ell\}_{\ell=0,\ldots,N-1}$ be a nested sequence of subdomains so that

$$\Omega^{\ell-1} \supseteq \Omega^\ell, \tag{1}$$

for $\ell = 1, \ldots, N - 1$. Each Ω^ℓ is defined as a collection of cells with respect to the tensor–product grid of level $\ell - 1$.

Example 1. Figures 2 and 3 show three subdomain hierarchies which will be used to demonstrate the performance of our algorithms and data structures:

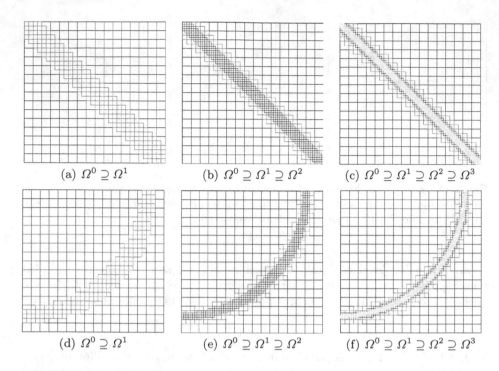

(a) $\Omega^0 \supseteq \Omega^1$ (b) $\Omega^0 \supseteq \Omega^1 \supseteq \Omega^2$ (c) $\Omega^0 \supseteq \Omega^1 \supseteq \Omega^2 \supseteq \Omega^3$

(d) $\Omega^0 \supseteq \Omega^1$ (e) $\Omega^0 \supseteq \Omega^1 \supseteq \Omega^2$ (f) $\Omega^0 \supseteq \Omega^1 \supseteq \Omega^2 \supseteq \Omega^3$

Fig. 3. Two nested sequences of subdomains — indicated as *linear* (top) and *curvilinear* (bottom) in Example 1. They satisfy relation (1) with respect to two (left), three (middle) and four (right) hierarchical levels.

- *rectangular* (refinement over rectangular–shaped regions);
- *linear* (refinement along a diagonal layer);
- *curvilinear* (refinement along a curvilinear trajectory).

By starting with an initial tensor–product configuration at level 0, the tensor–product grid associated with level $\ell + 1$ is obtained by subdividing any cell of the previous level into four parts. Each subdomain Ω^ℓ is then defined as a certain collection of cells with respect to the grid of level ℓ so that (1) is satisfied. Figure 2 illustrates an example of hierarchical refinement over rectangular–shape regions where the central area of the mesh is always refined up to the maximum level of detail. The other two subdomain hierarchies mentioned above are shown in Figure 3 up to four refinement levels so that $\Omega^0 \supseteq \ldots \supseteq \Omega^3$.

For each hierarchical level ℓ, with $\ell = 0, \ldots, N - 1$, let $\mathcal{B}^\ell$ be the normalized B–spline basis of the spline space V^ℓ with respect to a certain degree (d, d) defined on corresponding nested knot sequences. We say that

$$\beta \in \mathcal{B}^\ell \text{ is active } \Leftrightarrow \text{supp}^0 \beta \subseteq \Omega^\ell \wedge \text{supp}^0 \beta \nsubseteq \Omega^{\ell+1},$$

where $\text{supp}^0 \beta = \text{supp} \beta \cap \Omega^0$ is a slightly modified support definition which makes local refinements possible also along the boundaries of Ω^0. A B–spline

$\beta \in \mathcal{B}^\ell$ is then *active* if it is completely contained in Ω^ℓ but not in $\Omega^{\ell+1}$, and *passive* otherwise.

We may assume the initial domain Ω^0 to be an axis-aligned box[1]. By denoting with k the number of knot spans of level 0 along the edges of Ω^0, which is assumed to be the same for both directions, we define a *characteristic matrix* X^ℓ of size $s^\ell \times s^\ell$, with $s^\ell = (2^\ell k + d)$, for $\ell = 0, \dots, N-1$. These matrices collect the information about active/passive B–splines level by level, namely

$$X_{i,j}^\ell = \begin{cases} 1 & \text{if } \beta_{i,j}^\ell \text{ is active,} \\ 0 & \text{otherwise,} \end{cases}$$

where $\beta_{i,j}^\ell$ is a B–spline of level ℓ. The indices i, j are chosen such that exactly the B–splines $\beta_{i,j}^\ell$ with $i, j = 1, \dots, s^\ell$ are non–zero on Ω^0.

Definition 1 ([11,12], extended in [18]). *The hierarchical B–spline (HB–spline) basis $\mathcal{H}$ is defined as the set of all* active *B–splines defined over the tensor–product grid of each level,*

$$\mathcal{H} = \bigcup_{\ell=0,\dots,N-1} \{\beta_{i,j}^\ell \in \mathcal{B}^\ell : X_{i,j}^\ell = 1\}.$$

A spline function represented in the hierarchical B–spline basis is then defined as a linear combination of active B–splines from different levels in the hierarchy. In order to evaluate the considered spline in a given point of the domain, the contribution of all the active B–splines (from the minimum to the maximum level of basis functions whose support is non–zero on that point) has to be computed and then added together. The cost of the hierarchical evaluation algorithm is then quadratic and linear with respect to the degree (B–spline evaluation) and the number of levels, respectively.

Truncated hierarchical B–splines [5,6] form a different basis for the same multilevel B–spline space. The key idea behind this alternative hierarchical construction is to properly exploit the *refinable* nature of the B–spline basis which allows to express a B–spline of level ℓ in terms of $(d+2)^2$ functions which belong to level $\ell+1$ and of certain binomial coefficients scaled by a factor 2^{-d} with respect to any dimension. By using this subdivision rule, any function $\tau \in V^\ell$ can be represented according to a two–scale relation with respect to the basis $\mathcal{B}^{\ell+1}$ of $V^{\ell+1}$, namely

$$\tau = \sum_{\beta \in \mathcal{B}^{\ell+1}} c_\beta^{\ell+1}(\tau)\, \beta,$$

with certain coefficients $c_\beta^{\ell+1}(\tau) \in \mathbb{R}$. The *truncation* of $\tau \in V^\ell$ with respect to $\mathcal{B}^{\ell+1}$ and $\Omega^{\ell+1}$ is the function $\text{trunc}^{\ell+1}\tau \in V^{\ell+1}$ defined as:

$$\text{trunc}^{\ell+1}\tau = \sum_{\beta \in \mathcal{B}^{\ell+1}, \text{supp}\, \beta \not\subseteq \Omega^{\ell+1}} c_\beta^{\ell+1}(\tau)\, \beta.$$

---

[1] Different shapes are easily identified at subsequent levels as shown in Figure 3.

The overall truncation of a hierarchical B–spline $\beta \in \mathcal{B}^\ell \cap \mathcal{H}$ is defined by recursively applying the truncation with respect to the different levels,

$$\mathrm{trunc}\,\beta = \mathrm{trunc}^{N-1}(\mathrm{trunc}^{N-2}\ldots(\mathrm{trunc}^{\ell+1}\beta)).$$

By recursively discarding the contribution of active B–splines of subsequent levels from coarser B–splines, we obtain the definition of the truncated basis.

Definition 2 ([5]). *The truncated hierarchical B-spline (THB–spline) basis* $\mathcal{T}$ *is defined by*

$$\mathcal{T} = \{\mathrm{trunc}\,\beta_{i,j}^\ell : X_{i,j}^\ell = 1, \ell = 0, \ldots, N-2\} \cup \{\beta_{i,j}^{N-1} : X_{i,j}^{N-1} = 1\}.$$

Truncated hierarchical B–spline are linearly independent, non-negative, form a partition of unity and preserve the nested nature of the spline spaces [5]. Moreover, the construction of THB–splines is strongly stable with respect to the supremum norm provided that the knot vectors satisfy certain reasonable assumptions — see [6] for more details.

In addition to the characteristic matrices $\{X^\ell\}_{\ell=0}^{N-1}$, we consider another sequence of matrices $\{C^\ell\}_{\ell=0}^{N-1}$ of the same size and with the same sparsity pattern, i.e. $X_{i,j}^\ell = 0$ implies $c_{i,j}^\ell = 0$. These matrices store the coefficients associated to the (active) basis functions in the representation of a spline function with respect to the truncated basis. The following simple algorithm performs the evaluation of a hierarchical spline function which is represented in terms of THB–splines.

Algorithm EVAL_THB(seqmat X, seqmat C, int D, int LMAX, float U,V)

\\ seqmat X is the sequence of characteristic matrices, i.e., X[L] is the characteristic matrix of level L
\\ seqmat C is the sequence of coefficient matrices associated to the spline function f, i.e., C[L] is the coefficient matrix of level L
\\ int D is the degree in both directions
\\ int LMAX is the maximum refinement level $N - 1$
\\ float U,V are evaluation parameters
Identify the (D+1)×(D+1) sub–matrix M of C[0] which contains the coefficients of those B–splines of level 0 that are non–zero at (U,V)
for L = 1 to LMAX do {
    Generate the matrix S by applying one step of B–spline subdivision to M
    Identify the (D+1)×(D+1) sub–matrix T of S which contains the coefficients of those B–splines of level L that are non–zero at (U,V)
    for each pair of indices i,j in T do {
        if X[L](i,j) == 1 then T(i,j) = C[L](i,j) }
    M = T }
return the value f obtained by applying de Boor's algorithm to M

In this algorithm, the sub-matrices M, S, and T at a certain level are always accessed by global indices, i.e., indices with respect to the entire array of all tensor–product splines of that level. The following proposition clarifies the connection between the evaluation algorithm and the truncated hierarchical B–spline basis.

Theorem 1. *The value $f(u, v)$ computed by the algorithm is the value of a function represented in the THB–spline basis.*

This can be proved by applying the algorithm to Kronecker–type coefficient data, where exactly one coefficient is nonzero and equals 1. This corresponds to the evaluation of a truncated basis function.

The *cost* of the THB–spline evaluation algorithm EVAL_THB is equal to $N - 1$ times the application of the B–spline subdivision rule plus the cost due to the standard de Boor's algorithm. Consequently, it grows linearly with the number of levels and quadratically with the degree of the splines. This is similar to the costs needed for evaluating the classical (non-truncated) hierarchical B–splines. The computational cost could be further reduced

- by starting the **for** loop at the minimum level of functions which are active at the given point (u, v), and
- by stopping it at the maximum level of functions which are active at this point.

With this modification, the computational costs grows linearly with the number of levels which are active at the given point. This number can be controlled by choosing a suitable refinement strategy.

The following sections discuss data structures and algorithms for manipulating and storing the subdomain hierarchy and for representing the characteristic matrices and coefficient matrices.

3   Representing and Manipulating the Subdomain Hierarchy

The domain Ω^0 is now assumed to be a box consisting of $2^n \times 2^n$ cells of the coarsest tensor-product grid, where n is a non-negative integer. This assumption is made in order to facilitate the use of a quadtree data structure. Moreover, in order to simplify the implementation, the edges of the coarsest tensor–product grid should have the length 2^{M-1}, where M is the maximum number of levels, i.e. $N \le M$. Under this assumption, all coordinates of bounding boxes in the algorithms presented below are integers. Alternatively, one may use other exact data types than integers (e.g. rational numbers), thereby eliminating the restriction on the number of levels.

3.1   The Subdomain Hierarchy Quadtree

We represent the *entire* subdomain hierarchy by a *single* quadtree. Each node of the quadtree takes the form

```
struct qnode{
    aabb  box;
    int   level;
    *node nw;
    *node ne;
    *node sw;
    *node se; };
```

where the axis–aligned bounding box `aabb box` is characterized by coordinates of its upper left and lower right corner, `level` defines the highest level in which the box is completely contained and `nw`, `ne`, `sw`, `se` are pointers to the four children of the node. These children represent the northwestern, northeastern, southwestern and southeastern part of the box after the dyadic subdivision. All pointers to these children are set to null until the node is created during the insertion process, which is described by the `INSERTBOX` algorithm below.

Let $\Omega^\ell = \bigcup_i b_i^\ell$, where each b_i^ℓ is a collection of cells forming a rectangular box. During the creation of the quadtree which represents the subdomain hierarchy, for each level ℓ, we insert all boxes b_i^ℓ which define Ω^ℓ. The following recursive algorithm performs the insertion of a box b_i^ℓ into the quadtree:

Algorithm `INSERTBOX(box B, qnode Q, int L)`
```
    \\ box B is the box which will be inserted
    \\ qnode Q is the current node of the quadtree
    \\ int L is the level
    if B == Q.box then {
        Q.level = L
        visit all nodes in the subtree with root Q; if the level of a node is less
            than L then increase it to L }
    else {
        for child in {Q.nw, Q.ne, Q.sw, Q.se} do {
            if child != null then {
                if B∩Q.box ≠ ∅ then INSERTBOX(B∩Q.box, child, L) }
            else {
                create the box childbox of child
                if B∩childbox ≠ ∅ then
                    create the node child
                    set child.box to childbox, child.level to Q.level and the
                        four children to null
                    INSERTBOX(B∩childbox, child, L) } } }
```

After each box insertion we perform a cleaning step, visiting all sub–trees and deleting those where all nodes have the same level. This reduces the depth of the tree to a minimal value and optimizes the performance of all algorithms.

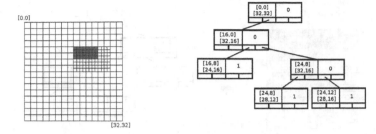

Fig. 4. Initial subdomain structure (left) and corresponding quadtree (right) which stores the boxes related to level 0 and 1 in the hierarchy. The box $b = [16,8] \times [24,12]$ (red) has to be inserted into the quadtree at level 2.

Example 2. To explain the **INSERTBOX** algorithm, we consider the subdomain hierarchy composed of three levels ($N = 3$), two of which (level 0 and 1) are initially present. This is shown in Figure 4, together with the related quadtree representation. The domain Ω^0 has $k = 16$ edges of length $2^{N-2} = 2$. The box $b = [16,8] \times [24,12]$ will be inserted at level 2 into the hierarchy. The cells with respect to the tensor–product grid of level 1 covered by b are depicted in red in Figure 4.

The execution of the algorithm is illustrated in Figure 5. At each step, we highlight the current node Q and the corresponding box in the subdomain hierarchy (Figure 5, right and left column, respectively). The insertion starts by considering the root of the tree, where the box b is compared with the axis–aligned bounding box stored in the root. Since these two boxes are not the same, the level of the root remains unchanged.

Subsequently, we have to identify which boxes between the ones stored in the four children of the root overlap with b, see Figure 5(a). In this case b is completely contained in the box represented by the **ne** child of the root. The recursive call of **INSERTBOX** is therefore applied to this child only. The situation in Figure 5(b) is similar to the previous case. After the split, the algorithm is recursively applied to the **sw** child.

In the third step shown in Figure 5(c) instead, the box b overlaps with the boxes related to two children (**nw** and **ne**) of the current node. Then, b is also subdivided and the recursion is called on both children.

Figure 5(d) shows the last step executed by the insertion of the box b. Two new nodes are created and inserted into the quadtree. Since these nodes coincide with the two parts of b, we set their level to 2. Clearly, the box to be inserted does not necessarily become a single node of the quadtree but it may be stored into several nodes.

3.2 Queries

In order to create the characteristic matrices introduced in Section 2, we define three query functions on the quadtree. These queries allow to understand if all

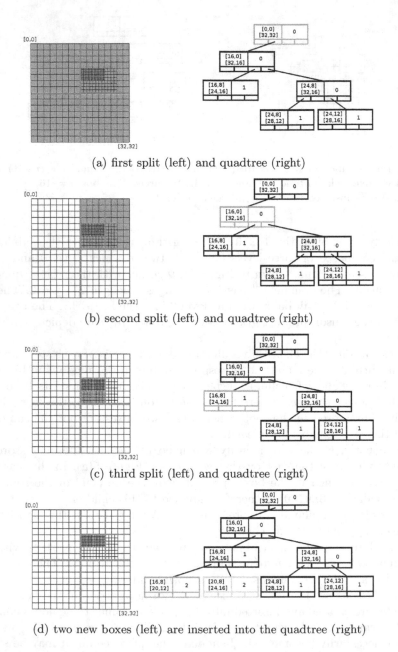

(a) first split (left) and quadtree (right)

(b) second split (left) and quadtree (right)

(c) third split (left) and quadtree (right)

(d) two new boxes (left) are inserted into the quadtree (right)

Fig. 5. Different steps performed by the INSERTBOX function to insert the box $b =$ $[16, 8] \times [24, 12]$ into the subdomain hierarchy of Figure 4. The subsequent splits are shown on the subdomain hierarchy (blue lines on the left) with respect to the visit of the quadtree (right).

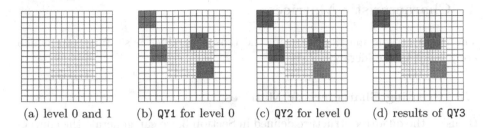

(a) level 0 and 1   (b) QY1 for level 0   (c) QY2 for level 0   (d) results of QY3

Fig. 6. Results of the three queries functions with respect to a subdomain hierarchy (a) with two levels. In case of QY1 (b) and QY2 (c), the green/red boxes correspond to a positive/negative answer. QY3 (d) returns 1 for the green boxes and 0 for the red ones.

basis functions β of a certain hierarchical level whose support is contained in a given box b are active or passive.

Given a box b defined as a collection of cells with respect to the tensor–product grid of level ℓ, the first query (QY1), returns true if

$$b \subseteq \Omega^\ell \quad \wedge \quad b \cap \Omega^i = \emptyset, \quad i > \ell. \tag{2}$$

Thus, if QY1 returns true, then all the basis functions of level ℓ whose support is completely contained in the box b are *active*, i.e., they are present in the hierarchical spline basis.

If the second query QY2 returns true then all the basis functions of level ℓ whose support is contained in the box b are *passive*, i.e., they are not present in the hierarchical spline basis. This is characterized by the following condition:

$$b \cap \Omega^\ell = \emptyset \quad \vee \quad b \subseteq \Omega^\ell, \quad \text{for some } i > \ell. \tag{3}$$

The third query QY3 returns the highest level ℓ with the property that Ω^ℓ contains the box b.

All the three queries can easily be implemented with the help of the quadtree structure described before. In particular, the structure of queries QY1 and QY2 is similar. We visit the quadtree until we find a leaf node or a node where the result of the query changes from to true to false. At that point, we can conclude the visit and return false. On the other hand, query QY3 requires a complete visit of the quadtree.

Example 3. Figure 6(b–d) shows the results of the three queries with respect to the subdomain hierarchy composed of two levels (level 0 and 1) shown on Figure 6(a) for four sampled boxes of level 0. Figures 6(b) and (c) display the results of QY1 and QY2 for $\ell = 0$, respectively. The boxes in green correspond to a positive answer to the query, the red boxes to a negative one. Finally, Figure 6(d) shows the results for QY3. The green boxes correspond to answer 1 and the red ones to answer 0.

4   Characteristic Matrices

The characteristic matrices identify the tensor–product basis functions which are present in the hierarchical basis.

4.1   Creating Characteristic Matrices

By using the quadtree structure defined in Section 3, we can generate the characteristic matrices introduced in Section 2 to represent and evaluate THB–splines. For the creation of these matrices we considered two different approaches:

- the *one–by–one approach* where we determine the entries of the characteristic matrices one by one by applying QY3 to each single basis function;
- the *all–at–once approach* where we try to set as many values as possible in one single step. This requires a more sophisticated algorithm.

During the creation of the characteristic matrices by the all–at–once approach, we try to set many entries of the matrices at the same time. In order to do this, the query functions are initially called for boxes which cover the initial domain Ω^0. The SETMAT algorithm below creates the characteristic matrices for all subdomains in the subdomain hierarchy.

Algorithm SETMAT(qnode Q, seqmat X)

> \\ qnode Q is the root of the quadtree which stores the subdomain hierarchy
> \\ seqmat X is the sequence of characteristic matrices, i.e. X[L] is the characteristic matrix of level L
> for all levels L do {
>   Create the index set I for all functions of level L acting on Ω^0. I is an
>   axis-aligned box in index space.
>   SETBOX(B,X[L]) }

SETMAT calls the algorithm SETBOX. When the answer active/passive cannot be given for the current call, the considered box is split into 4 disjoint axis–aligned bounding boxes and SETBOX function is recursively applied to them.

Algorithm SETBOX(aabbis I, mat XL)

> \\ aabbis I is an axis-aligned box in index space
> \\ mat XL is a characteristic matrix of level L
> The level L is a global variable
> Create the axis-aligned bounding box B covering all cells of level L which
>   belong to the supports of functions with indices in I
> if QY1(B, L) then {
>   for all indices (i,j) in I do XL[i,j]=1 }
> elseif QY2(B, L) then {
>   for all indices (i,j) in I do XL[i,j]=0 }
> elseif I is a single pair (i,j) then {

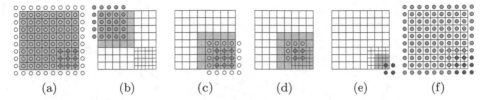

Fig. 7. A subdomain hierarchy with two levels and some of the boxes I in index space (shown as circles) along with the associated bounding boxes B in parameter space (grey) considered by SETBOX when creating the characteristic matrix X^0 for this subdomain hierarchy (a–e). Active (green) and passive (red) functions of level 0 (f).

```
    k =   QY3(B, L)
    if k == L then XL[i,j]=1
    else XL[i,j]=0 }
else {
    split I into 4 disjoint axis-aligned bounding boxes I1-I4 by subdividing
        each edge (approximately) into halves in index space.
    Apply SETBOX to I1-I4 and XL }
```

Example 4. Figure 7 shows a subdomain hierarchy with two levels, consisting of a square Ω^0 and a subdomain Ω^1 in the southeastern corner, which is shown in blue. The four pictures (a–e) visualize several index sets I (shown by circles) and the associated boxes B (grey) which are considered by SETBOX when creating X^0 for biquadratic splines.

Initially, SETBOX considers the entire set of basis functions (a) and concludes that it has to be subdivided. The northwestern subset is shown in (b). Query QY1 returns 1, therefore the functions are all active; no subdivision is needed. The northeastern and southwestern subsets (not shown) are dealt with similarly. The southeastern subset (c), however, has to be subdivided. Considering its northwestern subset (d) does not lead to a conclusion again, needing another subdivision. The functions in this index set have to be analyzed one-by-one (not shown). The northeastern and southwestern subsets (not shown) are dealt with similarly. For the southeastern subset (e), however, query QY2 returns 1, therefore the functions are all passive. Finally, the procedures arrives at the correct classification of basis functions of level 0 (f).

As Example 5 shows the all–at–once approach is not necessarily faster compared to the one–by–one mentioned at the beginning of this section. However, the approach becomes considerably faster with an increasing number of levels. This is demonstrated by the next example.

Example 5. Figure 8 compares the all–at–once setting with the one–by–one method. The number of queries called by the one-by-one approach is the same for the three hierarchical refinements in Figure 9 since it depends solely on the number of basis functions. This approach is faster for small numbers of basis functions, which typically correspond to a small numbers of hierarchical levels.

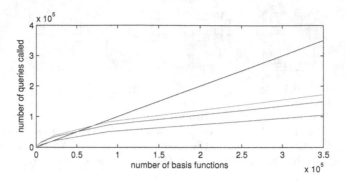

Fig. 8. The plot visualizes the number of queries needed to create the characteristic matrices for the three examples in Figure 2, 3 and 9. Compared to the all–at–once approach (cyan: linear, green: curvilinear, red: square-shaped refinement), the one–by–one approach (blue: same for all examples) is faster for small numbers of levels and basis functions, but it becomes slower for higher ones.

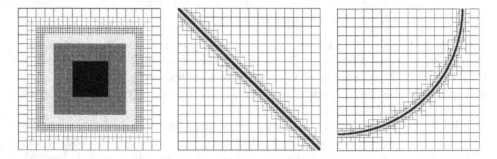

Fig. 9. The three subdomain hierarchies considered in Example 6: rectangular (left), linear (middle) and curvilinear (right) refinement, all with six levels.

On the other hand, the all–at–once approach becomes faster for higher numbers of basis functions in all the three considered cases since the number of queries grows sub–linearly with respect to the number of basis functions.

4.2   Using Sparse Data Structures

The representation of THB–splines in terms of characteristic matrices allows a fast look–up during the evaluation process and a simple update of the values when the underlying subdomain hierarchy changes. The drawback of this representation is the rather large memory consumption, which can exceed the available physical memory even for relatively small meshes and low numbers of levels. Indeed, it grows exponentially with the number of levels.

This problem can be solved by using a suitable sparse matrix data structure. For our experiments, we chose the *compressed sparse column* (CSC) data structure. The nonzero elements (read first by column) are stored in a one–dimensional array. A second array stores the row indices corresponding to these values and a third one collects the indices into the first two arrays of the leading entry in each column [8].

As detailed in the next section, the CSC structure significantly reduces the memory consumption of our approach (see Example 6). In fact, we will observe that the memory consumption grows linearly with the number of degrees of freedom, instead of exponentially with the number of levels. In addition, the price paid for reducing the memory requirements is only a small increase of the computational time (see Examples 7 and 8).

5 Examples

We implemented the proposed algorithms and data structures in C++. For the manipulation of the characteristic matrices we used the sparse MATLAB representation in terms of the *compressed sparse column* approach mentioned at the end of the previous section. The experiments have been performed on a laptop running the Windows 7 operating system (Intel Core I5-2520 2.5GHz, 4GB RAM, 64 bit).

Example 6. We compare the memory consumption of full characteristic matrices with the memory consumption of the matrices represented in the CSC structure for the three subdomain hierarchies in Figure 9 (rectangular, linear, and curvilinear).

The experimental results in Figure 10 show that the memory needed by the sparse matrix data structure is considerably smaller then the one related to the full matrix representation. Moreover, the memory consumption grows only *linearly* with the numbers of degrees of freedom instead of exponentially with the number of levels. This is the optimal result that one can expect, since a coefficient for each active basis function needs to be stored anyway.

We observe a difference between the results related to the rectangular–shaped refinement with respect to the linear and curvilinear case. The reason is the different nature of the refinement procedure. In the linear and curvilinear case, the refined area is reduced at each new level and the coarser levels do not change (see Figure 3). In the rectangular case, the refined area of the highest level is constant and the size of lower level subdomains increases (see Figure 2). Thus, using the sparse data structure does not decrease the *order* of memory consumption in this case, since the number of degrees of freedom grows exponentially with the number of levels.

The next example analyzes the influence of using the sparse data structures to the time needed to evaluate the multilevel spline functions using the algorithm EVAL_THB.

Example 7. Figure 11 visualizes the distribution of the computation times needed to evaluate the multilevel spline function at 1000 points with (blue bars in the

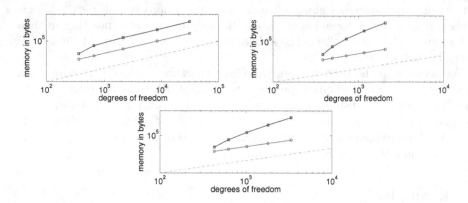

Fig. 10. Memory needed to represent the characteristic matrices without (blue) and with (green) the use of sparse data structures for different numbers of degrees of freedom related to the square (top left), the circle (top right) and the line refinement (bottom) refinement. The dashed red line has slope 1 and indicates linear growth.

plot) and without (red bars in the plot) the use of sparse data structures for the linear refinement shown in Figure 9. Two facts can be observed:

- the evaluation time does not depend significantly on the location of the point with respect to the subdomain hierarchy;
- using the sparse data structure increases the evaluation time only by a very small amount.

Note that the evaluation times in this example vary between 0.153 and 0.195 milliseconds.

Finally we analyze the relation between evaluation time and the number of levels in the hierarchy.

Example 8. We consider the curvilinear refinement shown on the right of Figure 9. Figure 12 compares the evaluation times for 10,000 parameters obtained by using either the full or the sparse matrix representation. We may note that the computational time grows *linearly* with the increasing level of refinement for both representations, with a small overhead caused by using the sparse data structure. The values do not include the time necessary for creating the corresponding data structures, only the evaluation algorithm EVAL_THB is considered.

6   Non–uniform Knots and General Refinement

In order to discuss more general knot configurations, we now describe the modifications of data structures and algorithms which are required to extend the framework to non–uniform knots and different multiplicities.

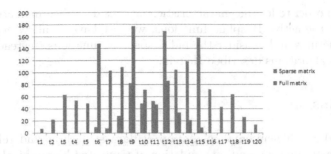

Fig. 11. The labels t1,...,t20 on the horizontal axis represent uniform time intervals between minimal (0.153 ms) and maximal (0.195 ms) time needed by the evaluation algorithm. The vertical axis indicates the number of points whose evaluation time falls into these intervals.

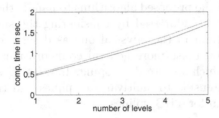

Fig. 12. Computational time needed to evaluate the multilevel spline function at 10,000 points for curvilinear refinement with various numbers of levels with (green) and without (blue) using the sparse data structure.

First of all, two vectors with strictly monotone knot values in both directions have to be stored. Two additional arrays will register the associated multiplicities. In this way, if we allow also knots with zero multiplicities, we can even perform non–dyadic refinements by simply considering to consecutive refinement levels where some of the newly inserted knots have multiplicity 0.

Concerning the modification of the THB–spline evaluation algorithm presented in Section 2, the only difference is in the use of B–spline subdivision with respect to non uniform knots, namely standard knot insertion, instead of uniform B–spline refinement. At each iteration, the knot insertion algorithm has to be applied to the proper sub–matrix of the corresponding matrix computed at the previous step.

In the quadtree data structure introduced in Section 3, the axis–aligned bounding box has now to be in index space and possibly of rectangular shape. In addition, the splitting into halves considered in the related insertion algorithm of Section 3.1 has to be rounded to the nearest integer. No modification are needed for the query functions described in Section 3.2.

Finally, in order to let the characteristic matrices described in Section 4 properly identify the active B–spline functions, we just have to allow general rectangular structures and consider the index space of functions (instead of knots) into the two algorithms described therein.

7   Conclusion

We proposed an efficient implementation of data structures and related algorithms for the evaluation and manipulation of truncated hierarchical B–splines. Several examples show the advantageous behavior of the data structures and algorithms in terms of memory overheads and computational costs. Indeed, the memory consumption grows only linearly with the number of degrees of freedom, but there is no significant increase of the time needed to evaluate the multilevel spline function.

The generalization of the proposed algorithms to handle the non–uniform case and multiple knots can be addressed by considering the subdomain hierarchy in index space rather than in the physical one as described in Section 6. We are currently working on these more general configurations in the frame of a new software library which we are developing. Interesting subjects for future research include the extension to multivariate splines and the identification of the refinement algorithm for THB–splines.

Acknowledgments. Gábor Kiss is supported by the Austrian Science Fund (FWF) through the Doctoral Program in Computational Mathematics (W1214, DK3). Carlotta Giannelli is a Marie Curie Postdoctoral Fellow within the 7th European Community Framework Programme under grant agreement n°272089 (PARADISE). Bert Jüttler has received support by the Austrian Science Fund (FWF) through the National Research Network Geometry + Simulation (S117). This research has also received funding from the Marie Curie Actions – Industry-Academia Partnerships and Pathways (IAPP) funding scheme under grant agreement n°324340 (EXAMPLE).

References

1. Bornemann, P.B., Cirak, F.: A subdivision–based implementation of the hierarchical B–spline finite element method. Comput. Methods Appl. Mech. Engrg. 253, 584–598 (2012)
2. Deng, J., Chen, F., Feng, Y.: Dimensions of spline spaces over T–meshes. J. Comput. Appl. Math. 194, 267–283 (2006)
3. Dokken, T., Lyche, T., Pettersen, K.F.: Polynomial splines over locally refined box-partitions. Comput. Aided Geom. Design 30, 331–356 (2013)
4. Forsey, D.R., Bartels, R.H.: Hierarchical B–spline refinement. Comput. Graphics 22, 205–212 (1988)
5. Giannelli, C., Jüttler, B., Speleers, H.: THB–splines: the truncated basis for hierarchical splines. Comput. Aided Geom. Design 29, 485–498 (2012)

6. Giannelli, C., Jüttler, B., Speleers, H.: Strongly stable bases for adaptively refined multilevel spline spaces. Adv. Comp. Math. (to appear, 2013)
7. Giannelli, C., Jüttler, B.: Bases and dimensions of bivariate hierarchical tensor–product splines. J. Comput. Appl. Math. 239, 162–178 (2013)
8. Gilbert, J.R., Moler, C., Schreiber, R.: Sparse matrices in MATLAB: design and implementation. SIAM J. Matrix Anal. Appl. 13, 333–356 (1992)
9. Gonzalez-Ochoa, C., Peters, J.: Localized–hierarchy surface splines (LeSS). In: Proceedings of the 1999 Symposium on Interactive 3D Graphics, pp. 7–15. ACM, New York (1999)
10. Greiner, G., Hormann, K.: Interpolating and approximating scattered 3D–Data with hierarchical tensor product B–splines. In: Méhauté, A.L., Rabut, C., Schumaker, L.L. (eds.) Surface Fitting and Multiresolution Methods. In Innovations in Applied Mathematics, pp. 163–172. Vanderbilt University Press, Nashville (1997)
11. Kraft, R.: Adaptive and linearly independent multilevel B–splines. In: Le Méhauté, A., Rabut, C., Schumaker, L.L. (eds.) Surface Fitting and Multiresolution Methods, pp. 209–218. Vanderbilt University Press, Nashville (1997)
12. Kraft, R.: Adaptive und linear unabhängige Multilevel B–Splines und ihre Anwendungen. PhD Thesis, Universität Stuttgart (1998)
13. Lee, S., Wolberg, G., Shin, S.Y.: Scattered data interpolation with multilevel B–splines. IEEE Trans. on Visualization and Computer Graphics 3, 228–244 (1997)
14. Schillinger, D., Dedè, L., Scott, M.A., Evans, J.A., Borden, M.J., Rank, E., Hughes, T.J.R.: An isogeometric design–through–analysis methodology based on adaptive hierarchical refinement of NURBS, immersed boundary methods, and T–spline CAD surfaces. Comput. Methods Appl. Mech. Engrg., 249–252, 116–150 (2012)
15. Schumaker, L.L., Wang, L.: Approximation power of polynomial splines on T–meshes. Comput. Aided Geom. Design 29, 599–612 (2012)
16. Sederberg, T.W., Zheng, J., Bakenov, A., Nasri, A.: T–splines and T–NURCCS. ACM Trans. Graphics 22, 477–484 (2003)
17. Stollnitz, E.J., DeRose, T.D., Salesin, D.H.: Wavelets For Computer Graphics: Theory and Application, 1st edn. Morgan Kaufmann Publishers, Inc. (1996)
18. Vuong, A.-V., Giannelli, C., Jüttler, B., Simeon, B.: A hierarchical approach to adaptive local refinement in isogeometric analysis. Comput. Methods Appl. Mech. Engrg. 200, 3554–3567 (2011)
19. Yvart, A., Hahmann, S.: Hierarchical triangular splines. ACM Trans. Graphics 24, 1374–1391 (2005)

The LIR Space Partitioning System Applied to Cartesian Grids

Sven Linden[1,2], Hans Hagen[1], and Andreas Wiegmann[3]

[1] University of Kaiserlautern, 67663 Kaiserlautern, Germany
[2] Fraunhofer ITWM, 67663 Kaiserslautern, Germany
[3] Math2Market GmbH, 67663 Kaiserslautern, Germany

Abstract. We introduce a novel multi-dimensional space partitioning method. A new type of tree combines the advantages of the Octree and the KD-tree without having their disadvantages. We present in this paper a new data structure allowing local refinement, parallelization and proper restriction of transition ratios between cells. Our technique has no dimensional restrictions at all. The tree's data structure is defined by a topological algebra based on the symbols $A = \{L, I, R\}$ that encode the partitioning steps. The set of successors is restricted such that each cell has the partition of unity property to partition domains without overlap. With our method it is possible to construct a wide choice of spline spaces to compress or reconstruct scientific data such as pressure and velocity fields and multidimensional images. We present a generator function to build a tree that represents a voxel geometry. The space partitioning system is used as a framework to allow numerical computations. This work is triggered by the problem of representing, in a numerically appropriate way, huge three-dimensional voxel geometries that could have up to billions of voxels.

1   Introduction

The goal of this work is to introduce and apply a novel mathematical model to partition n-dimensional domains. We give a detailed definition and description of the theoretical background and the algorithms. The space partitioning is done by a tree that can be seen as a hybrid of an Octree and a KD-tree. We combine the advantages while avoiding the disadvantages.

The basic idea of the tree is the definition of a ternary alphabet. This alphabet is applied recursively and dimensionally independently to a system of functions. We abstract from the geometrical properties and define an algebraic approach to efficiently partition domains and evaluate functions recursively. Evaluation and computation as well as proofs on these trees can be done by structural induction.

The tree can be used to compress geometries as well as scalar and multi-dimensional fields. It is possible to access different kinds of information and operators that are influenced by the structure of the tree, e.g. interpolation schemes, differential operators, subsets of the given domain and neighborhoods of cells. In this work we focus on voxel geometries. But it is also possible to

M. Floater et al. (Eds.): MMCS 2012, LNCS 8177, pp. 324–340, 2014.

partition different kinds of sets. In many applications it is a disadvantage if the tree degenerates in a single direction. Therefore we introduce an input function that is able to prevent degeneration of the tree where it leads to a disadvantage.

2   Related Work

There are many different kinds of space partitioning methods.

Regular grids are an easy commonly used way to discretize the two- and three-dimensional space. They allow an efficient alignment of the data in the memory and enables the user to easily formulate discretized differential operators. Data can be accessed in constant time, but the disadvantage of the regular grids is that the requirements for computational effort and memory grows at least with the power of the dimension. It is also not easy to describe smooth boundaries, e.g. the interfaces in two-phase flows and between different materials in solid mechanics.

A way to treat this problem is to use a body-fitted mesh of tetrahedra. But a mesh of tetrahedra requires a lot of overhead, e.g. you have to store positions, normal vectors and topological information. This overhead significantly decreases the number of cells that can be stored and processed.

Another way to treat this problem is to use an Octree [1] [2]. A disadvantage of the Octree is the limited choice of partitioning. You have just the choice to do no partitioning or to partition simultaneously in all dimensions. Therefore one is forced to increase the number of cells even in directions where it is not necessary. But the Octree is an important data structure in numerical mathematics. It received attention in isogeometric analysis [3] [4] within the last years.

The latter issue is addressed by the KD-tree [5]. It is able to increase the number of cells just in one direction. But a KD-tree is not well suited to use for numerical calculations as it is designed for partitioning point clouds. Another problem is the high number of interior nodes in higher dimensional settings.

A detailed description of the Octree and KD-tree can also be found in [6] and [7] covering the latest developments and applications.

The model we describe in this paper avoids the disadvantages of the KD-tree and Octree and combines their advantages.

3   Method

In this section we describe the basic theory of a tree structure that we call LIR-tree and the corresponding space partitioning system. The tree arises from dimensionally independent and recursive application of a ternary alphabet. We also introduce a number of auxiliary tree structures that are used to construct the LIR-tree.

3.1   Alphabet

For a one-dimensional finite interval there exist three choices: no partitioning, partition and take the left part of the interval, or partition and take the right part of the interval. Similar to partitioning there exist three choices for embedding intervals: no embedding, embed to the left or embed to the right. These choices are used to define the alphabet:

Definition 1.
$$A = \{L, I, R\} \tag{1}$$

*A is called the **alphabet** and contains three symbols that denote: L - left, R - right and I - identity. L and R are interpreted as **complementary** symbols and I as neutral symbol. We introduce a unary minus operator defined by*

$$-L := R \quad -R := L \quad -I := I \tag{2}$$

that is also used vector-wise and element-wise.

Definition 2. *The bold notation is used to see the symbols in A as sets by*

$$\mathbf{A} = \{\mathbf{L}, \mathbf{I}, \mathbf{R}\} \tag{3}$$

with the symbol sets

$$\mathbf{L} = \{L\} \quad \mathbf{I} = \{L, R\} \quad \mathbf{R} = \{R\} \tag{4}$$

and to introduce the conversion

$$v = (v_1, \cdots, v_n) \in A^n \Leftrightarrow \mathbf{v} = \mathbf{v_1} \times \cdots \times \mathbf{v_n} \in \mathbf{A}^n. \tag{5}$$

Definition 3. *The set of vectors of symbols defined by*

$$P := \{p \subseteq A^n : \bigcup_{v \in p} \mathbf{v} = \mathbf{I}^n \wedge \forall_{\substack{v,w \in p \\ v \neq w}} \mathbf{v} \cap \mathbf{w} = \emptyset\} \tag{6}$$

denotes all sets of vectors that are partitions of unity. The sets in P are the basis to construct the LIR-tree.

Definition 4. *We use S_n to denote the symmetric group and $\{1, -1\}^n$ as selective inversion. Let $p, q \in P$ be two partitions of unity. $p \sim q$ means they are equivalent with respect to rotation and inversion. That is*

$$p \sim q \quad \Leftrightarrow \quad \exists_{s \in S_n} \exists_{h \in \{1, -1\}^n} \{(v_{s_i})_{i=1}^n : v \in p\} = \{h \cdot v : v \in q\} \tag{7}$$

then $P/\sim$ describes the set of equivalence classes.

Tab. 1 shows the number of different unique and equivalent partitions of unity. Figure 1 and Fig. 2 illustrate a choice of partitions of unity for the two- and three-dimensional case. The number of different partitions grows very fast but is small until $n = 4$.

Table 1. Number of unique and equivalent partitions of unity for $n \in \mathbb{N}$

n	1	2	3	4	5
$|P|$	3	8	154	89512	71319425714
$|P/\sim|$	2	4	15	434	> 100000

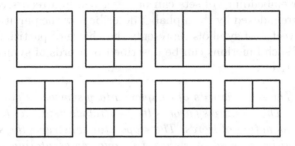

Fig. 1. 8 unique and 4 equivalent partitions of unity exist for the two-dimensional case. The last two partitions with two elements in the first row belong to the same equivalence class and the partitions with three elements in the second row belong to the same equivalence class.

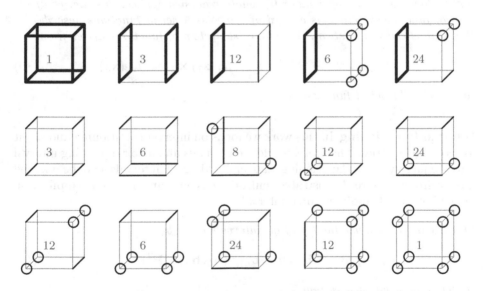

Fig. 2. 154 unique and 15 equivalent partitions of unity exist for the three-dimensional case. For a better visibility we use bold entities representing the different parts. The bold squares represent a vector of symbols with two identity symbols, a bold line represents a vector with one identity and a bold circle represents a vector with zero identity symbols. The numbers inside the cubes denote the cardinality of the equivalence classes.

3.2 Space Partitioning System

Partitioning (or embedding) of sets can be described in a recursive way. Sets of functions that are indexed by the alphabet describe how the partitioning is done. A symbol or a vector of symbols represents the choice of partitioning function. Composition of such functions can be described by words of symbols or vectors of words, respectively.

Definition 5. *The set of **words** of complementary symbols is given by the word monoid $(\mathbf{I}^*, \cdot, I)$. The identity symbol is the neutral element, i.e. the empty word and $\cdot$ is the concatenation of words. The $*$ operator constructs the set of all finite words. The definition is used to induce a unique representation for partitioned domains. Let $w \in \mathbf{I}^k$ be a word, then a minus operator for words is defined by $-w := (-w_{n-i+1})_{i=1}^n$ such that complementary symbols are inverted and the order is reversed.*

Definition 6. *Let $\mathfrak{D}$ be a set and $f_A = \{f_L, f_R, f_I\}$ be a set of functions such that $f_{a \in A} : \mathfrak{D} \to \mathfrak{D}$. Then we introduce the notation*

$$f_{w \in \mathbf{I}^k} := f_{w_k} \circ \cdots \circ f_{w_1} \tag{8}$$

*for the lower and the upper index to denote recursive applications, i.e. composition of functions where w is a word of symbols. A set of functions described in that way is called a **system**. A vector-wise system of functions arises from*

$$f_{v \in A^n} : \mathfrak{D}^n \to \mathfrak{D}^n \qquad \mathbf{x} \mapsto f_{v_1}(\mathbf{x}_1) \times \cdots \times f_{v_n}(\mathbf{x}_n) \tag{9}$$

to describe the set of functions f_{A^n}.

Interval Partitioning In this work we focus on interval partitioning and use it for voxel geometries. Therefore, we define intervals and a corresponding interval partitioning system. Partitioning and embedding of intervals can be merged into a group. Figure 3 illustrates multi-dimensional and recursive application of symbols to a two-dimensional interval.

Definition 7. *We use the set of all intervals given by*

$$\mathfrak{B} := \{\mathbf{b} = [\mathbf{b}_L, \mathbf{b}_L] : (\mathbf{b}_L, \mathbf{b}_R) \in \mathbb{R}^2\} \tag{10}$$

to define the combined system

$$\xi_A^A := \{\xi_L^L, \xi_I^L, \xi_R^L, \xi_L^I, \xi_I^I, \xi_R^I, \xi_L^R, \xi_I^R, \xi_R^R\} \tag{11}$$

such that $\xi_{b \in A}^{a \in A} : \mathfrak{B} \to \mathfrak{B}$ where $\xi_A^L = \{\xi_L^L, \xi_I^L, \xi_R^L\}$ denotes the system of partitioning

$$\xi_L^L(\mathbf{b}) = [\mathbf{b}_L, \frac{1}{2}\mathbf{b}_L + \frac{1}{2}\mathbf{b}_R] \qquad \xi_I^L(\mathbf{b}) = \mathbf{b} \qquad \xi_R^L(\mathbf{b}) = [\frac{1}{2}\mathbf{b}_L + \frac{1}{2}\mathbf{b}_R, \mathbf{b}_R] \tag{12}$$

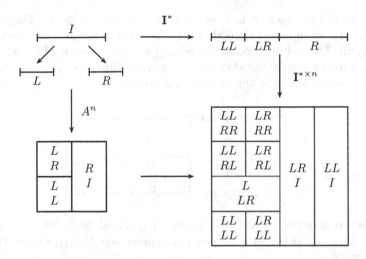

Fig. 3. Intent of the Alphabet A and its generalization to an n-dimensional domain and recursive application. L denotes left in x-direction and bottom in y-direction while R denotes right in x-direction and top in y-direction. The red lines indicate the partitioning in the first level and the blue lines indicate the partitioning in the second level.

with the neutral system $\xi_A^I = \{\xi_L^I, \xi_I^I, \xi_R^I\}$ which is defined by

$$\xi_L^I(\mathbf{b}) = \xi_I^I(\mathbf{b}) = \xi_R^I(\mathbf{b}) = \mathbf{b} \tag{13}$$

and $\xi_A^R = \{\xi_L^R, \xi_I^R, \xi_R^R\}$ denotes the system of embedding intervals defined by

$$\xi_L^R(\mathbf{b}) = [2\mathbf{b}_L - \mathbf{b}_R, \mathbf{b}_R] \qquad \xi_I^R(\mathbf{b}) = \mathbf{b} \qquad \xi_R^R(\mathbf{b}) = [\mathbf{b}_L, 2\mathbf{b}_R - \mathbf{b}_L] \tag{14}$$

The composition of multi-dimensional interval partitioning functions is defined by

$$\varXi := \{\xi_w^q : q \in \mathbf{I}^* \wedge w \in \mathbf{I}^{* \times n}\} \tag{15}$$

Theorem 1. $(\varXi, \circ)$ is a group.

Proof. It is sufficient to show the existence of the neutral and inverse elements. $\xi_I^{a \in \mathbf{I}} = \xi_{a \in A}^I$ is the neutral function with different notations. Let $\xi_w^q \in \varXi$ then $\xi_{-w}^{-q} \in \varXi$ is the inverse function.

Example 1. The vectors $(L, L), (I, R) \in A^2$ applied to $\mathbf{b} \in \mathfrak{B}^2$ yield

$$\xi_{(L,LR)}^L(\mathbf{b}) = \begin{pmatrix} \xi_L^L([\mathbf{b}_{L,L}, \mathbf{b}_{L,R}]) \\ \times \\ \xi_{LR}^L([\mathbf{b}_{R,L}, \mathbf{b}_{R,R}]) \end{pmatrix} = \begin{pmatrix} [\mathbf{b}_{L,L}, \frac{1}{2}\mathbf{b}_{L,L} + \frac{1}{2}\mathbf{b}_{L,R}] \\ \times \\ \xi_R^L([\mathbf{b}_{R,L}, \frac{1}{2}\mathbf{b}_{R,L} + \frac{1}{2}\mathbf{b}_{R,R}]) \end{pmatrix} \tag{16}$$

$$= \begin{pmatrix} [\mathbf{b}_{L,L}, \frac{1}{2}\mathbf{b}_{L,L} + \frac{1}{2}\mathbf{b}_{L,R}] \\ \times \\ [\frac{3}{4}\mathbf{b}_{R,L} + \frac{1}{4}\mathbf{b}_{R,R}, \frac{1}{2}\mathbf{b}_{R,L} + \frac{1}{2}\mathbf{b}_{R,R}] \end{pmatrix} \tag{17}$$

Vectors of words describe how interval partitioning is done. They can be understood as a set of instructions that point to a sub-interval with respect to an initial interval. The combination of embedding and partitioning ξ_{vv}^{RL} with $v \in A^n$ corresponds to the v-neighbored domain. Figure 4 illustrates the application of different vectors of words to the same initial domain.

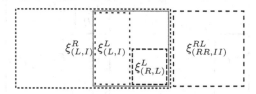

Fig. 4. Effects of Ξ on the cyan colored domain: $\xi_{(L,I)}^{R}$ embeds to the left, $\xi_{(L,I)}^{L}$ restricts to the left, $\xi_{(R,L)}^{L}$ restricts to the bottom right corner and $\xi_{(RR,II)}^{RL}$ defines the right neighbor domain.

3.3   The Oracle Tree

In this section we introduce a formalism to define tree structures. The approach is used in a recursive way to generate trees of higher order. Different kinds of trees are described that can be used for diverse applications.

Oracles. We define tree structures by structural induction with a function that maps the current location, i.e. the node to a set of succeeding edges. These function are called oracles and are motivated by the memory layout of trees inside the compuational memory.

Definition 8. *Let $(X, +)$ be a monoid with the generator set $0 \notin T \subseteq X$. The generator set is used to define the generator function*

$$\mho : X \to \mathcal{P}(T) = \{U : U \subseteq T\} \tag{18}$$

where the elements of X are mapped to a subset of the generator set ($\mathcal{P}$ denotes the power set). Then we define the tree that is constructed with respect to $\mho$ by

$$\mathcal{G} = (\mathcal{X}, \mathcal{E}) \qquad \mathcal{X} \subseteq X \qquad \mathcal{E} \subseteq \mathcal{X} \times \mathcal{X} \tag{19}$$

such that there exists a root node $0 \in \mathcal{X}$. We construct a tree by structural induction such that

$$x \in \mathcal{X} \Rightarrow \forall_{t \in \mho(x)} x + t \in \mathcal{X} \wedge (x, x + t) \in \mathcal{E}. \tag{20}$$

*We introduce the notation $\mathcal{X}(\mho)$ and $\mathcal{E}(\mho)$ to denote nodes and edges, respectively. The generator function $\mho$ generates a tree and is called an **oracle** if and only if every node has exactly one predecessor except the root node, i.e.*

$$(y, x) \in \mathcal{E}(\mho) \Rightarrow 0 \neq x \quad \wedge \quad 0 \neq x \in \mathcal{X}(\mho) \Rightarrow \exists!(y, x) \in \mathcal{E}(\mho). \tag{21}$$

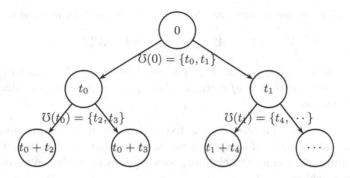

Fig. 5. Construction of an oracle tree. The oracle function returns the edges to succeeding nodes. The nodes are defined by adding the preceding edges.

We use the definition of oracles to construct trees using the symbols in A or vectors of symbols, respectively. An example for the general case is shown in Fig. 5.

Ternary Oracle. In many situations we want to be able to deal with arbitrary sets of vectors, e.g. partitions of unity or overlapping sets of vectors to store data. The number of different vectors and the number of sets of vectors is determined by the dimension. We introduce the ternary oracle and the range oracle that are suited to efficiently represent and access arbitrary subsets of A^n.

Definition 9. *Let $k \in \mathbb{N}$, $v \in A^k$ and $a \in A$ then we introduce the following $|\cdot|$ operators to count specified symbols in a vector:*

$$|v| := k \qquad |v|_{\mathbf{a}} := |\{i \in \mathbb{N} : v_i \in \mathbf{a}\}| \quad |v|_a := |\{i \in \mathbb{N} : v_i = a\}| \quad (22)$$

Definition 10. *A **ternary oracle** Υ is defined by the word monoid $(A^*, \cdot, \epsilon)$ such that*

$$\Upsilon : A^* \to \mathcal{P}(A) \quad (23)$$

where all leafs are at level n, i.e.

$$\forall v \in \mathcal{X}(\Upsilon) \quad |v| = n \Rightarrow \Upsilon(v) = \emptyset \quad (24)$$

The set of all ternary oracles is denoted by $\boldsymbol{\Upsilon}$.

The ternary oracle is used to describe subsets of A^n efficiently. Since the number of different partitions of unity is small until $n = 4$ (see Table 1) we use a look-up table in the implementation to eliminate any overhead in computational time and memory. An example of a ternary tree with three levels is shown in Fig. 6.

Definition 11. *Let $j, k \in \mathbb{N}$ with $j \le k \le n$ then we define the **range-oracle** $\Upsilon_j^k \in \boldsymbol{\Upsilon}$ by*

$$\Upsilon_j^k : A^* \to \mathcal{P}(A) \qquad v \mapsto \{\mathbf{I} : |v|_{\mathbf{I}} \le k\} \cup \{I : n - |v|_I \ge j\} \quad (25)$$

which is a subset of the ternary oracle such that the vectors defined by

$$V_j^k := \{v \in A^n : j \leq |v|_\mathbf{I} \leq k\} \subset \mathcal{X}(\Upsilon_j^k)$$

are the corresponding leaf nodes. The tree constructed by Υ_j^k is called range-tree and allows the representation of vertices, edges, faces or other higher dimensional entities in a cuboid.

Since the structure of a range tree is fixed and known at compilation time we can eliminate any overhead similar to the ternary tree. An important application of the range-tree is store neighborhoods of cells such that we can access a neighbor by a vector $v \in A^n$.

Vector-Oracle. In this section we describe a tree structure called vector-tree that uses vectors of symbols for construction of a tree that represents a set of vectors of words. By restriction to partitions of unity we finally get the LIR-tree.

Definition 12. *Let $n \in \mathbb{N}$ then a **vector-oracle** Ω is defined by the vector of words monoid $(\mathbf{I}^{*\times n}, \cdot, I)$ where $\cdot$ is the vector-wise concatenation of words and I the vector of empty words such that*

$$\Omega : \mathbf{I}^{*\times n} \to \mathcal{P}(A^n).$$

The set of all vector-oracles is denoted by $\boldsymbol{\Omega}$.

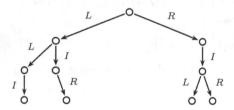

Fig. 6. Example of a ternary-tree encoding vectors in A^3

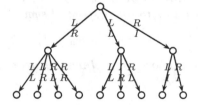

Fig. 7. Example of a vector-tree encoding vectors of words in $\mathbf{I}^{*\times 2}$

Definition 13. *By restricting the set of successors it is possible to introduce the partition of unity property. Therefore, the **LIR-oracle** is defined by*

$$\Omega_P : \mathbf{I}^{*\times n} \to P$$

*such that Ω_P constructs a vector tree and satisfies the partition of unity property in each node. The nodes in the tree are also called **cells**. The set of all LIR-oracles is denoted by $\boldsymbol{\Omega}_P$. The trees that can be constructed by LIR-oracles are called **LIR-trees**.*

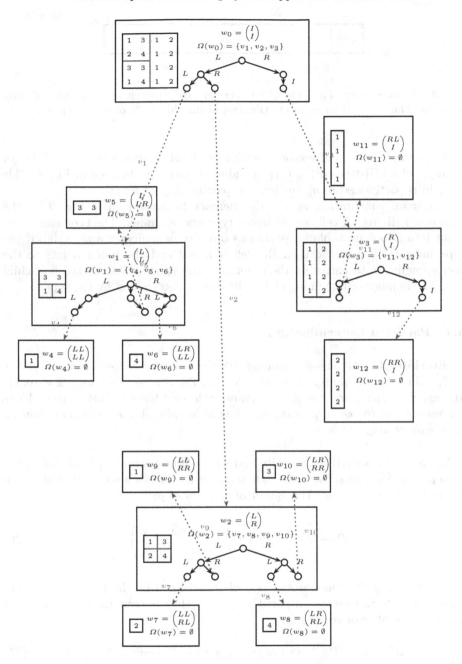

Fig. 8. Example of a LIR-tree that is defined by the set of cells $\mathcal{X}(\Omega_P) = \{w_0, \cdots, w_{12}\}$ and the set of links $\mathcal{E}(\Omega_P) = \{(w_0, w_1), \cdots, (w_3, w_{12})\}$. Note that the ternary trees are realized by a look-up table and do not occupy memory. The numbers inside the rectangles illustrate the domain and the corresponding data that is partitioned. The red lines show how the domains are partitioned and the red dotted lines show the links between cells.

type	index
8 bit	56 bit

Fig. 9. Memory layout of a cell in 64 bit. The type indicates either a partition of unity or a leaf. The index either points to the start of the child cells or a data entity.

An example of a small vector-tree with two levels is shown in Fig. 7. A larger example of a LIR-tree with a corresponding domain can be seen in Fig. 8. This domain is partitioned using the interval partitioning system.

In our implementation we use the memory layout shown in Fig. 9 for the cells in a LIR-tree. A cell is split into a type and an index. The type can either point into a look-up table for partitions of unity or indicates a material. If the type indicates a material then the cell is a leaf and the index points to the corresponding data entity. In the other case the index points to the first child cell. The number of remaining child cells can be determined by the type.

3.4 Partition Determination

Definition 14. *The one-dimensional edges can be represented by the range tree Υ_{n-1}^{n-1}. We assign a binary value to each edge, i.e. $S := \mathbb{B}^{n \cdot 2^{n-1}}$ with $\mathbb{B} = \{0, 1\}$. We use S to determine and modify appropriate partitions to build a tree. A one represents that the corresponding edge has to be split. A zero represents that an edge may or may not be split.*

Let $p \in P$ be a partition of unity and $s \in S$. If no vector of p contains a split then p is conform to s. We define a function ψ that takes a partition and return a zero where ever an edge is a subset of a vector of p:

$$\psi : P \to S \qquad p \mapsto \left(\begin{cases} 0 & \text{if } \exists_{v \in p} \mathbf{e} \subseteq \mathbf{v} \\ 1 & \text{else} \end{cases} \right)_{e \in E} . \qquad (26)$$

with $E := V_{n-1}^{n-1}$. The inverse function of ψ is defined by the ψ^{-1} function that takes a $s \in S$ and returns a partition $p \in P$ that is conform to s and has a minimum number of vectors.

$$\psi^{-1} : S \to P \qquad s \mapsto \arg\min\{|p| : p \in P \wedge \neg\psi(p) \wedge s = 0\} \qquad (27)$$

where $\wedge$ and $\neg$ denote bit-wise operators. Since the minimum partition is not unique, ψ^{-1} returns an arbitrary conforming partition that is minimal.

Similar to the ternary- and range-tree we use a look-up table for the evaluation of ψ and ψ^{-1} in our implementation. That way we get a $\mathcal{O}(1)$ runtime-complexity.

3.5 Input Function

In this section we describe an oracle that constructs a (non-degenerative) tree until a given threshold for partitioning is reached. It is also possible to use overlapping sets to increase the number of cells at interfaces.

Let $u \in \mathbb{R}^n$ be a vector of real numbers that denote the limit for partitioning. Then

$$\vartheta : \mathfrak{B} \to S \qquad \mathbf{b} \mapsto (|\mathbf{b}_i| > u_i)_{v \in E} \tag{28}$$

such that $v_i = I$ denotes a function that uses an interval and determines possible edges which are candidates to be split with respect to u.

The basic idea of the actual input function is to split edges where the corresponding segments of voxels contain different values. A segment of voxels is a finite straight chain of voxels that is used for analysis, see Fig. 10. We assume that $\omega : \mathbb{R}^n \to \mathfrak{D}$ is a given function that returns the type of material at each $x \in \mathbb{R}^n$ where $\mathfrak{D}$ denotes the domain of material. Then the next part of the input function is defined by

$$\theta^{k \in \mathbb{R}} : \mathfrak{B} \to S \qquad \mathbf{b} \mapsto \left(\exists x, y \in \nu_v^k(\mathbf{b}) \begin{cases} \omega(x) \neq \omega(y) \\ x_i \neq y_i \Rightarrow v_i = I \end{cases} \right)_{v \in E} \tag{29}$$

where we use the (overlapping) domains that are given by $\nu_{a \in A}^{k \in \mathbb{R}} : \mathfrak{B} \to \mathfrak{B}$ with

$$\nu_{a \in I}^k([\mathbf{b}_L, \mathbf{b}_R]) = [\mathbf{b}_L - k \cdot |\mathbf{b}|, \mathbf{b}_R + k \cdot |\mathbf{b}|] \tag{30}$$

$$\nu_I^k([\mathbf{b}_L, \mathbf{b}_R]) = [\mathbf{b}_L - \frac{k}{2} \cdot |\mathbf{b}|, \mathbf{b}_R + \frac{k}{2} \cdot |\mathbf{b}|] \tag{31}$$

The range of analysis is given by $k \in \mathbb{R}$. If only θ and ϑ are used for construction then we get non-degenerative trees.

In a context where the domain is strongly elongated in one direction, it might be a good idea to degenerate until a cube-like subdomain is given. Thus, we post-process θ and ϑ by

$$\zeta^k : \mathfrak{B} \to S \qquad \mathbf{b} \mapsto s \wedge \left(|\mathbf{b}_j| > \frac{1}{2} u \right)_{v \in E} \qquad v_j = I \tag{32}$$

$$u = \max\{|b_i| : \exists_{w \in E} w_i = I \wedge s_w\} \qquad s = \vartheta(\mathbf{b}) \wedge \theta^k(\mathbf{b}) \tag{33}$$

and define the oracle

$$\Omega_P(w) := \psi^{-1}(\zeta^k(\xi_w^L(\mathbf{b})))$$

where we assume that $\mathbf{b} \in \mathfrak{B}$ is the initial domain we want to partition.

4 Results

In the last section we show the results of experiments where we applied the LIR-tree to a complex dataset and compare it to the Octree and KD-tree. In our research we focus on voxel geometries, especially on fiber geometries, see Fig. 11. Therefore we chose a complex generated fiberglass dataset for our experiments. We restrict to a three-dimensional context since the LIR-tree is designed for higher dimensional partitioning.

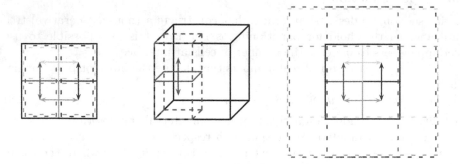

Fig. 10. The colored regions illustrate the domain and direction of analysis for the edges. In the left and center case we use ν^0 and in the right case we use ν^1. A color component shows the edges and its domains of analysis. The direction of the arrow corresponds to i in Eqn. 29.

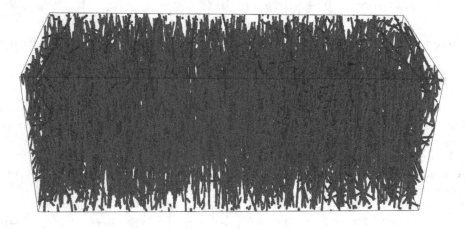

Fig. 11. 3d fibrous material used as porous media model and as composite material models proposed in [9] and available from [8]

4.1   Generated Fiberglass

For our experiments we generated a very complex fiberglass dataset with the GeoDict software suite [8]. The dataset is an approximation to a High-Efficiency Particulate Air (HEPA) filter medium. HEPA filters are composed of randomly arranged fibers on the micrometer scale.

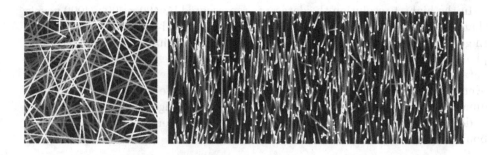

Fig. 12. 2D views of a fiberglass medium from [8]: (x,y)-view, (y,z)-view

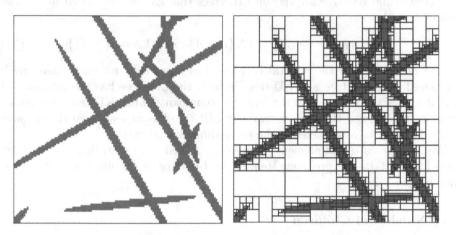

Fig. 13. Cut through the fiberglass

Fig. 14. LIR-tree with ν^0

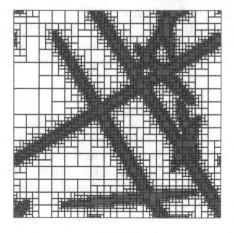

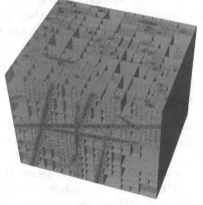

Fig. 15. LIR-tree with ν^1

Fig. 16. 3D-View of the LIR-tree

In the highest resolution the dataset is represented by $400 \times 400 \times 4000$ voxels that can either be empty or solid. The fibers have a solid-volume-fraction of 8%. In general they have an anisotropic spatial orientation that is isotropic in (x, y)-plane and layered in z-direction, see Fig. 12. A two-dimensional cut through the dataset is shown in Fig. 13. We use the input function with ζ^0, see Fig. 14. An overlapping input function is used in Fig. 15 to illustrate the behavior. A three-dimensional view of the tree structure is shown in Fig. 16. The partitioning is done by ξ_v^L with $v \in A^n$ where we modified the center to meet the voxel boundaries.

4.2 Tree Comparison

The Octree and KD-tree are special LIR-trees that can be generated by restriction of P with

$$P_{oct} = \{\emptyset, \mathbf{I}^3\} \qquad P_{kd}^i = \{\emptyset, \{v \in A^3 : v_i \in \mathbf{I} \land v_{j \neq i} = I\}\} \tag{34}$$

Figure 17 shows that the general LIR-tree has fewer cells for each voxellength compared to the Octree and KD-tree. In fact, the LIR-tree has the interior cell complexity of the Octree and the leaf cell complexity of the KD-tree. Numerous experiments showed that on average the LIR-tree has at least two times fewer cells compared to other trees in the three-dimensional case.

We are interessted in efficient numerically suited representations rather then pure binary data compression. Hence, the LIR-tree is an efficient method for that purpose.

4.3 Number of Children

In the next experiment we investigate the distribution of the number of children. Figure 18 shows the distribution for the fiberglass example. It turns out that in most cases a partitioned cell has two, three or four children. The cases of two and three children are changing their places with respect to the voxellength due to the non-cubic dataset. On average they make up 25% of the partitions. A five children partition occurs in 10% of the cases while the six, seven and eight children partitions occur in up to 4% of the cases.

4.4 Error Analysis

In our last experiment we used the highest resolution of the fiberglass medium as reference and compared a lower resolved LIR-tree with respect to the number of incorrect voxel values and volume defect. Figure 18 shows that the relative error has linear correlation to the voxellength. But it also shows that the relative error has a quadratic correlation to the number of cells. Hence, it makes sense to use the number of cells to investigate convergence orders instead of the voxellength. The error of the volume is much lower compared to the total error. That can be an important property for numerical computations where the same mass leads to the same behavior.

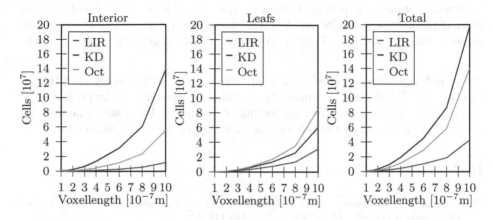

Fig. 17. Number of cells used by the LIR-tree, KD-tree and Octree. The interior, leaf and total number of cells are considered

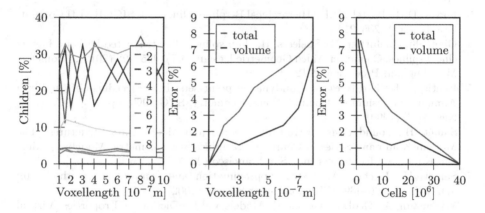

Fig. 18. Left: Number of children distribution, Center: Incorrect voxel values with respect to voxellength, Right: Incorrect voxel values with respect to number of cells

5   Conclusions

We presented a novel space partitioning system based on a ternary alphabet. Recursive and dimensional application to systems of functions leads to a tree structure that combines the advantages of the Octree and the KD-tree without having their disadvantages. This is also achieved by using different types of look-up tables and compile-time generated data structures.

We compared the LIR-tree to the Octree and KD-tree and observed that the LIR-tree needs at most half the number of cells in a three dimensional context. This is due to the fact that the LIR-tree is a generalization and can use more sophisticated methods to analyze and decompose the geometry. The LIR-tree has the interior cell complexity of the Octree and the leaf complexity of the

KD-tree. Numerous experiments showed that in a three dimensional context most of the partitioned cells have 2-4 children. For convergence analysis it makes sense to take the number of cells into account in addition to the spatial length.

Acknowledgements. This work is sponsored by Math2Market GmbH, the Fraunhofer ITWM and the University of Kaiserslautern under the project (P3) Applied System Modeling for Engineering of Stochastic Processes. The authors also would like to thank the referees for their suggestions and comments.

References

1. Jackins, C., Tanimoto, S.: Oct-trees and Their Use in Representing Three-Dimensional Objects. CGIP 14(3), 249–270 (1980)
2. Jackins, C., Tanimoto, S.: Quad-Trees, Oct-Trees, and K-Trees: A Generalized Approach to Recursive Decomposition of Euclidean Space. PAMI 5(5), 533–539 (1983)
3. Forsey, D.R., Bartels, R.H.: Hierarchical B-spline refinement. SIGGRAPH Comput. Graph. 22(4), 205–212 (1988)
4. Giannelli, C., Jüttler, B., Speleers, H.: Thb-splines: The truncated basis for hierarchical splines. Computer Aided Geometric Design 29(7), 485–498 (2012); Geometric Modeling and Processing 2012
5. Bentley, J.L.: K-d trees for semidynamic point sets. In: Proceedings of the Sixth Annual Symposium on Computational Geometry, SCG 1990, pp. 187–197. ACM, New York (1990)
6. Samet, H.: Foundations of Multidimensional and Metric Data Structures (The Morgan Kaufmann Series in Computer Graphics and Geometric Modeling). Morgan Kaufmann Publishers Inc., San Francisco (2005)
7. Samet, H., Kochut, A.: Octree approximation and compression methods. In: 3DPVT, pp. 460–469. IEEE Computer Society (2002)
8. Wiegmann, A.: GeoDict: Geometric Models and PreDictions of Properties. Virtual material laboratory (2001), http://www.geodict.com
9. Schladitz, K., Peters, S., Reinel-Bitzer, D., Wiegmann, A., Ohser, J.: Design of acoustic trim based on geometric modeling and flow simulation for non-woven. Computational Materials Science 38(1), 56–66 (2006)
10. Chen, H., Huang, T.: A Survey of Construction and Manipulation of Octrees. CVGIP 43(3), 409–431 (1988)
11. Hunter, G.M.: Efficient computation and data structures for graphics. PhD thesis, Princeton, NJ, USA (1978)
12. Bentley, J.L.: Multidimensional binary search trees used for associative searching. Commun. ACM 18(9), 509–517 (1975)
13. Duncan, C.A., Goodrich, M.T., Kobourov, S.G.: Balanced aspect ratio trees: Combining the advantages of k-d trees and octrees. J. Algorithms 38(1), 303–333 (2001)
14. Tobler, R.F.: The rkd-Tree: An Improved kd-Tree for Fast n-Closest Point Queries in Large Point Sets. In: Proceedings of Computer Graphics International 2011, CGI 2011 (2011)

Local Hierarchical h-Refinements in IgA Based on Generalized B-Splines

Carla Manni[1], Francesca Pelosi[1], and Hendrik Speleers[2]

[1] Dipartimento di Matematica, Università di Roma "Tor Vergata",
Via della Ricerca Scientifica, 00133 Roma, Italy
[2] Departement Computerwetenschappen, Katholieke Universiteit Leuven,
Celestijnenlaan 200A, B-3001 Leuven, Belgium

Abstract. In this paper we construct multilevel representations in terms of a hierarchy of tensor-product generalized B-splines. These representations combine the positive properties of a non-rational model with the possibility of dealing with local refinements. We discuss their use in the context of isogeometric analysis.

Keywords: Isogeometric analysis, Generalized B-splines, Hierarchical bases, Multilevel representations.

1 Introduction

Isogeometric Analysis (IgA) is a recent, but well established and successful, paradigm for the analysis of problems governed by partial differential equations [9,17]. Its main goal is to improve the connection between numerical simulation and Computer Aided Design (CAD) systems.

In its original formulation, the main idea in IgA is to use directly the geometry provided by CAD systems – which usually is expressed in terms of tensor-product Non-Uniform Rational B-Splines (NURBS) – and to approximate the unknown solutions of differential equations by the same type of functions. This results in three principal advantages of IgA with respect to classical FEM (Finite Element Methods). First, complicated geometries are represented more accurately and some common profiles as conic sections are exactly described. This exact or accurate description of the geometry has a beneficial influence on the numerical solution of the addressed differential problem. Second, the description of the geometry is incorporated exactly at the coarsest mesh level and mesh refinement does not modify the geometry. This greatly simplifies the refinement process because it eliminates the necessity of interacting with the CAD system when mesh refinement is carried out. Finally, B-spline and NURBS representations allow an easy treatment and refinement of spaces with high approximation order and an inherent higher smoothness than those in classical FEM. This has been proved to be superior in various applications, see [9,17], and references therein.

On the other hand, the three main advantages summarized above are not a distinguishing property of NURBS. Actually, NURBS suffer from some relevant

M. Floater et al. (Eds.): MMCS 2012, LNCS 8177, pp. 341–363, 2014.
© Springer-Verlag Berlin Heidelberg 2014

geometric drawbacks. For instance, they lack an exact description of transcendental curves of interest in applications, and their parametrization of conic sections does not correspond to natural arc length. In addition, NURBS poorly behave with respect to differentiation and integration which are crucial operators in analysis. On this concern, it is sufficient to have a look at the complicated structure of the derivative of a NURBS curve of a given order.

In the CAGD (Computer Aided Geometric Design) community several alternatives to the rational model have been proposed to overcome the above mentioned drawbacks. The so-called generalized B-splines are of relevant interest in this context, see [6,21,22,28,34] and references therein. They are piecewise functions with sections in more general spaces than algebraic polynomial spaces (like classical B-splines). Suitable selections of such spaces – typically including trigonometric or hyperbolic functions – allow an exact representation of polynomial curves, conic sections, helices and other profiles of salient interest in applications. Moreover, generalized B-splines possess all fundamental properties of algebraic B-splines (recurrence relation, compact minimum support, local linear independence, ...) which are shared by NURBS as well. Finally, contrarily to NURBS, they behave completely similar to B-splines with respect to differentiation and integration. Therefore, tensor-product generalized B-splines can offer an interesting alternative to NURBS in IgA as investigated in [8,23,24].

Adaptive local refinement is a crucial ingredient for obtaining, in an efficient way, an accurate solution of partial differential equations. However, NURBS rely on a tensor-product structure, so they do not allow adequate local refinements. This motivates the interest in alternative structures for IgA that permit local refinements, such as T-splines [11], hierarchical splines [33], LR (Locally Refined) splines [10], or B-splines over triangulations [32].

Hierarchical B-splines were introduced in [12] as an accumulation of tensor-product B-splines with nested knot vectors, and they were further investigated in [19] and [13,14,33]. Of course, the concept of hierarchical bases can be considered for more general spaces than tensor-product B-splines, see also [14]. In particular, a hierarchical structure can be built for tensor-product generalized B-splines, with suitable section spaces, as they suffer from the same drawbacks of NURBS with respect to local refinements.

Besides the classical hierarchical bases, in the literature there also exist so-called quasi-hierarchical bases and truncated hierarchical bases. They are both alternative bases for a hierarchical space build on a sequence of nested linear spaces. The notion of quasi-hierachical bases for hierarchical spaces has been introduced in [16,20] and also used in [29,30]. The concept of truncated bases has been described in [13,14].

In this paper we define a multilevel representation in terms of a hierarchy of tensor-product generalized B-splines, and we discuss its use in the context of IgA. In this way, we can combine the positive properties of a non-rational model with the possibility of dealing with local refinements. The proposed hierarchical construction extends the classical one, because it does not necessarily require nested sequences of spaces.

The remainder of the paper consists of four sections. Section 2 briefly summarizes the definition and main properties of generalized B-splines, while in Section 3 we define multilevel representations for them. Section 4 illustrates the performances of the generalized multilevel representations as a tool supporting local refinements in IgA in two benchmark problems taken from the literature. We end in Section 5 with some final comments.

2 Generalized B-Splines

To make the paper self contained, we devote this section to summarize the definition and basic properties of generalized B-splines. Further details can be found in the cited references and in [23, Section 3]. Assuming a sequence of knots is given,

$$\Xi := \{\xi_1 \leq \xi_2 \leq \cdots \leq \xi_{n+p+1}\}, \quad n \in \mathbb{N}, \tag{1}$$

classical B-splines of degree p defined over (1) are a basis for piecewise polynomial functions with a suitable smoothness, i.e. functions with sections in the space of algebraic polynomials of degree p,

$$\mathbb{P}_p := \langle 1, t, \ldots, t^{p-2}, t^{p-1}, t^p \rangle.$$

Functions with a given smoothness and belonging piecewisely to more general spaces as

$$\mathbb{P}_p^{u_i, v_i} := \langle 1, t, \ldots, t^{p-2}, u_i(t), v_i(t) \rangle, \quad t \in [\xi_i, \xi_{i+1}), \quad i = 1, \ldots, n+p, \tag{2}$$

can be considered as well, see [6] and references therein. In the section spaces (2) the functions u_i, v_i can be selected such that salient profiles of interest are exactly represented and/or special characteristics are obtained. Popular choices for (2) are:

$$\mathbb{E}_{p, \alpha_i} := \langle 1, t, \ldots, t^{p-2}, \exp(\alpha_i t), \exp(-\alpha_i t) \rangle, \quad 0 < \alpha_i \in \mathbb{R}, \tag{3}$$

$$\mathbb{T}_{p, \alpha_i} := \langle 1, t, \ldots, t^{p-2}, \cos(\alpha_i t), \sin(\alpha_i t) \rangle, \quad 0 < \alpha_i(\xi_{i+1} - \xi_i) < \pi, \tag{4}$$

which lead to exponential and trigonometric splines respectively. They allow an exact representation of conic sections as well as of some transcendental curves (helix, cycloid,...), see [4]. Exponential splines are often referred to as hyperbolic splines because the space (3) coincides with the space

$$\langle 1, t, \ldots, t^{p-2}, \cosh(\alpha_i t), \sinh(\alpha_i t) \rangle.$$

This alternative formulation shows more clearly the connection between spaces (3) and (4). Other interesting section spaces are

$$\mathbb{VD}_{p, \alpha_i} := \left\langle 1, t, \ldots, t^{p-2}, \left(\frac{\xi_{i+1} - t}{\xi_{i+1} - \xi_i} \right)^{\alpha_i}, \left(\frac{t - \xi_i}{\xi_{i+1} - \xi_i} \right)^{\alpha_i} \right\rangle, \quad p \leq \alpha_i \in \mathbb{R}, \tag{5}$$

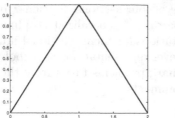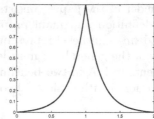

Fig. 1. Generalized B-spline $B^{(1)}_{i,\Xi}$ with knot sequence $\Xi = \{0, 1, 2\}$. Left: the classical polynomial case. Right: the exponential case with $\alpha_i = 5$.

which lead to the so-called variable degree splines. Exponential and variable degree splines are a powerful tool in shape-preserving approximation and/or interpolation.

Results on the approximation power of the spaces (2) can be found in [6, Section 3] (see also [5,25,28]). In particular, for fixed values of the involved parameters, the spaces (3) and (4) have the same approximation power of $\mathbb{P}_p$.

Remark 1. The section spaces in (2) may be different on each interval. Thus, generalized B-splines allow an exact representation of profiles composed by a sequence of curve segments of different kind: arcs of ellipses, hyperbolas, polynomial curves, etc.

It is well known that it is possible to construct B-spline-like functions with sections in spaces (2), see [5,18,21,26,27,28,34] and references therein. The so-called *generalized B-splines of degree p*, defined over the knot sequence (1), will be denoted by $B^{(p)}_{i,\Xi}$. To simplify the notation we omit the reference to the section spaces (2), even though this would be more appropriate. The specific section spaces will be clear from the context.

More precisely, we assume that[1]

$$u_i, v_i \in C^{p-1}[\xi_i, \xi_{i+1}],$$

and that $\{u_i^{(p-1)}, v_i^{(p-1)}\}$ is a Chebyshev system in $[\xi_i, \xi_{i+1}]$ and an Extended Chebyshev system in (ξ_i, ξ_{i+1}). Thus, without loss of generality we can assume

$$u_i^{(p-1)}(\xi_i) > 0, \quad u_i^{(p-1)}(\xi_{i+1}) = 0, \quad v_i^{(p-1)}(\xi_i) = 0, \quad v_i^{(p-1)}(\xi_{i+1}) > 0.$$

According to [21] (see also [26,34] and references therein), generalized B-splines can be defined with the following recurrence relation formula:

$$B^{(1)}_{i,\Xi}(t) := \begin{cases} \dfrac{v_i^{(p-1)}(t)}{v_i^{(p-1)}(\xi_{i+1})}, & \text{if } t \in [\xi_i, \xi_{i+1}), \\[2mm] \dfrac{u_{i+1}^{(p-1)}(t)}{u_{i+1}^{(p-1)}(\xi_{i+1})}, & \text{if } t \in [\xi_{i+1}, \xi_{i+2}), \\[2mm] 0, & \text{elsewhere}; \end{cases}$$

---

[1] For more general constructions with less restrictive hypotheses we refer to [26,27].

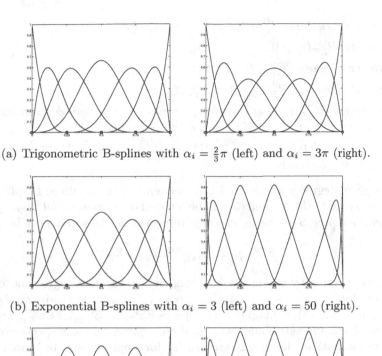

(a) Trigonometric B-splines with $\alpha_i = \frac{2}{3}\pi$ (left) and $\alpha_i = 3\pi$ (right).

(b) Exponential B-splines with $\alpha_i = 3$ (left) and $\alpha_i = 50$ (right).

(c) Variable degree B-splines with $\alpha_i = 6$ (left) and $\alpha_i = 30$ (right).

Fig. 2. Examples of generalized B-splines of degree 3 defined on the knot sequence $\Xi = \{0, 0, 0, 0, 1/4, 1/2, 3/4, 1, 1, 1, 1\}$.

$$B_{i,\Xi}^{(p)}(t) := \delta_{i,\Xi}^{(p-1)} \int_{-\infty}^{t} B_{i,\Xi}^{(p-1)}(s)\mathrm{d}s - \delta_{i+1,\Xi}^{(p-1)} \int_{-\infty}^{t} B_{i+1,\Xi}^{(p-1)}(s)\mathrm{d}s, \quad p \geq 2,$$

where $\delta_{i,\Xi}^{(p)} := [\int_{-\infty}^{+\infty} B_{i,\Xi}^{(p)}(s)\mathrm{d}s]^{-1}$ and fractions with zero denominators are considered to be zero. The knot sequence (1) allows us to define n generalized B-splines of degree p, namely $B_{1,\Xi}^{(p)}, \ldots, B_{n,\Xi}^{(p)}$. Two generalized B-splines of degree 1 are depicted in Figure 1, and some sets of cubic generalized B-splines are illustrated in Figure 2.

Generalized B-splines possess all desirable properties of classical polynomial B-splines [3,5,21].

Proposition 1. *Let $B_{i,\Xi}^{(p)}$, $i = 1, \ldots, n$, be generalized B-splines of degree $p \geq 2$ associated to the knot sequence (1). Then the following properties hold:*

- *piecewise structure:* $B_{i,\Xi}^{(p)}(t) \in \mathbb{P}_p^{u_j,v_j}$, $t \in [\xi_j, \xi_{j+1})$,
- *positivity:* $B_{i,\Xi}^{(p)}(t) \geq 0$,
- *partition of unity:* $\sum_{i=1}^n B_{i,\Xi}^{(p)}(t) \equiv 1$, $t \in [\xi_{p+1}, \xi_{n+1})$,
- *compact support:* $B_{i,\Xi}^{(p)}(t) = 0$, $t \notin [\xi_i, \xi_{i+p+1}]$,
- *smoothness:* $B_{i,\Xi}^{(p)}(t)$ *is at least* $p - \rho_j$ *times continuously differentiable at* ξ_j *being* ρ_j *the multiplicity of* ξ_j *in the knot sequence,*
- *local linear independence:* $B_{i-p,\Xi}^{(p)}(t), \ldots, B_{i-1,\Xi}^{(p)}(t), B_{i,\Xi}^{(p)}(t)$ *are linearly independent on* $[\xi_i, \xi_{i+1})$.

For a given degree p and a fixed knot sequence Ξ, generalized B-splines with section spaces as in (3), (4) and (5) will be referred to as *exponential, trigonometric* and *variable degree B-splines of degree* p, respectively. We will denote by

$$\mathbb{ES}_{\Xi,\alpha}^p, \quad \mathbb{TS}_{\Xi,\alpha}^p, \quad \mathbb{VDS}_{\Xi,\alpha}^p, \quad \mathbb{S}_{\Xi}^p$$

the spaces spanned by exponential, trigonometric, variable degree and classical B-splines of degree p, respectively. Here $\alpha = \{\ldots, \alpha_i, \ldots\}$ stands for the set of real parameters in (3), (4) and (5). The spaces $\mathbb{ES}_{\Xi,\alpha}^p$, $\mathbb{TS}_{\Xi,\alpha}^p$, $\mathbb{VDS}_{\Xi,\alpha}^p$ are called *exponential, trigonometric* and *variable degree spline spaces,* respectively.

Stable evaluation algorithms are crucial for applications. It is well known that a stable evaluation of exponential functions is a difficult task. On the other hand, variable degree splines possess graphical properties analogous to exponential splines as the parameters α_i increase, see Figure 2 (b)–(c), but they profit of an efficient evaluation. Indeed, variable degree splines can be obtained by a *geometric construction* consisting of simple corner cutting schemes which are applied to the (generalized) de Boor control polygon and produce the (generalized) Bézier control polygons associated with the Bernstein-like representation of the section spaces (5), see [5,7]. Thanks to this geometric construction, the evaluation of the spline results in a stable computation.

Tensor-product exponential, trigonometric and variable degree splines have been profitably used in the context of IgA, including the treatment of advection-diffusion problems, see [8,23,24]. However, as for classical B-splines and NURBS, the tensor-product structure prevents the possibility of local refinements. On this concern, in the next section we address the problem of constructing multilevel bases in terms of a hierarchy of generalized B-splines.

3   Generalized Multilevel Bases

In this section we construct multilevel bases in terms of a hierarchy of generalized B-splines. Hierarchical tensor-product B-splines have been introduced in [12] as an accumulation of tensor-product B-splines with nested knot vectors, in order to allow local editing of tensor-product spline surfaces. Nevertheless, the definition of hierarchical spaces and bases can be more general, and it is not confined to tensor-product algebraic B-splines, see [14].

The classical definition of hierarchical bases and spaces assumes to deal with a sequence of nested linear spaces. Here, we extend the usual construction to sequences of not necessarily nested spaces. We prove that, under the hypothesis of local linear independence for the basis functions considered in each space, the resulting set of functions still possesses the fundamental property of linear independence.

Let Ω_0 be the closure of a bounded domain in $\mathbb{R}^d$. We consider a sequence of finite-dimensional linear spaces of functions defined on Ω_0,

$$V^0, V^1, \ldots, V^\ell, \ldots \tag{6}$$

and let Ω_0^ℓ be the closure of domains in $\mathbb{R}^d$ such that

$$\Omega_0 \supset \Omega_0^1 \supset \cdots \supset \Omega_0^\ell \supset \cdots . \tag{7}$$

Setting $n_\ell := \dim(V^\ell)$, we assume

$$\mathcal{B}^\ell := \{b_{1,\ell}, \ldots, b_{n_\ell,\ell}\}$$

is a basis of V^ℓ. A multilevel basis can be obtained by considering first all the basis elements in $\mathcal{B}^0$ whose support overlaps Ω_0, and by considering then an iterative procedure which selects at each level ℓ all the basis functions of the previous level whose support is not entirely contained in Ω_0^ℓ and all the basis functions in $\mathcal{B}^\ell$ whose support is entirely contained in Ω_0^ℓ, see Figure 3.

More precisely, the multilevel space associated to a hierarchy of depth L is the space spanned by the set of functions constructed according to the following definition, see [13,14,33]. We denote by $\text{supp}(b_{i,\ell})$ the intersection of the support of $b_{i,\ell}$ with Ω_0.

Definition 1. *The multilevel set of basis functions $\mathcal{H}$ associated to a hierarchy of domains (7) of depth L is recursively constructed as follows:*

i) $\mathcal{H}^0 := \{b_{i,0} : \text{supp}(b_{i,0}) \neq \emptyset\}$;
ii) for $\ell = 0, \ldots, L-2$:

$$\mathcal{H}^{\ell+1} := \mathcal{H}_C^{\ell+1} \cup \mathcal{H}_F^{\ell+1},$$

where

$$\mathcal{H}_C^{\ell+1} := \{b_{i,j} \in \mathcal{H}^\ell : \text{supp}(b_{i,j}) \not\subset \Omega_0^{\ell+1}\},$$

$$\mathcal{H}_F^{\ell+1} := \{b_{i,\ell+1} \in \mathcal{B}^{\ell+1} : \text{supp}(b_{i,\ell+1}) \subset \Omega_0^{\ell+1}\};$$

iii) $\mathcal{H} := \mathcal{H}^{L-1}$.

We are focusing on (local) mesh refinements in the context of IgA – the so-called h-refinements. Therefore, in the following we assume that each space V^ℓ in (6) is a space of generalized B-splines with fixed degree p defined over a given knot sequence. For $\ell = 0, \ldots, L-1$, we assume that the knot vectors

$$\Xi_\ell := \{\xi_{1,\ell} \leq \xi_{2,\ell} \leq \cdots \leq \xi_{n_\ell+p+1,\ell}\}, \quad n_\ell \in \mathbb{N}, \tag{8}$$

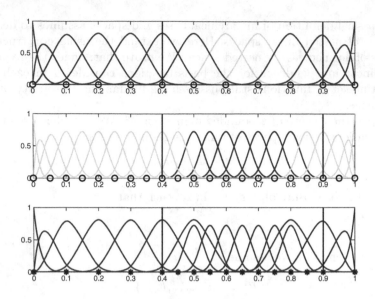

Fig. 3. Illustration of the construction of a multilevel representation of cubic exponential B-splines with $\alpha_{i,\ell} = 50$. The light blue functions are removed in the construction of the multilevel basis.

and the section spaces

$$\mathbb{P}^{u_i,v_i}_{p,\ell} := \langle 1, t, \ldots, t^{p-2}, u_{i,\ell}(t), v_{i,\ell}(t) \rangle, \ t \in [\xi_{i,\ell}, \xi_{i+1,\ell}), \quad i = 1, \ldots, n_\ell + p, \quad (9)$$

are given. Moreover, we assume that the domains Ω_0^ℓ in (7) are selected as the union of a number of knot intervals defined by Ξ_ℓ in (8).

We will refer to the elements of $\mathcal{H}$ as *generalized multilevel B-splines* and to the spaces they span as *generalized multilevel spline spaces*. This multilevel construction generalizes the classical definition of hierarchical B-splines (see [33]) in the following ways:

- the multilevel basis is constructed as a hierarchy of generalized B-splines;
- different section spaces may be chosen not only on the different knot intervals (see Remark 1) but also on the different levels;
- the different spaces V^ℓ are not necessarily nested.

Despite this flexibility, the generalized multilevel B-splines preserve many properties of the classical hierarchical B-splines. Thanks to the local linear independence of generalized B-splines, the construction in Definition 1 ensures that the obtained set $\mathcal{H}$ consists of linearly independent functions, as proved in the following theorem. Therefore, we will refer to them as *generalized multilevel bases*.

Theorem 1. *Let $\mathcal{H}$ be given as in Definition 1 where V^ℓ are spaces of generalized B-splines defined on the knot sequences (8) with section spaces as in (9).*

Then,
- *the elements of $\mathcal{H}$ are linearly independent;*
- *the elements of $\mathcal{H}$ are non-negative.*

Proof. Let us consider a linear combination of the elements of $\mathcal{H}$ which vanishes in Ω_0. Denoting by $I_\ell \subset \{1, 2, \ldots, n_\ell\}$ the set of indices of the elements of $\mathcal{B}^\ell$ belonging to $\mathcal{H}$, $\forall t \in \Omega_0$, we have

$$0 = \sum_{\ell=0}^{L-1} \sum_{k \in I_\ell} \beta_{k,\ell}\, b_{k,\ell}(t), \quad \beta_{k,\ell} \in \mathbb{R}.$$

In order to show the linear independence of the elements of $\mathcal{H}$, we have to prove that all the coefficients in this representation are zero. By (7) and by Definition 1, the only functions in $\mathcal{H}$ which are possibly not zero in $\Omega_0 \setminus \Omega_0^1$ belong to $\mathcal{B}^0$. Therefore, $\forall t \in \Omega_0 \setminus \Omega_0^1$, we have

$$0 = \sum_{\ell=0}^{L-1} \sum_{k \in I_\ell} \beta_{k,\ell}\, b_{k,\ell}(t) = \sum_{k \in I_0} \beta_{k,0}\, b_{k,0}(t).$$

From Proposition 1 we know that the elements of $\mathcal{B}^0$ are locally linearly independent, and so we obtain that $\beta_{k,0} = 0$, $k \in I_0$. We now proceed in a similar way for all levels $\ell = 1, \ldots, L-2$ sequentially. By considering all $t \in \Omega_0^\ell \setminus \Omega_0^{\ell+1}$ for a given ℓ, we obtain that $\beta_{k,\ell} = 0$, $k \in I_\ell$. Finally, imposing that the linear combination must vanish for all $t \in \Omega_0^{L-1}$, we obtain that $\beta_{k,L-1} = 0$, $k \in I_{L-1}$.

In addition, from Proposition 1 it follows that any $b_{i,\ell}$ is non-negative. Hence, the functions in $\mathcal{H}$ are non-negative as well. □

The procedure described in Definition 1 allows a great flexibility in the construction of generalized multilevel B-splines because spline spaces with different section spaces and different knot sequences are allowed at different levels. Nevertheless, when nested spaces are of interest, we can also consider nested spaces of generalized B-splines defined over nested knot vectors. More precisely, the following result can be easily proved.

Proposition 2. *Let $\mathcal{B}^\ell := \{B_{i,\Xi_\ell}^{(p)},\ i = 1, \ldots, n_\ell\}$. The spaces $V^\ell := \langle \mathcal{B}^\ell \rangle$ are nested, if the knot sequences are nested, i.e.,*

$$\Xi_\ell \subset \Xi_{\ell+1}, \tag{10}$$

and if

$$\langle 1, t, \ldots, t^{p-2}, u_{i,\ell}(t), v_{i,\ell}(t) \rangle = \langle 1, t, \ldots, t^{p-2}, u_{j,0}(t), v_{j,0}(t) \rangle, \tag{11}$$

for each $[\xi_{i,\ell}, \xi_{i+1,\ell}) \subset [\xi_{j,0}, \xi_{j+1,0})$.

The main advantage of nested spaces is the ensured reduction (not deterioration) of the error.[2] Moreover, as a particular case of the results in [14] (see also [33]), we obtain that the intermediate spaces $\langle \mathcal{H}^\ell \rangle$ are nested in the construction of Definition 1.

---

[2] This is not always the case for non-nested spaces, but when particular features (as layers) have to be detected, non-nested spaces can be powerful as well, see e.g. [32].

Proposition 3. *Let $\mathcal{H}$ be constructed as in Definition 1, where V^{ℓ} are spaces of generalized B-splines defined on the knot sequences (8) with section spaces as in (9). If (10) and (11) hold, then $\langle \mathcal{H}^{\ell} \rangle \subseteq \langle \mathcal{H}^{\ell+1} \rangle$.*

We now consider two multilevel spaces $\langle \mathcal{H} \rangle$ and $\langle \hat{\mathcal{H}} \rangle$ constructed by using the same nested generalized spline spaces V^{ℓ} over two different domain hierarchies $\{\Omega_0^{\ell}\}_{\ell=0,\ldots,L-1}$ and $\{\hat{\Omega}_0^{\ell}\}_{\ell=0,\ldots,L-1}$. From [14] (see also [33]) we know that, if the second domain sequence enlarges the first one, then this property is inherited by the corresponding generalized multilevel spline spaces.

Proposition 4. *Let $\{\Omega_0^{\ell}\}_{\ell=0,\ldots,L-1}$ and $\{\hat{\Omega}_0^{\ell}\}_{\ell=0,\ldots,L-1}$ be two domain hierarchies such that $\Omega^0 = \hat{\Omega}^0$ and $\Omega^{\ell} \subseteq \hat{\Omega}^{\ell}$ for $\ell = 1,\ldots,L-1$. Let $\mathcal{H}^{\ell}$ and $\hat{\mathcal{H}}^{\ell}$ be constructed as in Definition 1 on both domain hierarchies, where V^{ℓ} are spaces of generalized B-splines defined on the knot sequences (8) with section spaces as in (9). If (10) and (11) hold, then $\langle \mathcal{H}^{\ell} \rangle \subseteq \langle \hat{\mathcal{H}}^{\ell} \rangle$.*

Remark 2. Exponential and trigonometric splines easily provide nested sequences of spaces. As far as variable degree splines are concerned, we note that, if $[\xi_{i,\ell}, \xi_{i+1,\ell})$ is a proper subset of $[\xi_{j,0}, \xi_{j+1,0})$, but $\alpha_{j,0} \neq p$, then the space

$$\left\langle 1, t, \ldots, t^{p-2}, \left(\frac{\xi_{i+1,\ell} - t}{\xi_{i+1,\ell} - \xi_{i,\ell}} \right)^{\alpha_{j,0}}, \left(\frac{t - \xi_{i,\ell}}{\xi_{i+1,\ell} - \xi_{i,\ell}} \right)^{\alpha_{j,0}} \right\rangle$$

is not a subspace of

$$\left\langle 1, t, \ldots, t^{p-2}, \left(\frac{\xi_{j+1,0} - t}{\xi_{j+1,0} - \xi_{j,0}} \right)^{\alpha_{j,0}}, \left(\frac{t - \xi_{j,0}}{\xi_{j+1,0} - \xi_{j,0}} \right)^{\alpha_{j,0}} \right\rangle.$$

In general, the elements of $\mathcal{H}$ do not sum up to one. When a sequence of nested spaces is taken in (6) such that V_0 contains constants, different strategies can be considered to obtain a normalized basis in $\langle \mathcal{H} \rangle$. For example, the so-called truncated hierarchical bases presented in [13,14] are particularly interesting for their stability properties. For the sake of simplicity, we just consider the elements of $\mathcal{H}$ as basis elements and we are not concerned with normalization.

The results and the proofs of the previous theorem and propositions immediately extend to the multidimensional tensor-product setting.

4   Numerical Examples

For the sake of completeness, we first briefly summarize the formulation of the isogeometric analysis approach for numerical approximation of PDE solutions. For a more comprehensive presentation we refer to [9], see also [8,11,17] for a short summary. For the sake of simplicity, we explain the paradigm for linear stationary problems in two dimensions with Dirichlet boundary conditions.

Let $\mathcal{L}$ be a second order (elliptic) operator on the domain Ω with Lipschitz boundary $\partial\Omega$. Let us consider the problem

$$\begin{cases} \mathcal{L}u = f, & \text{in } \Omega, \\ u = g, & \text{on } \partial\Omega, \end{cases} \tag{12}$$

for the unknown $u : \Omega \to \mathbb{R}$. Assuming without loss of generality homogeneous Dirichlet boundary conditions (i.e. $g = 0$), the weak formulation of (12) is formulated as follows:

$$\text{find} \quad u \in \mathbb{V}_D, \quad \text{such that} \quad a(u,v) = F(v), \quad \forall\, v \in \mathbb{V}_D, \tag{13}$$

where

$a : \mathbb{V} \times \mathbb{V} \to \mathbb{R}$ is a bilinear form depending on $\mathcal{L}$,
$F : \mathbb{V} \to \mathbb{R}$ is a linear form depending on f,
$\mathbb{V} := H^1(\Omega)$, $\mathbb{V}_D := H^1_0(\Omega)$.

It is well known that the Galerkin approach to approximate the solution of (13) consists in selecting a suitable subspace

$$\mathbb{V}_h := \langle \phi_1, \phi_2, \ldots, \phi_{n_h}, \phi_{n_h+1}, \ldots, \phi_{n_h+n_h^b} \rangle \subset \mathbb{V},$$

and in looking for $u_h \in \mathbb{V}_{D_h} \subset \mathbb{V}_D$, such that

$$a(u_h, v_h) = F(v_h), \quad \forall v_h \in \mathbb{V}_{D_h}. \tag{14}$$

Here, we assume that an approximation (which is possibly exact) Ω_h of the physical domain Ω is given, and

$$\mathbb{V}_{D_h} := \{ v_h \in \mathbb{V}_h : v_{h|\partial\Omega_h} = 0 \}.$$

Assuming $\mathbb{V}_{D_h} = \langle \phi_1, \phi_2, \ldots, \phi_{n_h} \rangle$ and setting $u_h := \sum_{i=1}^{n_h} q_i \phi_i$, condition (14) gives rise to a linear system $A\mathbf{q} = \mathbf{F}$, where $A \in \mathbb{R}^{n_h \times n_h}$ is the stiffness matrix $A_{i,j} := a(\phi_j, \phi_i)$ and $\mathbf{F} \in \mathbb{R}^{n_h}$ is the data vector defined by $F_i := F(\phi_i)$, $i = 1, \ldots, n_h$. Different methods correspond to different choices of $\mathbb{V}_h$.

In IgA, assuming the parametric domain $\Omega_0 := [0,1] \times [0,1]$ is given, the physical domain, Ω, is represented by a *global geometry function*

$$\mathbf{G} : \Omega_0 \to \overline{\Omega}, \quad \mathbf{G}(\boldsymbol{\omega}) = \sum_{i=1}^{n_h+n_h^b} \mathcal{N}_i(\boldsymbol{\omega})\mathbf{P}_i, \quad \mathbf{P}_i \in \mathbb{R}^2, \ \boldsymbol{\omega} \in \Omega_0, \tag{15}$$

where the basis functions

$$\{ \mathcal{N}_1, \ldots, \mathcal{N}_{n_h+n_h^b} \} \tag{16}$$

have to be selected to produce an exact representation of the geometry. The space $\mathbb{V}_h$ is then defined by

$$\phi_i(\mathbf{x}) := \mathcal{N}_i \circ \mathbf{G}^{-1}(\mathbf{x}) = \mathcal{N}_i(\boldsymbol{\omega}), \quad i = 1, \ldots, n_h + n_h^b, \quad \mathbf{x} = \mathbf{G}(\boldsymbol{\omega}).$$

In IgA based on NURBS, the functions in (16) are (tensor-product) NURBS. Here we select the basis (16) as generalized multilevel B-spline bases constructed from tensor-product exponential, trigonometric and/or variable degree generalized B-splines.

In the next subsections we consider two examples illustrating the potential of these generalized multilevel spaces in IgA. For the sake of simplicity, we do not use advanced error estimators and corresponding automatic refinement strategies. Hence, the presented hierarchical meshes are constructed manually. For automatic refinement strategies we refer to the literature, see e.g. [1,11,33].

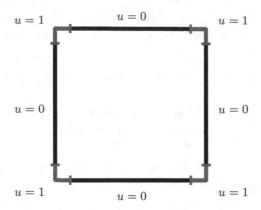

Fig. 4. Example 1. The domain with the boundary conditions.

4.1   Example 1: Reaction-Diffusion Problem

We consider the following problem, see [2]:

$$\begin{cases} -\kappa \Delta u + u = f, & \text{in } \Omega, \\ u = g, & \text{on } \partial\Omega, \end{cases} \tag{17}$$

with $\kappa = 10^{-3}$ and Dirichlet boundary conditions as depicted in Figure 4.

We consider the approximated solutions in the bivariate spaces spanned by the tensor-product of three different univariate bases, all with sections in four-dimensional spaces: cubic exponential B-splines (EXP$_3$ BSP), cubic variable degree B-splines (VDP$_3$ BSP) and classical cubic B-splines. Moreover, we also consider the solutions in the bivariate spaces spanned by generalized multilevel bases built from the same tensor-product spaces mentioned before, defined on the hierarchical meshes shown in Figure 5.

The discontinuous Dirichlet boundary conditions have been approximated by a continuous function: a Schoenberg-like quasi-interpolant obtained by using the values of the boundary data at the Greville abscissas of the cubic B-splines as depicted in Figure 5 (right). The "true" Greville abscissas could have been used (dealing with cubic-like spaces) but for the sake of simplicity the standard ones have been considered.

We compute in all cases the condition number of the corresponding stiffness matrices in the 2-norm and the minimum value of the solution evaluated on a regular 400×400 grid. We observe that all the solutions have 1.000 as maximum value on the same evaluation grid.

Table 1 shows the results obtained on uniform refinements of the initial rectangular 5×5 mesh, and the parameters of the spaces of generalized B-splines are fixed for all refinement levels: $\alpha_{i,\ell} = 50$ for all sections i and all levels ℓ in case of the exponential B-splines and $\alpha_{i,\ell} = 5$ for all i and all ℓ in case of the variable degree B-splines. On the other hand, the results in Table 2 are obtained by varying the values of the tension parameters for different refinement levels. Tables 3

Table 1. The results for Example 1 using tensor-products of cubic exponential B-splines (EXP$_3$ BSP) with $\alpha_{i,\ell} = 50$ for all sections i and all levels ℓ, cubic variable degree B-splines (VDP$_3$ BSP) with $\alpha_{i,\ell} = 5$ for all i and all ℓ, and cubic B-splines. They are all defined on uniformly refined meshes with h-length intervals per side.

Example 1, uniform mesh refinements

h	dof	EXP$_3$ BSP, $\alpha_{i,\ell} = 50$		VDP$_3$ BSP, $\alpha_{i,\ell} = 5$		Cubic BSP	
		min	cond	min	cond	min	cond
0.2	36	-4.982 10^{-3}	13.04	-1.048 10^{-2}	30.99	-8.352 10^{-2}	250.77
0.1	121	-6.452 10^{-4}	14.21	-6.648 10^{-4}	13.84	-1.603 10^{-3}	121.99
0.05	441	-9.312 10^{-5}	14.81	-1.017 10^{-4}	5.50	-1.061 10^{-4}	40.76
0.025	1681	-3.743 10^{-8}	21.34	0.000 $10^{\,0}$	4.95	-1.306 10^{-7}	29.67

Table 2. The results for Example 1 using tensor-products of cubic exponential B-splines (EXP$_3$ BSP) and cubic variable degree B-splines (VDP$_3$ BSP) with different values of the tension parameters $\alpha_{i,\ell}$ for each refinement level ℓ in Table 1.

Example 1, uniform mesh refinements

h	dof	EXP$_3$ BSP			VDP$_3$ BSP		
		$\alpha_{i,\ell}$	min	cond	$\alpha_{i,\ell}$	min	cond
0.2	36	35	-3.102 10^{-3}	20.30	6	-3.296 10^{-3}	22.02
0.1	121	35	-1.304 10^{-4}	25.95	4	-1.171 10^{-4}	26.41
0.05	441	60	-9.187 10^{-5}	10.96	4	-9.283 10^{-5}	9.14
0.025	1681	100	-9.080 10^{-18}	11.06	4	0.000 $10^{\,0}$	7.81

Table 3. The results for Example 1 using tensor-products of cubic exponential B-splines (EXP$_3$ BSP), cubic variable degree B-splines (VDP$_3$ BSP) and cubic B-splines, defined on the hierarchical meshes shown in Figure 5 and consisting of quadrilaterals with $h_{\min}$ as minimum side length.

Example 1, hierarchical mesh refinements

$h_{\min}$	dof	EXP$_3$ BSP, $\alpha_{i,\ell} = 50$		VDP$_3$ BSP, $\alpha_{i,\ell} = 5$		Cubic BSP	
		min	cond	min	cond	min	cond
0.2	36	-4.982 10^{-3}	13.04	-1.048 10^{-2}	30.99	-8.352 10^{-2}	250.77
0.1	56	-6.576 10^{-4}	489.46	-6.648 10^{-4}	172.44	-1.423 10^{-3}	484.56
0.05	92	-1.062 10^{-4}	957.39	-1.017 10^{-4}	352.45	-4.782 10^{-4}	1092.37
0.025	144	-2.382 10^{-6}	1784.63	-6.789 10^{-6}	520.76	-1.215 10^{-5}	1434.21

Table 4. The results for Example 1 using tensor-products of cubic exponential B-splines (EXP$_3$ BSP) and cubic variable degree B-splines (VDP$_3$ BSP) with different values of the tension parameters $\alpha_{i,\ell}$ for each refinement level ℓ and defined on the same four hierarchical meshes as in Table 3.

Example 1, hierarchical mesh refinements

$h_{\min}$	dof	EXP$_3$ BSP			VDP$_3$ BSP		
		$\alpha_{i,\ell}$	min	cond	$\alpha_{i,\ell}$	min	cond
0.2	36	35	-3.102 10^{-3}	20.30	6	-3.296 10^{-3}	22.02
0.1	56	35	-4.431 10^{-4}	556.82	4	-5.171 10^{-4}	172.71
0.05	92	60	-8.703 10^{-5}	1351.18	4	-9.581 10^{-5}	352.78
0.025	144	100	-9.881 10^{-7}	2085.02	4	-1.063 10^{-6}	522.41

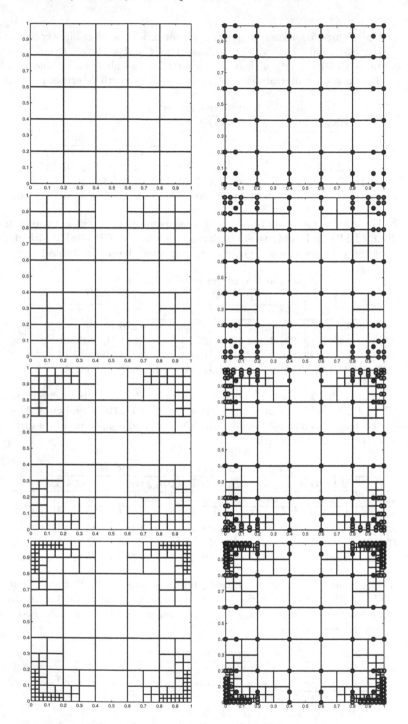

Fig. 5. Example 1. Hierarchical meshes (left) and Greville abscissas of cubic tensor-product B-splines (right) for different refinement levels.

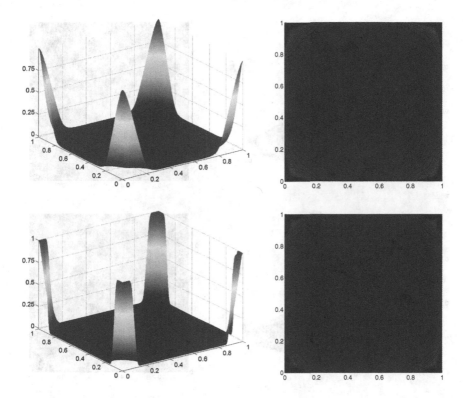

Fig. 6. Example 1. Generalized multilevel cubic exponential B-spline solutions on the first and last mesh shown in Figure 5 with tension parameters $\alpha_{i,\ell} = 35, 35, 60, 100$ at the levels $\ell = 0, 1, 2, 3$ for all i.

and 4 report the corresponding results obtained in the considered locally refined multilevel space setting.

Figures 6 and 7 show the computed approximate solutions and their contour plots using cubic exponential B-splines and cubic variable degree B-splines respectively, defined on the first and last hierarchical mesh given in Figure 5 and with the parameters given in Table 4. Figure 8 shows the reduction of undershoots versus the number of degrees of freedom (dof), in the cases of uniform and local hierarchical refinements with the different considered generalized spline spaces.

As illustrated by the numerical and graphical results, exponential and variable degree B-splines have a similar behavior, and they provide a good identification of the boundary layer with less undershoots than in the case of classical B-splines. Considering their similar performances, variable degree B-splines appear to be particularly attractive because of their reduced computational cost and the available stable evaluation algorithms.

Finally, we remark that the increase in the condition number of the stiffness matrices (see Tables 3 and 4) is a common behavior when dealing with local

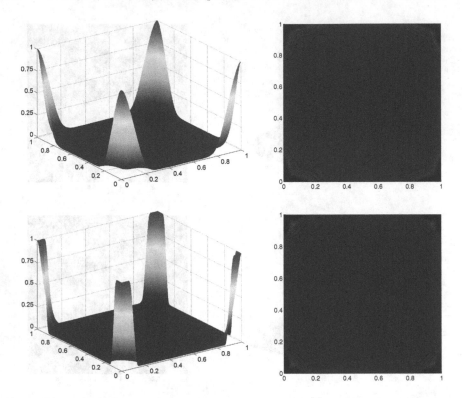

Fig. 7. Example 1. Generalized multilevel cubic variable degree B-spline solutions on the first and last mesh shown in Figure 5 with tension parameters $\alpha_{i,0} = 6$ at level 0 and $\alpha_{i,\ell} = 4$ at the other levels $\ell = 1, 2, 3$ for all i.

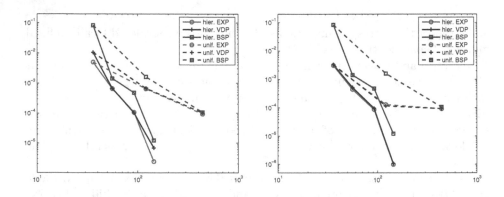

Fig. 8. Example 1. Min values (as absolute value) of the solution using cubic exponential B-splines (red), cubic variable degree B-splines (black) and classical cubic B-splines (blue) versus the number of degrees of freedom, defined on locally refined hierarchical meshes (solid) and uniformly refined meshes (dashed). Left: the tension parameters are fixed as in Tables 1 and 3. Right: see Tables 2 and 4 for the tension parameters used at each level.

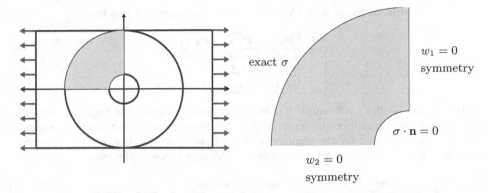

Fig. 9. Example 2. Elastic plate with a circular hole: problem setting.

refinements. Other results with alternative spaces and local refinement strategies for this problem can be found in [2,32].

4.2 Example 2: a Problem in Solid Mechanics

This example aims to illustrate the performances of local refinements based on multilevel trigonometric B-splines by solving a classical problem in solid mechanics [11,17,23,31]. We consider an infinite plate with a circular hole of radius r, subject to an in-plane uniform tension T_x in x-direction, see Figure 9 (left). For a homogeneous and isotropic material this problem features an exact solution which can be found in [15, Section 7.6]. The infinite plate is modeled by a finite circular domain with radius R. Due to the symmetry, the computational domain Ω is restricted to a quarter, see Figure 9 (right). We study the linear elastic behavior of the displacement field $\mathbf{w} : \Omega \to \mathbb{R}^2$ described by

$$\operatorname{div} \sigma(\mathbf{w}) = 0 \ \text{ in } \Omega.$$

The exact solution is applied as a Neumann boundary condition, see Figure 9 (right). For the sake of completeness we recall that $\sigma(\mathbf{w}) := \{\sigma_{ij}(\mathbf{w})\}_{i,j=1,2}$ with

$$\sigma_{ij}(\mathbf{w}) := \lambda \operatorname{div}(\mathbf{w})\delta_{ij} + 2\mu\epsilon_{ij}(\mathbf{w}), \quad \epsilon_{ij}(\mathbf{w}) := \frac{1}{2}\left(\frac{\partial w_i}{\partial x_j} + \frac{\partial w_j}{\partial x_i}\right), \quad i,j = 1,2,$$

$$\mathbf{w} := (w_1, w_2), \quad (x,y) := (x_1, x_2), \quad \lambda := \frac{E\nu}{(1+\nu)(1-2\nu)}, \quad \mu := \frac{E}{2(1+\nu)},$$

where E denotes the Young modulus and ν the Poisson ratio. In our computed example we have taken

$$r = 1, \ R = 4, \ E = 10^5, \ \nu = 0.3, \ T_x = 10.$$

Without the hole, the stress would be uniform

$$\sigma_{1,1} = T_x, \quad \sigma_{1,2} = \sigma_{2,2} = 0.$$

Table 5. Example 2. Control points of the geometry function.

k	$P_{k,1}$	$P_{k,2}$	$P_{k,3}$
1	(-1,0)	(-2.5,0)	(-4,0)
2	(-1,1)	(-2.5,2.5)	(-4,4)
3	(0,1)	(0,2.5)	(0,4)

Table 6. The results for Example 2 using tensor-product quadratic trigonometric B-splines and tensor-product quadratic B-splines defined on uniformly refined meshes. The error is computed for the displacement $\mathbf{w}$ of the exact solution in the L_2-norm, and the approximated value of $\sigma_{1,1}$ is given at the upper side of the hole.

Example 2, uniform mesh refinements

dof	TRG$_2$ BSP			Quadratic BSP		
	error of $\mathbf{w}$	$\sigma_{1,1}(0,1)$	cond	error of $\mathbf{w}$	$\sigma_{1,1}(0,1)$	cond
96	4.421 10^{-5}	26.722	25.02	8.564 10^{-5}	26.931	23.43
300	8.881 10^{-6}	28.921	33.24	1.893 10^{-5}	28.243	33.02
1044	2.612 10^{-6}	29.502	151.81	5.563 10^{-6}	28.905	151.35
3876	9.341 10^{-8}	29.996	693.45	1.371 10^{-7}	29.802	691.34

Table 7. The results for Example 2 using quadratic trigonometric B-splines and quadratic B-splines defined on the hierarchical meshes shown in Figure 10. The error is computed for the displacement $\mathbf{w}$ of the exact solution in the L_2-norm, and the approximated value of $\sigma_{1,1}$ is given at the upper side of the hole.

Example 2, hierarchical mesh refinements

dof	TRG$_2$ BSP			Quadratic BSP		
	error of $\mathbf{w}$	$\sigma_{1,1}(0,1)$	cond	error of $\mathbf{w}$	$\sigma_{1,1}(0,1)$	cond
96	4.421 10^{-5}	26.722	25.02	8.564 10^{-5}	26.931	23.43
230	5.025 10^{-6}	28.993	31.42	1.423 10^{-5}	28.762	29.34
574	6.703 10^{-7}	29.745	101.04	1.758 10^{-6}	29.243	96.13
1122	8.052 10^{-8}	30.012	245.25	2.132 10^{-7}	29.962	231.81

This distribution will alter only in the vicinity on the hole. More precisely, we get a peak stress concentration at the upper side of the hole: the stress component $\sigma_{1,1}$ takes the value 30 at the point $(0,1)$.

To exactly represent the geometry we construct a global geometry function as in (15) considering the tensor-product space $\mathbb{TS}^2_{\Xi,\alpha} \otimes \mathbb{S}^2_{\Xi}$ with $\alpha_i = \frac{\pi}{2}$. The control points $\mathbf{P}_i = \mathbf{P}_{k,j}$, $k,j = 1,2,3$, are depicted in Table 5 for a coarse grid consisting of one interval per edge. Then we approximate both components of the displacement in the bivariate tensor-product spaces $\mathbb{TS}^2_{\Xi,\alpha} \otimes \mathbb{S}^2_{\Xi}$ (TRG$_2$ BSP) with $\alpha_i = \frac{\pi}{2}$. Classical quadratic B-splines $\mathbb{S}^2_{\Xi} \otimes \mathbb{S}^2_{\Xi}$ (quadratic BSP) have also been considered for the sake of comparison. Next, we also consider bivariate spaces spanned by generalized multilevel bases built from the same tensor-product spaces mentioned before, defined on the hierarchical meshes shown in Figure 10. Note that all the considered spaces are nested. We compute in all cases

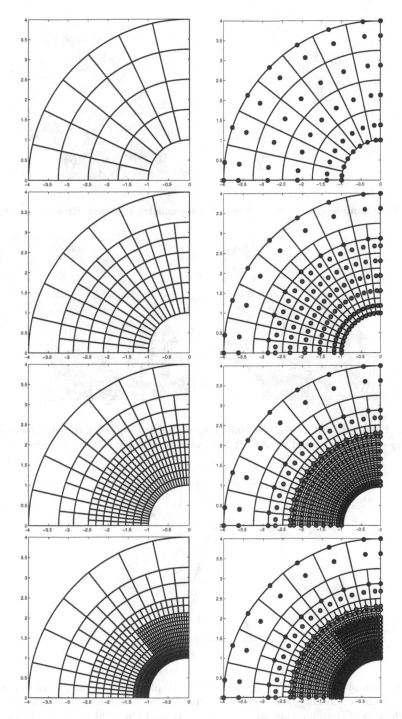

Fig. 10. Example 2. Hierarchical meshes (left) and Greville abscissas for quadratic tensor-product B-splines (right) for four refinement levels.

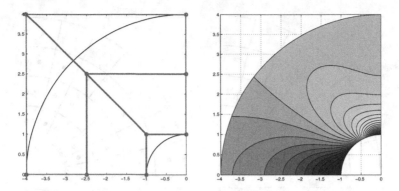

Fig. 11. Example 2. Left: control net of the geometry function. Right: exact $\sigma_{1,1}$.

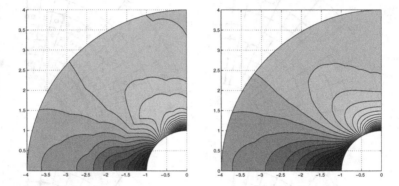

Fig. 12. Example 2. Multilevel quadratic trigonometric B-spline approximations of $\sigma_{1,1}$ with $\alpha_{i,\ell} = \pi/2$ on the first mesh (left) and last mesh (right) shown in Figure 10.

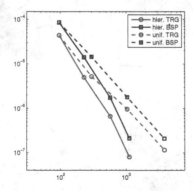

Fig. 13. Example 2. L_2-norm of the error for the displacement using quadratic trigonometric B-splines (red line) and classical quadratic B-splines (blue line) versus the number of degrees of freedom, computed on locally refined hierarchical meshes (solid) and uniformly refined meshes (dashed).

the condition number of the corresponding stiffness matrices in the 2-norm and the error with respect to the exact displacement in the L_2-norm.

Table 6 shows the results obtained on uniform refinements of the initial rectangular mesh, see Figure 10 (top). Table 7 reports the corresponding results obtained in the locally refined multilevel space setting using the meshes given in Figure 10. Figure 13 shows the reduction of the computed error versus the number of degrees of freedom, in the cases of uniform and local hierarchical refinements with the different considered spaces. Finally, Figure 12 shows the computed stress component $\sigma_{1,1}$ on the hierarchical meshes as in Figure 10.

Other results with different spaces and local refinement strategies for this problem can be found in [11,31,33].

5 Closure

Generalized B-splines have been used as an alternative to standard NURBS as an efficient problem-oriented tool in isogeometric analysis. However, they share with NURBS a tensor-product structure. This prevents the possibility of local refinements, which are a crucial ingredient for obtaining, in an efficient way, an accurate solution of partial differential equations.

Multilevel bases – originally introduced for tensor-product B-splines – offer a natural structure to obtain local refinements and can be extended to tensor-product generalized B-splines with suitable section spaces.

Moreover, multilevel bases can be constructed from not necessarily nested sequences of spaces. Combining this with the flexibility of generalized B-splines results in a variety of generalized multilevel spaces, whose characteristics can be tuned according to the considered problem.

In particular, generalized multilevel spaces based on variable degree B-splines present interesting performances because they couple the ability of modeling sharp features of exponential B-splines, with the low computational cost and stable evaluation algorithms of classical algebraic B-splines.

Of course, if non-nested spaces are considered to build the hierarchy, finer levels do not result in a nested sequence of (multilevel) discretization spaces. Nested spaces are not a necessary ingredient in IgA, see for example [31,32]. Nevertheless, nested hierarchical structures can be obtained with generalized B-splines as well, whenever preferred. In this case, at each level of the hierarchy, generalized B-spline spaces defined over nested sequences of knots and with sections in nested spaces have to be considered.

Finally, we have focused on local mesh refinements (h-refinements), but as for classical B-splines hierarchical p- and k-refinements can be considered as well. Since the section spaces (2) support degree-raising, nested hierarchical structures can be obtained also in this case if preferred.

References

1. Bank, R.E., Smith, R.K.: A posteriori error estimates based on hierarchical bases. SIAM J. Numer. Anal. 30, 921–935 (1993)
2. Bazilevs, Y., Calo, V.M., Cottrell, J.A., Evans, J.A., Hughes, T.J.R., Lipton, S., Scott, M.A., Sederberg, T.W.: Isogeometric analysis using T-splines. Comput. Methods Appl. Mech. Engrg. 199, 229–263 (2010)
3. de Boor, C.: A Practical Guide to Splines, revised edn. Springer (2001)
4. Carnicer, J.M., Mainar, E., Peña, J.M.: Critical length for design purposes and Extended Chebyshev spaces. Constr. Approx. 20, 55–71 (2004)
5. Costantini, P.: Curve and surface construction using variable degree polynomial splines. Comput. Aided Geom. Design 17, 419–446 (2000)
6. Costantini, P., Lyche, T., Manni, C.: On a class of weak Tchebycheff systems. Numer. Math. 101, 333–354 (2005)
7. Costantini, P., Manni, C.: Geometric construction of generalized cubic splines. Rend. Matem. Appl. 26, 327–338 (2006)
8. Costantini, P., Manni, C., Pelosi, F., Sampoli, M.L.: Quasi-interpolation in isogeometric analysis based on generalized B-splines. Comput. Aided Geom. Design 27, 656–668 (2010)
9. Cottrell, J.A., Hughes, T.J.R., Bazilevs, Y.: Isogeometric Analysis: Toward Integration of CAD and FEA. John Wiley & Sons (2009)
10. Dokken, T., Lyche, T., Pettersen, K.F.: Polynomial splines over locally refined box-partitions. Comput. Aided Geom. Design 30, 331–356 (2013)
11. Dörfel, M., Jüttler, B., Simeon, B.: Adaptive isogeometric analysis by local h-refinement with T-splines. Comput. Methods Appl. Mech. Engrg. 199, 264–275 (2010)
12. Forsey, D.R., Bartels, R.H.: Hierarchical B-spline refinement. Comput. Graph. 22, 205–212 (1988)
13. Giannelli, C., Jüttler, B., Speleers, H.: THB-splines: the truncated basis for hierarchical splines. Comput. Aided Geom. Design 29, 485–498 (2012)
14. Giannelli, C., Jüttler, B., Speleers, H.: Strongly stable bases for adaptively refined multilevel spline spaces. Adv. Comput. Math. (to appear, 2013)
15. Gould, P.L.: Introduction to Linear Elasticity. Springer, Berlin (1999)
16. Grinspun, E., Krysl, P., Schröder, P.: CHARMS: a simple framework for adaptive simulation. ACM Trans. Graphics 21, 281–290 (2002)
17. Hughes, T.J.R., Cottrell, J.A., Bazilevs, Y.: Isogeometric analysis: CAD, finite elements, NURBS, exact geometry and mesh refinement. Comput. Methods Appl. Mech. Engrg. 194, 4135–4195 (2005)
18. Koch, P.E., Lyche, T.: Interpolation with exponential B-splines in tension. In: Farin, G., Hagen, H., Noltemeier, H., Knödel, W. (eds.) Geometric Modelling, pp. 173–190. Springer (1993)
19. Kraft, R.: Adaptive and linearly independent multilevel B-splines. In: Le Méhauté, A., Rabut, C., Schumaker, L.L. (eds.) Surface Fitting and Multiresolution Methods, pp. 209–218. Vanderbilt University Press, Nashville (1997)
20. Krysl, P., Grinspun, E., Schröder, P.: Natural hierarchical refinement for finite element methods. Int. J. Numer. Meth. Eng. 56, 1109–1124 (2003)
21. Kvasov, B.I., Sattayatham, P.: GB-splines of arbitrary order. J. Comput. Appl. Math. 104, 63–88 (1999)
22. Mainar, E., Peña, J.M., Sánchez-Reyes, J.: Shape preserving alternatives to the rational Bézier model. Comput. Aided Geom. Design 18, 37–60 (2001)

23. Manni, C., Pelosi, F., Sampoli, M.L.: Generalized B-splines as a tool in isogeometric analysis. Comput. Methods Appl. Mech. Engrg. 200, 867–881 (2011)
24. Manni, C., Pelosi, F., Sampoli, M.L.: Isogeometric analysis in advection-diffusion problems: tension splines approximation. J. Comput. Appl. Math. 236, 511–528 (2011)
25. Marušic, M., Rogina, M.: Sharp error bounds for interpolating splines in tension. J. Comput. Appl. Math. 61, 205–223 (1995)
26. Mazure, M.L.: Chebyshev-Bernstein bases. Comput. Aided Geom. Design 16, 649–669 (1999)
27. Mazure, M.L.: How to build all Chebyshevian spline spaces good for geometric design? Numer. Math. 119, 517–556 (2011)
28. Schumaker, L.L.: Spline Functions: Basic Theory, 3rd edn., Cambridge U.P. (2007)
29. Speleers, H., Dierckx, P., Vandewalle, S.: Quasi-hierarchical Powell-Sabin B-splines. Comput. Aided Geom. Design 26, 174–191 (2009)
30. Speleers, H., Dierckx, P., Vandewalle, S.: On the local approximation power of quasi-hierarchical Powell-Sabin splines. In: Dæhlen, M., Floater, M., Lyche, T., Merrien, J.-L., Mørken, K., Schumaker, L.L. (eds.) MMCS 2008. LNCS, vol. 5862, pp. 419–433. Springer, Heidelberg (2010)
31. Speleers, H., Manni, C., Pelosi, F.: From NURBS to NURPS geometries. Comput. Methods Appl. Mech. Engrg. 255, 238–254 (2013)
32. Speleers, H., Manni, C., Pelosi, F., Sampoli, M.L.: Isogeometric analysis with Powell-Sabin splines for advection-diffusion-reaction problems. Comput. Methods Appl. Mech. Engrg. 221–222, 132–148 (2012)
33. Vuong, A.-V., Giannelli, C., Jüttler, B., Simeon, B.: A hierarchical approach to adaptive local refinement in isogeometric analysis. Comput. Methods Appl. Mech. Engrg. 200, 3554–3567 (2011)
34. Wang, G., Fang, M.: Unified and extended form of three types of splines. J. Comput. Appl. Math. 216, 498–508 (2008)

Exploring Matrix Generation Strategies
in Isogeometric Analysis

Angelos Mantzaflaris[1] and Bert Jüttler[2]

[1] RICAM, Austrian Academy of Sciences, Linz, Austria
Angelos.Mantzaflaris@oeaw.ac.at
[2] AG, Johannes Kepler University, Linz, Austria
Bert.Juettler@jku.at

Abstract. An important step in simulation via isogeometric analysis (IGA) is the *assembly step*, where the coefficients of the final linear system are generated. Typically, these coefficients are integrals of products of shape functions and their derivatives. Similarly to the finite element analysis (FEA), the standard choice for integral evaluation in IGA is Gaussian quadrature. Recent developments propose different quadrature rules, that reduce the number of quadrature points and weights used. We experiment with the existing methods for matrix generation. Furthermore we propose a new, quadrature-free approach, based on interpolation of the geometry factor and fast look-up operations for values of B-spline integrals. Our method builds upon the observation that exact integration is not required to achieve the optimal convergence rate of the solution. In particular, it suffices to generate the linear system within the order of accuracy matching the approximation order of the discretization space. We demonstrate that the best strategy is one that follows the above principle, resulting in expected accuracy and improved computational time.

Keywords: isogeometric analysis, stiffness matrix, mass matrix, numerical integration, quadrature.

1 Introduction

The advent of IGA by Hughes et al. [11] has motivated new approaches to the entire process of simulation and numerical solving of partial differential equations (PDEs). The benefits of the isogeometric paradigm include the exact representation of the geometry by using flexible B-spline representations as a basis for analysis. In this realm, the whole of the analysis process is revisited to exploit the new possibilities. Lately special focus has been given to the matrix generation step, since it is one of the sub-processes that is likely to admit considerable improvement in this new analysis environment. Indeed, e.g. in [8] the authors perform simulations on the deformation of turbine blades using both IGA and FEA and conclude that even though IGA has a clear advantage regarding the number of degrees of freedom, matrix generation (by means of quadrature) constitutes a bottleneck in the overall running times.

M. Floater et al. (Eds.): MMCS 2012, LNCS 8177, pp. 364–382, 2014.

During the analysis process, several approximate computational steps are executed, while computing an unknown field over the given geometry. Typically, given a geometry (or physical domain) and a boundary value problem, the unknown solution field is projected onto a finite-dimensional sub-space, i.e. we restrict ourselves to finding a solution in that space. Then a linear system is generated, consisting of a matrix with e.g. mass, stiffness terms, as well as a load vector containing the moments with respect to the right-hand side. The solution of the resulting linear system yields the coefficients of the unknown field in the chosen discretization space. In each of these steps, errors are introduced and accumulate in the final solution. In most cases the principal error sources during the process are the discretization error coming from projection of the solution and the integration error made in the generation step.

Typically, the discretization of a differential equation leads to matrices with entries being integrals of products of shape functions and their derivatives. These integrals over elements in the physical domain are transformed to integrals over the support of the basis functions, resulting in integrands involving the (inverse of the) Jacobian of the geometry map. The most we can hope for is a good approximation of these quantities, since the integrals of rational functions in the best case lead to non-rational expressions.

When it comes to convergence, a main parameter is the order of accuracy of the entire process. We shall confirm that a minimal order of accuracy has to be maintained throughout the analysis pipeline in order to obtain the expected convergence. Similarly, an intermediate step with a higher order of convergence is unnecessary, since a current super-convergence is likely to be canceled by a subsequent step.

Numerical integration by use of evaluations of the integrand alone is often referred to as quadrature in one dimension and as cubature in higher dimensions. The problem of deriving quadrature rules for integrals involving B-splines was first considered over thirty years ago. Indeed, in [10] the authors computed rules for the moments of (linear, quadratic and cubic) B-spline functions, in order to solve a parabolic PDE using Galerkin's method. The interest in the topic is revived lately, after the introduction of IGA.

In [12] the authors present optimal quadrature rules for the mass and stiffness of uniform B-spline discretizations, i.e. rules with the minimum number of nodes that are exact for the product of two B-splines, upto a fixed degree. The number of nodes (points) plus the number of weights in this minimal rule coincides with the dimension of the spline space of integrands, and this is why it is known as the *half-point rule*. The optimal rule is defined over the whole domain of the B-spline space, and the computation of the nodes and weights leads to a global, non-linear system of equations, which is tackled with a Newton iteration. This limits the practical ability to derive of the rule to small degree and to small number of elements. The authors anticipate this constraint by splitting big domains into macro-elements, thus resorting to a non-optimal strategy.

In [1] the spline space of the product of two uniform B-spline basis functions is further investigated, in order to produce a feasible, computable rule. The basis functions are grouped with respect to the size of their support. In particular, basis functions have support over at most two elements and are translates of a small group of distinct basis functionals. This allows to derive a rule which is defined over one or two elements, and can be obtained as the solution of a "local" non-linear system that, unlike [12], does not depend on the number of elements.

An experimental study of the Gauss rule, and the optimal rule on macro-elements of [12] is done in the recent work [15]. They perform experiments on a Poisson problem over a domain given by the identity mapping, with a unit Jacobian determinant. Their focus is on the degree of exactness of different rules as well as their practical computational cost. Since the parameterization is the identity, the shape functions are simply B-splines, therefore exact evaluation of the stiffness matrix is feasible when using quadrature rules that integrate exactly the respective integrands.

Throughout this paper we consider uniform knot vectors; we note that any mesh can be properly refined so that it becomes uniform almost everywhere. We focus on the univariate case, since for higher dimensions the tensor-product structure allows re-using the same technique coordinate-wise.

We set up a model Poisson problem, and use it firstly to briefly review the different available approaches for matrix generation in IGA. We elaborate on a new approach based on (quasi-)interpolation of the geometry factor in the integrand. An ingredient needed for our method is the exact evaluation of integrals of tri-products of B-splines, which can be done symbolically. We experiment with the different approaches, and demonstrate that the requirement for a method having high degree of exactness is not crucial, in the sense that this exactness does not propagate to the final solution, since the accuracy of the final solution is limited by the discretization error. Instead we verify that it suffices to adopt a method whose accuracy matches the discretization error, in order to maintain all essential information that is contained in the stiffness matrix regarding the problem. The proposed quadrature-free method has the above property, while avoiding the use Newton iteration for deriving quadrature nodes and weights. Contrarily to high-accuracy quadrature it requires less evaluations, therefore it partially overcomes a common bottleneck in terms of computational cost.

In the next section we describe the model problem, its discretization and the expected numerical error. Then we look at different quadrature-based assembly strategies in Section 3 and we introduce our method in Section 4. Experimental results are presented in Section 5 and short conclusions follow in Section 6.

2   The Model Problem

In this section we present the model problem that is used to present the different assembly methods and perform experiments in Section 5.

We consider a homogeneous bar of length $L = 5$, subject to a distributed load $f(x)$ that is acting along the $x-$axis (Figure 1). The longitudinal displacement $u(x)$ that is produced by the force is the solution of the one-dimensional Poisson

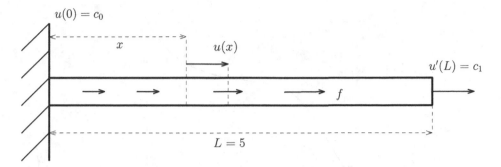

Fig. 1. A homogeneous bar problem with zero-displacement boundary on the left. A horizontal line force f is applied and an end condition.

equation

$$-u''(x) = f(x) \quad , \quad x \in \Omega, \tag{1}$$

with physical domain being a real interval $\Omega = [0, L]$. For an isogeometric model, we parameterize the bar by a B-spline geometry map $G : [0, 1] \rightarrow \Omega$,

$$G(t) = \sum_{i=0}^{n} g_i N_{i,p}(t),$$

supported on a uniform, open knot vector (Figure 2). The B-Spline basis functions $N_{i,p}$ are piece-wise polynomials of degree p, and have continuity C^{p-1} across the interior knots, provided that the knot vector has only simple knots. We refer the reader to standard textbooks, e.g. [6] for an introduction to spline theory. For parameterizing this one-dimensional problem, an identity map would be the best choice. However, when we use the tensor-product of univariate B-spline spaces for 2D or 3D problems, a linear geometry map for non-trivial geometries is no longer possible. Our aim is to simulate this fact, and therefore we shall consider non-trivial mappings (of several degrees) for the bar. In particular, the integrands that we aim at treating are rational functions, so that we shall access the full effects of a geometry mapping on the simulation process.

We impose a Dirichlet boundary condition on the left end and a Neumann condition on the right end of the bar,

$$u(0) = c_0 \quad \text{and} \quad u'(L) = c_1, \tag{2}$$

where c_0, c_1 are constants. The physical interpretation is that we have an initial displacement c_0 in the fixed end of the bar, and we know the magnitude c_1 of the force acting at the free end.

First we derive the weak formulation. We multiply (1) by a test function $v(x)$, and after integration by parts we get

$$\int_\Omega u'(x)v'(x)dx = \int_\Omega v(x)f(x)dx + c_1 v(1) \quad \text{or} \quad a(u,v) = \langle f, \phi_i \rangle + \phi_i(1)c_1 , \tag{3}$$

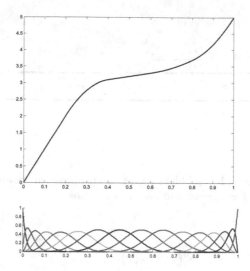

Fig. 2. Parameterization of the bar using B-splines of degree 6

using trial and test spaces

$$\mathcal{U} = \{u \ : \ u(0) = c_0\} \quad \text{and} \quad \mathcal{V} = \{v \ : \ v(0) = 0\},$$

for u and v respectively. One can show that $a(\cdot, \cdot)$ is bi-linear, symmetric, co-ercive, and bounded. Hence, the theory of abstract boundary value problems can be applied to show the existence of a unique weak solution in the infinite-dimensional Hilbert space $H^1(\Omega)$.

2.1   Discretization

We restrict test functions and trial solutions to the push forward of the (finite dimensional) B-spline space $V^h \subset H^1(\Omega)$, i.e. we apply Galerkin approximation:

$$v^h(x) = \sum_{i=1}^{n} v_i \phi_i(x) \in \mathcal{V}^h \quad \text{and} \tag{4}$$

$$u^h(x) = \sum_{i=1}^{n} u_i \phi_i(x) + c_0 \phi_0(x) \in \mathcal{U}^h, \tag{5}$$

with $\phi_i = N_{i,p} \circ G^{-1}$ and $\phi_i(0) = 1$ for $i = 0$, and 0 otherwise. Using the weak form (3) we arrive at a linear system of equations

$$\sum_{j=0}^{n} u_i a(\phi_i, \phi_j) = \ell(\phi_i) + \phi_i(1)c_1 - c_0 a(\phi_i, \phi_0) \quad , \quad i = 1, \dots, n \tag{6}$$

with $a(\phi_i, \phi_j) := \int_\Omega \phi_i'(x)\phi_j'(x)dx$ and $\ell(\phi_i) := \langle f, \phi_i \rangle = \int_\Omega \phi_i(t)f(t)dx$. The coefficients of the solution u^h are described by the linear system

$$Ku = b, \qquad (7)$$

where $K_{ij} = a(\phi_i, \phi_j)$, $b_i = \langle f, \phi_i \rangle + \phi_i(1)c_1 - c_0 a(\phi_i, \phi_0)$ and u stands for the vector of unknown coefficients $(u_1, \ldots, u_n)$. Plugging in B-splines of degree p, we arrive at the stiffness entries

$$K_{ij} = a(\phi_i, \phi_j) = \int_\Omega \phi_i'(x)\phi_j'(x)dx = \int_{\Omega_0} N_{i,p}'(t)N_{j,p}'(t)\frac{dt}{G'(t)}, \qquad (8)$$

whereas the right-hand side involves the inner product

$$\ell(\phi_i) = \langle f, \phi_i \rangle = \int_\Omega f(x)\phi_i(x)\,dx = \int_{\Omega_0} f(G(t))N_{i,p}(t)|G'(t)|\,dt . \qquad (9)$$

Finally, it is useful to mention the mass term, that may also appear in the variational form,

$$\langle \phi_i, \phi_j \rangle = \int_{\Omega_0} \phi_i(x)\phi_j(x)\,dx = \int_\Omega N_{i,p}(t)N_{j,p}(t)|G'(t)|\,dt . \qquad (10)$$

2.2 A Priori Error Considerations

In this section we recall some facts regarding the numerical error that is expected to appear during the analysis pipeline. Looking at every individual step of the process, we see that the following error factors are likely to occur:

1. The *approximation error* for the geometry representation, e.g. in case of polynomial B-spline discretization of planar domains with circular boundary, or the use of linear splines on domains with curved boundaries.
2. The *discretization error*, coming from the projection of the unknown solution field onto the solution space V^h.
3. The error coming from interpolating non-constant Dirichlet boundary conditions.
4. The numerical integration or *consistency error* appearing during the generation of K and b in (7), involving integrals of products of shape functions and moments of the source function respectively.
5. The possible numerical inaccuracy introduced when solving the linear system (7).

Therefore, the quality of the final solution is determined as a combination of all intermediate strategies chosen for these steps.

1. Geometry approximation error in the context of IGA can be safely assumed zero, since we shall be using for analysis (a refinement of) the same basis in which the geometry is given in the first place.

2. Discretization error is of order $p + 1$ if measured in the L^2−norm, when B-splines of degree p are used [2,21]. Consequently, this is the optimal convergence rate that we expect to obtain in the final solution.

3. The error factor coming from incorporating Dirichlet boundary is usually zero, e.g. for constant or polynomial conditions on the boundary; in particular it is always zero in the 1D case we are considering [3].

4. The consistency error influences greatly the quality of the numerical solution, since the computed values of the bi-linear form $a(\cdot, \cdot)$ and inner product $\langle \cdot, \cdot \rangle$ can deviate significantly from theoretical values. Indeed, the consistency error stems from approximating the stiffness matrix and the load vector by some computed (perturbed) versions $a_h \cong a$ and $\ell_h \cong \ell$, e.g. we actually compute $\tilde{K}_{ij} = a_h(\phi_i, \phi_j)$ instead of (7). These quantities are in general rational and could be evaluated exactly only if the geometry transformation is linear (i.e. the Jacobian determinant is a constant) and the PDE has constant coefficients.

Consequently, when approximating the (matrix of the) bilinear form a by a numerically computed a_h, we replace the original problem (3) by the perturbed one $a_h(u_h, v_h) = \ell_h(v_h)$. The effect of this perturbation on the solution of the original problem is captured by Strang's first lemma Strang's first lemma [16,17]:

$$\|u - u_h\|_V \leq C \left(\inf_{v_h \in V_h} \left\{ \|u - v_h\| + \sup_{w_h \in V_h} \frac{a(v_h, w_h) - a_h(v_h, w_h)}{\|w_h\|_V} \right\} + \right.$$
$$\left. + \sup_{w_h \in V_h} \frac{\ell(w_h) - \ell_h(w_h)}{\|w_h\|_V} \right), \tag{11}$$

with $a(\cdot, \cdot)$ and $\ell(\cdot)$ as in (6) and C being a constant that does not depend on the discretization step-size h.

The lemma describes (under reasonable assumptions, e.g. smoothness and V_h-ellipticity) the accumulation of discretization and consistency errors on the solution. It states that the variation of the computed solution is bounded by the a best approximation error $\inf_{v_h \in V_h} \|u - v_h\|$ and the consistency error present in the bi-linear form and load vector. We expect to observe the same type of behavior in our isogeometric setting.

5. Finally, the error introduced while solving the resulting linear system $\tilde{K}u = b$ is negligible, when iterative solvers are utilized. Indeed, the quality of the solution will depend on the number of iteration steps executed, therefore in practice at the point where we have generated the linear system we can approach its solution up to high precision using stable solvers with preconditioning and sufficiently many iteration steps. We also mention that the condition number of the stiffness matrix for uniform knot-meshes is known to be of order $O\left(h^{-2}\right)$ for mesh-size h, which is analogous to the case of the traditional finite element method (see [7] for more details).

3   Quadrature-Based Approaches

We now consider our main topic of interest, the assembly step. We need to compute the quantities (8)–(10) in order to form (7). In the present section we discuss existing quadrature methods in order to prepare the ground for a quadrature-free approach that follows right after.

As already mentioned, we focus on the 1D case, since the derivation of quadrature rules for 2D or 3D is done by taking the tensor product of univariate rules. We introduce two notations; we denote by $\mathbb{P}^m$ the space of polynomials of degree m and by $\mathbb{S}_q^m$ the space of piece-wise polynomial (spline) functions of degree m and continuity q at the knots.

Numerical integration is typically based on a quadrature (or, in higher dimension, cubature) formula:

$$\int_{-1}^{1} g(x)\,dx = \sum_{i=1}^{r} w_i g(x_i) + e_g \ , \tag{12}$$

with weights w_i, nodes (or integration points) x_i, and error term e_g. A quadrature formula is specified by providing weights w_i and nodes x_i for the integration domain $[-1, 1]$. Then the weights and nodes can be mapped to any other interval with a linear change of variables.

When designing or choosing quadrature rules for a problem, an important parameter is the trade-off between the number of evaluations of the integrand used and the quality of the approximation of the integral in question. In the frame of IGA, locality of the support of basis functions favors quadrature approaches: at a given point, it suffices to evaluate only those functions whose support contains the point.

3.1   Gauss Quadrature

The most commonly used quadrature rule is the Gauss integration rule [18]. We shall briefly describe its advantages.

We define the degree of exactness (or algebraic precision) of a rule to be equal to p, if all polynomial functions of degree at most p are exactly integrated by the rule, i.e. in (12) $e_g = 0$ for all g in the polynomial space $\mathbb{P}^p$ and there exists $g \in \mathbb{P}^{p+1}$ such that $e_g \neq 0$. Under certain smoothness assumptions on the integrand, applying a rule of degree of exactness p guarantees an approximation error of $p + 1$, where p is its degree of exactness. A formula of the form (12) can be chosen to be exact on a polynomial space of degree p if it has at least $p + 1$ "quadrature degrees of freedom", ie. $2r \geq p + 1$.

The nodes are the at roots of the r-th Legendre polynomial mapped onto the integration interval, and the weights follow from a simple formula. This choice of nodes (and weights) is minimal (or optimal) with respect to the degree of exactness; they provide the unique rule with r nodes that is exact on $\mathbb{P}^{2r-1}$.

The approximation power for sufficiently regular integrands is $2r$ [13]; this error bound is based on the fact that all the weights are positive. This means that if we are interested in approximating an integral within an order of accuracy $p+1$, then it suffices to evaluate the integrand on $r = \lceil (p+1)/2 \rceil$ Gauss nodes.

3.2 The Half-point Rule

The half-point rule of [12] is an attempt to specify minimal exact rules for $\mathbb{S}_r^m$, i.e. the analogue of Gauss rules for $\mathbb{P}^m$. For a spline space of dimension n, an optimal quadrature formula is one with $\lceil n/2 \rceil$ nodes. Note the limit case of discontinuous splines, where $\dim \mathbb{S}_{-1}^m = k \dim \mathbb{P}^m$, with k being the number of knot-spans; this equality implies that the optimal rule for $\mathbb{S}_{-1}^m$ is the Gauss rule, as expected. But if the continuity is bigger then the spline space dimension drops, and we should be able to do better.

The derivation of such rules requires solution of a global non-linear problem, expressing the exactness of the rule on the basis functions. Since the Newton solver is only applicable for a limited number of unknowns, this computation can only be carried out for small n, or equivalently for a small number of elements. However, the authors succeeded in using the rule in an non-optimal way by computing the node values for problems with a small number of elements and then tiling the rule along larger meshes, that is, they consider rules on so called macro-elements. The rules where derived numerically for B-spline discretizations of degree up to three. In addition, the node values for degree four are computed in [15]. Questions regarding uniqueness and stability of these numerically computed rules (for example weight positivity, approximation power) are still open.

3.3 A Local, Feasible Rule

In the recent work [1] the authors explore quadrature rules for specific product-spline spaces, where the stiffness or mass integrands belong to. The B-spline space (of higher degree) of the product of two uniform B-spline basis functions (or derivatives) is further investigated, in order to produce a feasible, computable rule. It is observed that the basis functions of the product-spline space $\mathbb{S}_{p-2}^{2p}$ are supported in at most two knot-spans, and the basis functions are grouped with respect to the size of their support.

Using the periodicity of uniform basis functions supported on one or two elements, they derive a system of equations expressing exactness on the basis, this time mapped back to a reference interval. This allows to setup a rule which can be obtained as the solution of a "local" non-linear system that, unlike [12], does not depend on the number of degrees of freedom. An additional system of equations is set up for deriving a rule for boundary basis functions, where multiple knots are present. The number of variables in these Newton systems depends on the degree and the smoothness, but not on the number of elements. Moreover, the number of quadrature nodes taken for this rule is close to the half-point rule. The quadrature nodes and weights can be computed for any degree using GeoPDEs [5].

4   Quadrature-free Assembly

In this section we explore a quadrature-free method for the assembly task. We shall replace the quadrature by an approximation, by means of interpolation or quasi-interpolation, of the part of the integrands (8), (9) or (10) that contains the geometry mapping G and its derivatives. This strategy aims at reducing the number of evaluations needed in quadrature-based approaches as well as avoiding the need to solve non-linear systems in order to derive quadrature rules. The proposed method consists in an initial approximation of the *geometry factor* that appears in the integrand (for example G' or $1/G'$) and a fast look-up operation for the resulting integrals, involving only products of B-splines. The idea is to approximate the integrand within the order of accuracy that matches the discretization error and consequently exploit the periodic nature of uniform B-spline bases. Similarly to quadrature-based approaches, with our approach the tensor product of 1D instances can be used to apply our technique to 2D or 3D patches. The only change is that the interpolation of the geometry factor needs to be carried out in higher dimension, which is also done by the tensor-product of 1D interpolation operators.

We shall use the stiffness term (8) for presentation purposes. The scheme is as follows:

1. First we approximate the geometry factor G by projecting it onto $\mathbb{S}^{p-1}_{p-2}$. Applying an interpolation operator Q to $1/G'$, we get

$$K_{ij} \cong \int_\Omega N'_{i,p} N'_{j,p} Q\left(\frac{1}{G'}\right) dt = \sum_{k=0}^{n} q_k \int_\Omega N'_{i,p} N'_{j,p} N_{k,p-1}\, dt . \tag{13}$$

   The geometry factor is thereafter expressed in the B-spline space and stiffness entries break down to a sum of tri-product integrals of B-splines.
2. Consequently we construct (or load) a look-up table $I^{p-1,p-1,p-1}_{i,j,k}$ for the integrals of tri-products $N_{i,p-1} \cdot N_{j,p-1} \cdot N_{k,p-1}$ of basis functions of $\mathbb{S}^{p-1}_{p-2}$ that appear in (13), after eliminating derivatives.
3. At this point we are able to assemble the matrix K by summing up contributions from the look-up tables.

A similar formula can be deduced for the load vector, by applying Q on $f \circ G$.

Note that the approximation of the "geometry factor" does not interfere with the exactness of the geometry representation, that is, the preservation of the boundary of the physical domain after discretization, as known in IGA. Indeed, this factor refers to the contribution of the integral transformation from the physical to the parameter domain and of possibly non-constant coefficients of the PDE. For the numerical evaluation of integrals of rational function approximating is inevitable, and usually some kind of interpolation takes place, e.g. polynomial interpolation in the case of Gaussian quadrature. In our approach we restrict this interpolation to the actual non-polynomial part of the integrand.

When using open knot vectors, a number of special B-Splines appear at the boundaries of the parametric domain, due to the multiplicity of the boundary knots. We may incorporate these cases in our setting by including the corresponding integral values in a lookup table that is used for all boundaries. Their values may be obtained exactly by means of Gaussian quadrature. In the one-dimensional case, however, we employ directly Gaussian quadrature for the two boundaries, since the potential savings in this case are negligible. Note that having only uniform B-Splines also on the boundaries would still need special treatment, since the integration domain would be in that case a genuine subset of the intersection of the supports of the B-Splines that are involved.

The ingredients required are a suitable interpolation operator Q and a look-up table for the integral of B-spline tri-products. We address these two requirements in the following paragraphs.

4.1   Approximating the Geometry Factor

The first ingredient of the proposed method is an approximation operator Q to be applied on the geometry factor denoted hereafter $\eta(t)$.

There are at least two options for Q; one can use a global interpolation scheme, that would require solving a linear system, or a local quasi-interpolation scheme [4,14]. The interpolation points used in Q are directly available, for instance one can use the Greville abscissae or any other point-set given by the quasi-interpolation scheme. This way we replace $\eta(t)$ by a B-spline function

$$Q\eta(t) = \sum_{i=1}^{n} \eta_i N_{i,m}(t) . \tag{14}$$

Strang-Fix conditions (cf. [17]) hold for the B-spline basis, which implies that quasi-interpolation can be applied to approximate η within order $m + 1$ when $\mathbb{S}_{m-1}^m$ is used for interpolation, that is, halving the knot-spans should decrease the error by 2^{m+1} .

The geometry map (and the geometry factor) is almost everywhere smooth except from the knots of the coarse mesh. Therefore approximation on the fine grid is expected to behave nice; this is confirmed by the experiments in Section 5. Since in the case of the stiffness matrix the denominator of $\eta(t)$ is a B-Spline function of degree $p-1$, we choose $m = p-1$ to match the continuity of η at the knots. Another strategy is to add double knots and use $\mathbb{S}_{p-2}^p$, but experiments indicate that the degree $m = p - 1$ is sufficient.

4.2   Exact Integrals of Products of B-Splines

The scheme requires look-up tables for the integral over their support of product of two or three B-spline functions. For this, we consider uniform splines of degree p over the infinite knot vector $h\mathbb{Z}$, for some length $h \in \mathbb{R}$. Note that all formulas presented in this section refer to B-Splines without repeated knots; integrals that involve "boundary" B-Splines, e.g. the few ones appearing at the two endpoints of the basis shown in Figure 2 will need special care and are not covered.

The computation of integrals of inner products of B-splines is studied in [19]. Furthermore, a formula for the inner product $\langle N_{i,p}(t), N_{i+j,p} \rangle$ of two uniform B-splines of the same degree (i.e. entries of the Gramian matrix), the second being a shift by j knots, seems to be known to the B-spline community, even though we were not able to locate a proof of that formula in the literature. More generally, for different degrees we discovered that the integral (Figure 3 left):

$$\int_0^{mh} N_{i,m}(t)N_{i+j,n}(t)\, dt = hI_j^{m,n} \quad , \quad j = 1-n, \ldots, m-1 , \tag{15}$$

where $I_j^{m,n}$ corresponds to $h = 1$ can be computed explicitly by the formula

$$I_j^{m,n} = \frac{E(m+n-1, n+j-1)}{(m+n-1)!} , \tag{16}$$

for all $j = 1-n, \ldots, m-1$. Here $E(i,j)$ denotes the so-called *Eulerian numbers* (cf. [20]):

$$E(i,j) = \sum_{k=0}^{j} (-1)^k \binom{i+1}{k}(j-k+1)^i , \quad i,j \in \mathbb{Z} .$$

Eulerians can be computed by the recursion:

$$E(i,j) = (i-j)E(i-1,j-1) + (j+1)E(i-1,j),$$

which is in fact quite close to the B-spline recursion. The relation between Eulerian numbers and B-splines seems to be a deep one, see [9,20] for more information. By symmetry $I_j^{p,p} = I_{-j}^{p,p}$, therefore assembling the Gramian matrix of uniform B-splines involves essentially computing p distinct integrals, namely $I_0^{p,p}, I_1^{p,p}, \ldots, I_{p-1}^{p,p}$.

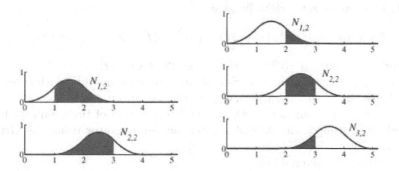

Fig. 3. Overlap of the support of 2 and 3 uniform B-splines after shifting

For the tri-product integral we write down the factors in terms of shifts, and analogously to (15) we have (Figure 3 right):

$$\int_0^{mh} N_{i,m}(t)N_{i+j,n}(t)N_{i+j+k,p}(t)\, dt = hI_{j,k}^{m,n,p}. \tag{17}$$

In this case there is no closed formula available, however, since these are just integrals of piece-wise polynomials, we can compute their values by an exact Gauss rule in every knot-span, or by using symbolic integration. A closed formula for $I_{j,k}^{m,n,p}$ might also exist; nevertheless up to now computing these values symbolically has not been a problem. In particular, a small Mathematica worksheet can produce the values for degrees up to 10 in less than one second. Table 1 provides these data for degrees 2 to 4.

Table 1. Values for B-spline tri-products for B-spline degrees 2 to 4. The rows correspond to the degree and the values correspond to $I_{j,k}^{p,p,p}$.

j,k	0,0	0,1	0,2	0,3	0,4	1,1	1,2	1,3	2,2
2	$\frac{12}{35}$	$\frac{43}{420}$	$\frac{1}{840}$	0	0	$\frac{1}{168}$	0	0	0
3	$\frac{1979}{7560}$	$\frac{18871}{181440}$	$\frac{31}{6480}$	$\frac{1}{181440}$	0	$\frac{85}{6048}$	$\frac{17}{181440}$	0	0
4	$\frac{4393189}{20756736}$	$\frac{3465461}{34594560}$	$\frac{129119}{14152320}$	$\frac{13411}{155675520}$	$\frac{1}{88957440}$	$\frac{6474701}{311351040}$	$\frac{376723}{622702080}$	$\frac{349}{622702080}$	$\frac{251}{155675520}$

5   Experimental Results

We use the problem of Section 2, considering a bar described by the geometry map $G : [0,1] \to [0,5]$, using B-splines of degrees up to 10, and exact solution

$$u = \left(7t + 2t^2 - 3t^3\right) \sin(t) \cos(t).$$

For this solution we get a right-hand side

$$f = \left(18t^2 - 8t - 14\right) \cos(2t) - \left(6t^3 - 4t^2 - 23t + 2\right) \sin(2t),$$

an boundary conditions $u(0) = 0$, $u'(1) = \sin(2) + 6\cos(2)$.

The control points for the experiments are chosen randomly, with some care to avoid singular parametrizations, e.g. for degree 6 we used the geometry function of Figure 2. For our new method, the interpolation of the geometry factor was performed using a simple global interpolation operator using the Greville abscissae as interpolation points.

The methods considered are

- Gauss(m): Gauss rule with m nodes.
  We take $m = p + 1$ ("full" Gauss), as well as $m = p - 1, p - 2, p - 3$.
- ACHRS: The exact integration rule of [1].
- Quadrature-free: The method of Section 4.

Full Gauss quadrature Gauss($p + 1$), is the exact rule for mass and stiffness integrands, since they have degree upto $2p$. Also, the Gauss rule matching the approximation order of the discretization has $p - 1$ quadrature points. We choose

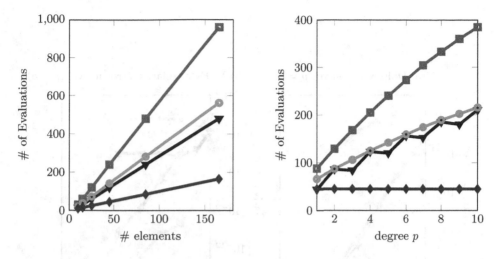

Fig. 4. Number of evaluations for fixed degree $p = 5$ (left) and for fixed number of elements (right). Legend: ● ACHRS, ■ Gauss($p + 1$), ▼ Gauss($p - 1$), ♦ Quad-free.

to employ the ACHRS rule of [1] in our experiments, since it is defined for any degree, and an implementation to derive the rules is available in GeoPDEs [5]. Lastly, we test the strategy presented here-in using interpolation of a part of the integrand.

Convergence depends greatly on the assembly of the load vector. To access the effect of the quadrature on the stiffness matrix alone, we used a high precision rule for the load vector in all experiments. We present plots for degrees 5 and 6 in figures 5 and 6 respectively. We have experimented with degrees up to 10, and in all cases the results follow the same pattern.

Our first task is to investigate the order of convergence and the *convergence threshold* using different assembly strategies. Monotonic convergence of order 6 (resp. 7) for degree 5 (resp. 6) of the overall error is confirmed for Gauss($p + 1$), ACHRS and Quadrature-free, as seen in 5(a) and 6(a). Since the discretization error is known, the result we get is in agreement with the prediction of Strang's lemma (11). Note that using Gauss quadrature with $\lceil (p + 1)/2 \rceil$ points or less, the order or convergence drops, as expected.

The second experiment studies the consistency error. We look at a central entry of the stiffness matrix in terms of numerical accuracy, and the results are presented in figures 5(b) and 6(b). Interestingly, the accuracy in terms of the relative error, in which we estimate the bi-linear form $K_{ij} = a(\phi_i, \phi_j)$ using Gauss($p-1$), Gauss($p - 2$) or Gauss($p - 3$) remains constant under h−refinement. This can be justified by the fact that refinement shrinks both the integrand and the integration interval, i.e. this is not an adaptive quadrature, where the integrand is fixed and the integration interval is subdivided. The rule ACHRS behaves very similarly to full Gauss quadrature with $p + 1$ points. Using quadrature-free assembly, we gain adaptivity, since the function that is approximated is fixed

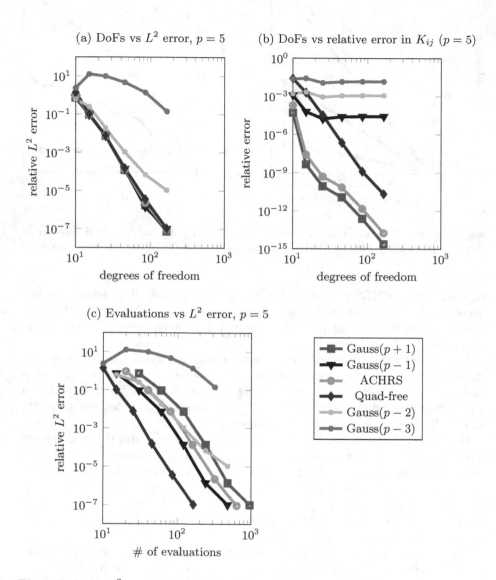

Fig. 5. Relative L^2 error of the solution (a) and relative numerical error of an entry of K (b) plotted against the DoFs of the problem. In (c), we plot the relative L^2 error of the solution against the number of evaluations performed during stiffness generation, for different strategies. Discretization is done using uniform B-splines of degree 5.

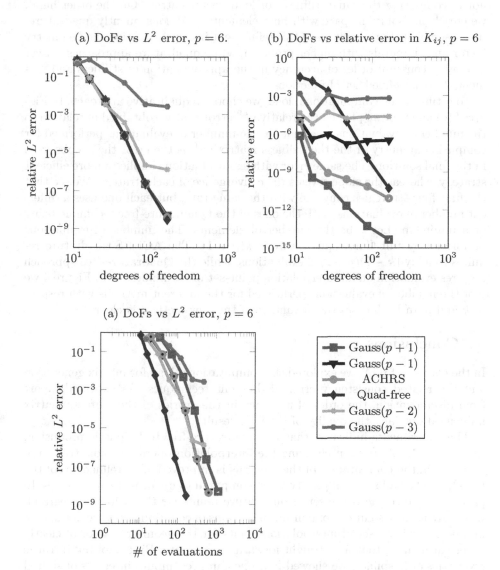

Fig. 6. Relative L^2 error of the solution (a) and relative numerical error of an entry of K (b) plotted against the DoFs of the problem. In (c), we plot the relative L^2 error of the solution against the number of evaluations performed during stiffness generation, for different strategies. Discretization is done using uniform B-splines of degree 6.

(because the geometry map is fixed), therefore refinement improves the approximation accuracy in the stiffness matrix. Indeed, when refining the mesh, we are not re-computing the same stiffness, or load vector entries. On the other hand, we enrich the solution space with finer elements, and consequently quadrature-free assembly takes advantage of this refinement in approximating the (geometry factor and) integrals with higher precision, while quadrature approaches deliver a roughly constant order of accuracy in the approximation of stiffness and moments for the refined set of functions.

To estimate the computational load, we choose a qualitative approach. In Figures 5(c) and 6(c) we plotted the relative L^2 error that we obtained in relation to the number of evaluations invested, i.e. the number of evaluations performed for computing an entry of the the stiffness matrix versus the error that is observed in the final solution. The same error with less evaluations implies a more efficient strategy, whereas the slope reveals the convergence of each strategy. We see that the first four strategies converge with the same rate, but each one uses a different number of evaluations, with the ones of the Quadrature-free technique being at a minimum. Let k be the number of elements. The number of evaluations performed by full Gauss quadrature is $k(p + 1)$. The ACHRS quadrature requires roughly $(k - 2)(p + 2)/2$ evaluations, while the Quadrature-free approach requires evaluation on an interpolation point-set of cardinality k. In Figure 4 we plot the number of evaluations performed for the different methods with respect to k and p. In both cases the advantage of the new method is clear.

6   Conclusions

In the context of IGA, we explored the computational load for matrix generation and the related consistency error of different techniques. As expected, apart from discretization, the order of accuracy in the entries of the stiffness matrix influenced crucially the quality of the final result.

The experiments indicated that the quadrature-free technique is promising, since it is flexible (e.g. in choosing the interpolation operator), adaptive (the accuracy in the computation of the integrals is improved after refinement of the basis), and has a lower complexity when compared to quadrature approaches. In particular, we replaced the set of quadrature points for the stiffness integrand, which are usually solutions of non-linear systems of equations that are not known in advance, with a set of interpolation points for the geometry factor of clearly lower cardinality, that are straight-forward to compute. By exploiting intrinsic properties of B-splines, we showed how one can precompute integrals of shifted tri-products and therefore avoid duplicated computations. We believe that there exist closed formulas, or even better, recurrences, that compute these integral values, at least in the uniform case; this is a topic for further research.

The behaviour of Gauss quadrature rules with decreasing number of points, that we experimented with, demonstrates that an optimal assembly strategy with respect to computational load is to generate the system within the order of accuracy implied by the discretization error. The quadrature-free approach presented herein follows this principle while using a minimum number of evaluations.

Finally, we think that it is a worthy challenge to extend the rules of [1,12] to spaces such as $\mathbb{S}^p_{p-2}$ and prove good approximation properties, as in the case of Gauss quadrature where the degree of exactness is closely connected to the approximation order.

Acknowledgement. This research was supported by the National Research Network "Geometry + Simulation" (NFN S117, 2012–2016), funded by the Austrian Science Fund (FWF).

References

1. Auricchio, F., Calabrò, F., Hughes, T., Reali, A., Sangalli, G.: A simple algorithm for obtaining nearly optimal quadrature rules for NURBS-based isogeometric analysis. Comput. Meth. Appl. Mech. Eng. (2012)
2. Beirão da Veiga, L., Buffa, A., Rivas, J., Sangalli, G.: Some estimates for h-p-k-refinement in isogeometric analysis. Numerische Mathematik 118, 271–305 (2011)
3. Costantini, P., Manni, C., Pelosi, F., Sampoli, M.L.: Quasi-interpolation in isogeometric analysis based on generalized B-splines. Comput. Aided Geom. Des. 27(8), 656–668 (2010)
4. de Boor, C., Fix, G.: Spline approximation by quasi-interpolants. J. Approx. Theory 8, 19–45 (1973)
5. de Falco, C., Reali, A., Vázquez, R.: GeoPDEs: A research tool for isogeometric analysis of PDEs. Advances in Engineering Software 42(12), 1020–1034 (2011)
6. Farin, G.: Curves and surfaces for CAGD: A practical guide. Morgan Kaufmann Publishers Inc., San Francisco (2002)
7. Gahalaut, K., Tomar, S.: Condition number estimates for matrices arising in the isogeometric discretizations. Technical Report RR-2012-23, Johann Radon Institute for Computational and Applied Mathematics, Linz (December 2012), https://www.ricam.oeaw.ac.at/publications/reports/12/rep12-23.pdf
8. Großmann, D., Jüttler, B., Schlusnus, H., Barner, J., Vuong, A.-V.: Isogeometric simulation of turbine blades for aircraft engines. Comput. Aided Geom. Des. 29(7), 519–531 (2012)
9. He, T.-X.: Eulerian polynomials and B-splines. J. Comput. Appl. Math. 236(15), 3763–3773 (2012)
10. Hopkins, T., Wait, R.: Some quadrature rules for Galerkin methods using B-spline basis functions. Comput. Meth. Appl. Mech. Eng. 19(3), 401–416 (1979)
11. Hughes, T., Cottrell, J., Bazilevs, Y.: Isogeometric analysis: CAD, finite elements, NURBS, exact geometry and mesh refinement. Comput. Meth. Appl. Mech. Eng. 194(39-41), 4135–4195 (2005)
12. Hughes, T., Reali, A., Sangalli, G.: Efficient quadrature for NURBS-based isogeometric analysis. Comput. Meth. Appl. Mech. Eng. 199(5-8), 301–313 (2010)
13. Kahaner, D., Moler, C., Nash, S.: Numerical methods and software. Prentice-Hall, Inc., Upper Saddle River (1989)
14. Lyche, T., Schumaker, L.: Local spline approximation methods. J. Approx. Theory 25, 266–279 (1979)
15. Patzák, B., Rypl, D.: Study of computational efficiency of numerical quadrature schemes in the isogeometric analysis. In: Proc. of the 18th Int'l Conf. Engineering Mechanics, EM 2012, pp. 1135–1143 (2012)

16. Strang, G.: Approximation in the finite element method. Numerische Mathematik 19, 81–98 (1972)
17. Strang, G., Fix, G.J.: An Analysis of the Finite Element Method. Prentice-Hall, Englewood Cliffs (1973)
18. Stroud, A., Secrest, D.: Gaussian quadrature formulas. Prentice-Hall Series in Automatic Computation. Prentice-Hall, Englewood Cliffs (1966)
19. Vermeulen, A.H., Bartels, R.H., Heppler, G.R.: Integrating products of B-splines. SIAM J. on Sci. and Stat. Computing 13(4), 1025–1038 (1992)
20. Wang, R.-H., Xu, Y., Xu, Z.-Q.: Eulerian numbers: A spline perspective. J. Math. Anal. and Appl. 370(2), 486–490 (2010)
21. Beirão da Veiga, B.Y.L., Cottrell, J.A., Hughes, T.J.R., Sangalli, G.: Isogeometric analysis: Approximation, stability and error estimates for h-refined meshes. Math. Models Methods Appl. Sci. 16(07), 1031–1090 (2006)

Commutator Estimate for Nonlinear Subdivision

Peter Oswald and Tatiana Shingel[*]

SES, Jacobs University, 28759 Bremen, Germany
p.oswald@jacobs-university.de

Abstract. Nonlinear multiscale algorithms often involve nonlinear perturbations of linear coarse-to-fine prediction operators S (also called subdivision operators). In order to control these perturbations, estimates of
the "commutator" $SF - FS$ of S with a sufficiently smooth map F are
needed. Such estimates in terms of bounds on higher-order differences of
the underlying mesh sequences have already appeared in the literature,
in particular in connection with manifold-valued multiscale schemes. In
this paper we give a compact (and in our opinion technically less tedious)
proof of commutator estimates in terms of local best approximation by
polynomials instead of bounds on differences covering multivariate S
with general dilation matrix M. An application to the analysis of normal multiscale algorithms for surface representation is outlined.

Keywords: subdivision operators, polynomial reproduction, local
polynomial best approximation, nonlinear multiscale transforms.

1 Introduction

Throughout this paper, we assume familiarity with the basics of the theory and
applications of subdivision schemes [1,5,12,17]. Let $M \in \mathbb{Z}^{s \times s}$ be a dilation
matrix, i.e., M has integer entries and every eigenvalue is of modulus > 1. We
consider linear subdivision operators $S : \ell_\infty(\mathbb{Z}^s) \to \ell_\infty(\mathbb{Z}^s)$ with dilation matrix
M acting on sequences $x = (x_\phi)_{\phi \in \mathbb{Z}^s} \in \ell_\infty(\mathbb{Z}^s)$ according to

$$(Sx)_\psi = \sum_{\phi \in \mathbb{Z}^s} a_{\psi - M\phi} x_\phi, \qquad \psi \in \mathbb{Z}^s, \tag{1}$$

where $(a_\phi)_{\phi \in \mathbb{Z}^s}$ is a fixed finitely supported sequence, called the mask of S. The
case $M = 2\mathrm{Id}$, where Id is the $s \times s$ identity matrix, corresponds to dyadic
subdivision and is important for many applications. If we consider vector-valued
sequences $\mathbf{x} = (x^{(1)}, \dots, x^{(n)}) \in \ell_\infty^n(\mathbb{Z}^s)$ then S acts on them componentwise, i.e.,
$S\mathbf{x} = (Sx^{(1)}, \dots, Sx^{(n)})$. Other situations, such as vector subdivision operators
and non-shift-invariant mesh topologies, are not considered.

We are interested in estimating the commutator term $SF(\mathbf{x}) - F(S\mathbf{x})$ in terms
of "smoothness" properties of the vector-valued sequence $\mathbf{x} \in \ell_\infty^n(\mathbb{Z}^s)$, where

[*] The authors gratefully acknowledge that their work was supported by Deutsche
Forschungsgemeinschaft grant OS-122/3-1.

M. Floater et al. (Eds.): MMCS 2012, LNCS 8177, pp. 383–402, 2014.

$\mathbf{x} = (x^{(1)}, \ldots, x^{(n)})$, under generic assumptions on S and the nonlinear map F : $\mathbb{R}^n \to \mathbb{R}$ belonging to the Hölder class $C^{K,\rho}$. The estimates of the commutator map are local in nature due to the finite support of the mask associated with S, and should therefore be formulated using local smoothness measures of $\mathbf{x}$. We prefer to use *local polynomial best approximations* to the vector-valued sequence $\mathbf{x}$ restricted to finite index sets $I_\alpha \subset \mathbb{Z}^s$ (so-called invariant neighborhoods of S) associated with each $\alpha \in \mathbb{Z}^s$. The role of the family of invariant neighborhoods $\{I_\alpha\}_{\alpha \in \mathbb{Z}^s}$ for a given S is roughly speaking as follows: Fix a set Γ_M of representatives for the factor space $\mathbb{Z}^s/M\mathbb{Z}^s$, obviously $|\Gamma_M| = |\det M|$. Then the restriction of $x \in \ell_\infty(\mathbb{Z}^s)$ to I_α fully determines the values of Sx restricted to any of the $|\det(M)|$ invariant neighborhoods $I_{M\alpha+\epsilon}$, $\epsilon \in \Gamma_M$. Throughout the paper, it is assumed that $\{I_\alpha\}_{\alpha \in \mathbb{Z}^s}$ is shift-invariant, i.e., $I_\alpha = I_0 + \alpha$. This defines a bijection on $\{I_\alpha\}_{\alpha \in \mathbb{Z}^s}$, and recursively a $|\det(M)|$-ary tree which can be conveniently used to study the subdivision process $S^j x$, $j \geq 0$, in terms of local subdivision maps $S_\epsilon : \ell_\infty(I_0) \to \ell_\infty(I_0)$ acting according to $S_\epsilon(x|_{I_0}) = (Sx)_{I_\epsilon}$, $\epsilon \in \Gamma_M$. E.g., the joint spectral radius of the family $\{S_\epsilon\}_{\epsilon \in \Gamma_M}$ (restricted to certain subspaces of $\ell_\infty(I_0)$) will determine the convergence of the subdivision process to a continuous limit function and the Hölder smoothness class the latter belongs to.

The paper is organized as follows. In Section 2, we collect auxiliary results on S, F, and the smoothness measures based on local polynomial best approximations. In particular, the notion of the order of exact polynomial reproduction P_e of a subdivision operator S is emphasized and related to other notions of polynomial reproduction or generation used in the literature. Our main results, local and global estimates for the commutator $SF(\mathbf{x}) - F(S\mathbf{x})$ in terms of the introduced smoothness measures for $\mathbf{x}$, and their dependence on P_e, are formulated and proved in Section 3. In various generality, similar results can be found in [3,4,6,7,10,13,16,18,19], where estimates are stated in terms of finite difference norms of $\mathbf{x}$. Our contributions are in emphasizing local estimates via local polynomial best approximations as an alternative approach in multivariate situations (which will also be useful in the case of semi-regular, non-shift-invariant mesh topologies and adaptive subdivision), in formulating results for $F \in C^{K,\rho}$ rather than for $F \in C^\infty$, and in clarifying the role of the order of exact polynomial reproduction P_e of a subdivision operator (see Section 2.1).

In the concluding Section 4 we discuss an application of the commutator estimate to the smoothness analysis of normal multiscale transforms, a nonlinear wavelet-type transform for the representation and re-parametrization of smooth two-dimensional surfaces in $\mathbb{R}^3$ introduced in [8,11]. This analysis is conditional on the well-posedness of the transform on regular, shift-invariant topologies, which will be investigated in a forthcoming paper [15].

2   Definitions and Preliminary Facts

2.1   Properties of S

We recall the properties of linear subdivision operators S given by (1) which will be of use.

Definition 1. *The operator S has* polynomial reproduction order P *if for any polynomial p of degree $\deg p < P$ there exists a polynomial $p^S := p + q$, where q is a polynomial of degree $\deg q < \deg p$, such that*

$$S\left(p(M\cdot)|_{\mathbb{Z}^s}\right) = p^S|_{\mathbb{Z}^s}. \tag{2}$$

This definition is standard (see [1,5]), P is also called order of polynomial generation of S, see, e.g., [2]. It is equivalent to assuming that S leaves the space of polynomial sequences $p|_{\mathbb{Z}^s}$ of total degree $< P$ invariant. Any meaningful subdivision operator must preserve constant sequences, i.e., $S\mathbf{1} = \mathbf{1}$. Hence, we will always assume that $P \geq 1$.

Definition 2. *The operator S has* exact polynomial reproduction order P_e *with shift $c_S \in \mathbb{R}^s$ if for any polynomial p of degree $\deg p < P_e$ we have*

$$S\left(p(M\cdot)|_{\mathbb{Z}^s}\right) = p(\cdot + c_S)|_{\mathbb{Z}^s}. \tag{3}$$

In special cases, this definition appeared in [10], and independently in [2], where it is called step-wise polynomial reproduction. It is also equivalent to the notion of polynomial generation degree $(0, P_e)$ introduced in [7], see below. Introducing a shift $c_S \neq 0$ is useful to cover certain dual subdivision schemes (also called cell-centered in contrast to vertex-centered subdivision schemes). It is assumed that P and P_e denote the largest possible order for a given S. It is easy to see that $1 \leq P_e \leq P$. The equality $P_e = P$ holds for instance if $P = 1$ or if S is interpolating, i.e., if $(Sx)_{M\phi} = x_\phi$, $\phi \in \mathbb{Z}^s$ (in this case $c_S = 0$). Furthermore, we observe that if $P \geq 2$, then $P_e \geq 2$. Indeed, let $P \geq 2$. Then taking the monomials $p(\cdot) = x_i$ in (2), we have

$$S\left(x_i|_{(M\mathbb{Z})^s}\right) = x_i|_{\mathbb{Z}^s} + c_i, \quad i = 1, \ldots, s.$$

Letting $c_S = (c_1, \ldots, c_s)$, we see that (3) holds for all linear functions, i.e., $P_e \geq 2$. It turns out that for the commutator estimates considered in this paper, the exact polynomial reproduction order P_e, and not P, is relevant.

Definition 3. *The operator S is of* class (d, f) *if there exists $c_S \in \mathbb{R}^s$ such that*

$$S\left(p(M\cdot)|_{\mathbb{Z}^s}\right) - p(\cdot + c_S)|_{\mathbb{Z}^s} \in \Pi_d|_{\mathbb{Z}^s} \tag{4}$$

for all $p \in \Pi_f$. Here $0 \leq d < f \leq P$.

Note that in order to establish the class (d, f) condition, it is sufficient to verify it on monomials of total degree $< f$. Indeed, let

$$q_\gamma|_{\mathbb{Z}^s} := S\left((M\mathbf{x})^\gamma|_{\mathbb{Z}^s}\right), \quad |\gamma| < f, \tag{5}$$

satisfy (4). In this subsection, $\mathbf{x}$ stands for points in $\mathbb{R}^s$ which should not cause confusion. As to multivariate exponents $\gamma = (\gamma_1, \ldots, \gamma_s) \in \mathbb{N}_0^s$, where $\mathbb{N}_0$ denotes the set of all non-negative integers, we use the usual notation such as $|\gamma| = \gamma_1 + \ldots + \gamma_s$, $\gamma! = \gamma_1! \ldots \gamma_s!$, and $\mathbf{x}^\gamma = x_1^{\gamma_1} \ldots x_s^{\gamma_s}$, while for $\mathbf{x} \in \mathbb{R}^n$, we set $|\mathbf{x}| = \max_{i=1,\ldots,s} |x_i|$.

Then for any polynomial $p(\mathbf{x}) = \sum\limits_{|\gamma| < f} c_\gamma \mathbf{x}^\gamma$,

$$S\left(p(M\mathbf{x})|_{\mathbb{Z}^s}\right) = S\left(\sum_{|\gamma| < f} c_\gamma (M\mathbf{x})^\gamma|_{\mathbb{Z}^s}\right) = \sum_\gamma c_\gamma q_\gamma(\mathbf{x})|_{\mathbb{Z}^s}$$

$$= \sum_\gamma c_\gamma (\mathbf{x} + c_S)^\gamma|_{\mathbb{Z}^s} + r(\mathbf{x})|_{\mathbb{Z}^s} = p(\mathbf{x} + c_S)|_{\mathbb{Z}^s} + r(\mathbf{x})|_{\mathbb{Z}^s},$$

where r is a polynomial of total degree $< d$. The next theorem demonstrates the relation between the class (d, f) condition for the subdivision operator S and the preceding definitions of polynomial reproduction. In particular, we will see that the quantity P_e plays a key role in verifying the class (d, f) condition.

Theorem 1. *The subdivision operator is of class* (d, f), $f \leq P$, *if and only if* $0 < f - d \leq P_e$, *and* $\min(2, P) \leq P_e \leq P$. *Here,* P_e *is the order of exact polynomial reproduction for the subdivision operator S and P is the order of polynomial reproduction according to* (2), (3).

Proof. We need the following two lemmas. The first one is a simple consequence of Definitions 2 and 3, and stated without proof.

Lemma 1. *The operator S is of class* $(0, f)$ *if and only if* $f \leq P_e$.

Lemma 2. *For* $0 < f < P$, *the subdivision operator S is of class* (d, f) *if and only if it is of class* $(d + 1, f + 1)$.

Proof. The proof easily follows from the following auxiliary result. Recall formula (5) for q_γ. For $\{e_i\}_{i=1}^s$ denoting the standard unit coordinate basis in $\mathbb{R}^s$, and $\gamma \in \mathbb{N}_0^s$, $|\gamma| < P$, the following relation holds,

$$\frac{\partial}{\partial x_i} q_\gamma(\mathbf{x}) = \gamma_i q_{\gamma - e_i}(\mathbf{x}), \qquad i = 1, \ldots, s. \tag{6}$$

By shift-invariance,

$$q_\gamma(\mathbf{x} + M\mathbf{y})|_{\mathbb{Z}^s} = S\left((M(\mathbf{x} + \mathbf{y}))^\gamma|_{\mathbb{Z}^s}\right) = \sum_{0 \leq \beta \leq \gamma} \binom{\gamma}{\beta} (M\mathbf{y})^{\gamma - \beta} q_\beta(\mathbf{x})|_{\mathbb{Z}^s}.$$

The above equality can be regarded as a polynomial identity on the lattice $M\mathbb{Z}^s$. The identity still holds when $\mathbf{x}|_{\mathbb{Z}^s}$, $M\mathbf{y}|_{\mathbb{Z}^s}$ are replaced by $\mathbf{x}|_{\mathbb{R}^s}$, $\mathbf{t}|_{\mathbb{R}^s}$ respectively. Thus,

$$q_\gamma(\mathbf{x} + \mathbf{t}) - q_\gamma(\mathbf{x}) = \sum_{0 \leq \beta < \gamma} \binom{\gamma}{\beta} \mathbf{t}^{\gamma - \beta} q_\beta(\mathbf{x}).$$

Consequently, taking $\mathbf{t} = h e^i$, $h \to 0$, we establish (6).

Next, assume that the subdivision operator is of class (d, f). We would like to show that it is of class $(d+1, f+1)$. Let $c_S = (c_1, \ldots, c_s)$ and $\mathbf{x}' = (x_1, \ldots, v_i, \ldots, x_s)$

with v_i varying and the remaining x_j with $j \neq i$ being fixed. We integrate (6) from $-c_i$ to x_i. Thus,

$$\int\limits_{-c_i}^{x_i} \frac{\partial}{\partial v_i} q_\gamma(\mathbf{x}')dv_i = \int\limits_{-c_i}^{x_i} \gamma_i q_{\gamma - e_i}(\mathbf{x}')dv_i$$

$$= \gamma_i \int\limits_{-c_i}^{x_i} (\mathbf{x}' + c_S)^{\gamma - e_i} dv_i + \int\limits_{-c_i}^{x_i} r(\mathbf{x}')dv_i.$$

Hence,

$$q_\gamma(\mathbf{x}) = (\mathbf{x} + c_S)^\gamma + \tilde{r}(\mathbf{x}), \tag{7}$$

where $\tilde{r}(\mathbf{x})$ is a polynomial of degree $< d + 1$ by integration rules. Since the above condition holds for any monomial of degree $|\gamma| < P$, the class $(d+1, f+1)$ condition is established. The inverse statement, that $(d+1, f+1)$ implies (d, f), follows easily by differentiating (7) and again invoking formula (5). Thus, the statement of Lemma 2 is fully proved. $\square$

With these lemmas at hand, the proof of Theorem 1 is now immediate. It was noted earlier that if $P \geq 2$, then $P_e \geq 2$, and otherwise, $P = P_e = 1$. Thus, $\min(2, P) \leq P_e \leq P$. Further by Lemma 2, the class (d, f) condition is equivalent to the class $(d - 1, f - 1)$ condition, and proceeding by induction, to the class $(0, f - d)$ condition. It remains to apply Lemma 1, by which the latter is equivalent to $0 < f - d \leq P_e$. $\square$

The class (d, f) condition appeared before in the context of proximity analysis. In particular, in [7], a polynomial generation/reproduction property for S, referred to as *polynomial generation degree* (d, f), was introduced by requiring that whenever $p = \prod_{j=1}^m q_j \in \Pi_f$ is in factorized form with polynomials of lower degree then

$$p^S - \prod_{j=1}^m q_j^S \in \Pi_d|_{\mathbb{Z}^s}, \tag{8}$$

where p^S, q_j^S are as in Definition 1. We note that our Definition 3 is equivalent to (8), and the reader is referred to Lemma 3.5 in [7] for the proof of this fact. To be self-contained, here is the simple proof that (4) implies (8), a fact which we need later. By linearity of S, it is again enough to verify the implication for monomials, i.e., let us consider $p(\mathbf{x}) = \mathbf{x}^\gamma$ and $q_j(\mathbf{x}) = \mathbf{x}^{\gamma_j}$, where $|\gamma| = |\gamma_1| + \ldots + |\gamma_m| < f$ for some $0 \neq \gamma \in \mathbb{N}_0^s$, $j = 1, \ldots, m$. By (4) we can assume that

$$p^S(\mathbf{x}) = (\mathbf{x} + c_S)^\gamma + r(\mathbf{x}), \qquad q_j^S(\mathbf{x}) = (\mathbf{x} + c_S)^{\gamma_j} + r_j(\mathbf{x}),$$

where $r \in \Pi_d$ and $r_j \in \Pi_{|\gamma_j|+1-(f-d)}$ if $|\gamma_j| \geq f - d$, and $r_j(\mathbf{x}) = 0$ otherwise, compare to Lemma 2. Consequently,

$$p^S(\mathbf{x}) - \prod_{j=1}^m q_j^S(\mathbf{x}) = r(\mathbf{x}) - \Sigma(\mathbf{x}),$$

where $\Sigma(\mathbf{x})$ represents a large linear combination of products of m polynomial factors each, consisting of shifted monomials $(\mathbf{x} + c_S)^{\gamma_j}$ and (at least one!) non-zero $r_j(\mathbf{x})$. It is easy to see that all these products must have overall degree

$$\leq |\gamma_1| + \ldots + |\gamma_m| - (f - d) = |\gamma| - (f - d) < f - (f - d) = d.$$

This proves (8) for monomials, and thus in the general case.

2.2   Local Smoothness Measures

Estimates of the commutator map are local in nature due to the finite support of the mask associated with S, and should therefore be formulated using local smoothness measures of vector-valued sequences $\mathbf{x} \in \ell_\infty^n(\mathbb{Z}^s)$. We prefer to use *local polynomial best approximations* to $\mathbf{x}$ restricted to the finite index sets $I_\alpha \subset \mathbb{Z}^s$, $\alpha \in \mathbb{Z}^s$, of a shift-invariant family of invariant neighborhoods of S as introduced before. More precisely, let $\mathbf{p} = (p^{(1)}, \ldots, p^{(n)})$ denote a vector-polynomial with entries from Π_m, by which we denote the set of all algebraic polynomials p of total degree $< m$ on $\mathbb{R}^s$, assuming $\Pi_0 = \{0\}$. We will use the quantities

$$E_m(\mathbf{x}, I_\alpha) := \inf_{p_1, \ldots, p_s \in \Pi_m} \|\mathbf{x}|_{I_\alpha} - \mathbf{p}|_{I_\alpha}\|, \qquad m \geq 0, \qquad (9)$$

as local polynomial best approximations, where

$$\|\mathbf{x}|_{I_\alpha}\| = \max_{i=1,\ldots,n} \max_{\phi \in I_\alpha} |x_\phi^{(i)}|.$$

Obviously,

$$E_0(\mathbf{x}, I_\alpha) = \|\mathbf{x}|_{I_\alpha}\| \geq E_1(\mathbf{x}, I_\alpha) \geq E_2(\mathbf{x}, I_\alpha) \geq \ldots, \qquad \alpha \in \mathbb{Z}^s. \qquad (10)$$

Note that for any S with order P of polynomial reproduction, $m = 0, \ldots, P$, and $\epsilon \in \Gamma_M$, we have

$$E_m(Sx, I_{M\alpha+\epsilon}) \leq C E_m(x, I_\alpha), \qquad \alpha \in \mathbb{Z}^s, \qquad x \in \ell_\infty(\mathbb{Z}^s). \qquad (11)$$

This follows from (2), according to which for any $p \in \Pi_m$ we have $(Sp|_I)_{I_\epsilon} = p^S|_{I_\epsilon}$ for some polynomial $p^S \in \Pi_m$, and the boundedness of the linear maps $S_\epsilon : \mathbb{R}^I \to \mathbb{R}^{I_\epsilon}$ (called local subdivision maps). Indeed, for $\alpha = 0$

$$E_m(Sx, I_\epsilon) \leq \|(Sx)|_{I_\epsilon} - p^S|_{I_\epsilon}\| = \|(S(x|_{I_0} - p|_{I_0}))|_{I_\epsilon}\| \leq \|S_\epsilon\| \|x|_{I_0} - p|_{I_0}\|.$$

It remains to take the infimum with respect to all $p \in \Pi_m$ to establish (11), by shift-invariance, the result remains true for arbitrary $\alpha \in \mathbb{Z}^s$.

Due to the consistent use of maximum norms, in the following proofs it is enough to consider a scalar-valued sequence $x \in \ell_\infty(\mathbb{Z}^s)$ in place of $\mathbf{x}$. As before, we denote the restriction of a polynomial $p \in \Pi_m$ and a sequence $x \in \ell_\infty(\mathbb{Z}^s)$ to an index set $J \subset \mathbb{Z}^s$ by $p|_J$ and $x|_J$ respectively. Let $\mathbb{R}^J$ denote the space of all finite sequences with index set J, equipped with the maximum norm which we also denote by $\|\cdot\|$.

We start with some simple observations on equivalent norms.

Lemma 3. *Let $m = 0, 1, \ldots$ be given. If Z is a finite-dimensional normed space with norm $\|\cdot\|_Z$, and $T : \mathbb{R}^J \to Z$ is a linear operator with $\ker T = \Pi_m|_J$, then*

$$c\|T(x|_J)\|_Z \leq E_m(x, J) \leq C\|T(x|_J)\|_Z, \qquad x \in \ell_\infty(\mathbb{Z}^s), \tag{12}$$

where c, C do not depend on x. More generally, if $\ker T \subset \Pi_m|_J$ then the upper estimate in (12) holds, if $\Pi_m|_J \subset \ker T$ then the lower estimate in (12) holds.

The proof of this lemma is a simple exercise on factor norms on finite-dimensional spaces, and is omitted. Examples of operators to which Lemma 3 applies are $T = \mathrm{Id} - P$, where $P : \mathbb{R}^J \to \mathbb{R}^J$ is any projector onto $\Pi_m|_J \subset \mathbb{R}^J$. In particular, if $p|_J = 0$ implies $p = 0$ for any $p \in \Pi_m$ (we call such a J a unicity set for Π_m), then Π_m is unisolvent on a subset $J' \subset J$, and defining $P(x|_J)$ by the values on J of the interpolation polynomial to the data $(J', x|_{J'})$ is a possible choice. Alternatively, taking as P the least-squares projector onto $\Pi_m|_J$ will do as well. Equivalent norms in terms of the maximum norm of the vector of all m-th order differences of $x|_J$ that are well-defined on J follow under the same assumption.

For this paper, we need only the following consequence of Lemma 3 which is at the core of the technical part of our main result. Without loss of generality, from now on we will assume that $0 \in \Gamma_M$ and $0 \in I_0$, where I_0 is a set of unicity for Π_f. Let S be of class (d, f), where $0 \leq d < f \leq P$, as defined in (4). For given $x \in \mathbb{R}^{I_0}$, consider a sequence of polynomials p_m of degree $< m$, $m = 1, \ldots, f$, such that $p_1(\cdot) = x_0$, and

$$\|x - p_m|_{I_0}\| \leq CE_m(x, I_0), \qquad m = 1, \ldots, f.$$

E.g., the polynomials p_m could be defined by interpolation on appropriate subsets of I_0 containing 0. Set $q_1(M \cdot -c_S) = p_1(\cdot) = x_0$, and $q_m(M \cdot -c_S) = p_m(\cdot) - p_{m-1}(\cdot)$, $m = 2, \ldots, f$, and

$$d_\phi := \sum_{m=2}^{f} q_m(M\phi - c_S), \qquad r_\phi := x_\phi - x_0 - d_\phi, \qquad \phi \in I_0. \tag{13}$$

Since $d, r \in \mathbb{R}^{I_0}$, the vectors $(Sd)_\psi$ and $(Sr)_\psi$ are well-defined for all $\psi \in I_\epsilon$, $\epsilon \in \Gamma_M$. The following formulas and estimates are needed in the sequel:

Lemma 4. *Under the assumptions outlined above, we have for all $x \in \mathbb{R}^{I_0}$, $m = 2, \ldots, f$, and $\epsilon \in \Gamma_M$*

$$(q_m(M \cdot -c_S))^S - q_m \in \Pi_{\max(0, m-(f-d))}, \tag{14}$$

$$\|S(q_m(M \cdot -c_S)|_{I_0})|_{I_\epsilon}\| \leq C\|q_m(M \cdot -c_S)|_{I_0}\|$$
$$\leq CE_{m-1}(x, I_0) \leq CE_1(x, I_0), \tag{15}$$

and

$$\|(Sr)|_{I_\epsilon}\| \leq C\|r\| \leq CE_f(x, I_0) \leq CE_1(x, I_0). \tag{16}$$

The constants $C > 0$ do not depend on x, m, and ϵ. $\square$

Proof. The first and last estimates in (15) and (16) trivially follow from the boundedness of the local subdivision operators S_ϵ, and (10), respectively. In order to establish (15), we use the definition of $q_m(M \cdot -c_S) = p_m - p_{m-1} \in \Pi_m \subset \Pi_f$ and the fact that any two norms on the finite-dimensional space Π_f are equivalent:

$$\|q_m(M \cdot -c_S)|_{I_0}\| = \|(p_m - p_{m-1})|_{I_0}\|$$
$$\leq \|x - p_m|_{I_0}\| + \|x - p_{m-1}|_{I_0}\| \leq CE_{m-1}(x, I_0),$$

The last step in the above inequality follows from the definition of the polynomials p_m and (11). For the estimate in (16), observe that $r = x - p_f|_{I_0}$ by definition of r and the polynomials p_m, q_m. □

2.3   Properties of F

From now on, we assume without loss of generality that $F : \mathbb{R}^n \to \mathbb{R}$ belongs to the Hölder smoothness class $C^{K,\rho}$, where $K = 0, 1, \ldots$, and $0 < \rho \leq 1$. By this we mean that F possesses continuous partial derivatives $D^\gamma F$ up to order K, and that $D^\gamma F$ of order $|\gamma| = K$ are Lipschitz continuous on $\mathbb{R}^n$ with exponent ρ, with uniform bounds on all occurring derivatives and Lipschitz constants. As a consequence, by Taylor's formula we have

$$F(\mathbf{t}) = \sum_{k=0}^{K} D_{\mathbf{a}}^k F(\mathbf{t} - \mathbf{a}) + R_{\mathbf{a}}^K(\mathbf{t}, \mathbf{a}), \qquad \mathbf{t}, \mathbf{a} \in \mathbb{R}^n, \tag{17}$$

where

$$D_{\mathbf{a}}^0 F(\mathbf{t} - \mathbf{a}) = F(\mathbf{a}), \qquad D_{\mathbf{a}}^k F(\mathbf{t} - \mathbf{a}) = \sum_{\gamma \in \mathbb{Z}_+^n : |\gamma|=k} \frac{D^\gamma F(\mathbf{a})}{\gamma!} (\mathbf{t} - \mathbf{a})^\gamma,$$

are multi-linear forms of degree $k = 0, \ldots, K$ with argument $\mathbf{t} - \mathbf{a}$, and

$$|D_{\mathbf{a}}^k F(\mathbf{t} - \mathbf{a})| \leq C|\mathbf{t} - \mathbf{a}|^k, \quad k = 0, \ldots, K, \qquad |R_{\mathbf{a}}^K(\mathbf{t}, \mathbf{a})| \leq C|\mathbf{t} - \mathbf{a}|^{K+\rho}, \tag{18}$$

uniformly in $\mathbf{t}, \mathbf{a} \in \mathbb{R}^n$, respectively. The assumption that the constant C in (18) is independent of $\mathbf{t}, \mathbf{a} \in \mathbb{R}^n$ is not as stringent as it looks, as in typical applications one can always show that values $\mathbf{t}, \mathbf{a}$ of interest belong to a compact subset of $\mathbb{R}^n$.

3   Main Result

For given S of class (d, f) with $0 \leq d < f \leq P$, where $P > 0$ is the order of polynomial reproduction, and $F \in C^{K,\rho}(\mathbb{R}^n)$, define

$$\tilde{K} := \min(K, f - 1), \qquad \tilde{\rho} := \min(1, K + \rho - \tilde{K}) \geq \rho, \tag{19}$$

and

$$\Omega_{f,K,\rho}(\mathbf{x}, I) := E_1(\mathbf{x}, I)^{\tilde{K}+\tilde{\rho}} + \sum_{(\nu_1,\dots,\nu_{f-1})\in\Xi_{f,\tilde{K}}} \prod_{l=1}^{f-1} E_l(\mathbf{x}, I)^{\nu_l}, \qquad (20)$$

where $\Xi_{f,\tilde{K}} = \emptyset$ if $\tilde{K} < 2$, and

$$\Xi_{f,\tilde{K}} := \{(\nu_1,\dots,\nu_{f-1}) \in \mathbb{N}_0^{f-1} : 2 \leq \sum_{m=1}^{f-1} \nu_m \leq \tilde{K}, \ \sum_{m=1}^{f-1} m\nu_m = f\} \qquad (21)$$

for $\tilde{K} \geq 2$ (in this case $f \geq 3$). Note that $\tilde{K} + \tilde{\rho} = \min(K + \rho, f)$ by definition (19).

The main result of this paper is the following theorem. Even though the exact polynomial reproduction order P_e of S does not explicitly surface in its formulation, it is implicit due to Theorem 1.

Theorem 2. *Let $F : \mathbb{R}^n \to \mathbb{R}$ belong to the Hölder smoothness class $C^{K,\rho}$, where $K = 0, 1, \dots$, and $0 < \rho \leq 1$. Let S be a linear subdivision operator (1) of class (d, f) with a finitely supported mask, as defined in (4). Assume that the associated family of invariant neighborhoods $\{I_\alpha\}_{\alpha\in\mathbb{Z}^s}$ satisfies the properties listed in the previous sections, in particular I_0 contains 0 and is a set of unicity for Π_f. Then, with $\tilde{K}$, $\tilde{\rho}$, and $\Omega_{f,K,\rho}(\mathbf{x}, I)$ defined as above, for any $\mathbf{x} \in \ell_\infty^n(\mathbb{Z}^s))$ and $\alpha \in \mathbb{Z}^s$ we have the estimate*

$$E_d\left(SF(\mathbf{x}) - F(S\mathbf{x}), I_{M\alpha+\epsilon}\right) \leq C\Omega_{f,K,\rho}(\mathbf{x}, I_\alpha), \quad \epsilon \in \Gamma_M. \qquad (22)$$

Proof. The proof is modeled after results for $s = n = 1$ given in [3] for interpolating S, and later in [10] for general dyadic S. Similar results obtained in the context of manifold subdivision are contained in [4,6,7,10,13,16,18,19], all these earlier results use smoothness measures in terms of finite differences of sequences rather than local polynomial best approximations. Due to shift-invariance, it is sufficient to consider the case $\alpha = 0$. To simplify the notation, let $I := I_0$ and $\tilde{I} := I_\epsilon$, where $\epsilon \in \Gamma_M$ is arbitrarily fixed. Consider the difference $SF(\mathbf{x})_\psi - F((S\mathbf{x})_\psi)$ for arbitrarily fixed $\psi \in \tilde{I}$. In order to estimate the error of best approximation of this quantity by polynomials of degree $< d$, we use the Taylor formula (17) for F with arguments $\mathbf{a} = \mathbf{x}_0 \in \mathbb{R}^n$ and $\mathbf{t} = \mathbf{x}_\phi$, $\phi \in I$, and with parameters $\tilde{K} \leq K$ and $\tilde{\rho}$ from (19) instead of K and ρ, respectively. This gives with $\mathbf{t} = \mathbf{x}_\phi$, $\phi \in I$,

$$F(\mathbf{x}_\phi) = \mathbf{x}_0 + D_{\mathbf{a}}^1 F(\mathbf{x}_\phi - \mathbf{x}_0) + \sum_{k=2}^{\tilde{K}} D_{\mathbf{a}}^k F(\mathbf{x}_\phi - \mathbf{x}_0) + R^{\tilde{K}}(\mathbf{x}_\phi, \mathbf{x}_0),$$

and with $\mathbf{t} = S\mathbf{x}_\psi$ for $\psi \in I$

$$F(S\mathbf{x}_\psi) = \mathbf{x}_0 + D_{\mathbf{a}}^1 F(S\mathbf{x}_\psi - \mathbf{x}_0) + \sum_{k=2}^{\tilde{K}} D_{\mathbf{a}}^k F(S\mathbf{x}_\psi - \mathbf{x}_0) + R^{\tilde{K}}(S\mathbf{x}_\psi, \mathbf{x}_0).$$

Next, we apply the linear subdivision operator S to the first equation, and use the fact that S reproduces constants and that $D_a^1 F(\cdot)$ is a linear form. This shows that the first two terms in the sum expression for $SF(\mathbf{x})_\psi$ and $F(S\mathbf{x}_\psi)$ coincide. Consequently, for all $\psi \in \tilde{I}$

$$(SF(\mathbf{x}) - F(S\mathbf{x}))_\psi = \sum_{k=2}^{\tilde{K}} \{(SD_a^k F(\mathbf{x}|_I - \mathbf{x}_0)_\psi - D_a^k F(S\mathbf{x}_\psi - \mathbf{x}_0)\} + \tilde{R}_\psi. \quad (23)$$

By (18), and the boundedness of the local subdivision operators S_ϵ, the remainder term

$$\tilde{R}_\psi = SR^{\tilde{K}}(\mathbf{x}, \mathbf{x}_0)_\psi - R^{\tilde{K}}(S\mathbf{x}_\psi, \mathbf{x}_0)$$

can be bounded by

$$|\tilde{R}_\psi| \leq C(\|\mathbf{x}|_I - \mathbf{x}_0\| - \|(S\mathbf{x})|_{\tilde{I}} - \mathbf{x}_0\|^{\tilde{K}+\tilde{\rho}}) \leq CE_1(\mathbf{x}, I)^{\tilde{K}+\tilde{\rho}},$$

where the last step follows by observing that $(S\mathbf{x})|_{\tilde{I}} - \mathbf{x}_0 = S_\epsilon(\mathbf{x}|_I - \mathbf{x}_0)$ and $\mathbf{p}_1(\cdot) = \mathbf{x}_0$ (see Lemma 4, when applied componentwise to $\mathbf{x}$). Note that if $\tilde{K} < 2$, we would not have the summation term in (23), and the proof would be complete. Thus, from now on we consider $\tilde{K} > 2$. Recall from (13) and Lemma 4 applied componentwise to $\mathbf{x}$ that

$$\mathbf{x}_\phi - \mathbf{x}_0 = \mathbf{d}_\phi + \mathbf{r}_\phi = \sum_{m=2}^{f} \mathbf{q}_m(M\phi - c_S) + \mathbf{r}_\phi, \qquad \phi \in I,$$

and consider the terms in

$$\sum_{k=2}^{\tilde{K}} \{(SD_a^k F(\mathbf{d} + \mathbf{r}))_\psi - D_a^k F(S(\mathbf{d} + \mathbf{r})_\psi)\}. \quad (24)$$

Since $D_a^k F(\cdot)$ are k-linear forms on the vector space $\mathbb{R}^n$, where $k = 2, \ldots, \tilde{K}$, the sum (24) can be expressed as a large, yet finite sum of terms of the form

$$S(g_1|_I \ldots g_k|_I)_\psi - S(g_1|_I)_\psi \cdot \ldots \cdot S(g_k|_I)_\psi$$

of differences of products with $2 \leq k \leq \tilde{K}$ functions resp. sequences g_l of the form either $g_l = q_m^{(i)}(M \cdot - c_S)$ or $g_l = r^{(i)}$, where $i = 1, \ldots, n$, $m = 2, \ldots, f$ vary arbitrarily within the respective limits. The coefficients in this linear combination do not depend on $\psi \in \tilde{I}$, they can also be bounded independently of $\mathbf{x}$ under the assumptions on F, outlined in Section 2.3. We group these products into three cases:

i) Products that contain at least one remainder term $g_l = r_\phi^{(i)}$: Due to (15) and (16), all these products are uniformly bounded from above by

$$C\sum_{k=2}^{\tilde{K}} E_f(\mathbf{x}, I) E_1(\mathbf{x}, I)^{k-1} \leq CE_f(\mathbf{x}, I)\left(E_1(\mathbf{x}, I) + E_1(\mathbf{x}, I)^{\tilde{K}-1}\right). \quad (25)$$

ii) Products that contain only "polynomial" terms $g_l := q_{m_l}^{(i_l)}(M \cdot -c_S)$, $l = 1, \ldots, k$, with overall degree $\deg(q_{m_1}^{(i_1)} \ldots q_{m_k}^{(i_k)}) = \sum_{l=1}^{k}(m_l - 1) < f$. Since S is of class (d, f), Lemma 2 holds, and (4) implies (8), we have

$$(g_1 \ldots g_k)^S = \prod_{l=1}^{k} q_{m_l}^{(i_l)}(\cdot) + \tilde{g}$$

where $\tilde{g}$ is a polynomial of degree $< d$, and

$$g_l^S = (q_{m_l}^{(i_l)}(M \cdot -c_S))^S = q_{m_l}^{(i_l)}(\cdot) + \tilde{g}_l,$$

where $\tilde{g}_l$ is a polynomial of degree $< \deg q_{m_l}^{(i_l)} - (f - d)$, $l = 1, \ldots, k$. Obviously, a certain number of these polynomials may vanish, this happens whenever $m_l - (f - d) \leq 0$. If $\tilde{g}_l$ does not vanish then $\deg \tilde{g}_l \leq \deg g_l - (f - d) = (m_l - 1) - (f - d)$. Thus

$$\begin{aligned}
S((g_1 \ldots g_k)|_I))_\psi &- S(g_1|_I)_\psi \ldots S(g_k|_I)_\psi \\
&= (g_1 \ldots g_k)^S(\psi) - g_1^S(\psi) \ldots g_k^S(\psi) \\
&= \tilde{g}(\psi) - \Sigma(g_1^S, \ldots, g_k^S, \tilde{g}_1, \ldots, \tilde{g}_k)(\psi),
\end{aligned}$$

where $\Sigma(\cdot)$ is a finite linear combination of all possible products of at most k polynomials of the form g_l^S or $\tilde{g}_l$, with the condition, that at least one factor is of the form $\tilde{g}_l \not\equiv 0$ (if all $\tilde{g}_l$ vanish then $\Sigma(\cdot) = 0$). Therefore, each of these products has degree

$$\leq \sum_{l=1}^{k}(m_l - 1) - (f - d) = \deg g_1 \ldots g_k - (f - d) < f - (f - d) = d.$$

Consequently, this group of products contributes an overall term $\tilde{p}(\psi)$ to (24), where $\tilde{p} \in \Pi_d$ (which can be neglected in the overall estimation).

iii) The last group corresponds to products of only "polynomial" terms $g_l := q_{m_l}^{(i_l)}(M \cdot -c_S)$, $l = 1, \ldots, k$, with overall degree $\sum_{l=1}^{k}(m_l - 1) \geq f$. Since by (15),

$$|S((g_1 \ldots g_k)|_I))_\psi - S(g_1|_I)_\psi \ldots S(g_k|_I)_\psi|$$
$$\leq C \| \prod_{l=1}^{k} q_{m_l}^{(i_l)}(M \cdot -c_S)|_I \| \leq C \prod_{l=1}^{k} E_{m_l-1}(\mathbf{x}, I),$$

the sum of all these terms is bounded from above by the expression

$$\leq C \sum_{k=2}^{\tilde{K}} \sum_{\substack{1 \leq \mu_l < f \\ \sum_{l=1}^{k} \mu_l \geq f}} \prod_{l=1}^{k} E_{\mu_l}(\mathbf{x}, I),$$

where we have set $\mu_l = m_l - 1$ for convenience. By (10), it is evident that this bound can be simplified by including only terms with $\sum_{l=1}^{k} \mu_l = f$ in the summation. Then it can be rewritten as

$$\leq C \sum_{\substack{2 \leq \sum_{m=1}^{f-1} \nu_m \leq \tilde{K} \\ \sum_{m=1}^{f-1} m\nu_m = f}} \prod_{l=1}^{f-1} E_l(\mathbf{x}, I)^{\nu_l} = C \sum_{(\nu_1, \ldots, \nu_{f-1}) \in \Xi_{f, \tilde{K}}} \prod_{l=1}^{f-1} E_l(\mathbf{x}, I)^{\nu_l}. \quad (26)$$

The desired estimate for $\tilde{K} > 2$ follows from

$$E_d \left(SF(\mathbf{x}) - F(S\mathbf{x}), I_e \right) \leq \| (SF(\mathbf{x}) - F(S\mathbf{x}))|_{I_e} - \tilde{p}_{I_e} \|,$$

where the polynomial $\tilde{p}$ of degree $< d$ addresses the contributions of group ii) products to (24), if one takes into consideration the estimates obtained in i) and iii) for the remaining terms contributing to (24). Indeed, by definition of $\tilde{K}$ and $\tilde{p}$, the terms in the bound (25) are bounded either by terms appearing in (26), or by $CE_1(\mathbf{x}, I)^{\tilde{K}+\tilde{p}}$. $\square$

For the ease of referencing, we formulate the "global" version of Theorem 2 which follows immediately if we define global smoothness measures of sequences $\mathbf{x} \in \ell_\infty^n(\mathbb{Z}^s)$ according to

$$\|\mathbf{x}\|_m := \sup_{\alpha \in \mathbb{Z}^s} E_m(\mathbf{x}, I_\alpha). \quad (27)$$

We have suppressed the dependence on the choice of the family of invariant neighborhoods $\{I_\alpha\}$ in the notation, it is not difficult to check by invoking Lemma 3 that under the assumptions of Theorem 2 these smoothness measures are equivalent for different families.

Theorem 3. *Under the conditions of Theorem 2, for any $\mathbf{x} \in \ell_\infty^n(\mathbb{Z}^s)$, we have the estimate*

$$\|SF(\mathbf{x}) - F(S\mathbf{x})\|_d \leq C\tilde{\Omega}_{f,K,\rho}(\mathbf{x}), \quad (28)$$

where, similar to (20),

$$\tilde{\Omega}_{f,K,\rho}(\mathbf{x}) := \|\mathbf{x}\|_1^{\tilde{K}+\tilde{p}} + \sum_{(\nu_1, \ldots, \nu_{f-1}) \in \Xi_{f, \tilde{K}}} \prod_{l=1}^{f-1} \|\mathbf{x}\|_l^{\nu_l}. \quad (29)$$

If, as mentioned after Lemma 3, best approximations $E_m(\mathbf{x}, I_\alpha)$ are replaced by equivalent expressions $\max |\Delta^\gamma \mathbf{x}|$ using finite differences of order $m = |\gamma|$ (the maximum is taken with respect to all those finite differences defined on I_α) then we recover the results implicitly or explicitly proved in, e.g., [16,19], and in greater generality in [7].

4   Application to Normal Multiscale Transforms

4.1   Smoothness of Normal Reparameterization

Roughly speaking, normal multiscale transforms of a given two-dimensional surface Σ embedded in $\mathbb{R}^3$ start from a coarse mesh $\mathbf{v}^0$ of points on Σ, with a mesh topology for which a surface subdivision process governed by a linear subdivision operator S is well-defined. The analysis step of the transform recursively creates finer meshes $\mathbf{v}^j$, $j \geq 1$, by predicting a mesh of base points $S\mathbf{v}^{j-1}$ (usually not on Σ) and associated unit normal vectors $\mathbf{n}^j$ using only information from the previous mesh $\mathbf{v}^{j-1}$, and then intersecting the resulting lines through the base points in the direction of the associated normal vectors with Σ. For short, this process can be written as

$$\mathbf{v}^j = S\mathbf{v}^{j-1} + d^j \mathbf{n}^j, \qquad j \geq 0, \tag{30}$$

where the vectors d^j are scalar sequences of signed distances from the base points in $S\mathbf{v}^{j-1}$ to the associated surface intersection points in $\mathbf{v}^j$, called details (they correspond to wavelet coefficients). Here, the formal product $d^j \mathbf{n}^j$ is understood entry-wise. In other words, the resulting transform

$$(\Sigma \quad \leftrightarrow) \quad \{\mathbf{v}^0, \mathbf{v}^1, \mathbf{v}^2, \ldots\} \quad \leftrightarrow \quad \{\mathbf{v}^0, d^1, d^2, \ldots\}$$

needs storage for $\mathbf{v}^0$ and scalar rather than vector-valued detail sequences d^j, $j \geq 1$. The reconstruction from the latter can be done using (30) and the fixed rules for predicting base points and approximate unit normals. We refer to [8,11] for details on the algorithms and surface compression applications.

   In this section, our goal is to illustrate how the above commutator estimate is used to analyze the smoothness of the re-parametrization of Σ induced by the meshes $\mathbf{v}^j$. We do this under the assumption that the two-dimensional dilation matrix M is isotropic, i.e., all its eigenvalues have the same modulus $\sqrt{|\det(M)|}$, and that the mesh topologies are regular, i.e., equivalent to $\mathbb{Z}^2$. We consider a graph surface $\Sigma_F = \{(\mathbf{x}, z) : z = F(\mathbf{x}), \mathbf{x} \in \mathbb{R}^2\}$, where $F : \mathbb{R}^2 \to \mathbb{R}$ belongs to the Hölder class $C^{K,\rho}$ with some $K \geq 1$, see Subsection 2.3. We also assume that (under certain conditions on the density and regularity of the coarsest mesh $\mathbf{v}^0$) the existence of a normal multiscale transform for all $j \geq 1$ with properties which will be detailed below has been settled (this step, called well-posedness of the transform, is non-trivial; we refer to our forthcoming paper [15]). To define the smoothness question, it is natural to identify the mesh sequences $\mathbf{v}^j = (\mathbf{x}^j, F(\mathbf{x}^j))$ generated by the normal multiscale transform with grid functions on the lattices $M^{-j}\mathbb{Z}^2$, and define convergence of this transform by the existence of an at least continuous vector function $\mathbf{v}^\infty : \mathbb{R}^2 \to \mathbb{R}^3$, such that

$$\lim_{j \to \infty} \|\mathbf{v}^\infty|_{M^{-j}\mathbb{Z}^2} - \mathbf{v}^j\| = 0.$$

How much Hölder smoothness does $\mathbf{v}^\infty$ possess? Since $\mathbf{v}^j = (\mathbf{x}^j, F(\mathbf{x}^j))$, we can equivalently talk about the Hölder smoothness of the limit function $\mathbf{x}^\infty$ for $\{\mathbf{x}^j\}$ mapping from $\mathbb{R}^2$ into $\mathbb{R}^2$.

The answer depends on three parameters: The Hölder smoothness $s_F := K + \rho$ of the surface Σ_F, the Hölder smoothness exponent s_S of the linear subdivision scheme associated with S, and the order of exact polynomial reproduction P_e of S. What we want to establish is the fact that

$$\mathbf{v}^\infty, \mathbf{x}^\infty \in C^{s_N-}, \qquad s_N := \min(P_e, s_\Sigma, s_S), \qquad (31)$$

where we denoted $C^{s-} := \cup_{K+\rho<s} C^{K,\rho}$. Thus, limit functions of the normal multiscale transform are guaranteed to be in $C^{K,\rho}$ whenever $K + \rho < s_N$ but not necessarily for $K + \rho = s_N$. The restrictions by s_Σ and s_S are natural, the appearance of P_e is also expected due to similar results for the curve case, see [10].

4.2   Assumed Facts

Here are the assumptions and facts that we need to prove (31). Let $a := \sqrt{|\det(M)|}$, and assume that the invariant neighborhoods of S used to define mesh norms in Section 2.2 are large enough. Also, assume $2 \le P_e \le P$ for the orders of polynomial reproduction of S, and $s_S > 1$. Finally, concerning the graph surface Σ_F, we assume $s_F > 1$ as well.

(a) From the well-posedness proof we need that the sequences $\mathbf{v}^j = (\mathbf{x}^j, F(\mathbf{x}^j))$, and d^j are well-defined for all $j \ge 1$ and satisfy (30). For $\mathbf{x}^j$, we need

$$\|\mathbf{x}^j\|_1 \le C_{\mathbf{x}^0} a^{-j}, \qquad j \ge 0. \qquad (32)$$

For (32) to hold, the coarsest mesh $\mathbf{v}^0$ must satisfy certain density and regularity requirements (see [3], [10], [14] for examples in the curve and surface case) which also determine the exact dependency of the constant on $\mathbf{x}^0$.

For the approximate normals we require (similar to the curve case [10]) that the $\mathbf{n}^j_\alpha$ coincide with the true unit normal to Σ_F at a nearby point: There exists a sequence $\tilde{\mathbf{x}}^j$ of points in $\mathbb{R}^2$ such that for all $\alpha \in \mathbb{Z}^2$

$$\mathbf{n}^j_\alpha = \mathbf{n}_\Sigma(\tilde{\mathbf{x}}^j_\alpha) := (1 + |\nabla F|^2)^{-1/2}(-\nabla F, 1)|_{\tilde{\mathbf{x}}^j_\alpha}, \quad |\mathbf{x}^j_\alpha - \tilde{\mathbf{x}}^j_\alpha| \le C\|\mathbf{x}^j\|_1. \; (33)$$

The standard construction of approximate normals that satisfies these conditions under mild conditions on the meshes $\mathbf{x}^j$ can be found in [14]. Note that $\|\mathbf{x}^j\|_1$ represents (within constant factors) the largest distance between neighboring $\mathbf{x}^j_\alpha$, i.e., the mesh-size of the mesh $\mathbf{x}^j$ in $\mathbb{R}^2$.

(b) We have

$$\|d^j\| \le C\|SF(\mathbf{x}^{j-1}) - F(S\mathbf{x}^{j-1})\|. \qquad (34)$$

The proof uses (a), in particular the assumption (33) on the approximate normals, the uniform boundedness of ∇F, and repeats the reasoning of Lemma 3.3 in [10] almost without change.

(c) For $m = 1, \ldots, P$, we can define numbers

$$\gamma_m := \min(m, \sup\{\gamma > 0 : \|S^j x\|_m = O(a^{-\gamma j}), \ j \to \infty, \ \forall x \in \ell_\infty(\mathbb{Z}^2)\}).$$

It is well known that $\gamma_1 > 0$ is a necessary condition for convergence, and that the γ_m provide lower bounds for the limit smoothness of the linear subdivision scheme governed by S, moreover, under mild additional conditions on S, γ_P coincides with the Hölder exponent of S. For the purpose of this paper, set

$$s_S := \max_{m=1,\ldots,P} \gamma_m = \gamma_P.$$

The numbers γ_m, and in particular $s_S = \gamma_P$, can be identified as joint spectral radii of families of operators $\{S_{\epsilon,m}\}_{\epsilon \in \Gamma_M}$ representing restrictions of the local subdivision operators S_ϵ to certain finite-dimensional operators acting on subspaces of $\mathbb{R}^{I_0}$. As detailed in [15], using the theory of joint spectral radii and the connection of the $S_{\epsilon,m}$ with the definition of $\|\cdot\|_m$, we have the following two facts of relevance for our derivation of (31):

$$\gamma_m = \min(m, s_S) = \min(m, \gamma_P), \qquad m = 1, \ldots, P, \tag{35}$$

and for every $\delta > 0$, there exists a semi-norm $\|\cdot\|_m^*$ depending on S and δ such that for all $x \in \ell_\infty(\mathbb{Z}^2)$ and $m = 1, \ldots, P$,

$$\|Sx\|_m^* \leq a^{-(\gamma_m - \delta)} \|x\|_m^*, \qquad \|x\|_m^* \approx \|x\|_m. \tag{36}$$

Naturally, the statement of (36) also holds for vector-valued meshes $\mathbf{x}$.

(d) To link with the limit smoothness problem, the following theorem is helpful. It is proved in [15] and originated from a similar perturbation result established in [3].

Theorem 4. *Let S be a fixed subdivision operator, and assume $\gamma_m > 0$ for some $m = 1, \ldots, P$. Assume that $\{x^j \in \ell_\infty(\mathbb{Z}^2)\}_{j \geq 0}$ is such that*

$$\|x^j - Sx^{j-1}\| \leq Ca^{-j\beta}, \qquad j \geq 1,$$

for some $\beta > 0$. Then the limit function x^∞ for the mesh sequences x^j exists and belongs to $C^{\min(\gamma_m, \beta)-}$. Applied componentwise, the statement holds for vector-valued mesh sequences $\{\mathbf{x}^j\}_{j \geq 0}$ as well.

4.3 Proof of (31)

The proof of (31) proceeds by induction in $m = 2, \ldots, m_0$, where m_0 is the smallest integer $\geq s_N$. Since $1 < s_N \leq P_e$, we have $2 \leq m_0 \leq P_e$, and thus we have that S is of class $(0, m)$ for all $m = 2, \ldots, m_0$. Consequently, if $m = 2$, by (34), and Theorem 3 with $d = 0$ and $f = 2$, we have

$$\|\mathbf{x}^j - S\mathbf{x}^{j-1}\| \leq \|d^j\| \leq C\|x^{j-1}\|_1^{\min(s_F, 2)} \leq C_{\mathbf{x}^0} a^{-j \min(s_F, 2)}, \qquad j \geq 1.$$

By Theorem 4 this implies $\mathbf{x}^\infty \in C^{\min(s_F,s_S,2)-}$ which is the desired result of (31) if $P_e = 2$ or if $\min(s_F, s_S) \leq 2$ which is equivalent to $s_N \leq m_0 = 2$.

Before we can proceed with induction in m, we need an estimate for $\|\mathbf{x}^j\|_2$ in the case $s_N > 2$ (such an estimate is also integral part of the proof of Theorem 4). Invoking (36), (34), and Theorem 3 with $d = 0$ and $f = 2$, we get

$$\|\mathbf{x}^j\|_2^* \leq \|S\mathbf{x}^{j-1}\|_2^* + C\|d^j\| \leq a^{-(\gamma_2-\delta)}\|\mathbf{x}^{j-1}\|_2^* + C\|x^{j-1}\|_1^{\min(s_F,2)}$$
$$\leq a^{-(\gamma_2-\delta)}\|\mathbf{x}^{j-1}\|_2^* + C_{\mathbf{x}^0}a^{-j\min(s_F,2)}, \qquad j \geq 1,$$

where the last step is justified by (32). Recursing in j and using the norm equivalence in (36), we get

$$\|\mathbf{x}^j\|_2 \leq C\|\mathbf{x}^j\|_2^*$$
$$\leq C_{\mathbf{x}^0}\left(a^{-(\gamma_2-\delta)j}\|\mathbf{x}^0\|_2^* + \sum_{k=0}^{j-1} a^{-k(\gamma_2-\delta)-(j-k)\min(s_F,2)}\|\mathbf{x}^0\|_1^{\min(s_F,2)}\right)$$
$$\leq C_{\mathbf{x}^0}a^{-j\min(\gamma_2-\delta,s_F)} = C_{\mathbf{x}^0}a^{-j(2-\delta)}, \qquad j \geq 0,$$

where $\delta > 0$ is arbitrary. The simplification of the exponent in the last estimation step is justified since from $s_N = \min(s_F, s_S, P_e) > 2$ and (35), we have

$$\min(\gamma_2 - \delta, s_F) = \min(s_S, 2) - \delta = 2 - \delta.$$

The correct formulation of the induction step is as follows. Suppose that for a given $m \geq 2$ satisfying $2 \leq m < m_0 \leq P_e$ and $m < s_N$ we have established that

$$\|\mathbf{x}^j\|_r \leq C_{\mathbf{x}^0}a^{-j(r-\delta)}, \qquad j \geq 0, \tag{37}$$

holds for all $r = 2, \ldots, m$ and any $\delta > 0$ (naturally, the constants also depend on the concrete value of δ, and may generally grow as $\delta \to 0$). For $r = 1$, we have the similar, slightly stronger estimate (32). With this at hand, we repeat the two steps shown above for $m = 2$ for $m + 1$. First, we use (34) and apply Theorem 4 with $d = 0$ and $f = m + 1$:

$$\|\mathbf{x}^j - S\mathbf{x}^{j-1}\| = \|d^j\| \leq C\tilde{\Omega}_{m+1,K,\rho}(\mathbf{x}^{j-1}).$$

Examining the definition (29), and plugging in (37) and (32), we see that

$$\tilde{\Omega}_{m+1,K,\rho}(\mathbf{x}^{j-1}) \leq C_{\mathbf{x}^0}\left(a^{-j\min(s_F,m+1)} + \sum_{(\nu_1,\ldots,\nu_m)\in\Xi_{m+1,\tilde{K}}} \prod_{l=1}^m a^{-j(l-\delta)\nu_l}\right)$$
$$\leq C_{\mathbf{x}^0}\left(a^{-j\min(s_F,m+1)} + C_m a^{-j(m+1-\tilde{K}\delta)}\right)$$
$$\leq C_{\mathbf{x}^0}a^{-j\min(s_F,m+1-\tilde{\delta})}, \qquad j \geq 1,$$

where $\tilde{\delta} = \tilde{K}\delta$ is again any positive constant. Thus, substituting into the bound for $\|d^j\|$, by Theorem 4 we can conclude that $\mathbf{x}^\infty \in C^{\min(s_F,s_S,m+1)-}$. This proves the claim of (31) for $m_0 = m + 1$ which is equivalent to $m < s_N \leq m + 1$.

To finish the induction step, we consider now the case $s_N > m+1$, and establish (37) for $r = m+1$ repeating the same strategy as demonstrated above for $m = 2$: By using (36) and repeating the application of Theorem 3, we arrive at

$$
\begin{aligned}
\|\mathbf{x}^j\|^*_{m+1} &\leq \|S\mathbf{x}^{j-1}\|^*_{m+1} + C\|d^j\| \\
&\leq a^{-(\gamma_{m+1}-\delta)}\|\mathbf{x}^{j-1}\|^*_{m+1} + C\tilde{\Omega}_{m+1,K,\rho}(\mathbf{x}^{j-1}) \\
&\leq a^{-(m+1-\delta)}\|\mathbf{x}^{j-1}\|^*_{m+1} + C_{\mathbf{x}^0}a^{-j\min(s_F,m+1-\tilde\delta)} \\
&\leq \ldots \leq C_{\mathbf{x}^0}a^{-j(m+1-\tilde\delta)}, \qquad j \geq 1,
\end{aligned}
$$

since $s_N > m+1$ implies both $s_F > m+1$ and $\gamma_{m+1} = \min(m+1,s_S) = m+1$. This shows (37) for $r = m+1$ because $\|\mathbf{x}^j\|_{m+1} \leq C\|\mathbf{x}^j\|^*_{m+1}$ holds according to (36). This concludes the induction, and establishes (31) under the mentioned assumptions on the well-posedness of the normal multiscale transform.

4.4   Remarks

First, the above considerations based on Theorem 4 also deliver an asymptotic estimate for the detail decay,

$$
\|d^j\| \leq C_{\mathbf{x}^0}a^{-j(\min(s_F,s_S+1,P_e)-\delta)}, \qquad j \geq 1, \tag{38}
$$

where $\delta > 0$ is arbitrary (as can be seen from the argument below, in some cases we can set $\delta = 0$ as well). Indeed, the above proof delivers the estimate

$$
\|d^j\| \leq C_{\mathbf{x}^0}a^{-j\min(s_F,m_0-\tilde\delta)}, \qquad j \geq 1,
$$

with arbitrarily small $\tilde\delta > 0$, where m_0 was the smallest integer $\geq s_N$, in other words $m_0 - 1 < s_N \leq m_0$. If the minimum defining the exponent in (38) is attained for P_e or s_F then the latter estimate is identical with (38). Suppose now that $s_S + 1 < \min(s_F, P_e)$. In this case, $m_0 - 1 < s_N = s_S = \gamma_{m_0} \leq m_0 < P_e$ by definition of s_N and m_0. This implies two things: An estimate of the form

$$
\|\mathbf{x}^j\|_{m_0} \leq C_{\mathbf{x}^0}a^{-j(s_S-\tilde\delta)}, \qquad j \geq 1,
$$

from the last induction step (semi-norms $\|\mathbf{x}^j\|_r$ with $r < m_0$ satisfy (37) and (32), respectively), and the possibility of applying Theorem 3 with $d = 0$, $f = m_0+1$. This gives

$$
\begin{aligned}
\|d^j\| &\leq C\tilde{\Omega}_{m_0+1,K,\rho}(\mathbf{x}^{j-1}) \\
&\leq C_{\mathbf{x}^0}\Big(a^{-j\min(s_F,m_0+1)} + a^{-j(1+s_S-\delta)} + \sum_{\nu\in\Xi_{m_0+1,K}:\nu_{m_0}=0}\prod_{l=1}^{m}a^{-j(l-\delta)\nu_l}\Big) \\
&\leq C_{\mathbf{x}^0}a^{-j(s_S+1-\tilde\delta)}, \qquad j \geq 1.
\end{aligned}
$$

In the estimation of $\tilde{\Omega}_{m_0+1,K,\rho}(\mathbf{x}^{j-1})$, we have used the fact that there is a single multi-index $(\nu_1,\ldots,\nu_{m_0}) \in \Xi_{m_0+1,\tilde{K}}$ with $\nu_{m_0} \neq 1$, namely $(1,0,\ldots,0,1)$.

The corresponding term in $\tilde{\Omega}_{m_0+1,K,\rho}(\mathbf{x}^{j-1})$ is treated by substituting the estimate for $\|\mathbf{x}^j\|_{m_0}$ and results in the separate, asymptotically dominant term $C_{\mathbf{x}^0}a^{-j(1+s_S-\delta)}$ in the above estimation, all other product terms are served by (37) and (32). This establishes (38).

Secondly, the statement (31) on the smoothness of the re-parametrization and the detail decay estimate (38) of normal multiscale transforms are backed by numerical experiments reported earlier [14] for some standard S, such as Loop, Butterfly, and linear subdivision schemes. Here, topologies are triangular, and refinement is by quadrisection (thus, $M = 2\mathrm{Id}$ and $a = 2$). On the one hand, they show that under the formulated conditions no better results should be expected, on the other hand, for certain combinations of S and the prediction scheme for approximate normals they point towards possible improvements in the smoothness of re-parametrization. To illustrate this fact, we have applied the Loop-based normal multiscale transform to a non-polynomial C^∞ graph surface with two different versions of approximate normals, and monitored the behavior of the norms of discrete derivatives $\Delta_m^j := 2^{-jm}\sup_{|\gamma|=m}\|\Delta^\gamma \mathbf{v}^j\|$ over a compact region of $\mathbb{R}^2$ of order ≤ 3 (in all tests, the initial $\mathbf{v}^0$ was obtained by sampling the underlying function F on a random perturbation of the lattice $\mathbb{Z}^2$). Since for the Loop subdivision operator S we have $P_e = 2$ and $s_S = 3$, our result (31) guarantees smoothness in the Hölder scale up to $s_N = 2$ only. In other words, we expect the Δ_m^j to stay bounded as j increases for $m = 1, 2$, but not necessarily for $m = 3$.

The standard version of normal prediction for triangular subdivision schemes is to compute first normal vectors $\mathbf{n}_t$ to each triangular face t from its vertices given by $\mathbf{v}^{j-1}$ (obviously, these $\mathbf{n}_t$ have length proportional to the area of t). Normals $\mathbf{n}_v$ and $\mathbf{n}_e$ at vertices v and midpoints of edges e are created by averaging the $\mathbf{n}_t$ of triangles adjacent to vertices and edges, respectively. Depending on the type of topology refinement, the sequence of approximate normals $\mathbf{n}_j$ for determining $\mathbf{v}^j$ is then obtained by normalizing to unit length an appropriate subset of these triangle-, vertex-, and edge-attached normals. To find $\mathbf{v}^j$ from $\mathbf{v}^{j-1}$ in the Loop-based normal multiscale transform, this would be the $\mathbf{n}_v$ for relocating the old point positions given by $\mathbf{v}^{j-1}$, and the $\mathbf{n}_e$ for inserting new vertices associated with edge midpoints into the mesh represented by $\mathbf{v}^j$. Interpolating transforms with triangle quadrisection as topology refinement, e.g., the one based on the Butterfly subdivision scheme as advocated in [8,11], need only normalized $\mathbf{n}_e$ since the old point positions from $\mathbf{v}^{j-1}$ are kept in $\mathbf{v}^j$ (this also leads to reduced storage for the detail sequences d^j). The first 3 columns of Table 1 below show the values for Δ_m^j for $j = 3, \ldots, 8$, and $m = 1, 2, 3$, obtained for a smooth C^∞ graph surface (given by a non-polynomial F) with this standard choice of approximate normals. As one can see, norms of first and second order discrete derivatives stay bounded, as expected by theory, but also third-order discrete derivatives seem to behave nicely. This observation is independent of the particular choice of F, as long as $s_F > 3$. That we should not expect higher than smoothness order 3 is clear since the normal multiscale transform coincides with the linear Loop subdivision scheme if F is linear for which $s_S = 3$ is known to be the best possible Hölder smoothness exponent for regular mesh topologies.

Table 1. Growth of norms of discrete derivatives for Loop-based normal multiscale transform

Level	Standard normals			Perturbed normals		
j	Δ_1^j	Δ_2^j	Δ_3^j	Δ_1^j	Δ_2^j	Δ_3^j
3	0.9429	2.9408	14.0567	2.3954	7.2715	12.11
4	0.9550	3.4326	17.9330	2.4835	11.8531	32.82
5	0.9587	3.5028	19.9322	2.4310	14.0084	112.74
6	0.9601	3.5235	20.6469	2.4106	17.1809	168.35
7	0.9603	3.5299	20.9226	2.4095	17.6150	262.57
8	0.9603	3.5323	21.0369	2.4086	17.7224	332.21

To check if this observed extra smoothness is related to the specific choice of standard normals, or if we missed out on a possible improvement of (31), we have run a test with perturbed normals. More precisely, instead of the $\mathbf{n}^j$ containing the standard unit normals we have used $\tilde{\mathbf{n}}^j$ whose elements are given by $\tilde{\mathbf{n}}_\alpha^j = \mathbf{n}_{\alpha'}^j$, where the index α' is taken randomly from a small neighborhood $|\alpha' - \alpha| \leq C$ of $\alpha \in \mathbb{Z}^2$, with C independent of α. Since $\mathbf{n}^j$ satisfies the property (33) needed for the proof of (31) then so does $\tilde{\mathbf{n}}^j$. As can be seen from the results in the last 3 columns of Table 1, this time we record a significant growth of discrete derivative norms Δ_3^j of order 3 which indicates (but not necessarily implies!) that we probably cannot expect improvements in (31) under the stated assumptions. It is an interesting open question to theoretically understand for which choices of approximate normals $\mathbf{n}^j$ improved smoothness of the normal re-parametrization can be expected. In particular, if $s_F \geq 3$, do we always obtain C^{3-} smooth limits for the Loop-based normal multiscale transform if we use the standard normals? Similar questions arose in connection with B-spline based normal multiscale transforms for curves, see [10], and have been partially answered in [9]. Addressing this question might also be key to explaining the numerically observed improved detail decay rates for the combined normal multiscale transforms proposed in [14].

References

1. Cavaretta, A.S., Dahmen, W., Micchelli, C.A.: Stationary Subdivision. In: Memoirs AMS, vol. 453, Amer. Math. Soc., Providence (1991)
2. Charina, M., Conti, C.: Polynomial reproduction of multivariate scalar subdivision schemes. J. Comput. Appl. Math. 240, 51–61 (2013)
3. Daubechies, I., Runborg, O., Sweldens, W.: Normal multiresolution approximation of curves. Constr. Approx. 20, 399–463 (2004)
4. Duchamp, T., Xie, G., Yu, T.P.Y.: Single basepoint subdivision schemes for manifold-valued data: time-symmetry without space-symmetry. Found. Comput. Math. (to appear, 2013)
5. Dyn, N.: Subdivision schemes in CAGD. In: Light, W. (ed.) Advances in Numerical Analysis, vol. 2, pp. 36–104. Oxford Univ. Press, Oxford (1992)

6. Grohs, P.: Smoothness equivalence properties of univariate subdivision schemes and their projection analogues. Numer. Math. 113, 163–180 (2009)
7. Grohs, P.: A general proximity analysis of nonlinear subdivision schemes. SIAM J. Math. Anal. 42, 729–750 (2010)
8. Guskov, I., Vidimče, K., Sweldens, W., Schröder, P.: Normal meshes. In: Proceedings 27th Annual Conference on Computer Graphics and Interactive Techniques, pp. 95–102. ACM Press/Addison-Wesley (2000)
9. Harizanov, S.: Analysis of nonlinear subdivision and multi-scale transforms. PhD Thesis, Jacobs University Bremen (2011)
10. Harizanov, S., Oswald, P., Shingel, T.: Normal multi-scale transforms for curves. Found. Comput. Math. 11, 617–656 (2011)
11. Khodakovsky, A., Guskov, I.: Compression of normal meshes. In: Geometric Modeling for Scientific Visualization, pp. 189–206. Springer, Berlin (2004)
12. Latour, V., Müller, N.W.: Stationary subdivision for general scaling matrices. Math. Z 227, 645–661 (1998)
13. Navayazdani, E., Yu, T.P.Y.: On Donoho's Log-Exp subdivision scheme: choice of retraction and time-symmetry. Multiscale Model. Sim. 9, 1801–1828 (2011)
14. Oswald, P.: Normal multi-scale transforms for surfaces. In: Boissonnat, J.-D., Chenin, P., Cohen, A., Gout, C., Lyche, T., Mazure, M.-L., Schumaker, L. (eds.) Curves and Surfaces 2011. LNCS, vol. 6920, pp. 527–542. Springer, Heidelberg (2012)
15. Oswald, P.: Normal multiscale transforms for surfaces: Shift-invariant topologies (in preparation)
16. Wallner, J., Navayazdani, E., Grohs, P.: Smoothness properties of Lie group subdivision schemes. Multiscale Model. Sim. 6, 493–505 (2007)
17. Warren, J., Weimer, H.: Subdivision Methods for Geometric Design: A Constructive Approach. Morgan Kaufmann Publ., San Francisco (2002)
18. Weinmann, A.: Subdivision schemes with general dilation in the geometric and nonlinear setting. J. Approx. Th. 164, 105–137 (2012)
19. Xie, G., Yu, T.P.Y.: Smoothness equivalence properties of general manifold-valued data subdivision schemes. Multiscale Model. Sim. 7, 1073–1100 (2008)

Joining Primal/Dual Subdivision Surfaces

Sergey Podkorytov[1], Christian Gentil[1],
Dmitry Sokolov[2], and Sandrine Lanquetin[1]

[1] LE2I - Université de Bourgogne
[2] LORIA - Universite Nancy I

Abstract. In this article we study the problem of constructing an intermediate surface between two other surfaces defined by different iterative construction processes. This problem is formalised with Boundary Controlled Iterated Function System model. The formalism allows us to distinguish between subdivision of the topology and subdivision of the mesh. Although our method can be applied to surfaces with quadrangular topology subdivision, it can be used with any mesh subdivision (primal scheme, dual scheme or other.) Conditions that guarantee continuity of the intermediate surface determine the structure of subdivision matrices. Depending on the nature of the initial surfaces and coefficients of the subdivision matrices we can characterise the differential behaviour at the connection points between the initial surfaces and the intermediate one. Finally we study the differential behaviour of the constructed surface and show the necessary conditions to obtain an almost everywhere differentiable surface.

Keywords: Iterative function system, attractor, surface, subdivision, differentiability.

1   Introduction

The global objective of our work is to develop a geometric modeller based on the paradigm of fractal geometry. More precisely we aim for modelling shapes with iterative processes. Our formalism covers traditional models like NURBS and subdivision surfaces and also a new world of shapes, not accessible by the polynomial models and having a particular aesthetic.

In this article we focus on the problem of connecting two shapes built by two different iterative processes. This problem often arises in context of subdivision curves and surfaces. Different biregular subdivision schemes have been studied for primal schemes ([SL03], [LL03], [SW05], [JLZ09], [BLND11]). Still now, no biregular subdivision have been performed on primal/dual schemes because the subdivision surface process is based on the mesh subdivision. The study of [KSD12] to transform a primal scheme into a dual scheme may open new perspectives for subdivision surfaces.

Our approach is based on the BCIFS model that explicitly code (implement) the topological subdivision of the limit surface. We describe the way to deal

M. Floater et al. (Eds.): MMCS 2012, LNCS 8177, pp. 403–424, 2014.
© Springer-Verlag Berlin Heidelberg 2014

with surfaces that has quadrangular topological subdivision, but any mesh subdivision (primal scheme, dual scheme, etc.) We are not limited by any type of meshes (quads, triangles, etc.) since with the BCIFS model the limit surface is defined as an attractor. As with classical Iterated Function System we can choose any starting shape (or type of mesh) to construct a precise enough approximation. We also study the differential properties of the surfaces generated by our method. We use the formalism of Boundary Controlled Iterated Function System [TBSG+06],[SGB12] to describe two surfaces and construct the intermediate surface. An automaton describes the construction process. Adjacency and incidence equations guarantee the continuity and define the structure of the subdivision matrices.

2   Background on IFS, CIFS and BCIFS

2.1   IFS and Projected IFS

Given a complete metric space (E, d) an Iterated Function System (*IFS*) is a finite set of contractive operators $\mathbb{T} = \{T_i\}_{i=0}^{N-1}$ acting on points of E.

It is possible to define an operator $\mathbb{T} : \mathcal{H}(E) \to \mathcal{H}(E)$, called Hutchinson operator. It maps non-empty compact subset K to $\bigcup_{i=0}^{N-1} T_i(K)$.

If operators T_i are contracting in the space (E, d), then operator $\mathbb{T}$ is contracting in $(\mathcal{H}(E), d_{\mathcal{H}})$, the space of non-empty compact subsets in E with the Hausdorff distance [Bar88].

According to [Hut81], there exists a unique compact A such that $\mathbb{T}(A) = A$, i.e., the fixed point of $\mathbb{T}$. Furthermore, due to the contractivity of $\mathbb{T}$, A can be calculated as the limit: $A = \lim_{i\to\infty} \mathbb{T}^i(K)$. This limit does not depend on the initial compact K as long as it is not empty. This property is illustrated by the sequence of images in figure 2 (top).

The idea of projective IFS was introduced by Zair and Tosan [ZT96]. By separating the iterative space from the modelling space, it is possible to construct fractal shapes with control points. Similar to splines determined by the basic functions defined in a barycentric space, attractors are defined in barycentric space whose dimensions correspond to the number of control points: $A \subset BI^n = \{\lambda \in \mathbb{R}^n | \sum_{i=0}^{n-1} \lambda_i = 1\}$, where n is the number of control points. Then the attractor is projected to the modelling space with the transformation defined by control points $PA = \{\sum_{i=0}^{n-1} P_i \lambda_i | \lambda_i \in A\}$, where $P = [P_0 \ P_1 \cdots P_{n-1}]$ is the vector composed of control points.

We limit our study to linear operator acting in the barycentric spaces. The operators on the barycentric space can be written as linear operators on $\mathbb{R}^n$, with a specific constraint on its matrix representation: each column of the matrix must have the sum of its elements equal to 1. The fixed point of such operator, as well as an attractor of the IFS composed of such operators, always belong to the barycentric plane.

2.2   Controlled IFS (CIFS)

It is possible to extend this model by adding rules controlling the iterative process. These rules can described with an automaton [PH94], [TT95], [MW88].
States of the automaton are associated with iterative spaces and arcs represent
transformation applicable at the current state. This gives a new way to control
the shape of the attractor. The left scheme in figure 1 represents the automaton
for the classic IFS which generates the BARNSLEY fern. The fern is self-similar,
i.e., it is built from an infinite number of copies of itself.

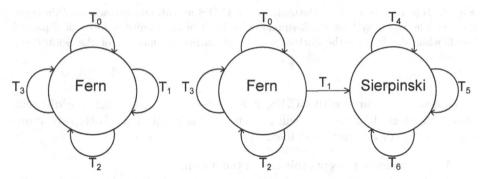

Fig. 1. The automatons giving the transformation applying rules. Left: simple automaton corresponding to the ordinary IFS. Right: Modified automaton with two states. The
attractor of the Fern state include the attractor of the Sierpinski state.

We can add a new state with three transformations defining the SIERPINSKI
triangle. We also change the destination of the arc T_1 (see right part in figure 1).
After the transformation T_1 is applied once, next steps will then follow the
SIERPINSKI subdivision. The attractor of the new controlled IFS is composed of
an infinite number of the SIERPINSKI triangles (see bottom row in figure 2).

2.3   Boundary Controlled IFS (BCIFS)

The Boundary Controlled IFS (BCIFS) model enhances the CIFS model by
adding the B-Rep notion [TBSG$^+$06], [SGB12]. It gives a way to explicitly state
the face-edge-vertex structure of the attractor [Gen92]. We can also write the
incidence and adjacency constraints on the subdivision process and thus control
the topology: classic (curve, surface, ...) or fractal topology.

B-Rep concepts used here are more general than the classical B-Rep concepts. Topological elements can be fractal objects. For example, a face can be a
SIERPINSKI triangle, or an edge can be a CANTOR set, but the B-Rep structure
remains consistent. This approach differs from the traditional model by the ability to clarify the relation of incidence and adjacency with the subdivision process
of the given topological structure. For the sake of simplicity we will present its
application to curves.

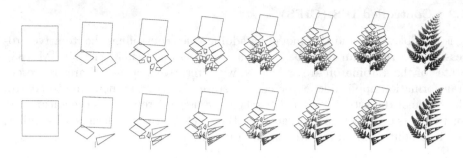

Fig. 2. Top: BARNSLEY fern; Bottom: with a C-IFS we can mix attractors of different nature. The Barnsley fern is self-similar, it is built of an infinite number of copies of itself, while the fern on the bottom is built of an infinite number of the SIERPINSKI triangles.

To describe a curve with BCIFS, it is necessary to distinguish the different spaces in which different cells will be defined. Each cell of the B-Rep structure (here an edge or a vertex) is defined by:

- A state which represents this cell in the automaton
- An iterative space associated with it, more specifically a barycentric space. The dimension of this space is equal to the number of control points of the cell
- An IFS reflecting the subdivision of the cell

For the curve defined by n control points and whose vertices depend on p control points we obtain the following structure:

- For the edge:
  - a state called e
  - an iterative space = a barycentric space of dimension n
  - an IFS = a set of at least two matrices ($n \times n$) representing edge subdivision
- For each vertex (which can be different):
  - two states v^l and v^r (for the left and right vertices respectively)
  - an iterative space = a barycentric space of dimension p
  - an IFS = a set of one matrix ($p \times p$) representing vertex subdivision

At this point we have a set of iterated function systems, where each IFS describes a cell of the B-Rep structure. If the IFS are composed of arbitrary operators there is no guarantee that the edge is really bordered by vertices and that subdivision of the edge does result in continuous curve. To address this issue we will use additional constraints on BCIFS matrices, but before that we need to add relations between different cells.

To do so we introduce boundary operators. In our example, different IFS associated with the edge and the vertices are defined in the barycentric spaces.

Boundary operators create a link between the space defining the system of the nested subspaces, i.e., the space for the vertex is a subspace of the space for the edge. As an edge has two vertices we use two boundary operators.

For example, consider a curve with three control points. If the first vertex depends on the first two control points and while the second one depends on the last two, then their respective boundary operators are:

$$b_0 = \begin{pmatrix} 1 & 0 \\ 0 & 1 \\ 0 & 0 \end{pmatrix}, b_1 = \begin{pmatrix} 0 & 0 \\ 1 & 0 \\ 0 & 1 \end{pmatrix}.$$

The general automaton for the curve with an edge subdivided in two parts and two different vertex subdivisions is presented in figure 3.

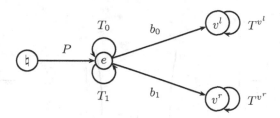

Fig. 3. An automaton representing a curve with two edge subdivision $T_{0,1}$ and two different vertex subdivisions T^{v^l}, T^{v^r}.

2.4   Topological Constraints

Incidence and adjacency constraints can be easily identified from the graph representing the progression of the automaton. The first level of subdivision is depicted in figure 4.

Adjacency constraints The edge is subdivided in two parts, so the "left" part has to be connected to the "right" one through the intermediate vertices (see figure 4). First of all the states v^l and v^r have to be identical, or in other word the "left" and the "right" vertices need to have the same "nature" and be subdivided by the same operator, otherwise the continuity is not ensured. So equality $T^{v^l} = T^{v^r} = T^v$ is necessary.

Another condition is deduced by writing the equivalence of paths in the graph: the left vertex of the right subdivision has to correspond to the right vertex of the left subdivision:

$$T_0 b_1 = T_1 b_0.$$

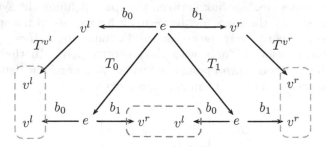

Fig. 4. The unfolding of the automaton generating subdivision system. This system is built in such way, that it guarantees the topological structure. The incidence constraints on the sides (in yellow) and adjacency constraints in the middle (in blue.)

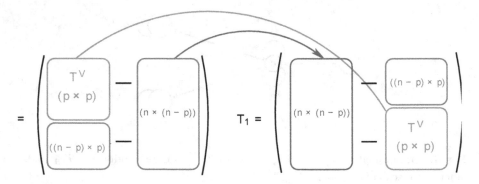

Fig. 5. General structure of the subdivision matrices for the curve with n control points, whose vertices are controlled by p control points.

Incidence constraints In the same manner, incidence constraints express the fact that vertices must remain at the ends of the edge during the subdivision process: subdivision of the left vertex of the edge has to correspond to the left vertex of the left subdivision of the edge (see figure 4). Writing down equivalences between paths gives the following equations:

$$b_0 T^v = T_0 b_0,$$

$$b_1 T^v = T_1 b_1.$$

Resolving the constraints, adjacency and incidence determine structures in the subdivision matrices in the form of equalities between columns and sub-matrices (see figure 5). Two examples of curves that can be described by a BCIFS are presented in figure 6.

Fig. 6. Left: a cubic spline with 4 control points. Right: a fractal curve with 3 control points.

2.5   Regular Quadrangular Surface Subdivision

The same principle can be used to describe the subdivision process for quadrangular surfaces. An automaton for such surface has three states: F for the face, E for the edges that are bordering the face (in this case all the edges have the same nature, i.e., are defined by the same iterative procedure) and V for the vertices (see figure 7).

For surfaces a larger number of the incidence and adjacency constraints is required than for curves, but the underlying principle is the same. Subdividing an edge should be equal to an edge of the patch subdivision, while neighbouring patches should share common edges. All the constraints can be illustrated by figure 8. Full list of constraints can be found in appendix A.

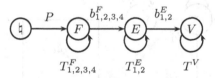

Fig. 7. An automaton for the surface with four edges

2.6   Irregular Quadrangular Surface Subdivision

Although in this article we focus on regular patches, our method can be modified work with irregular patches as well. Here we show how to modify the automaton to change the regular patch into a patch with one irregular vertex. Consider a quadrangular patch with one irregular vertex. Such patch is subdivided into three regular patches and one irregular. The corresponding automaton is presented in figure 9. Note that the irregular patch (F^{ir}) has two irregular edges (E^{ir}) as well as two regular ones (E). The irregular edge is subdivided ($T_{1,2}^{E^{ir}}$) into a regular and irregular edge. Conditions that guarantee the continuity of the irregular patch can be deduced in the same manned as for its regular counterpart.

In this article we focus only on regular subdivisions, but the example above shows that our approach can be tailored to irregular surfaces as well.

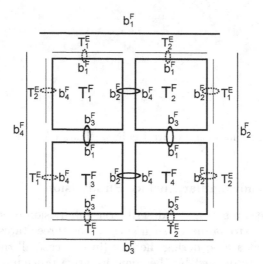

Fig. 8. Incidence and adjacency constraints for quadrangular surface subdivision. Each subdivided face has to be connected with neighbouring faces by sharing respective edges (represented by ellipses in solid), while edge subdivision must be an edge for one of the subdivided faces (dotted ellipses).

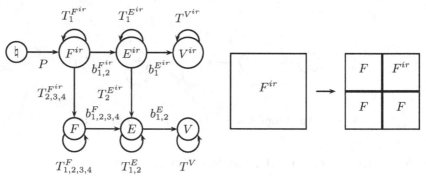

Fig. 9. Left: An automaton for the irregular patch. Right: a schematic subdivision of the irregular patch.

3   Construction of the Intermediate Surface.

At first we consider the automaton that defines two initial surfaces. To keep thing simple we consider the initial surface to be a square patches with four edges and four vertices, but the same method can be easily applied to surfaces with more complex configuration. The initial automaton is presented in figure 10. The initial state is denoted by ♮, while L and R denote state corresponding to the respective respective initial surfaces.

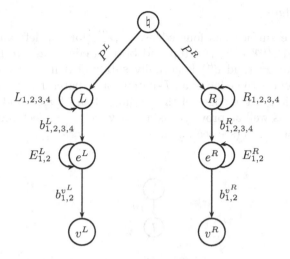

Fig. 10. Initial BCIFS. Each branch of the automaton corresponds to an initial surface.

Now we introduce a state I for the intermediate surface. And for this surface we propose the following subdivision. The intermediate surface is subdivided into six parts: two of them have the same nature as the first initial surface, another two as the second initial surface and the last two are similar to the intermediate surface itself. This subdivision and a graph representing it is shown in figure 11. Here we use the BCIFS to deduce the condition on the transformations of the intermediate patch to ensure its continuity. To do this we need to guarantee that the edges of the patch are continuous curves and that patches obtained after the subdivision are connected with each other, as described below.

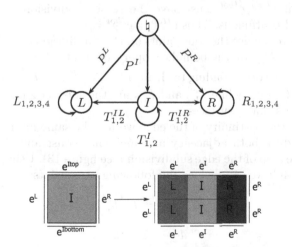

Fig. 11. Top: Automaton describing the subdivision of the intermediate patch. Note that edge and vertex states are omitted for conserving space. Bottom: Schematic representation of the same subdivision.

3.1 Outer Edges

The intermediate surface has four edges: e^L, e^R (for the left and right edges respectively) and e^{Itop}, e^{Ibot} (for top and bottom edges) as can be seen in figure 11. Note that e^L (and e^R) is actually subdivided into two smaller edges, while each of them is the edge of an L-patch (or R-patch respectively). We can use the subdivision of the edges of the initial patches for e^L and e^R guarantee their continuity, as well as provided us an easy we to connect the intermediate surface with initial ones (see section 3.3).

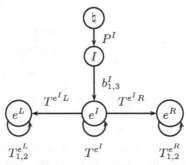

Fig. 12. The automaton depicting top and bottom edges subdivision of the intermediate patch.

Edges e^{Itop}, e^{Ibot} differ from e^L and e^R as they are subdivided into three parts different instead of two uniform parts. Also note that after one iteration two intermediate are produced and they share a common edge (see figure 11). This means that e^{Itop}, e^{Ibot} must have the same subdivision, so that that the surface is indeed continuous. Thus $e^{Itop} = e^{Ibot} = e^I$.

Now we want to deduce the constraints on the subdivision of e^I. It is presented in figure 12. Here $b^I_{1,3}$ are the boundary operators that selects the edges in question, e^I is the state corresponding to them, $T^{e^I L}, T^{e^I R}$ and T^{e^I} are the operators of the edge subdivision, while $T^{e^R}_{1,2}$ and $T^{e^L}_{1,2}$ are the subdivision operators for the edges of respected initial surfaces.

To guarantee the continuity of the edge we use the same method as in section 2.4. We can deduce both adjacency and incidence constraints from the graph representing one step of the edge subdivision (see figure 13). Using the equivalent paths in this graph we can deduce the following constraints:

$$T^{e^I L} b^{e^L}_2 = T^{e^I} b^{e^I}_1,$$

$$T^{e^I} b^{e^I}_2 = T^{e^I R} b^{e^R}_1,$$

$$b^{e^I}_1 T^v_1 = T^{e^I L} b^L_1,$$

$$b^{e^I}_2 T^v_2 = T^{e^I R} b^R_2.$$

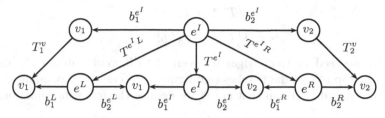

Fig. 13. Part of the graph representing one iteration of the automaton.

3.2 Inner Edges

There are 7 shared edges between the patches obtained after the first iteration (see figure 11). For each of them we need to write a constraint on the subdivision operators.

At first we consider two R-patches. One step of the subdivision that produces this two patches is presented in figure 14. (Note that nodes not related to the considered patches or to the common edge are not presented). From the

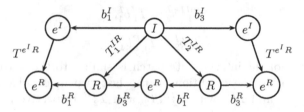

Fig. 14. Part of the graph representing one step of the subdivision. This part focuses on two R patches and the common edge between them.

equivalent paths in this graph we can deduce the following equations:

$$T_1^{IR} b_3^R = T_2^{IR} b_1^R,$$

$$b_1^I T^{e^I R} = T_1^{IR} b_1^R,$$

$$b_3^I T^{e^I R} = T_2^{IR} b_3^R.$$

In the similar ways the condition on edges between the I-patches:

$$T_1^I b_3^I = T_2^I b_1^I,$$

$$b_1^I T^{e^I} = T_1^I b_1^I,$$

$$b_3^I T^{e^I} = T_2^I b_3^I.$$

And between the L-patches:

$$T_1^{IL} b_3^L = T_2^{IL} b_1^L,$$

$$b_1^I T^{e^I L} = T_1^{IL} b_1^L,$$

$$b_3^I T^{e^I L} = T_2^{IL} b_3^L.$$

Now we proceed to treat edges between the different patches. At first we consider the top row of the patches (see figure 11). The corresponding graph is presented in figure 15. From that graph we deduce the following equations:

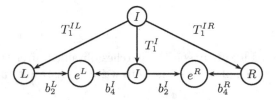

Fig. 15. Part of the graph representing one step of the subdivision. This part focuses on top row of three patches and edges between them.

$$T_1^{IL} b_2^L = T_1^I b_4^I,$$

$$T_1^I b_2^I = T_1^{IR} b_4^R.$$

Note the top row of the intermediate patch is similar to the bottom one. This means that to obtain the condition on the last two inner edges, we need to substitute T_1^{IL}, T_1^I and T_1^{IR} with T_2^{IL}, T_2^I and T_2^{IR} respectively. This way we obtain the following equations:

$$T_2^{IL} b_2^L = T_2^I b_4^I,$$

$$T_2^I b_2^I = T_2^{IR} b_4^R.$$

3.3   Connection to the Initial Surfaces

Finally we need to ensure that the intermediate patch actually connects two initial surfaces. Since we chose to use the subdivision of the initial surfaces edges for e^L and e^R, we can connect the initial surface to the intermediate one by using the same control points for respective edges. This can be written as following:

$$P^I b_2^I = P^R b_4^R,$$

$$P^I b_4^I = P^L b_2^L,$$

where P^I, P^R, P^L are vectors of control points for respective surfaces. From these condition we can also deduce the minimum dimension of BI^I. That is a sum of dimension for initial surfaces respective edges. The said dimension can be increased, therefore introducing additional control points to the intermediate patch.

3.4   Approximation of the Limit Surface

The adjacency and incidence condition described above only guarantee the continuity of the limit shape. Hence the iterations with arbitrary compact set as a starting point are not necessary continuous. However with correct starting set, we can obtain a continuous surface that approximate the limit shape as the number of iterations increases. For each face state a quad is chosen as a compact set to be transformed accordingly with the rules of the BCIFS. Said quads of course lie within the corresponding barycentric space. The vertices of each quad must be chosen specifically that incidence and adjacency condition are conserved with each iteration. This is true if each vertex of the tetragon lies within barycentric subspace which corresponds to that vertex iterative subspace and have the same coordinates within the subspaces. One way to satisfy that condition is to choose the fixed points of respective transformations as quads vertices.

Different steps of approximation of the intermediate surface is presented in figure 16. Different quads are color coded to show their "origin" from different barycentric spaces. Green is for an image of the quad from BI^I (barycentric space corresponding to the state I), red is for BI^R and blue is for BI^L.

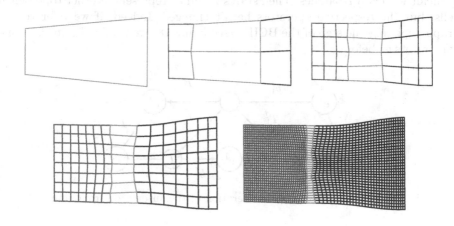

Fig. 16. Continuous approximation of the limit surface

Here we choose quads for convenience, but choosing any other configuration will yield the sequence of approximations that converge to the same limit surface.

4   Differentiability

In this section we study the differential properties of the surfaces that was described in the previous section.

It is easy to see that as the subdivision process goes on, the intermediate patch is tasselled with patches that has the same subdivision as the initial surfaces.

In this paragraph we show that there exists a self similar curve between two different areas tiled with images of the first initial patch or the second one. If we discard the states L, R, e^R, e^L and corresponding transitions, we obtain the following automaton (see figure 17). Note that state e^I lost two of three

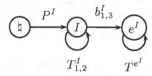

Fig. 17. BCIFS without states L, R, e^R, e^L and respective transitions

transition leading from it. This means that it has become a vertex state. Similar thing has happened with state I. This state now represents not a face, but an edge. So we obtained an edge, that is bounded by two vertices, and now we show that it is continuous.

Previously in section 3.1 we have already derived constraints on $T_{1,2}^I$, when we dealt with two I-patches. Then states I and e^I represented other topological cells, but the respective operators hasn't changed. Indeed, if we refer to the graph depicting one step of the BCIFS (see figure 18) we would obtain the same constraints as before.

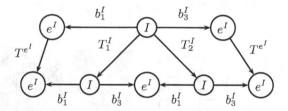

Fig. 18. Graph representing one iteration of the automaton from figure 17

So with exception of one continuous curve, the whole intermediate surface is tiled with images of the initial surfaces. This means, that said area inherits its differential properties from the initial surfaces.

Now we proceed to study the differential properties of the constructed surface along and across the special curve we described previously. At first we study the differential behaviour at the endpoint of this curve. Note that the said endpoints are the fixed points of T_1^I and T_2^I respectively.

At first we consider the fixed point of T_1^I. Let us denote the intermediate surface with S. Without the loss of generality we can assume that the fixed point of T_1^I is 0, i.e., $T_1^I(0) = 0$. Let $\{\lambda_j\}_{j=0}^n$ be a set of eigenvalues of T_1^I such that $1 = \lambda_0 > |\lambda_1| = |\lambda_2| > |\lambda_3| \geq \ldots \geq |\lambda_n|$, $\lambda_0 = 1$ and λ_1, λ_2 are positive. We want to prove the hypothesis, that the eigenvectors v_1 and v_2 are the basis of tangent space. To prove that we need to demonstrate, that for any sequence

$\{y_n\}_{n=0}^{\infty}$ such that $y_n \to 0$, $\forall n$ $y_n \in S$ the following is true: $f_n = f(y_n) \to 0$ where $f(x) = \frac{dist(x,p)}{|Pr_p(x)|} \to 0$, $p = \{y | y = a\mathbf{v_1} + b\mathbf{v_2}, \forall a, b \in \mathbb{R}\}$ and Pr_p is a projection onto plane p.

Let us consider a sequence of compact sets A_n where $A_0 = S$ and $A_n = T_1^I(A_{n-1})$, $\forall n > 0$. Let us also define another sequence $B_n = \overline{A_n \setminus A_{n+1}}$, a closure of the difference between A_n and A_{n+1}. Note that $\forall n \geq 0$ B_n is also a compact set and $0 \notin B_n$. Also the following is true:

$$\cup_{n=0}^{\infty} B_n = A_0.$$

Here is the more general version of the previous equality:

$$\cup_{n=k}^{\infty} B_n = A_k$$

Without the loss of generality we can consider the plane p to be the plane Oxy. Then:

$$f(x) = \frac{|x - Pr_p(x)|}{|Pr_p(x)|} = \frac{|(0,0,0,x_3,\ldots,x_n)|}{|(0,x_1,x_2,0,\ldots,0)|}.$$

Let $x_1 \in B_1$, then $\exists x_0 \in B_0$ such that $T_0(x_0) = x_1$.

$$f(x_1) = f(T_1^I(x_0)) = \frac{|(0,0,0,\lambda_3 x_3,\ldots,\lambda_n x_n)|}{|(0,\lambda_1 x_1,\lambda_2 x_2,0,\ldots,0)|} \leq$$

$$\leq \frac{|\lambda_3 \cdot (0,0,0,x_3,\ldots,x_n)|}{|\lambda_1 \cdot (0,x_1,x_2,0,\ldots,0)|} = \frac{|\lambda_3|}{|\lambda_1|} f(x_0).$$

So for any n we can write the following:

$$\max_{x \in B_n} f(x) = \frac{|\lambda_3|}{|\lambda_1|} \max_{x \in B_{n-1}} f(x) = \left(\frac{|\lambda_3|}{|\lambda_1|}\right)^n \max_{x \in B_0} f(x).$$

Since $\frac{|\lambda_3|}{|\lambda_1|} < 1$:

$$\lim_{n \to \infty} \max_{x \in B_n} f(x) = 0.$$

This also implies that $\max_{x \in A_n} f(x) \to 0$. Therefore $\forall y_n$ such that $\lim_{n \to \infty} f(y_n) = 0$.

So the sufficient condition for existence of tangent plane at the fixed point of T_1^I are the following two statements:

− T_1^I have two equal positive sub-dominant eigenvalues.
− The corresponding eigenspaces are not the same.

Same conditions can be applied for the fixed point of T_2^I.

Let us assume, that tangent planes exist at the fixed point of T_1^I and T_2^I. Since the curve in question is continuous, applying T_1^I to the fixed point of T_2^I yields the same result as applying T_2^I to the fixed point of T_1^I. If the results of applying these operator to the normals of the respective tangent planes are collinear, then the tangent plane also exists at the "middle point".

Applying different finite combinations of $T_{1,2}^I$ we can obtain a set of points along the curve such that the tangent plane to the surface exists at any such point. Such set is dense within the curve (see [Ben09].)

5   Refining the BCIFS

So far we have deduced a BCIFS that produces a variety of limit surfaces satisfying the initial conditions. A specific surface can be chosen by specifying the parameter values for the BCIFS operators. Unfortunately, specifying parameter directly can have a somewhat unpredictable effect on the final shape. In the previous section we have shown, that the shape of the surface depends on the sub-dominant eigenvectors of the operators $T_{1,2}^I$. Now we are going to show how control points can be added to provide more intuitive way of designing the intermediate surface.

As was already noted before a special curve (an attractor of $T_{1,2}^I$) runs across the intermediate surface. We are going to reflect this in the BCIFS and assign new control points to directly influence the shape of this curve and therefore the surface itself. Explicit specification of the special curve allows us to view the intermediate surface as two quadrangular patches that share the common edge. This allows us to redefine its subdivision as presented in figure 19. As these two patches are mirrored copies of each other we can use the same automaton to resolve the adjacency and incidence constraints.

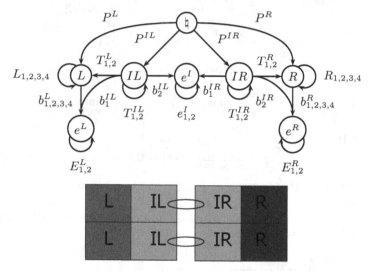

Fig. 19. Top: Refined BCIFS with two intermediate states. Bottom: Corresponding subdivision scheme.

But before we can resolve the constraints we need to specify the subdivision of the common edge. Virtually any subdivision may be chosen, but our goal is to build a smooth surface, so some restrictions apply. First of all the curve needs to be C^1 itself. For more studies on differentiability of self-similar curves please refer to [BGN09], [PGSL13] and [SGB12].

In the previous section we showed that existence of the tangent quarter-planes depend on the eigenvalues of $T_{1,2}^{IL}$ or $T_{1,2}^{IR}$ respectively. If two positive equal sub-dominant eigenvalues exist then the tangent plane will be collinear with respective eigenspace. In turn, tangent plane position in the modelling space depends on those control points which correspond to the smallest subspace that contains the tangent plane. So to be able to guarantee C^1-continuity between two patches the tangent planes must depend on the control points shared by two patches - control points of the common edge.

These can be further exploited if we consider the structure of the subdivision matrices. When we consider two matrices that are constrained by a boundary condition, e.g., a face and an edge subdivision, the face subdivision will always have block-triangular structure, with one block equal to the edge subdivision. For example:

$$T_1^{IL} = \begin{pmatrix} e_1^I & A_1 \\ 0 & B_1 \end{pmatrix}, \quad T_2^{IL} = \begin{pmatrix} e_2^I & A_2 \\ 0 & B_2 \end{pmatrix}$$

where e_1^I and B_1, B_2 are square blocks.

The eigenvalues of the edge subdivision are also eigenvalues of the face sub-division, while the corresponding eigenvector are embedded into the higher dimensional space. Note that if two patches share the same edge and hence the same edge subdivision, their face subdivision will have common blocks. In our case:

$$T_1^{IR} = \begin{pmatrix} e_1^I & A_3 \\ 0 & B_3 \end{pmatrix}, \quad T_2^{IR} = \begin{pmatrix} e_2^I & A_4 \\ 0 & B_4 \end{pmatrix}.$$

So to guarantee the C^1-continuity the edge subdivision block must provide two sub-dominant eigenvalues, so that the same tangent plane can be obtain on both sides of the special curve.

6   Examples

In this section we provide some examples of the surfaces constructed by the proposed method. The first example (see figure 20) is an intermediate surface between a Doo-Sabin and Catmull-Clark surfaces. The intermediate curve subdivision is obtained by tensor product of cubic b-spline subdivision and quadratic b-spline vertex subdivision. Hence $4 \times 2 = 8$ control points for the intermediate curve. The subdivision matrices used to generate this image can be found in appendices B and C. Note that in this particular case a complete basis of eigenvectors does not exist, so to check the differentiability generalised eigenvectors were used. Because proof is similar to one presented is section 4 we omit it to preserve space.

Our method can also be applied to fractal surfaces, and even spline patches can be connected to fractal surfaces (see figure 21).

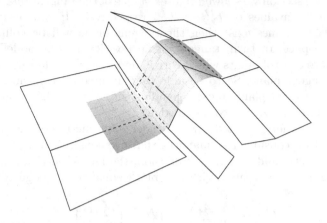

Fig. 20. An intermediate surface between the Doo-Sabin and Catmull-Clark surfaces with control polygons for initial edges of initial surfaces and intermediate curve

Fig. 21. An intermediate surface between a fractal surfaces and spline patch

7   Conclusion

Using the BCIFS model we have proposed a method of constructing the intermediate surface between two surfaces build by different iterative processes. We used the fact, that BCIFS distinguishes the topology and mesh subdivision, and were able to work around the differences between primal and dual subdivision schemes. With the chosen topology subdivision of the intermediate surface we guaranteed continuity with adjacency and incidence constraints.

After that we proceeded to analyse the differential behaviour of the intermediate surface. We were able to show, that the surface can be split into two different areas, that shares the differential behaviour with respective initial surface. We showed that the border between these two areas has specific differential behaviour, and that existence of the tangent plane at the curves points depend on the eigenvalues and eigenvector of the subdivision operators.

Finally we modified the automaton and explicitly specify the special curve on the intermediate surface. New control were added to control the shape of the curve and therefore shape of the surface. Condition that guarantee the C^1-continuity along that curve were also derived.

A   Appendix 1

List of conditions for quadrangular subdivision from figure 8.
   Adjacency constraints:

- Top edge:
$$b_1^F T_1^E = T_1^F b_1^F$$
$$b_1^F T_2^E = T_2^F b_1^F$$

- Right edge:
$$b_2^F T_1^E = T_2^F b_2^F$$
$$b_2^F T_2^E = T_4^F b_2^F$$

- Bottom edge:
$$b_3^F T_1^E = T_3^F b_3^F$$
$$b_3^F T_2^E = T_4^F b_3^F$$

- Left edge:
$$b_4^F T_2^E = T_1^F b_4^F$$
$$b_4^F T_1^E = T_3^F b_4^F$$

Incidence constraints:

$$T_1^F b_2^F = T_2^F b_4^F$$
$$T_2^F b_3^F = T_4^F b_1^F$$
$$T_4^F b_4^F = T_3^F b_2^F$$
$$T_3^F b_1^F = T_1^F b_3^F$$

B   Doo-Sabin Side Subdivision Matrices

The following matrices were used to generate Doo-Sabin side of the surface from figure 20:

$$T_1^L = \begin{pmatrix}
9/16 & 3/16 & 0 & 3/16 & 1/16 & 0 & a & b & 0 \\
3/16 & 9/16 & 9/16 & 1/16 & 3/16 & 3/16 & b & a & a \\
0 & 0 & 3/16 & 0 & 0 & 1/16 & 0 & 0 & b \\
3/16 & 1/16 & 0 & 9/16 & 3/16 & 0 & d & e & 0 \\
1/16 & 3/16 & 3/16 & 3/16 & 9/16 & 9/16 & e & d & d \\
0 & 0 & 1/16 & 0 & 0 & 3/16 & 0 & 0 & e \\
0 & 0 & 0 & 0 & 0 & 0 & g & i & 0 \\
0 & 0 & 0 & 0 & 0 & 0 & h & h & g \\
0 & 0 & 0 & 0 & 0 & 0 & i & g & h \\
0 & 0 & 0 & 0 & 0 & 0 & 0 & 0 & i \\
0 & 0 & 0 & 0 & 0 & 0 & k & m & 0 \\
0 & 0 & 0 & 0 & 0 & 0 & l & l & k \\
0 & 0 & 0 & 0 & 0 & 0 & m & k & l \\
0 & 0 & 0 & 0 & 0 & 0 & 0 & 0 & m
\end{pmatrix}$$

$$T_2^L = \begin{pmatrix}
3/16 & 0 & 0 & 1/16 & 0 & 0 & b & 0 & 0 \\
9/16 & 9/16 & 3/16 & 3/16 & 3/16 & 1/16 & a & a & b \\
0 & 3/16 & 9/16 & 0 & 1/16 & 3/16 & 0 & b & a \\
1/16 & 0 & 0 & 3/16 & 0 & 0 & e & 0 & 0 \\
3/16 & 3/16 & 1/16 & 9/16 & 9/16 & 3/16 & d & d & e \\
0 & 1/16 & 3/16 & 0 & 3/16 & 9/16 & 0 & e & d \\
0 & 0 & 0 & 0 & 0 & 0 & i & 0 & 0 \\
0 & 0 & 0 & 0 & 0 & 0 & h & g & i \\
0 & 0 & 0 & 0 & 0 & 0 & g & h & h \\
0 & 0 & 0 & 0 & 0 & 0 & 0 & i & g \\
0 & 0 & 0 & 0 & 0 & 0 & m & 0 & 0 \\
0 & 0 & 0 & 0 & 0 & 0 & l & k & m \\
0 & 0 & 0 & 0 & 0 & 0 & k & l & l \\
0 & 0 & 0 & 0 & 0 & 0 & 0 & m & k
\end{pmatrix}$$

$$T_1^{IL} = \begin{pmatrix}
3/16 & 1/16 & 0 & a & b & 0 & 0 & 0 & 0 & 0 & 0 & 0 & 0 \\
1/16 & 3/16 & 3/16 & b & a & a & 0 & 0 & 0 & 0 & 0 & 0 & 0 \\
0 & 0 & 1/16 & 0 & 0 & b & 0 & 0 & 0 & 0 & 0 & 0 & 0 \\
9/16 & 3/16 & 0 & d & e & 0 & 0 & 0 & 0 & 0 & 0 & 0 & 0 \\
3/16 & 9/16 & 9/16 & e & d & d & 0 & 0 & 0 & 0 & 0 & 0 & 0 \\
0 & 0 & 3/16 & 0 & 0 & e & 0 & 0 & 0 & 0 & 0 & 0 & 0 \\
0 & 0 & 0 & g & i & 0 & 3/8 & 3/32 & 0 & 0 & 1/8 & 1/32 & 0 & 0 \\
0 & 0 & 0 & h & h & g & 3/8 & 9/16 & 3/8 & 3/32 & 1/8 & 3/16 & 1/8 & 1/32 \\
0 & 0 & 0 & i & g & h & 0 & 3/32 & 3/8 & 9/16 & 0 & 1/32 & 1/8 & 3/16 \\
0 & 0 & 0 & 0 & 0 & i & 0 & 0 & 0 & 3/32 & 0 & 0 & 0 & 1/32 \\
0 & 0 & 0 & k & m & 1/8 & 1/32 & 0 & 0 & 3/8 & 3/32 & 0 & 0 \\
0 & 0 & 0 & l & l & k & 1/8 & 3/16 & 1/8 & 1/32 & 3/8 & 9/16 & 3/8 & 3/32 \\
0 & 0 & 0 & m & k & l & 0 & 1/32 & 1/8 & 3/16 & 0 & 3/32 & 3/8 & 9/16 \\
0 & 0 & 0 & 0 & 0 & m & 0 & 0 & 0 & 1/32 & 0 & 0 & 0 & 3/32
\end{pmatrix}$$

$$T_2^{IL} = \begin{pmatrix}
1/16 & 0 & 0 & b & 0 & 0 & 0 & 0 & 0 & 0 & 0 & 0 & 0 \\
3/16 & 3/16 & 1/16 & a & a & b & 0 & 0 & 0 & 0 & 0 & 0 & 0 \\
0 & 1/16 & 3/16 & 0 & b & a & 0 & 0 & 0 & 0 & 0 & 0 & 0 \\
3/16 & 0 & 0 & e & 0 & 0 & 0 & 0 & 0 & 0 & 0 & 0 & 0 \\
9/16 & 9/16 & 3/16 & d & d & e & 0 & 0 & 0 & 0 & 0 & 0 & 0 \\
0 & 3/16 & 9/16 & 0 & e & d & 0 & 0 & 0 & 0 & 0 & 0 & 0 \\
0 & 0 & 0 & i & 0 & 0 & 3/32 & 0 & 0 & 0 & 1/32 & 0 & 0 & 0 \\
0 & 0 & 0 & h & g & i & 9/16 & 3/8 & 3/32 & 0 & 3/16 & 1/8 & 1/32 & 0 \\
0 & 0 & 0 & g & h & h & 3/32 & 3/8 & 9/16 & 3/8 & 1/32 & 1/8 & 3/16 & 1/8 \\
0 & 0 & 0 & 0 & i & g & 0 & 0 & 3/32 & 3/8 & 0 & 0 & 1/32 & 1/8 \\
0 & 0 & 0 & m & 0 & 0 & 1/32 & 0 & 0 & 0 & 3/32 & 0 & 0 & 0 \\
0 & 0 & 0 & l & k & m & 3/16 & 1/8 & 1/32 & 0 & 9/16 & 3/8 & 3/32 & 0 \\
0 & 0 & 0 & k & l & l & 1/32 & 1/8 & 3/16 & 1/8 & 3/32 & 3/8 & 9/16 & 3/8 \\
0 & 0 & 0 & 0 & m & k & 0 & 0 & 1/32 & 1/8 & 0 & 0 & 3/32 & 3/8
\end{pmatrix}$$

The following values were substituted to obtain a C^1-continuous surface: $a = 0, b = 0, d = 1/4, e = 1/4, g = 1/4, h = 1/4, i = 0, k = 0, l = 0, m = 0$.

C   Catmull-Clark Side Subdivision Matrices

The following matrices were used to generate Catmull-Clark side of the surface from figure 20:

$$T_1^R = \begin{pmatrix}
1/4 & 1/16 & 0 & 0 & 1/16 & 1/64 & 0 & 0 & 0 & 0 & 0 & 0 & ma & na & 0 & 0 \\
1/4 & 3/8 & 1/4 & 1/16 & 1/16 & 3/32 & 1/16 & 1/64 & 0 & 0 & 0 & 0 & mb & nb & ma & na \\
0 & 1/16 & 1/4 & 3/8 & 0 & 1/64 & 1/16 & 3/32 & 0 & 0 & 0 & 0 & 0 & nc & mb & nb \\
0 & 0 & 0 & 1/16 & 0 & 0 & 0 & 1/64 & 0 & 0 & 0 & 0 & 0 & 0 & 0 & nc \\
1/4 & 1/16 & 0 & 0 & 3/8 & 3/32 & 0 & 0 & 1/4 & 1/16 & 0 & 0 & me & ne & 0 & 0 \\
1/4 & 3/8 & 1/4 & 1/16 & 3/8 & 9/16 & 3/8 & 3/32 & 1/4 & 3/8 & 1/4 & 1/16 & mf & nf & me & ne \\
0 & 1/16 & 1/4 & 3/8 & 0 & 3/32 & 3/8 & 9/16 & 0 & 1/16 & 1/4 & 3/8 & 0 & ng & mf & nf \\
0 & 0 & 0 & 1/16 & 0 & 0 & 0 & 3/32 & 0 & 0 & 0 & 1/16 & 0 & 0 & 0 & ng \\
0 & 0 & 0 & 0 & 1/16 & 1/64 & 0 & 0 & 1/4 & 1/16 & 0 & 0 & mi & ni & 0 & 0 \\
0 & 0 & 0 & 0 & 1/16 & 3/32 & 1/16 & 1/64 & 1/4 & 3/8 & 1/4 & 1/16 & mj & nj & mi & ni \\
0 & 0 & 0 & 0 & 0 & 1/64 & 1/16 & 3/32 & 0 & 1/16 & 1/4 & 3/8 & 0 & nk & mj & nj \\
0 & 0 & 0 & 0 & 0 & 0 & 0 & 1/64 & 0 & 0 & 0 & 1/16 & 0 & 0 & 0 & nk \\
0 & 0 & 0 & 0 & 0 & 0 & 0 & 0 & 0 & 0 & 0 & 0 & mm & nm & 0 & 0 \\
0 & 0 & 0 & 0 & 0 & 0 & 0 & 0 & 0 & 0 & 0 & 0 & mn & nn & mm & nm \\
0 & 0 & 0 & 0 & 0 & 0 & 0 & 0 & 0 & 0 & 0 & 0 & 0 & no & mn & nn \\
0 & 0 & 0 & 0 & 0 & 0 & 0 & 0 & 0 & 0 & 0 & 0 & 0 & 0 & 0 & no \\
0 & 0 & 0 & 0 & 0 & 0 & 0 & 0 & 0 & 0 & 0 & 0 & mq & nq & 0 & 0 \\
0 & 0 & 0 & 0 & 0 & 0 & 0 & 0 & 0 & 0 & 0 & 0 & mr & nr & mq & nq \\
0 & 0 & 0 & 0 & 0 & 0 & 0 & 0 & 0 & 0 & 0 & 0 & 0 & ns & mr & nr \\
0 & 0 & 0 & 0 & 0 & 0 & 0 & 0 & 0 & 0 & 0 & 0 & 0 & 0 & 0 & ns
\end{pmatrix}$$

$$
T_2^R =
\begin{pmatrix}
1/16 & 0 & 0 & 0 & 1/64 & 0 & 0 & 0 & 0 & 0 & 0 & 0 & na & 0 & 0 & 0 \\
3/8 & 1/4 & 1/16 & 0 & 3/32 & 1/16 & 1/64 & 0 & 0 & 0 & 0 & 0 & nb & ma & na & 0 \\
1/16 & 1/4 & 3/8 & 1/4 & 1/64 & 1/16 & 3/32 & 1/16 & 0 & 0 & 0 & 0 & nc & mb & nb & ma \\
0 & 0 & 1/16 & 1/4 & 0 & 0 & 1/64 & 1/16 & 0 & 0 & 0 & 0 & 0 & 0 & nc & mb \\
1/16 & 0 & 0 & 0 & 3/32 & 0 & 0 & 0 & 1/16 & 0 & 0 & 0 & ne & 0 & 0 & 0 \\
3/8 & 1/4 & 1/16 & 0 & 9/16 & 3/8 & 3/32 & 0 & 3/8 & 1/4 & 1/16 & 0 & nf & me & ne & 0 \\
1/16 & 1/4 & 3/8 & 1/4 & 3/32 & 3/8 & 9/16 & 3/8 & 1/16 & 1/4 & 3/8 & 1/4 & ng & mf & nf & me \\
0 & 0 & 1/16 & 1/4 & 0 & 0 & 3/32 & 3/8 & 0 & 0 & 1/16 & 1/4 & 0 & 0 & ng & mf \\
0 & 0 & 0 & 0 & 1/64 & 0 & 0 & 0 & 1/16 & 0 & 0 & 0 & ni & 0 & 0 & 0 \\
0 & 0 & 0 & 0 & 3/32 & 1/16 & 1/64 & 0 & 3/8 & 1/4 & 1/16 & 0 & nj & mi & ni & 0 \\
0 & 0 & 0 & 0 & 1/64 & 1/16 & 3/32 & 1/16 & 1/16 & 1/4 & 3/8 & 1/4 & nk & mj & nj & mi \\
0 & 0 & 0 & 0 & 0 & 0 & 1/64 & 1/16 & 0 & 0 & 1/16 & 1/4 & 0 & 0 & nk & mj \\
0 & 0 & 0 & 0 & 0 & 0 & 0 & 0 & 0 & 0 & 0 & 0 & nm & 0 & 0 & 0 \\
0 & 0 & 0 & 0 & 0 & 0 & 0 & 0 & 0 & 0 & 0 & 0 & nn & mm & nm & 0 \\
0 & 0 & 0 & 0 & 0 & 0 & 0 & 0 & 0 & 0 & 0 & 0 & no & mn & nn & mm \\
0 & 0 & 0 & 0 & 0 & 0 & 0 & 0 & 0 & 0 & 0 & 0 & 0 & 0 & no & mn \\
0 & 0 & 0 & 0 & 0 & 0 & 0 & 0 & 0 & 0 & 0 & 0 & nq & 0 & 0 & 0 \\
0 & 0 & 0 & 0 & 0 & 0 & 0 & 0 & 0 & 0 & 0 & 0 & nr & mq & nq & 0 \\
0 & 0 & 0 & 0 & 0 & 0 & 0 & 0 & 0 & 0 & 0 & 0 & ns & mr & nr & mq \\
0 & 0 & 0 & 0 & 0 & 0 & 0 & 0 & 0 & 0 & 0 & 0 & 0 & 0 & ns & mr
\end{pmatrix}
$$

$$
T_1^{IR} =
\begin{pmatrix}
1/16 & 1/64 & 0 & 0 & 0 & 0 & 0 & 0 & ma & na & 0 & 0 & 0 & 0 & 0 & 0 & 0 & 0 & 0 \\
1/16 & 3/32 & 1/16 & 1/64 & 0 & 0 & 0 & 0 & mb & nb & ma & na & 0 & 0 & 0 & 0 & 0 & 0 & 0 \\
0 & 1/64 & 1/16 & 3/32 & 0 & 0 & 0 & 0 & 0 & nc & mb & nb & 0 & 0 & 0 & 0 & 0 & 0 & 0 \\
0 & 0 & 0 & 1/64 & 0 & 0 & 0 & 0 & 0 & 0 & 0 & nc & 0 & 0 & 0 & 0 & 0 & 0 & 0 \\
3/8 & 3/32 & 0 & 0 & 1/4 & 1/16 & 0 & 0 & me & ne & 0 & 0 & 0 & 0 & 0 & 0 & 0 & 0 & 0 \\
3/8 & 9/16 & 3/8 & 3/32 & 1/4 & 3/8 & 1/4 & 1/16 & mf & nf & me & ne & 0 & 0 & 0 & 0 & 0 & 0 & 0 \\
0 & 3/32 & 3/8 & 9/16 & 0 & 1/16 & 1/4 & 3/8 & 0 & ng & mf & nf & 0 & 0 & 0 & 0 & 0 & 0 & 0 \\
0 & 0 & 0 & 3/32 & 0 & 0 & 0 & 1/16 & 0 & 0 & 0 & ng & 0 & 0 & 0 & 0 & 0 & 0 & 0 \\
1/16 & 1/64 & 0 & 0 & 1/4 & 1/16 & 0 & 0 & mi & ni & 0 & 0 & 0 & 0 & 0 & 0 & 0 & 0 & 0 \\
1/16 & 3/32 & 1/16 & 1/64 & 1/4 & 3/8 & 1/4 & 1/16 & mj & nj & mi & ni & 0 & 0 & 0 & 0 & 0 & 0 & 0 \\
0 & 1/64 & 1/16 & 3/32 & 0 & 1/16 & 1/4 & 3/8 & 0 & nk & mj & nj & 0 & 0 & 0 & 0 & 0 & 0 & 0 \\
0 & 0 & 0 & 1/64 & 0 & 0 & 0 & 1/16 & 0 & 0 & 0 & nk & 0 & 0 & 0 & 0 & 0 & 0 & 0 \\
0 & 0 & 0 & 0 & 0 & 0 & 0 & 0 & mm & nm & 0 & 0 & 3/8 & 3/32 & 0 & 0 & 1/8 & 1/32 & 0 \\
0 & 0 & 0 & 0 & 0 & 0 & 0 & 0 & mn & nn & mm & nm & 3/8 & 9/16 & 3/8 & 3/32 & 3/16 & 1/8 & 1/32 \\
0 & 0 & 0 & 0 & 0 & 0 & 0 & 0 & 0 & no & mn & nn & 3/32 & 3/8 & 9/16 & 0 & 1/32 & 1/8 & 3/16 \\
0 & 0 & 0 & 0 & 0 & 0 & 0 & 0 & 0 & 0 & no & 0 & 0 & 3/32 & 0 & 0 & 0 & 1/32 \\
0 & 0 & 0 & 0 & 0 & 0 & 0 & 0 & mq & nq & 0 & 0 & 1/8 & 1/32 & 0 & 3/8 & 3/32 & 0 & 0 \\
0 & 0 & 0 & 0 & 0 & 0 & 0 & 0 & mr & nr & mq & nq & 1/8 & 3/16 & 1/8 & 1/32 & 3/8 & 9/16 & 3/8 \\
0 & 0 & 0 & 0 & 0 & 0 & 0 & 0 & 0 & ns & mr & nr & 0 & 1/32 & 1/8 & 3/16 & 0 & 3/32 & 3/8 \\
0 & 0 & 0 & 0 & 0 & 0 & 0 & 0 & 0 & 0 & ns & 0 & 0 & 1/32 & 1/8 & 0 & 0 & 3/32
\end{pmatrix}
$$

$$
T_2^{IR} =
\begin{pmatrix}
1/64 & 0 & 0 & 0 & 0 & 0 & 0 & 0 & na & 0 & 0 & 0 & 0 & 0 & 0 & 0 & 0 & 0 & 0 \\
3/32 & 1/16 & 1/64 & 0 & 0 & 0 & 0 & 0 & nb & ma & na & 0 & 0 & 0 & 0 & 0 & 0 & 0 & 0 \\
1/64 & 1/16 & 3/32 & 1/16 & 0 & 0 & 0 & 0 & nc & mb & nb & ma & 0 & 0 & 0 & 0 & 0 & 0 & 0 \\
0 & 0 & 1/64 & 1/16 & 0 & 0 & 0 & 0 & 0 & 0 & nc & mb & 0 & 0 & 0 & 0 & 0 & 0 & 0 \\
3/32 & 0 & 0 & 0 & 1/16 & 0 & 0 & 0 & ne & 0 & 0 & 0 & 0 & 0 & 0 & 0 & 0 & 0 & 0 \\
9/16 & 3/8 & 3/32 & 0 & 3/8 & 1/4 & 1/16 & 0 & nf & me & ne & 0 & 0 & 0 & 0 & 0 & 0 & 0 & 0 \\
3/32 & 3/8 & 9/16 & 3/8 & 1/16 & 1/4 & 3/8 & 1/4 & ng & mf & nf & me & 0 & 0 & 0 & 0 & 0 & 0 & 0 \\
0 & 0 & 3/32 & 3/8 & 0 & 0 & 1/16 & 1/4 & 0 & 0 & ng & mf & 0 & 0 & 0 & 0 & 0 & 0 & 0 \\
1/64 & 0 & 0 & 0 & 1/16 & 0 & 0 & 0 & ni & 0 & 0 & 0 & 0 & 0 & 0 & 0 & 0 & 0 & 0 \\
3/32 & 1/16 & 1/64 & 0 & 3/8 & 1/4 & 1/16 & 0 & nj & mi & ni & 0 & 0 & 0 & 0 & 0 & 0 & 0 & 0 \\
1/64 & 1/16 & 3/32 & 1/16 & 1/16 & 1/4 & 3/8 & 1/4 & nk & mj & nj & mi & 0 & 0 & 0 & 0 & 0 & 0 & 0 \\
0 & 0 & 1/64 & 1/16 & 0 & 0 & 1/16 & 1/4 & 0 & 0 & nk & mj & 0 & 0 & 0 & 0 & 0 & 0 & 0 \\
0 & 0 & 0 & 0 & 0 & 0 & 0 & 0 & nm & 0 & 0 & 0 & 3/32 & 0 & 0 & 0 & 1/32 & 0 & 0 \\
0 & 0 & 0 & 0 & 0 & 0 & 0 & 0 & nn & mm & nm & 0 & 9/16 & 3/8 & 3/32 & 0 & 3/16 & 1/8 & 1/32 \\
0 & 0 & 0 & 0 & 0 & 0 & 0 & 0 & no & mn & nn & mm & 3/32 & 3/8 & 9/16 & 3/8 & 1/32 & 1/8 & 3/16 \\
0 & 0 & 0 & 0 & 0 & 0 & 0 & 0 & 0 & no & mn & nn & 0 & 0 & 3/32 & 3/8 & 0 & 0 & 1/32 \\
0 & 0 & 0 & 0 & 0 & 0 & 0 & 0 & nq & 0 & 0 & 0 & 1/32 & 0 & 0 & 0 & 3/32 & 0 & 0 \\
0 & 0 & 0 & 0 & 0 & 0 & 0 & 0 & nr & mq & nq & 0 & 3/16 & 1/8 & 1/32 & 0 & 9/16 & 3/8 & 3/32 \\
0 & 0 & 0 & 0 & 0 & 0 & 0 & 0 & ns & mr & nr & mq & 1/32 & 1/8 & 3/16 & 1/8 & 3/32 & 3/8 & 9/16 \\
0 & 0 & 0 & 0 & 0 & 0 & 0 & 0 & 0 & ns & mr & nr & 0 & 0 & 1/32 & 1/8 & 0 & 0 & 3/32
\end{pmatrix}
$$

The following values were substituted to obtain a C^1-continuous surface:

$$ma = 0, mb = 0, me = 0, mf = 0, mi = 1/8, mj = 1/8, mm = 3/8, mn = 3/8,$$

$$mq = 0, mr = 0, na = 0, nb = 0, nc = 0, ne = 0, nf = 0, ng = 0, ni = 1/32,$$

$$nj = 3/16, nk = 1/32, nm = 3/32, nn = 9/16, no = 3/32, nq = 0, nr = 0,$$

$$ns = 0.$$

References

[Bar88]   Barnsley, M.: Fractals everywhere. Academic Press Professional, Inc., San Diego (1988)

[Ben09]   Bensoudane, H.: Etude differentielle des formes fractales. PhD thesis, Université de Bourgogne (2009)

[BGN09]   Bensoudane, H., Gentil, C., Neveu, M.: Fractional half-tangent of a curve described by iterated function system. Journal Of Applied Functional Analysis 4(2), 311–326 (2009)

[BLND11]   Boumzaid, Y., Lanquetin, S., Neveu, M., Destelle, F.: An approximating-interpolatory subdivision scheme. Int. J. Pure Appl. Math. 71(1), 129–147 (2011)

[Gen92]   Gentil, C.: Les fractales en synthèse d'image: le modèle IFS. PhD thesis, Université LYON 1 (1992)

[Hut81]   Hutchinson, J.: Fractals and self-similarity. Indiana University Journal of Mathematics 30(5), 713–747 (1981)

[JLZ09]   Jiang, Q., Li, B., Zhu, W.: Interpolatory quad/triangle subdivision schemes for surface design. Computer Aided Geometric Design 26(8), 904–922 (2009)

[KSD12]   Kosinka, J., Sabin, M., Dodgson, N.: Cubic subdivision schemes with double knots. Computer Aided Geometric Design (2012)

[LL03]   Levin, A., Levin, D.: Analysis of quasi uniform subdivision. Applied and Computational Harmonic Analysis 15(1), 18–32 (2003)

[MW88]   Daniel Mauldin, R., Williams, S.C.: Hausdorff dimension in graph directed constructions. Transactions of the American Mathematical Society 309(2), 811–829 (1988)

[PGSL13]   Podkorytov, S., Gentil, C., Sokolov, D., Lanquetin, S.: Geometry control of the junction between two fractal curves. Computer-Aided Design 45(2), 424–431 (2012); Solid and Physical Modeling (2012)

[PH94]   Prusinkiewicz, P., Hammel, M.: Language-Restricted Iterated Function Systems, Koch Constructions, and L-systems (1994)

[SGB12]   Sokolov, D., Gentil, C., Bensoudane, H.: Differential behaviour of iteratively generated curves. In: Boissonnat, J.-D., Chenin, P., Cohen, A., Gout, C., Lyche, T., Mazure, M.-L., Schumaker, L. (eds.) Curves and Surfaces 2011. LNCS, vol. 6920, pp. 663–680. Springer, Heidelberg (2012)

[SL03]   Stam, J., Loop, C.: Quad/triangle subdivision. Computer Graphics Forum 22(1), 79–85 (2003)

[SW05]   Schaefer, S., Warren, J.: On c2 triangle/quad subdivision. ACM Trans. Graph. 24(1), 28–36 (2005)

[TBSG⁺06]   Tosan, E., Bailly-Sallins, I., Gouaty, G., Stotz, I., Buser, P., Weinand, Y.: Une modélisation géométrique itérative basée sur les automates. In: GTMG 2006, Journées du Groupe de Travail en Modélisation Géométrique, Cachan, Mars 22-23, pp. 155–169 (2006)

[TT95]   Thollot, J., Tosan, E.: Constructive fractal geometry: constructive approach to fractal modeling using language operations (1995)

[ZT96]   Zair, C.E., Tosan, E.: Fractal modeling using free form techniques. Comput. Graph. Forum 15(3), 269–278 (1996)

Reparametrization and Volume Mesh Generation for Computational Fluid Dynamics Using Modified Catmull-Clark Methods

Michael Rom[1,2] and Karl-Heinz Brakhage[1]

[1] Institut für Geometrie und Praktische Mathematik, RWTH Aachen, Germany
{rom,brakhage}@igpm.rwth-aachen.de
[2] German Research School for Simulation Sciences, Jülich, Germany

Abstract. A new technique is presented for using the Catmull-Clark subdivision method, modified for modeling sharp creases, to generate volume meshes used in computational fluid dynamics. Given a target surface of arbitrary genus, e.g., defined by a collection of trimmed B-spline patches, which represents an object in a flow, a simple polyhedron is constructed roughly approximating this target surface. After one Catmull-Clark subdivision, the polyhedron exclusively consists of quadrilaterals and its Catmull-Clark limit surface can be pre-computed. Points of the limit surface are projected onto the target surface and the control points of the polyhedron are adjusted by approximating the projected points. An iterative process of alternating subdivisions, projections and approximations leads to a watertight mesh consisting of untrimmed surface patches matching the given target surface. By attaching an offset mesh and a far-field mesh, a block-structured volume mesh is obtained, being well-suited for adaptive flow solvers.

Keywords: modified Catmull-Clark subdivision, B-spline surfaces and volumes, reparametrization, block-structured volume mesh generation, computational fluid dynamics.

1 Introduction

Numerical flow simulations require high-quality volume meshes. The generation of such meshes as well as the construction of the object around or through which the flow should be simulated can be very difficult and time-consuming. There is a wide variety of methodologies for the generation of meshes which can be categorized in terms of the resulting mesh, e.g., triangle/quadrilateral meshes, tetrahedral/hexahedral meshes or structured/unstructured meshes. Surveys of these methodologies can for instance be found in [1,2,3].

We have developed a new promising approach for the generation of volume meshes which can be used for the simulation of flows around given geometries, extending the ideas presented in [4]. The overall process is illustrated in Fig. 1. The dashed lines frame the steps we focus on in this paper. Given a target surface representing the object in the flow, the user only has to construct a simple initial

M. Floater et al. (Eds.): MMCS 2012, LNCS 8177, pp. 425–441, 2014.

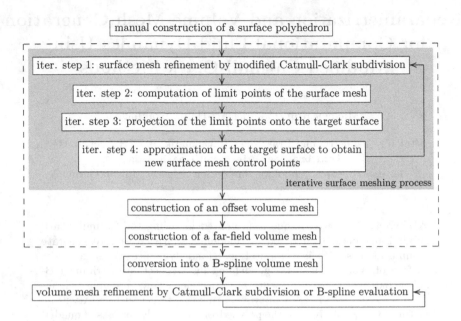

Fig. 1. Process of volume mesh generation

surface polyhedron which should be roughly approximating the target surface. This is done by defining the coordinates of the control points (= vertices), the face connectivity and optionally edges which should be treated as creases. Regarding the surface mesh generation in the iterative process, the target surface can be of arbitrary genus. However, the automation of the generation of a volume mesh based on such a surface mesh is difficult. Our goal is to build up a template library for recurring geometries, starting with a wing-fuselage configuration and an airplane engine which are presented in this paper.

For the two-dimensional subdivisions of the surface mesh (iteration step 1), we use the scheme presented by Catmull and Clark [5] which is applicable to surfaces of arbitrary genus. Due to the convergence of the Catmull-Clark limit surface, which is the limit of repeated subdivision, to uniform bi-cubic B-spline patches, see [5], we get a smooth C^2-surface with the exception of points where no tensor-product topology is given (C^1 there). The results of Stam's analysis of the subdivision matrix [6] with an extension for modeling creases by de Rose et al. [7] allow for the pre-calculation of limit points of the surface mesh at each refinement level (iteration step 2). The limit points can then be used for the approximation of a given target surface which in our practical applications usually is given as a trimmed B-spline surface: the Nelder-Mead optimization algorithm [8] is used for the projection of the limit points onto such a B-spline surface (iteration step 3) and, subsequently, the projected limit points are approximated to obtain new surface mesh control points (iteration step 4) by applying the CGLS method, called CGNR in [9]. The iterative surface meshing process is stopped if the approximation of the target surface is satisfying.

After or during the iterative surface meshing process, a block-structured volume mesh can be constructed, being attached to the surface mesh. If the iterative process is to be continued afterwards, the Catmull-Clark surface subdivision rules are replaced by an extension to volume rules. For that purpose, we apply a combination of the schemes presented by Joy and MacCracken [10] and Bajaj et al. [11]. The advantage of using a structured mesh for a numerical flow simulation instead of an unstructured one is the computational efficiency regarding speed and memory requirements. Our volume mesh consists of a body-fitted offset mesh and a far-field mesh. It can be converted to a B-spline volume mesh. Hence, it is possible to apply further mesh refinement by spline evaluation.

For the visualization of our meshes and B-spline surfaces, we have implemented an OpenGL interface which allows for user interaction, e.g., the iterative surface meshing process is applied step by step by the user such that intermediate interventions are possible, for instance for mesh smoothing or feature detection.

The rest of the paper is structured as follows: Section 2 gives an overview of the Catmull-Clark subdivision rules for surfaces and describes why the Catmull-Clark scheme is the method of choice for our purpose. In Sect. 3, we briefly explain how points of the limit surface can be calculated and demonstrate how they are used for the approximation of a target surface given by a B-spline representation. The results of the iterative surface meshing process depicted in Fig. 1 are illustrated for an example in Sect. 4. Section 5 describes a method for the construction of a body-fitted offset mesh, whereas Sect. 6 demonstrates how a far-field mesh can be attached to the offset.

All numerical computations which are mentioned in this paper are for error control done in double precision.

2   Background on Catmull-Clark Subdivision

In iteration step 1 of our mesh generation process, see Fig. 1, two-dimensional Catmull-Clark subdivision [5], extended by the possibility to model creases [7], is applied to the surface mesh, whereas three-dimensional subdivision can be applied to the volume mesh. A description of the extension of the Catmull-Clark subdivision rules to three-dimensional rules, see [10,11], is omitted in this work.

Catmull and Clark published their descriptions of quadratic and cubic subdivision surfaces in 1978. Contrary to tensor-product splines, their scheme can be applied to meshes that are not regular rectangular grids. A refinement step is defined by the following rules:

1. For each face, add a point given by the average of the N face vertices:

$$F = \frac{1}{N} \sum_{i=0}^{N-1} P_i \ . \tag{1}$$

2. For each edge, add a point given by a weighted average of the two new adjacent face points F_{left}, F_{right} and the edge midpoint E_c:

$$E = \frac{1}{4} \left(F_{\text{left}} + 2\,E_c + F_{\text{right}} \right) \ . \tag{2}$$

3. Move each old vertex to a new position given by a weighted average of the vertex P, the average $\tilde{F}$ of the new adjacent face points and the average $\tilde{E}_c$ of the adjacent edge midpoints:

$$P_{new} = \frac{1}{N}\left(\tilde{F} + 2\,\tilde{E}_c + (N-3)P\right) , \tag{3}$$

where N denotes the valence of P, i.e., the number of edges connected to P.

4. Build the new edges by splitting the old ones and connecting the new face points to the new adjacent edge points. For one old face with N vertices, this leads to N new faces instead of the old one.

On the boundary curve or on inner sharp creases, the edge point insertion rule 2 is replaced by the insertion of the edge midpoint, i.e.,

$$E = E_c \tag{4}$$

and the vertex recomputation rule 3 is changed to

$$P_{new} = \frac{1}{8}\left(P_{left} + 6\,P + P_{right}\right) , \tag{5}$$

where P_{left} and P_{right} denote the two boundary/crease vertices adjacent to P. The two replacement rules originate from the coefficients for midpoint knot insertion for a uniform cubic B-spline curve. Figure 2 illustrates the four subdivision rules for a simple mesh with one pentagon, one quadrilateral and two triangles.

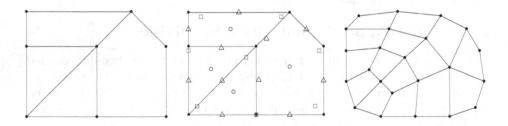

Fig. 2. Illustration of the rules of one Catmull-Clark subdivision, left: simple initial mesh, middle (rules 1-3): insertion of face points ($\circ$), edge points ($\triangle$) and recomputation of vertex positions ($\square$), right (rule 4): subdivided mesh after the reconnection of the points

In our modeling and mesh generation concepts, see [12], we want to end up with smooth untrimmed B-spline patches. These can be provided by an easily implementable conversion of the Catmull-Clark limit surface. Hence, this scheme is the method of choice. We summarize its crucial properties, cf. [13,14]:

– The surfaces can be of arbitrary genus since the subdivision rules can be carried out to a mesh of arbitrary topological type.

- After one subdivision, all faces are quadrilaterals.
- Except at extraordinary vertices (vertices of valence $N \neq 4$), the limit surface converges to uniform bi-cubic B-spline patches. Hence, the surface is C^2-continuous except at extraordinary vertices where it still is C^1-continuous.
- The number of extraordinary vertices is fixed after the first subdivision so that non-regular regions are scaled down with each further subdivision. This is illustrated in Fig. 4.
- After two subdivisions, each face can contain one extraordinary vertex at most. This allows us to easily compute points of the Catmull-Clark limit surface, see Sect. 3.
- The subdivision rules can be modified in such a way that they generate infinitely sharp as well as semi-sharp creases, i.e., creases for which the sharpness can vary from zero (meaning smooth) to infinite, see de Rose et al. [7].

Figure 3 shows an example for the application of the modified Catmull-Clark subdivision scheme. Using a simple initial polyhedron for modeling a wing-fuselage configuration (left), we end up with a smooth surface mesh after only three subdivisions (right). The bold edges are marked as sharp creases for the initial polyhedron and stay sharp during the subdivision process.

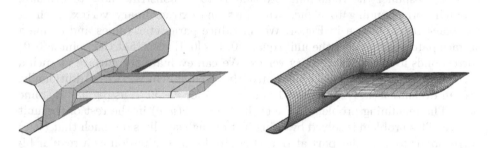

Fig. 3. Initial polyhedron (left) and polyhedron after three subdivisions (right) using the modified Catmull-Clark method

3   Limit Points of Catmull-Clark Subdivision Surfaces, Projection onto B-Spline Surfaces and Approximation

Stam [6] gave an algorithm for evaluating the Catmull-Clark scheme and its derivatives at arbitrary points. He chose an ordering for the control vertices such that the main part of the subdivision matrix has a cyclical structure. Hence, the discrete Fourier transform can be used to compute its eigenstructure. For the modified rules given in [7], this analysis is very technical. Details of the investigation and the implementation can be found in [15].

Due to the existence of extraordinary points, a surface mesh cannot be evaluated everywhere at each subdivision level by well-known B-spline algorithms because the control vertex structure near an extraordinary point is not a simple rectangular grid. Hence, all faces that contain extraordinary vertices cannot be evaluated as uniform B-splines. For our mesh, we assume that each face is a quadrilateral one and contains one extraordinary vertex at most, which both is fulfilled after two subdivisions at the latest. Figure 4 shows that the region in which the surface cannot be evaluated with standard methods (black) is scaled down with every subdivision.

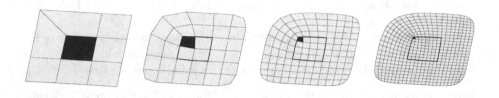

Fig. 4. Behavior near an extraordinary vertex of valence $N = 3$ when subdividing

Since we can evaluate the surface away from extraordinary vertices as a regular bi-cubic B-spline, the remaining problem is to demonstrate how to evaluate a patch corresponding to a face with just one extraordinary vertex, such as the black region shown in Fig. 4. We introduce parameter values and define a surface patch $\boldsymbol{x}(u,v)$ over the unit square $[0,1] \times [0,1]$ such that the point $\boldsymbol{x}(0,0)$ corresponds to the extraordinary vertex. We can evaluate the surface at such a vertex $(\boldsymbol{x}(0,0))$ as a linear combination of the circumfluent vertices. Additionally, we can evaluate $\boldsymbol{x}(u,1)$ for $u \in [0,1]$ and $\boldsymbol{x}(1,v)$ for $v \in [0,1]$ as regular B-spline part. The remaining problem is the evaluation of $\boldsymbol{x}(u,v)$ in the rest of the unit square. This problem is solved by doing just enough subdivisions such that (u,v) corresponds to a regular part at that stage to do the evaluation as a regular bi-cubic B-spline. Applying this technique enables us to pre-compute points of the Catmull-Clark limit surface.

3.1 Catmull-Clark Limit Surface

We can pre-compute points $\boldsymbol{L}_i$ of the Catmull-Clark limit surface, see iteration step 2 in Fig. 1, by using Stam's above-mentioned algorithm for evaluating the Catmull-Clark scheme at arbitrary points. For each face, we determine nine limit points as illustrated in Fig. 5 in which the vertex with the index 0 represents an extraordinary vertex. The four limit points corresponding to the face vertices $\boldsymbol{P}_0, \ldots, \boldsymbol{P}_3$ are given by

$$\boldsymbol{L}_0 = \boldsymbol{x}(0,0), \quad \boldsymbol{L}_1 = \boldsymbol{x}(0,1), \quad \boldsymbol{L}_2 = \boldsymbol{x}(1,1), \quad \boldsymbol{L}_3 = \boldsymbol{x}(1,0) , \qquad (6)$$

the four points corresponding to the edges by

$$\boldsymbol{L}_4 = \boldsymbol{x}\left(\frac{1}{2},1\right), \quad \boldsymbol{L}_5 = \boldsymbol{x}\left(1,\frac{1}{2}\right), \quad \boldsymbol{L}_6 = \boldsymbol{x}\left(\frac{1}{2},0\right), \quad \boldsymbol{L}_7 = \boldsymbol{x}\left(0,\frac{1}{2}\right) \qquad (7)$$

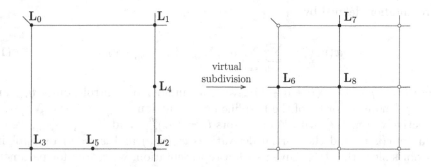

Fig. 5. Example for the computation of nine limit points for a face with one extraordinary vertex (index 0), left: six limit points can be calculated immediately, right: the remaining three limit points can be calculated after one subdivision which is only done implicitly for the evaluation

and the one corresponding to the face by

$$L_8 = x\left(\frac{1}{2}, \frac{1}{2}\right) . \tag{8}$$

The six limit points marked in the left part of Fig. 5 can be calculated immediately since they belong to $x(0,0)$, $x(u,1)$ for $u \in [0,1]$ or $x(1,v)$ for $v \in [0,1]$, see above. The computation of the remaining three limit points requires one subdivision of the mesh so that the new vertices with the indices 6, 7 and 8 belong to $x(u,1)$ for $u \in [0,1]$ or $x(1,v)$ for $v \in [0,1]$ at the new refinement level. For the whole mesh, the number of limit points sums up to the number of vertices plus the number of faces and edges of the current mesh. All the computations can be applied without explicitly subdividing and lead to coefficients for each vertex participating in the limit point computation. Hence, the limit points L_i can be written as a linear combination

$$L_i = c_i^T P^i , \tag{9}$$

where we have collected the involved vertices P_j in the vicinity of L_i in the vector P^i and the weights c_j of the associated vertices P_j in the vector c_i. All coefficients c_j are calculated prior to the first iteration and stored such that they do not have to be computed again for later iterations. Details of the computation and the choice of the associated vertices P_j can be found in [15]. The vertex and edge limit points are only calculated if they have not already been computed by considering an adjacent face before. If a control point belongs to a special edge, i.e., a crease, we need other rules for the evaluation. Therefore, we use our extension [15] in these cases.

3.2 Projection onto B-Spline Surfaces

Given a surface s, we can project a limit point L_i onto it: $L_i \rightarrow L_i^s$, see iteration step 3 in Fig. 1. Usually, the surface s or a patch of it is given by a B-spline representation defined by

$$\boldsymbol{x}(u,v) = \sum_{r=0}^{m-p} \sum_{s=0}^{n-q} \boldsymbol{y}_{r,s} N_{r,p,U}(u) N_{s,q,V}(v) \ , \tag{10}$$

where $(m - p + 1) \times (n - q + 1)$ gives the number of control points $\boldsymbol{y}_{r,s}$ and p and q denote the order of the B-spline basis functions $N_{r,p,U}(u)$ and $N_{s,q,V}(v)$, respectively. Together with knot vectors $U = (u_r)_{r=0}^{m}$ and $V = (v_s)_{s=0}^{n}$, we can evaluate $\boldsymbol{x}(u,v)$ and the partial derivatives $\boldsymbol{x}_u(u,v)$ and $\boldsymbol{x}_v(u,v)$ by applying de Boor's algorithm [16]. Given such a representation, we search for parameter values u_{L_i} and v_{L_i} for each limit point $\boldsymbol{L}_i$ such that

$$\|\boldsymbol{x}(u_{L_i}, v_{L_i}) - \boldsymbol{L}_i\|_2 = \min_{u,v} \|\boldsymbol{x}(u,v) - \boldsymbol{L}_i\|_2 \ . \tag{11}$$

To solve this distance minimization problem, we use the Nelder-Mead optimization algorithm [8] which is simple to understand and to implement. In addition, it does not need any derivatives, is very robust and can be adapted to the use with other than B-spline surfaces easily. For each limit point, a simplex, which in this two-dimensional case is a triangle in (u,v)-space, is built around the initial values (u_0, v_0) which have to be set once prior to the first projection. Due to the robustness of the algorithm, an inaccurate choice of initial values does not destroy the convergence of the algorithm. For each further projection, we can calculate far better initial values by taking averages of values from the previous projection. In addition, the relation between the B-spline surface and the vertices of the initial polyhedron has to be defined once if the B-spline representation contains more than one patch. The patch correlation can for instance be set depending on ranges of coordinate values in a particular direction.

In the first step of the Nelder-Mead algorithm, the distances $d_j = \|\boldsymbol{x}(u_j, v_j) - \boldsymbol{L}_i\|_2$, $j = 0, 1, 2$, between the B-spline surface points $\boldsymbol{x}(u_j, v_j)$, which are calculated for the triangle points $\boldsymbol{z}_j = (u_j, v_j)^T$, and the limit point $\boldsymbol{L}_i$ are computed. This determines the best, the second-best and the worst point of the triangle. Depending on this classification, the triangle is transformed and moved by a reflection, expansion or contraction of a single triangle point, or it is shrinked. The next iteration again starts with the calculation of the distances d_j. Special care has to be taken at the patch boundaries of the target surface. Depending on the type of such a boundary, which can for instance be a border (e.g., at a symmetry plane), a transition to another patch or periodic, the Nelder-Mead triangle has to be adjusted when it approaches a boundary. Another issue is the occurrence of sharp crease curves on the B-spline surface: in the first projection step, the points which are projected the closest to such a curve are moved to match the curve by resetting the corresponding parameter value u or v to the value of the curve. For further projections, this parameter value is fixed. Edge points, being inserted by later subdivisions between two vertices which are corresponding to a fixed curve, are also set to lie on that curve.

The algorithm is terminated if all three side lengths of the Nelder-Mead triangle are smaller than a threshold. The finer the mesh is, i.e., the smaller the distances between the vertices and, hence, the limit points are, the smaller the threshold value has to be. However, since a prediction for a good choice of the threshold value is difficult, we have set it to 10^{-10}. The additional computational cost compared to larger values is negligible due to the good initial values for u and v.

3.3   Approximation of the Target Surface

Finally, we approximate the projected limit points $L_i^s = x(u_{L_i}, v_{L_i})$ to obtain new control point positions with better approximation properties for the next iteration loop. Using (9), we get the equation

$$c_i^T P^i = L_i^s + e_i \qquad (12)$$

for a single point, where e_i is the error vector resulting from the approximation. Since we want to minimize the approximation error, we rewrite the problem for the whole mesh in the least-squares matrix-vector formulation

$$\|e\|_2 = \|CP - L^s\|_2 \rightarrow \min_P \ , \qquad (13)$$

where all coefficients are collected in the matrix C, all control points in the vector P and all projected limit points in the vector L^s. Due to computing a limit point for each vertex, each edge and each face, we have more limit points and consequentially more projection points than vertices. Hence, we end up with an over-determined sparse linear system for approximation, see iteration step 4 in Fig. 1. We apply the conjugate gradient method for linear least squares (CGLS) [9] which for a matrix $A \in \mathbb{R}^{m \times n}$, $m \geq n$, with full rank solves the normal equations $A^T A x = A^T b$ while taking care of the problems that $A^T A$ often is badly conditioned and that for a sparse A usually $A^T A$ is not sparse. The full rank of the coefficient matrix C is guaranteed since we have nine limit points for each quadrilateral face. The cost for one CGLS step is linear in the number of control points. The CGLS algorithm is stopped if the residual of the normal equations, for the iteration l given by $\|s^{(l)}\|_2 = \|A^T b - A^T A x^{(l)}\|_2$, is smaller than a threshold ε. The 2-norm condition number of $C^T C$, evaluated by Matlab, is always approximately 300, independent of the meshing problem and the current iteration of the surface mesh generation process. Hence, $C^T C$ is well-conditioned such that we expect a fast convergence without preconditioning and set ε to the rather small value 10^{-6}.

3.4   Iterative Surface Meshing Process: Summary

With $\#V$ denoting the number of vertices, $\#F$ the number of faces and $\#E$ the number of edges, the iterative surface meshing process regarding the points involved in the procedure can be formulated as follows:

1. The subdivision which is applied to the surface mesh control points (= vertices) $P_i^{(k)}$ of subdivision level k, $i = 0, \ldots, \#V_k - 1$, leads to a refined

mesh consisting of the vertices $\boldsymbol{P}_i^{(k+1)}$, $i = 0, \ldots, \#V_k + \#F_k + \#E_k - 1 = 0, \ldots, \#V_{k+1} - 1$.

2. The evaluation of the Catmull-Clark limit surface as done in this work results in the limit points $\boldsymbol{L}_i^{(k+1)}$, $i = 0, \ldots, \#V_{k+1} + \#F_{k+1} + \#E_{k+1} - 1$.

3. The projection of the limit points onto the target surface leads to the projected limit points $\boldsymbol{L}_i^{s,(k+1)}$, $i = 0, \ldots, \#V_{k+1} + \#F_{k+1} + \#E_{k+1} - 1$.

4. By approximating the target surface, the recomputed surface mesh vertices $\tilde{\boldsymbol{P}}_i^{(k+1)}$, $i = 0, \ldots, \#V_{k+1} - 1$ are obtained. These are the $\boldsymbol{P}_i$ for the next iteration loop.

4   Results of the Iterative Surface Meshing Process

From the sub-project *High Reynolds Number Aero-Structural Dynamics (HIRE-NASD)* of the collaborative research center SFB 401 *Flow Modulation and Fluid-Structure Interaction at Airplane Wings* at RWTH Aachen, see [17], we have a B-spline surface representing a wing-fuselage configuration. This is the gray surface depicted in Fig. 6. It consists of three patches: the simplified fuselage, the wing and the wing tip. An IGES file for this geometry can be found on http://www.igpm.rwth-aachen.de/brakhage/SFBmodel.

The polyhedron surrounding the surface has been constructed manually as a starting point for the generation of a surface mesh. It consists of quadrilaterals exclusively. Since it contains faces with more than one extraordinary vertex, the polyhedron has to be subdivided once prior to the first evaluation of the Catmull-Clark limit surface, as described above. The bold edges are marked to be sharp creases.

In a first step, we project the vertices of the initial polyhedron directly onto the B-spline surface to improve the initial polyhedron. After that, the iterative surface meshing process according to Fig. 1 is started. The result after two iterations is depicted in Fig. 7. Already at this stage, the limit surface of the mesh well matches the given B-spline surface: the average and maximum distance between the projected limit points of iteration step 3 and the corresponding limit points which can be computed for the control mesh shown in Fig. 7, which has been obtained by iteration step 4, are about $5.89 \cdot 10^{-5}$ and $9.33 \cdot 10^{-4}$, respectively. For comparison: the wing span is 1.29 and the length of the fuselage 1.63.

The convergence behavior of the overall process can be evaluated by computing the maximum of the distances between the limit points of iteration step 2 and the B-spline surface points of iteration step 3. Hence, for the current subdivision level k, we define the error

$$e^{(k)} = \max_i \|\boldsymbol{L}_i^{s,(k)} - \boldsymbol{L}_i^{(k)}\|_2 = \max_i \|\boldsymbol{x}(u_{L_i^{(k)}}, v_{L_i^{(k)}}) - \boldsymbol{L}_i^{(k)}\|_2 . \qquad (14)$$

For $k > 1$, the order of convergence p can then be estimated by

$$(h^p)^{k-1} = \frac{e^{(k)}}{e^{(1)}} \qquad \Leftrightarrow \qquad p = \frac{\log\left(\left(\frac{e^{(k)}}{e^{(1)}}\right)^{\frac{1}{k-1}}\right)}{\log(h)} . \qquad (15)$$

For the Catmull-Clark method, h is approximately 0.5. The B-spline patch for the wing was modeled to be C^2-continuous and afterwards a smoothing was applied. Hence, we expect a numerical order of convergence of $p \gtrsim 2$ for the wing. We get the errors $e^{(1)} = 3.01 \cdot 10^{-3}$ and $e^{(6)} = 1.59 \cdot 10^{-7}$, leading to $p = 2.84$. The B-spline patch for the fuselage also is C^2-continuous, but no smoothing has been used in this case. Hence, the expectation in this case is $p \approx 2$. The fuselage errors are $e^{(1)} = 5.75 \cdot 10^{-3}$ and $e^{(6)} = 1.89 \cdot 10^{-6}$ such that $p = 2.31$.

Instead of applying an iterative process, the initial polyhedron could be subdivided several times such that the projection of limit points and the successive approximation both would only be done once. However, the projection results are of much better quality using the iterative process since the initial values for u and v are improved with each further projection, see above. In addition, the subdivision in iteration step 1 leads to a smoothing of the approximated mesh obtained by the previous loop.

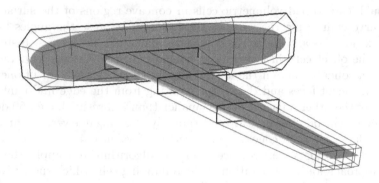

Fig. 6. Wing-fuselage configuration: given B-spline surface and manually constructed initial template polyhedron

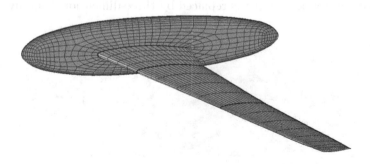

Fig. 7. Wing-fuselage configuration: given B-spline surface and surface mesh after two process iterations

5   Construction of a Body-Fitted Offset Mesh

So far, we have shown how to construct a surface mesh matching a given surface which for a flow simulation represents an object in the flow field. For the numerical simulation of that flow field, we need a volume mesh. We demonstrate our method for the construction of such a volume mesh using the example of the wing-fuselage configuration introduced in Sect. 4.

Viscous flows require a particularly high mesh quality near the object to resolve the boundary layer that may become very thin, e.g., for high Reynolds number flows. Hence, in a first step of our volume mesh generation, we construct a body-fitted offset mesh. The volumetric cells of that offset mesh should be as orthogonal to the surface as possible. The first idea to construct the offset would be to introduce one offset mesh point for each vertex of the surface mesh by marching away from the surface mesh along the vertex normal, which can be computed as the average or a weighted average of the adjacent face normals. This would lead to bad volumetric cells in concave regions of the surface, e.g., at the curve connecting the wing to the fuselage, because the angles between the vertex normals and the surface are approximately 45 degrees. Instead, we generate the offset cells in layers along concave curves. An edge is detected to belong to a concave curve if the angle between the normal vector of one of the two edge-adjacent faces and the vector pointing from the edge midpoint to the midpoint of the other adjacent face is smaller than a threshold, e.g., 60 degrees. The cells along a concave curve are generated by inserting one vertex, three edges and three faces for each concave curve vertex of valence 4. If a concave curve contains extraordinary vertices, we stop our algorithm and apply the vertex normal approach for the generation of the remaining cells which successively are attached to previously constructed cells.

For the wing-fuselage configuration, we can construct the offset mesh for instance after the first iteration of approximating the given surface. The resulting offset mesh is depicted in Fig. 8 in a translucent view. If the iterative surface meshing process is continued after the offset mesh generation, two-dimensional subdivision in iteration step 1 is replaced by three-dimensional subdivision.

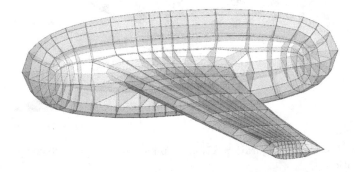

Fig. 8. Wing-fuselage configuration: offset mesh in a translucent view

6   Construction of a Far-Field Mesh

In addition to the offset mesh, the numerical simulation of the flow field around an object requires a far-field mesh. The intermediate and final results of the generation of a far-field mesh for the wing-fuselage configuration are illustrated in Fig. 9. Since we want to end up with a block-structured B-spline volume mesh, the first step is to partition the outer surface of the offset mesh into regular regions without extraordinary vertices, providing a starting point for the construction of the far-field blocks. This is done automatically. For our wing-fuselage configuration, the resulting 15 blocks are depicted in Fig. 9 (*i*). The bold lines mark the block borders.

The user can give the order in which the far-field blocks are constructed, define the coordinates of the outer block corners and decide whether a new block gets a new block number or is merged to its underlying block.

The first eight far-field blocks could for instance be attached to the fuselage blocks 1-8 along the symmetry plane, see Fig. 9 (*ii*). The user can decide how many layers of cells are to be created (in this case four) and whether the vertices of each layer are computed by applying transfinite interpolation, using an equidistant spacing or using the spacing of an adjacent block. Each outer offset mesh vertex is connected to its corresponding outer far-field vertex by computing the parabola or the straight line between these two points. For the uniqueness of the parabola calculation, we use the normal vector of the offset vertex.

The next step of the far-field mesh construction is filling the dent at the fuselage blocks 9-14 by cells, resulting in the mesh depicted in Fig. 9 (*iii*). The dent-filling cells get the same block number as their adjacent offset cells. This necessitates a change of the first eight far-field blocks: the first layer of cells of each of these eight blocks is allocated to its underlying offset block. Hence, all offset blocks which have been connected to far-field blocks so far consist of two layers of cells and all far-field blocks of three layers of cells. The reallocation of cell layers to underlying offset blocks due to filling dents is done automatically.

Then, four new blocks around the wing are attached to the blocks 11-14. Since these blocks contain concave curves, we have to treat them as a special case: the inner patch to which the block is to be attached is divided into two parts along the concave curve. The offset cells at the wing are added to the four new blocks. Together with one layer of far-field cells at the wing tip which belong to the same block as their underlying offset cells (block 15) and two blocks connected to the fuselage at the blocks 9 and 10 we obtain the mesh illustrated in Fig. 9 (*iv*).

The next eight blocks nearly complete the far-field mesh, as Fig. 9 (*v*) shows. We just need another three blocks such that the outer surface of the far-field mesh has a planar or partly even Cartesian structure, see Fig. 9 (*vi*). This makes the extension of the far-field mesh with further blocks easy and provides optimal mesh quality. Figure 10 shows the inner surface mesh together with a selection of volume mesh blocks to give an idea of the block partitioning inside the volume mesh. The whole mesh generation procedure, including the iterative surface mesh generation, takes less than a minute on a usual office workstation.

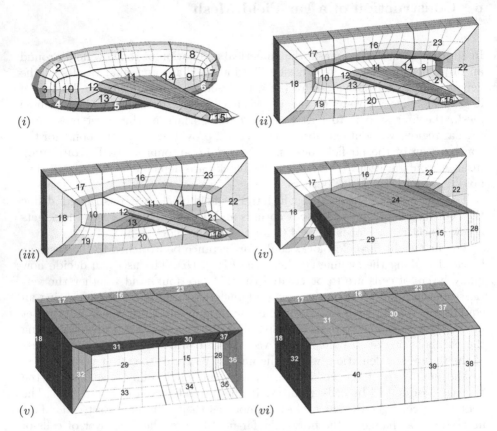

Fig. 9. Volume mesh generation: (*i*) offset mesh after partition into regular regions (15 blocks), (*ii*) 23 blocks, (*iii*) dent-filling, (*iv*) 29 blocks, (*v*) 37 blocks, (*vi*) 40 blocks

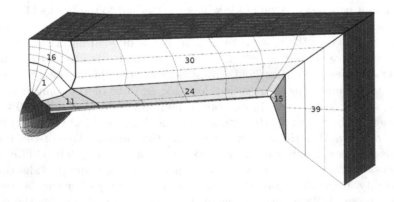

Fig. 10. Inner surface mesh and selected offset and far-field blocks

The final step is the conversion of the whole mesh into a B-spline mesh. Even though the extraordinary vertices of the surface mesh have become extraordinary edges in the volume mesh, resulting in C^0-continuity away from the inner surface, the meshes are suitable for the purpose of adaptive flow simulations since a nested mesh hierarchy can be created. For our flow solver Quadflow [18], this is required due to the adaptation strategy based on a multiscale analysis, see [19]. Of course, the usage of our meshes with other adaptive flow solvers after a conversion of the data structure is imaginable.

Since the construction of a far-field mesh needs some user input, the automation is difficult. Our goal is to build up a library of templates for recurring geometries like the wing-fuselage configuration and to provide tools for the creation of single blocks or cells as demonstrated above by creating normal blocks, blocks at patches with concave curves or dent-filling cells.

Another example for which we have constructed a template is the airplane engine depicted in Fig. 11 (i). The same techniques as the ones described above have been applied for the generation of an offset mesh as well as a far-field mesh. The final volume mesh contains 15 blocks. Figure 11 (ii) shows a selection of these blocks.

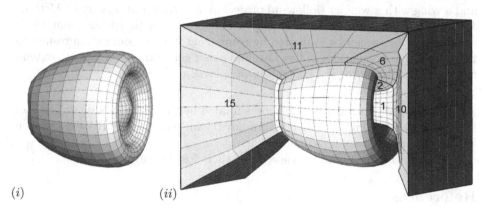

(i)   (ii)

Fig. 11. Template for an airplane engine: (i) surface mesh, (ii) selected blocks of the volume mesh

7 Conclusion and Future Work

We have presented a fast semi-automatic procedure for the generation of block-structured volume meshes with an inner surface mesh well matching a given target surface. These volume meshes are of high quality regarding the smoothness of the inner surface, the cell orthogonality (especially for the boundary-conforming offset mesh) and the suitability for (adaptive) flow solvers. Most of the procedure operates automatically, only an initial polyhedron has to be generated manually, a few parameters according to the given B-spline surface have to be set and the construction of a far-field volume mesh needs user input regarding the block sizes, the shape and spacing of the blocks and the (re)allocation of cell layers to other blocks. We use Catmull-Clark surface subdivision rules, modified to allow for the modeling of creases, to produce smooth surface meshes and an extension of the Catmull-Clark rules to volumes for the refinement of the attached volume mesh. Points of the inner Catmull-Clark limit surface can be pre-computed after each subdivision and projected onto a given B-spline surface. New control points of the mesh can then be obtained by approximating the projected limit points. This linear least-squares problem is solved by applying the CGLS method. The extension to a volume mesh is performed in two steps: at first a body-fitted offset mesh with high-quality volumetric cells is constructed automatically to resolve the boundary layer in the case of a viscous flow simulation. Additionally, a far-field mesh is attached to the offset mesh. Defining a few parameters, the far-field mesh can be constructed by choosing from template geometries, e.g., a wing-fuselage configuration, or single blocks can be built from a selection of template blocks, e.g., a block with a concave curve. Having these templates leads to a minimization of user input and, hence, allows for a fast generation of high-quality block-structured volume meshes.

Since experimental data from wind tunnel readings are available for the wing-fuselage configuration presented throughout the paper (HIRENASD project [17]) and a wing with a winglet (follow-up project *Aero-Structural Dynamics Methods for Airplane Design (ASDMAD)* [20], a cooperation with Airbus), our mesh generation process can be evaluated by comparing the flow simulation results to these data. For simulations of flows around other geometries, we want to extend the template library for the generation of volume mehses.

Acknowledgments. The first author is funded by the German Research School for Simulation Sciences, Jülich, Germany (joint venture of RWTH Aachen University and Forschungszentrum Jülich).

We thank the anonymous reviewers for their fruitful suggestions.

References

1. Owen, S.J.: A Survey of Unstructured Mesh Generation Technology. In: 7th International Meshing Roundtable, pp. 239–267 (1998)
2. Thompson, J.F., Soni, B.K., Weatherill, N.P. (eds.): Handbook of Grid Generation. CRC Press (1999)
3. Frey, P.J., George, P.L.: Mesh Generation. John Wiley & Sons (2008)
4. Rom, M., Brakhage, K.H.: Volume Mesh Generation for Numerical Flow Simulations using Catmull-Clark and Surface Approximation Methods. In: Research Notes Proceedings of the 20th International Meshing Roundtable (2011)
5. Clark, J., Catmull, E.: Recursively generated B-spline surfaces on arbitrary topological meshes. CAD 10(6), 350–355 (1978)
6. Stam, J.: Exact Evaluation of Catmull-Clark Subdivision Surfaces at Arbitrary Parameter Values. In: Proceedings of SIGGRAPH, pp. 395–404 (1998)
7. DeRose, T., Kass, M., Truong, T.: Subdivision Surfaces in Character Animation. In: Proceedings of the 25th Annual Conference on Computer Graphics and Interactive Techniques, pp. 85–94. ACM SIGGRAPH (1998)

8. Nelder, J.A., Mead, R.: A simplex method for function minimization. Computer Journal 7, 308–313 (1965)
9. Saad, Y.: Iterative Methods for Sparse Linear Systems, 2nd edn. SIAM (2003)
10. Joy, K., MacCracken, R.: The Refinement Rules for Catmull-Clark Solids. Technical Report CSE-96-1, Department of Computer Science, University of California, Davis (1996)
11. Bajaj, C., Schaefer, S., Warren, J., Xu, G.: A subdivision scheme for hexahedral meshes. The Visual Computer 18, 343–356 (2002)
12. Brakhage, K.H., Lamby, P.: Application of B-Spline Techniques to the Modeling of Airplane Wings and Numerical Grid Generation. CAGD 25(9), 738–750 (2008)
13. Brakhage, K.H.: Modified Catmull-Clark Methods for Modelling, Reparameterization and Grid Generation. In: Proceedings of the 2. Internationales Symposium Geometrisches Modellieren, Visualisieren und Bildverarbeitung, pp. 109–114. HfT Stuttgart (2007)
14. Brakhage, K.H.: Grid Generation and Grid Conversion by Subdivision Schemes. In: Proceedings of the 11th International Conference on Numerical Grid Generation in Computational Field Simulations (2009)
15. Rom, M.: Oberflächenreparametrisierung und Gittererzeugung für numerische Strömungssimulationen mit Hilfe von Catmull-Clark-Methoden. Diplomarbeit, RWTH Aachen (2009)
16. de Boor, C.: A Practical Guide to Splines. Springer (1978)
17. Ballmann, J., Dafnis, A., Baars, A., Boucke, A., Brakhage, K.-H., Braun, C., Buxel, C., Chen, B.-H., Dickopp, C., Kämpchen, M., Korsch, H., Olivier, H., Ray, S., Reimer, L., Reimerdes, H.-G.: Aero-structural Dynamics Experiments at High Reynolds Numbers. In: Schröder, W. (ed.) Summary of Flow Modulation and Fluid-Structure Interaction Findings. NNFM, vol. 109, pp. 389–424. Springer, Heidelberg (2010)
18. Bramkamp, F., Lamby, P., Müller, S.: An adaptive multiscale finite volume solver for unsteady and steady state flow computations. J. Comput. Phys. 197(2), 460–490 (2004)
19. Müller, S.: Adaptive Multiscale Schemes for Conservation Laws. LNCSE, vol. 27. Springer (2003)
20. Ballmann, J., Boucke, A., Chen, B.H., Reimer, L., Behr, M., Dafnis, A., Buxel, C., Buesing, S., Reimerdes, H.G., Brakhage, K.H., Olivier, H., Kordt, M., Brink-Spalink, J., Theurich, F., Büscher, A.: Aero-Structural Wind Tunnel Experiments with Elastic Wing Models at High Reynolds Numbers (HIRENASD - ASDMAD). In: 49th AIAA Aerospace Sciences Meeting, Orlando, Florida (2011)

A Theoretical Analysis of an Improved Rational Scheme for Spherical Camera Motions

Maria Lucia Sampoli[1], Alessandra Sestini[2], Gašper Jaklič[3], and Emil Žagar[3]

[1] Department of Information Engineering and Mathematics, University of Siena, Italy
marialucia.sampoli@unisi.it
[2] Department of Mathematics and Computer Science, University of Florence, Italy
alessandra.sestini@unifi.it
[3] Faculty of Mathematics and Physics, Institute of Mathematics,
Physics and Mechanics, University of Ljubljana, Slovenia
{gasper.jaklic,emil.zagar}@fmf.uni-lj.si

Abstract. A rational C^1 Hermite interpolation scheme on the sphere is introduced, improving the method proposed in [15]. On the base of a careful asymptotic analysis, a new selection of the free parameters is suggested, leading to a fourth order approximation scheme. The resulting curve is also endowed with a C^1 rational rotation-minimizing directed frame, which interpolates prescribed end orientations. The spline extension of the scheme is investigated. This is useful, for instance, in the description of smoothly varying camera motions, when a fixed target object is being imaged. Several examples are considered in order to show the performance of the proposed approach.

Keywords: Camera–orientation, spherical motion, rotation–minimizing frame, directed frame, Hermite interpolation.

1 Introduction

Camera planning is a very active applicative research area due to its connections to several fields such as cinematography [5], robotics [1] and also medical practice of endoscopic surgery [2]. In particular, applications where it is required that the camera moves along a spatial path while imaging a stationary target object are of great interest. The camera orientation can be prescribed by specifying the variation of an orthonormal *directed* frame embedded within the camera, which means that the instantaneous direction of one of its orientation vectors is given by the optical axis – the straight line joining the origin of a fixed coordinate system and the corresponding point on the curve describing the path of the camera, typically its center of mass. Furthermore, in order to avoid unnecessary rotations of the frame in the image plane which is orthogonal to the optical axis, we are interested in camera orientation having a *Rotation–Minimizing* Directed Frame (RMDF) associated to its path. So we deal with constrained rigid body motions. For simplicity, we deal only with spherical motions. Observe however that the control of the camera distance from the imaged object can be easily added later

M. Floater et al. (Eds.): MMCS 2012, LNCS 8177, pp. 442–455, 2014.
© Springer-Verlag Berlin Heidelberg 2014

as a separated task by using the spherical motion as polar projection of a spatial camera motion on the unit sphere ([16]). From an analytic point of view, we are interested in finding a rational form both of a curve o and of the associated RMDF $\mathbf{e}_j, j = 1, 2, 3$. Two different interpolation schemes for prescribing this kind of rigid body motions have already been developed in [9] and in [15], either using the quaternion algebra, which leads to a compact mathematical representation of the considered motions, and defining the spherical rational curve as the *polar indicatrix* on the unit sphere of a polynomial space curve with polynomial distance from the origin of the reference system. The algorithm developed in [9] produces a rational interpolant of degree 4 with an associated RMDF of degree 8. Unfortunately, it is not fully satisfactory because the considered data are not completely symmetric. In [15], this problem is overcome by raising the degree of the rational curve to 8 which is also the degree of the RMDF. The input data considered in such an approach are the end camera positions, orientations and angular velocities, which are data ensuring the possibility of defining a globally C^1 rigid body motion when a sequence of analogous data are assigned and the method is used locally. A suitable rational distance function interpolating assigned end distances is defined in [16] and combined to the above scheme in order to extend it to general spatial motions.

The theory of directed frames and in particular of RMDFs on a space curve has been developed in [7] where it is shown that it can be derived by a suitable reformulation of adapted frames and specifically of Rotation–Minimizing Adapted Frames (RMAFs), previously introduced in the literature to control different kind of rigid body motions. As a consequence, we observe that all the important theoretical results proved in the literature for RMAFs can be suitably re–read in terms of RMDFs. Since it was proved in [13] that there are no non–planar polynomial curves of degree 3 with rational RMAFs, quintics were studied in [8] and specific algebraic conditions on their quaternion coefficients, ensuring the rationality of the associated RMAFs, were derived – see also [12] for a full classification of this family of quintics. Thus, these results can also be interpreted in terms of polynomial curves with rational RMDFs, respectively of degree 2 and 4 and having rational polar indicatrices on the unit sphere of the same degree. Unfortunately, degree 4 rational spherical curves are not flexible enough to deal with a fully symmetric C^1 Hermite interpolation problem. On the other hand, even if in [11] a polynomial divisibility condition was obtained for expressing the rationality requirement for the RMAFs associated to a polynomial space curve of any degree, a constructive characterization of degree 7 polynomial curves with rational RMAFs (which would correspond also to that of polynomial space curves of degree 6 with rational RMDFs) is a rather difficult task and it is not yet available in the literature. In [15], rational spherical curves of degree 8, having a rotation–minimizing directed frame coincident with its (directed) Euler–Rodrigues Frame (ERF), which is rational by definition [4], were used. Conditions ensuring that the ERF is rotation–minimizing are easier to be obtained and actually they have been first derived in [4] and almost simultaneously, but independently, rewritten in [15] and in [10], in quaternion and Hopf map form respectively.

The aim of this paper is to improve the performance of the method introduced in [15] and to develop a theoretical analysis of its approximation power. In particular, we reduce the number of the interpolants produced by the previous method from four to two, we improve the related free parameter selection strategy and we test the C^1 spline implementation of the scheme. Furthermore we investigate both theoretically and numerically the approximation power of the modified scheme, proving that in general it has fourth order. The presented numerical results aim to show the approximation power of the scheme.

Observe that the considered interpolation problem has two degrees of freedom and, for each selection of them, it in general admits four distinct solutions. Since the beginning, numerical experiments indicated that it is necessary to suitably select the two free parameters in order to get at least one interpolant with a reasonable shape (even for reasonable input data). So in [15] a preliminary choice was proposed which is easy to implement and guarantees the possibility to get a good shape of the interpolant. The new choice of the free parameters given in this paper is based on asymptotic analysis. It keeps the simplicity of the older scheme and it allows us to get even better shapes for some input data.

The paper is organized as follows. In Section 2 we first summarize the main steps for deriving our scheme with the help of quaternion algebra, and then Section 3 discusses in detail the two free parameter selection, proposing new formulation for both of them. Section 4 is devoted to the asymptotic analysis study while in Section 5 we first introduce the extension of the scheme to splines and then we show the numerical results. Section 6 concludes the paper.

2   The Spherical Rational Interpolation Scheme

In this section we briefly summarize the problem and the scheme proposed in [15]. The main problem is to determine a rational curve $o : [0,1] \to \mathbb{S}^2$ with an associated rational RMDF,

$$\{e_i\}_{i=1}^3, \quad e_i : [0,1] \to \mathbb{R}^3, \quad i = 1,2,3,$$

such that the following interpolation conditions are fulfilled,

$$
\begin{array}{llll}
e_1(0) = o_i, & e_2(0) = v_i, & e_3(0) = z_i, & \omega(0) = \omega_i, \\
e_1(1) = o_f, & e_2(1) = v_f, & e_3(1) = z_f, & \omega(1) = \omega_f, \quad (1)
\end{array}
$$

with $\{o_i, v_i, z_i\}$ and $\{o_f, v_f, z_f\}$ denoting assigned end frames and ω_i and ω_f the corresponding assigned end angular velocities. Note that

$$e_1 \equiv o,$$

since the frame is directed, and also

$$\omega(t) \cdot e_1(t) \equiv 0, \quad t \in [0,1], \tag{2}$$

since the frame is rotation-minimizing too. Here $\boldsymbol{\omega}$ denotes the angular velocity vector defined by

$$e'_j(t) = \boldsymbol{\omega}(t) \times e_j(t), \qquad j = 1, 2, 3. \tag{3}$$

Observe that the condition (2) implies the following compatibility conditions on the end angular velocity input data,

$$\boldsymbol{\omega}_i \cdot \boldsymbol{o}_i = 0, \quad \boldsymbol{\omega}_f \cdot \boldsymbol{o}_f = 0,$$

and also

$$|\boldsymbol{\omega}(t)| \equiv |\boldsymbol{o}'(t)| = \sigma(t),$$

where, as usual, σ denotes the parametric speed of the curve.

In [15], the curve $\boldsymbol{o} = e_1$ and the frame are defined as

$$e_j(t) = \frac{\mathcal{A}(t)\, \mathbf{i}_j\, \mathcal{A}^*(t)}{\mathcal{A}(t)\, \mathcal{A}^*(t)}, \qquad j = 1, 2, 3,$$

with $\mathbf{i}_j, j = 1, 2, 3,$ denoting a triple of mutually orthogonal right–handed unit vectors and $\mathcal{A}$ a quaternion polynomial of degree 4 which can be suitably represented in Bernstein-Bézier form, [6], i.e.,

$$\mathcal{A}(t) = \sum_{i=0}^{4} \mathcal{A}_i\, b_i^4(t),$$

where $b_i^4, i = 0, \ldots, 4$ denote the Bernstein basis of the space of polynomials of degree less or equal to 4. Thus both the curve and the frame are rational of degree 8.

The end frame interpolation conditions given in (1) then imply

$$\mathcal{A}_0 = \lambda_i \mathcal{U}_i, \quad \mathcal{A}_4 = \lambda_f \mathcal{U}_f, \tag{4}$$

where $\lambda_i, \lambda_f > 0$ are free parameters and

$$\mathcal{U}_i\, \mathbf{i}_1\, \mathcal{U}_i^* = \boldsymbol{o}_i, \qquad \mathcal{U}_i\, \mathbf{i}_2\, \mathcal{U}_i^* = \boldsymbol{v}_i, \qquad \mathcal{U}_i\, \mathbf{i}_3\, \mathcal{U}_i^* = \boldsymbol{z}_i,$$

$$\mathcal{U}_f\, \mathbf{i}_1\, \mathcal{U}_f^* = \boldsymbol{o}_f, \qquad \mathcal{U}_f\, \mathbf{i}_2\, \mathcal{U}_f^* = \boldsymbol{v}_f, \qquad \mathcal{U}_f\, \mathbf{i}_3\, \mathcal{U}_f^* = \boldsymbol{z}_f.$$

Note that the unit quaternions $\mathcal{U}_i$ and $\mathcal{U}_f$ satisfying the above conditions are unique up to their sign (see the Appendix in [15] for their analytic expressions), so four pairs of solutions are available for $(\mathcal{U}_i, \mathcal{U}_f)$, i.e.,

$$(\hat{\mathcal{U}}_i, \hat{\mathcal{U}}_f), \quad (-\hat{\mathcal{U}}_i, \hat{\mathcal{U}}_f), \quad (\hat{\mathcal{U}}_i, -\hat{\mathcal{U}}_f), \quad (-\hat{\mathcal{U}}_i, -\hat{\mathcal{U}}_f). \tag{5}$$

The interpolation of the angular velocities further implies

$$\mathcal{A}_1 = \frac{\lambda_i}{8}\, (\boldsymbol{\omega}_i + \mu_i)\, \mathcal{U}_i, \quad \mathcal{A}_3 = \frac{\lambda_f}{8}\, (-\boldsymbol{\omega}_f + \mu_f)\, \mathcal{U}_f, \tag{6}$$

where μ_i and μ_f are two additional free parameters.
Introducing the unit quaternion

$$\mathcal{H} = \begin{pmatrix} h_s \\ \boldsymbol{h} \end{pmatrix} = \mathcal{U}_i \, \mathbf{i}_1 \, \mathcal{U}_f^*, \tag{7}$$

considering rotation-minimizing conditions and following [15], we are able to express the central quaternion coefficient $\mathcal{A}_2$ with the help of

$$D = (\boldsymbol{h} \cdot \boldsymbol{\omega}_i)(\boldsymbol{h} \cdot \boldsymbol{\omega}_f) + h_s^2(\boldsymbol{\omega}_i \cdot \boldsymbol{\omega}_f) + h_s[\boldsymbol{\omega}_i, \boldsymbol{h}, \boldsymbol{\omega}_f], \tag{8}$$

where $[\cdot, \cdot, \cdot]$ is the standard mixed product. We end up with

$$\mathcal{A}_2 = a_0 \, \mathcal{A}_0 + a_1 \, \mathcal{A}_1 + a_3 \, \mathcal{A}_3 + a_4 \, \mathcal{A}_4, \tag{9}$$

where the coefficients a_j, $j = 0, 1, 3, 4$, are defined by

$$a_0 = -\frac{(h_s \mu_i - \boldsymbol{h} \cdot \boldsymbol{\omega}_i)^2}{2D}, \quad a_1 = \frac{4 h_s}{D} \left(h_s \mu_i - \boldsymbol{h} \cdot \boldsymbol{\omega}_i \right),$$

$$a_4 = -\frac{(h_s \mu_f - \boldsymbol{h} \cdot \boldsymbol{\omega}_f)^2}{2D}, \quad a_3 = \frac{4 h_s}{D} \left(h_s \mu_f - \boldsymbol{h} \cdot \boldsymbol{\omega}_f \right).$$

From the requirement that the directed ERF is a rotation–minimizing frame, we obtain also

$$8 \, h_s = -\left(\mu_i \, \mu_f \, h_s - \boldsymbol{h} \cdot (\mu_f \, \boldsymbol{\omega}_i + \mu_i \, \boldsymbol{\omega}_f) - ((\boldsymbol{\omega}_i \cdot \boldsymbol{\omega}_f) \, h_s + [\boldsymbol{\omega}_i, \boldsymbol{h}, \boldsymbol{\omega}_f]) \right). \tag{10}$$

If we define

$$\mu = \frac{\mu_f}{\mu_i}, \tag{11}$$

we finally get a quadratic equation for μ_i,

$$(\mu \, h_s) \, \mu_i^2 - (\mu \, d_i + d_f) \, \mu_i + (8 \, h_s - d_c - \omega_c \, h_s) = 0, \tag{12}$$

where for brevity we have put

$$d_i = \boldsymbol{h} \cdot \boldsymbol{\omega}_i, \quad d_f = \boldsymbol{h} \cdot \boldsymbol{\omega}_f, \quad d_c = [\boldsymbol{\omega}_i, \boldsymbol{h}, \boldsymbol{\omega}_f], \quad \omega_c = \boldsymbol{\omega}_i \cdot \boldsymbol{\omega}_f.$$

In order to have a real solution for μ_i in (12), the positivity of its discriminant Δ has to be imposed, i.e.,

$$\Delta = \Delta(\mu) = d_i^2 \, \mu^2 + 2 \, \mu(d_i \, d_f - 2 \, h_s(8 \, h_s - d_c - \omega_c \, h_s)) + d_f^2.$$

The above expression is a quadratic polynomial in μ which for some data can be positive on the whole real axis, but for others it is positive only on specific ranges. However we may observe that when the discriminant is negative, simply reversing the sign of μ makes it positive. This was actually the strategy adopted in [15].

Thus we have proved the following lemma.

Lemma 1. *For any set of data, provided that D in (8) does not vanish, and for any chosen μ, such that $\Delta(\mu) > 0$, there always exist two admissible pairs (μ_i, μ_f) satisfying (11) and solving (10).*

There are still two free remaining (positive) parameters λ_i and λ_f occurring in (4) to be determined. However, as rational forms are considered, one of them, for instance λ_i, can be assumed to be equal to 1 without loss of generality. Hence, we can conclude that the scheme has two free parameters in total, the positive λ_f (denoted in the following just as λ) and the real μ introduced in (11).

Remark 1. Our choice of using curves and frames of degree eight allows us to always find a solution to the considered interpolation problem, but it gives two free parameters that have to be reasonably determined. On the other hand, it was already noted in [15] that degree reduction to six for both the curve and the frame is not possible in general, unless C^1 continuity is relaxed to G^1 continuity.

Actually, the numerical experiments have clearly suggested that, in order to get plausible shapes, the following additional condition, which was already introduced in [14], has to be fulfilled,

$$\mathcal{U}_i \cdot \mathcal{U}_f > 0. \tag{13}$$

Observe that such a choice is also confirmed by the developed asymptotic analysis of the scheme reported in Section 4. Hence the above requirement implies that the method, for each admissible choice of the parameters λ and μ, produces only two distinct interpolating curves. It might look like that there are four distinct solutions, since there are two different pairs of solutions for $(\mathcal{U}_i, \mathcal{U}_f)$ in (5) fulfilling (13) and, for each of them, two admissible pairs (μ_i, μ_f) verifying (11) and solving (10) exist. However, we observe that the coefficients in (10) do not change when the signs of $\mathcal{U}_i$ and $\mathcal{U}_f$ are both reversed. Consequently, (4), (6) and (9) imply that the curve and the frame produced by the scheme do not change.

3 Selection of Free Parameters

In order to make the presented procedure automatic and also to get a solution with a reasonable shape, a strategy for fixing the free parameters λ and μ has to be developed.

In [15] these parameters were determined according to the possibility of degree reduction when the data allows it. The choice for μ was proposed as

$$\mu = \frac{2h_s + \mathbf{h} \cdot \boldsymbol{\omega}_f}{2h_s + \mathbf{h} \cdot \boldsymbol{\omega}_i} \tag{14}$$

which is assumed to be equal to 1 if both the numerator and the denominator in the above expression vanish and is equal to $+\infty$ if only the denominator vanishes. Furthermore, as mentioned in the previous section, the sign of μ could be reversed to fulfill the positivity of the discriminant. The other parameter (λ) was set to

$$\lambda = \sqrt{\frac{|\boldsymbol{\omega}_i|}{|\boldsymbol{\omega}_f|}}. \tag{15}$$

Here we propose a different selection of λ and μ, which gives better results confirmed by the numerical experiments. This choice of μ makes the algorithm also more robust when (14) is not directly usable. Indeed, in such cases, instead of reversing the sign of μ, we set $\mu = \mu_1/2$, where μ_1 and μ_2 are the real solutions of the quadratic equation $\Delta(\mu) = 0$ and $|\mu_1| < |\mu_2|$. Such a selection was suggested by the observation that reversing the sign of μ was producing non-satisfactory results in some experiments and hence the new choice keeps the sign given by (14).

Concerning the parameter λ, we propose

$$\lambda = \mu^2 \frac{|\mu_i + \boldsymbol{\omega}_i|^2}{|\mu_f + \boldsymbol{\omega}_f|^2}, \tag{16}$$

which is preferable than the choice in (15) since it can produce a significant reduction of the approximation error. The robustness of such a selection is enforced by the analysis developed in the next section, where it is proved that it guarantees the fourth approximation order of the scheme when the data are sampled from a general smooth analytic curve.

4   Asymptotic Analysis

Let $\mathbf{X} : [a\,,\,b] \to \mathbb{S}^2$ be a smooth assigned analytic curve on the unit sphere which can also be expressed in terms of its arc length s, that is $\mathbf{X} = \mathbf{X}(s), s \in [0\,,\,L]$, where L is the total curve length. This simplifies the analysis due to $\mathbf{X} \cdot \ddot{\mathbf{X}} \equiv 0$ and $\mathbf{X} \cdot \dot{\mathbf{X}} \equiv 0$, where the upper dot denotes derivation with respect to the arc–length. Let us further denote by $\mathbf{E}_j(s), j = 1, 2, 3$, an associated directed rotation–minimizing frame, with an usual assumption $\mathbf{X}(s) = \mathbf{E}_1(s)$. We are interested in establishing how the curve on the unit sphere $\mathbf{o}(t), t \in [0\,,\,1]$ produced by our method is capable to approximate the arc with infinitesimal arc–length Δs on the curve $\mathbf{X}(s), s \in [0, \Delta s]$ when $\Delta s \to 0$, assuming that the data for our algorithm are sampled from the end points of the arc. More precisely, assuming that the polar torsion [7] of the reference curve $\mathbf{X}$ doesn't vanish at $s = 0$, we are interested in proving that for each $t \in [0\,,\,1]$, the difference vector $\mathbf{o}(t) - \mathbf{X}(t\Delta s)$ belongs to $O(\Delta s^4)$, where

$$\mathbf{X}(t\Delta s) = \sum_{j=0}^{3} \frac{d^j \mathbf{X}}{d^j s}(0) t^j \frac{\Delta s^j}{j!} + O(\Delta s^4).$$

Then the following input data for the symbolic implementation of our algorithm are assumed,

$$\mathbf{o}_i = \mathbf{X}(0), \quad \mathbf{v}_i = (0, \cos\theta, \sin\theta)^T, \quad \mathbf{z}_i = \mathbf{o}_i \times \mathbf{v}_i, \quad \boldsymbol{\omega}_i = \Delta s\,\mathbf{X}(0) \times \dot{\mathbf{X}}(0),$$
$$\mathbf{o}_f = \mathbf{X}(\Delta s), \quad \mathbf{v}_f = \mathbf{E}_2(\Delta s), \qquad \mathbf{z}_f = \mathbf{o}_f \times \mathbf{v}_f, \quad \boldsymbol{\omega}_f = \Delta s\,\mathbf{X}(\Delta s) \times \dot{\mathbf{X}}(\Delta s),$$

where θ is a free angular parameter which allows us to specify arbitrarily any admissible orientation of $\mathbf{E}_2(0)$ (and, consequently, of $\mathbf{E}_3(0)$). We note that the angular velocities $\boldsymbol{\omega}_i$ and $\boldsymbol{\omega}_f$ are obtained by multiplying both the extreme angular velocities of $\mathbf{X}$ with respect to the arc–length by the factor Δs, because in the algorithm they denote the angular velocities with respect to the local parameter t which varies in $[0\,,1]$.

Now, in order to simplify the analysis, without loss of generality, we can assume

$$\mathbf{X}(0) = (1,0,0)^T, \quad \dot{\mathbf{X}}(0) = (0,0-1)^T$$

which implies that $\boldsymbol{\omega}_i = \Delta s\,(0,1,0)^T$ and $\mathcal{U}_i = (\cos\frac{\theta}{2}, \sin\frac{\theta}{2}, 0, 0)^T$, where the sign of $\mathcal{U}_i$ can be arbitrarily chosen. Observe that the hypothesis of a non vanishing polar torsion of $\mathbf{X}$ at $s = 0$ implies that $\ddot{\mathbf{X}}\cdot\mathbf{j} \neq 0$, where, as usual, $\mathbf{j} = (0,1,0)^T$.

Moreover the expressions of $\mathbf{X}(\Delta s)$, $\dot{\mathbf{X}}(\Delta s)$, $\mathbf{E}_2(\Delta s)$ are replaced with their asymptotic Taylor expansions at $s = 0$ which are assumed to be of order 9 because it is necessary to prove the fourth approximation order of the scheme. Starting from the Taylor expansions of these data, with the help of a computer algebra system, we have obtained the associated expansions of the unit quaternions $\mathcal{U}_f$, satisfying (13), and $\mathcal{H}$ defined in (7) which, together with the expansion of the angular velocity $\boldsymbol{\omega}_f$, have allowed us to formulate the following preliminary result.

Lemma 2. *If the polar torsion of the curve $\mathbf{X}$ at $s = 0$ is not vanishing, then the coefficient μ, defined in (14), admits the following asymptotic expansion,*

$$\mu = \sum_{j=0}^{3} \mu^{(j)} \Delta s^j + O(\Delta s^4),$$

where in particular $\mu^{(0)} = 1$ and $\mu^{(1)} = \frac{1}{4}\frac{\ddot{\mathbf{X}}(0)\cdot\mathbf{j}}{\dot{\mathbf{X}}(0)\cdot\mathbf{j}}$.

Using this expansion for μ, and by the help of a computer algebra system again we can obtain corresponding expansions for the two admissible couples (μ_i, μ_f), fulfilling (10) and (11). Selecting between them the one associated to greater value for μ_i, we are able to prove the following intermediate result.

Lemma 3. *If the polar torsion of the curve $\mathbf{X}$ at $s = 0$ is not vanishing, then the coefficient λ defined in (16) admits the following asymptotic expansion,*

$$\lambda = \sum_{j=0}^{3} \lambda^{(j)} \Delta s^j + O(\Delta s^4),$$

where in particular $\lambda^{(0)} = 1$, $\lambda^{(1)} = -\frac{1}{2}\frac{\ddot{\mathbf{X}}(0)\cdot\mathbf{j}}{\dot{\mathbf{X}}(0)\cdot\mathbf{j}}$.

From the expansions of μ_i, μ_f and λ we can derive the asymptotic expansions up to order 4 of the quaternion coefficients $\mathcal{A}_i, i = 0,\ldots,4$, and so it is possible to obtain a symbolic expression of the curve $\mathbf{o}(t)$ which is rational with respect to t of degree 8 but also depends on Δs as a rational function,

$$\mathbf{o}(t) = \mathbf{o}(t, \Delta s) = \frac{\sum_{j=0}^{3} \mathbf{b}_j(t)\Delta s^j + O(\Delta s^4)}{\sum_{j=0}^{3} d_j(t)\Delta s^j + O(\Delta s^4)},$$

where the $\mathbf{b}_j(t), d_j(t), j = 0, \ldots, 4$ are respectively vector and scalar polynomials of degree less or equal to 8 and, in particular, $d_0(t) \equiv 1$. Thus $\mathbf{o}(t, \Delta s)$ admits the following Taylor expansion with respect to Δs,

$$\mathbf{o}(t, \Delta s) = \sum_{j=0}^{3} \boldsymbol{\gamma}_j(t)\frac{\Delta s^j}{j!} + O(\Delta s^4),$$

where

$$\boldsymbol{\gamma}_j(t) = \frac{d^j \mathbf{X}}{d^j s}(0)t^j, \quad j = 0, \ldots, 3.$$

Hence, we can formulate the following proposition,

Proposition 1. *Given an analytic smooth curve on the unit sphere, $\mathbf{X}(\xi), \xi \in [a, a + \Delta\xi]$ with non vanishing polar torsion at $\xi = a$, it is approximated with order 4 by one of the two rational curves $\mathbf{o}(t), t \in [0, 1]$ with rational directed rotation–minimizing frame produced by the presented method.*

In Table 3 reported in the next section, the numerical approximation order for two test cases confirms the above analysis.

5   Extension to Splines and Numerical Examples

Before presenting the results obtained with some examples which illustrate the procedure, we consider the extension of the scheme to splines. Hence the input data are a sequence of frames $(\mathbf{o}_j, \mathbf{v}_j, \mathbf{z}_j), j = 0, 1, \ldots, N$, and related compatible angular velocities $\boldsymbol{\omega}_j, j = 0, 1, \ldots, N$, together with an associated set of parameter values (knots) $a = \xi_0 < \xi_1 < \cdots < \xi_{N-1} < \xi_N = b$. The scheme defines a C^1 rational spline $\mathbf{R} : [a, b] \rightarrow \mathbb{S}^2$ on the sphere, with an associated rational rotation-minimizing directed frame

$$\{\mathbf{e}_1, \mathbf{e}_2, \mathbf{e}_3\}, \quad \mathbf{e}_j : [a, b] \rightarrow \mathbb{R}^3, \quad j = 1, 2, 3,$$

and angular velocity $\boldsymbol{\omega} = \mathbf{e}_1 \times \frac{d\mathbf{e}_1}{d\xi}$, such that

$$\mathbf{R}(\xi_j) = \mathbf{o}_j, \quad \mathbf{e}_1(\xi) = \mathbf{o}_j, \quad \mathbf{e}_2(\xi) = \mathbf{v}_j, \quad \mathbf{e}_3(\xi) = \mathbf{z}_j, \quad \boldsymbol{\omega}(\xi_j) = \boldsymbol{\omega}_j,$$

for $j = 0, \ldots, N$. The presented method can be used locally in each subinterval $[\xi_j, \xi_{j+1}]$ by linearly mapping it to $[0, 1]$. This gives the following local data:

$$
\begin{aligned}
(\mathbf{o}_i, \mathbf{v}_i, \mathbf{z}_i) &= (\mathbf{o}_j, \mathbf{v}_j, \mathbf{z}_j) & \boldsymbol{\omega}_i &= \Delta\xi_j \, \boldsymbol{\omega}_j, \\
(\mathbf{o}_f, \mathbf{v}_f, \mathbf{z}_f) &= (\mathbf{o}_{j+1}, \mathbf{v}_{j+1}, \mathbf{z}_{j+1}), & \boldsymbol{\omega}_f &= \Delta\xi_j \, \boldsymbol{\omega}_{j+1}, \quad j = 0, 1, \ldots, N-1,
\end{aligned}
$$

where $\Delta \xi_j := \xi_{j+1} - \xi_j$. In our experiments, the data points are always taken from analytical curves in order to stress the approximation power of the method. Hence, in all cases the sequence of curve positions, along with the corresponding tangents (providing also the angular velocities) are sampled from the analytical curve. The knots considered to define the spline curve are just the parameter values used for the sampling. The associated sequence of sampled rotation-minimizing frame orientations is computed by numerical solution of a first order differential equation (for instance the second or the third equation in (3)), with a compatible initial condition.

For the examples, we take the data from arcs of the Viviani curve, which has the following standard parameterization,

$$X(\xi) = \left(\frac{1}{2} \left(1 + \cos \xi\right), \; \frac{1}{2} \sin \xi, \; \sin \frac{\xi}{2} \right)^T, \; \xi \in [0, 2\pi],$$

or from arcs of a curve $X = (x, y, z)^T$ on the unit sphere, where

$$
\begin{aligned}
x(\xi) &= 2 \cos \alpha \; \cos \xi \; \cos k\xi + \; \sin \xi \; \sin k\xi, \\
y(\xi) &= 2 \cos \alpha \; \sin \xi \; \cos k\xi + \cos \xi \; \sin k\xi, \\
z(\xi) &= 2 \sin \alpha \; \cos k\xi,
\end{aligned}
\tag{17}
$$

with α and k denoting two free parameters. The first curve is well known in classical geometry, but it is also important in applications since it is related to the projection of a satellite path on the Earth. The curves in (17) are obtained as trajectories of a point on a great circle of a sphere, rotating about one of its axes. Such curves are again related with satellite motions and are called *satellite curves* (see [3], e.g.).

The three figures shown in the sequel are obtained by using the spline extension of the scheme. For each spline segment, the best interpolant between the two possible ones could be determined by some fairness criterion. Here, having the analytical curve as a reference, the choice is based on a comparison of the interpolant with the curve at some chosen points.

In Fig. 1, the Viviani test is considered. On the left, the resulting spline curve, made of 8 segments, along with the analytical curve is shown, while on the right the resulting curve and its rotation-minimizing frame are depicted.

In Fig. 2 the data are sampled from (17) with $\alpha = 7/8 \, \pi$, $k = 4/3$ and $\xi \in [0, 6\pi]$. The resulting spline curve is made of 16 segments. In Fig. 3, the data are taken again from (17), but with $\alpha = 3/4 \, \pi$ and $k = 2/3$. In this case the curve exhibits some points with very big tangent variation. In order to focus on the good approximation of such kind of curves, we concentrate only on the part given by $\xi \in [0, 3\pi]$. Although this part of the curve behaves like having a cusp, 12 segments are already enough to approximate it in a good way.

Note that in the first two examples the results are referring to a sampling which ensures that the splitting of the analytical curve is uniform with respect to its cumulative arc length. On the other hand, in the third example, the sampling allows a better reconstruction of the area where the shape of the curve is very sharp.

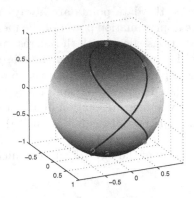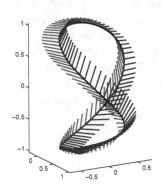

Fig. 1. Example 1. Left: the obtained spline curve (black) on the sphere along with the analytical curve (blue). Stared circles denote the junction points of the spline segments. Right: the resulting rotation-minimizing directed frame of the curve (only $\mathbf{e}_2$ and $\mathbf{e}_3$ are shown).

We conclude this section by presenting three sets of tables in order to check numerically the behavior of the approximation error for two test curves, the Viviani curve and the satellite curve from the second example. Such an error is obtained by considering the maximum Hausdorff distance ϵ_N between $\boldsymbol{X}$ and $\mathbf{R}$ computed in each subinterval $[\xi_j, \xi_{j+1}]$, $j = 0, \ldots, N - 1$:

$$\epsilon_N = \max_{j=0,\ldots,N-1} \left(d_{H_1}(\boldsymbol{X}_{|[\xi_j,\xi_{j+1}]}, \mathbf{R}_{|[\xi_j,\xi_{j+1}]}), d_{H_2}(\boldsymbol{X}_{|[\xi_j,\xi_{j+1}]}, \mathbf{R}_{|[\xi_j,\xi_{j+1}]}) \right),$$

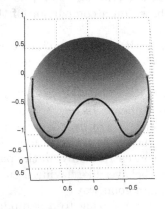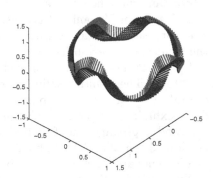

Fig. 2. Example 2. Left: the obtained spline curve (black) on the sphere along with the analytical curve (blue). Stared circles denote the junction points of the spline segments. Right: the resulting rotation-minimizing directed frame of the curve (only $\mathbf{e}_2$ and $\mathbf{e}_3$ are shown).

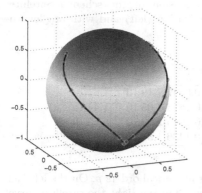 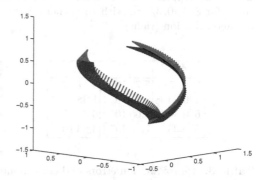

Fig. 3. Example 3. Left: the obtained spline curve (black) on the sphere along with the analytical curve (blue). Stared circles denote the junction points of the spline segments. Right: the resulting rotation-minimizing directed frame of the curve (only $\mathbf{e}_2$ and $\mathbf{e}_3$ are shown).

where

$$d_{H_1}(\mathbf{X}_{|[\xi_j,\xi_{j+1}]}, \mathbf{R}_{|[\xi_j,\xi_{j+1}]}) = \max_{\xi \in [\xi_j,\xi_{j+1}]} \left(\min_{\hat{\xi} \in [\xi_j,\xi_{j+1}]} |\mathbf{X}(\xi) - \mathbf{R}(\hat{\xi})| \right),$$

$$d_{H_2}(\mathbf{X}_{|[\xi_j,\xi_{j+1}]}, \mathbf{R}_{|[\xi_j,\xi_{j+1}]}) = \max_{\hat{\xi} \in [\xi_j,\xi_{j+1}]} \left(\min_{\xi \in [\xi_j,\xi_{j+1}]} |\mathbf{X}(\xi) - \mathbf{R}(\hat{\xi})| \right).$$

For both analytical curves we consider two different parameterizations, the standard one and the arc–length one. So the quantity

$$h_N = \max_{0 \le j \le N-1} \Delta \xi_j$$

for the arc–length parameterization is just the ratio between the curve length and N.

Table 1. Approximation errors of the spline extension of the scheme. Viviani curve (half curve) with standard parameterization (left) and with arc–length parameterization (right).

N	h_N	$\epsilon_N^{(1)}$	$\epsilon_N^{(2)}$
2	3.142	$3.43\ 10^{-2}$	$2.34\ 10^{-2}$
4	1.737	$3.44\ 10^{-3}$	$1.42\ 10^{-3}$
6	1.208	$1.00\ 10^{-3}$	$3.67\ 10^{-4}$
8	0.924	$4.24\ 10^{-4}$	$1.62\ 10^{-4}$

N	h_N	$\epsilon_N^{(1)}$	$\epsilon_N^{(2)}$
2	1.910	$5.26\ 10^{-2}$	$2.88\ 10^{-2}$
4	0.955	$5.97\ 10^{-3}$	$1.47\ 10^{-3}$
6	0.637	$1.55\ 10^{-3}$	$3.40\ 10^{-4}$
8	0.478	$7.27\ 10^{-4}$	$1.55\ 10^{-4}$

Tables 1–2 are aimed to confirm the benefit we get by selecting the free parameter λ with (16), instead of (15) introduced in [15]. Therefore the error $\epsilon_N^{(1)}$ obtained using (15) is compared with $\epsilon_N^{(2)}$, arising from the choice (16).

Table 2. Approximation errors of the spline extension of the scheme. Satellite curve for $\xi \in [0, 3/4\pi]$, with the generic parameterization (left) and with arc–length parameterization (right).

N	h_N	$\epsilon_N^{(1)}$	$\epsilon_N^{(2)}$
2	1.780	$5.72\ 10^{-3}$	$3.17\ 10^{-3}$
4	0.6458	$5.73\ 10^{-4}$	$1.98\ 10^{-4}$
6	0.4472	$1.65\ 10^{-4}$	$6.43\ 10^{-5}$
8	0.3413	$7.08\ 10^{-5}$	$3.96\ 10^{-5}$

N	h_N	$\epsilon_N^{(1)}$	$\epsilon_N^{(2)}$
2	0.5747	$7.19\ 10^{-3}$	$1.61\ 10^{-3}$
4	0.2873	$1.05\ 10^{-3}$	$1.44\ 10^{-4}$
6	0.1915	$2.83\ 10^{-4}$	$5.65\ 10^{-5}$
8	0.1436	$1.36\ 10^{-4}$	$3.82\ 10^{-5}$

Table 3. Approximation errors and corresponding numerical orders p. On the left the Viviani curve for $a = 5\pi/8$ and b reducing as shown. On the right the satellite curve of example 2, with $b = 3\pi/16$ and a increasing as shown. The total curve length L is also reported.

b	L	$\epsilon_1^{(2)}$	p
$13\pi/16$	0.3208	$6.75\ 10^{-5}$	
$23\pi/32$	0.1642	$5.19\ 10^{-6}$	3.83
$43\pi/64$	0.0832	$3.38\ 10^{-7}$	4.01
$83\pi/128$	0.0418	$2.16\ 10^{-8}$	4.01
$163\pi/256$	0.0210	$1.36\ 10^{-9}$	4.01

a	L	$\epsilon_1^{(2)}$	p
0	0.2592	$6.04\ 10^{-5}$	
$3\pi/32$	0.1360	$1.55\ 10^{-5}$	2.11
$9\pi/64$	0.0701	$1.10\ 10^{-6}$	3.99
$21\pi/128$	0.0356	$7.30\ 10^{-8}$	4.00
$45\pi/256$	0.0179	$4.70\ 10^{-9}$	4.00

Analyzing the error in the tables, we can conclude that the new choice of λ is always preferable, in particular when the arc–length parameterization is considered.

Finally in Table 3 the basic one segment implementation of the scheme is performed again for the Viviani and satellite. Successively reducing the length of the analytic curve defining the data for our scheme, we are able to compute the numerical approximation order p of the scheme which nicely confirms the developed theoretical asymptotic analysis.

6   Conclusions

A new method for the construction of rational C^1 Hermite interpolating curves on the sphere, endowed by a rational directed rotation–minimizing frame, is presented. Its spline extension for defining general spherical camera motions has been also considered. Starting from the method introduced in [15], new and better selection for the two free parameters have been studied, with the help of the asymptotic analysis. The resulting scheme has order of approximation 4. Several examples illustrate the effectiveness of the proposed method.

Acknowledgements. This research was partially supported by a Bilateral Research Project within the Executive Programme Italy-Slovenia 2011–2013. The INDAM support is also gratefully acknowledged.

References

1. Belghith, K., Kabanza, F., Bellefeuille, P., Hartman, L.: Automated camera planning to film robot operations. Artif. Intell. Rev. 37, 313–330 (2012)
2. Cao, C.G.L.: Guiding navigation in colonoscopy. Surgical Endoscopy 21, 408–484 (2007)
3. Capderou, M.: Satellites: Orbites Et Missions. Springer, Heidelberg (2003)
4. Choi, H.I., Han, C.Y.: Euler–Rodrigues frames on spatial Pythagorean–hodograph curves. Comput. Aided Geom. Design 19, 603–620 (2002)
5. Christie, M., Machap, R., Normand, J.-M., Olivier, P., Pickering, J.H.: Virtual camera planning: A survey. In: Butz, A., Fisher, B., Krüger, A., Olivier, P. (eds.) SG 2005. LNCS, vol. 3638, pp. 40–52. Springer, Heidelberg (2005)
6. Farin, G.: Curves and Surfaces for CAGD: A practical Guide, 5th edn. Academic Press, San Diego (2003)
7. Farouki, R.T., Giannelli, C.: Spatial camera orientation control by rotation–minimizing directed frames. Comput. Anim. Virtual World 20, 457–472 (2009)
8. Farouki, R.T., Giannelli, C., Manni, C., Sestini, A.: Quintic space curves with rational rotation–minimizing frames. Comput. Aided Geom. Design 26, 580–592 (2009)
9. Farouki, R.T., Giannelli, C., Sestini, A.: An interpolation scheme for designing rational rotation–minimizing camera motions. Advances in Computational Mathematics 38(1), 63–82 (2013)
10. Farouki, R.T., Han, C.Y., Dospra, P., Sakkalis, T.: Rotation–minimizing Euler–Rodrigues rigid–body motion interpolants. Comput. Aided Geom. Design 30, 653–671 (2013)
11. Farouki, R.T., Sakkalis, T.: Rational rotation–minimizing frames on polynomial space curves of arbitrary degree. Journal of Symbolic Computation 45, 844–856 (2010)
12. Farouki, R.T., Sakkalis, T.: A complete classification of quintic space curves with rational rotation–minimizing frames. Journal of Symbolic Computation 47, 214–226 (2012)
13. Han, C.Y.: Nonexistence of rational rotation–minimizing frames on cubic curves. Comput. Aided Geom. Design 25, 298–304 (2007)
14. Jüttler, B.: Rotation–minimizing spherical motions. In: Lenarčič, J., Husty, M. (eds.) Advances in Robot Kinematics: Analysis and Control, pp. 413–422. Kluwer, Dordrecht (1998)
15. Jaklič, G., Sampoli, M.L., Sestini, A., Žagar, E.: C^1 rational interpolation of spherical motions with rational rotation–minimizing directed frames. Comput. Aided Geom. Design 30, 159–173 (2013)
16. Sampoli, M.L., Sestini, A.: Rational rotation–minimizing polar oriented rigid body motions. In: Proceedings of the 21th Int. Workshop on Robotics in Alpe–Adria–Danube Region, vol. 4, pp. 284–291. Ed. Sci. Art, Italy (2012) ISBN: 978-88-95430-45-4

Construction and Analysis of Zone-dependent Interpolatory/Non-interpolatory Stochastic Subdivision Schemes for Non-regular Data

Xiaoyun Si[1], Jean Baccou[2], and Jacques Liandrat[1]

[1] Centrale Marseille, LATP, UMR 7353, CNRS, Aix-Marseille univ.,
13451 Marseille, France
[2] Institut de Radioprotection et de Sûreté Nucléaire(IRSN),
PSN-RES/SEMIA/LIMAR, CE Cadarache, 13115 Saint Paul Les Durance, France
`xiaoyun.si@centrale-marseille.fr`, `jean.baccou@irsn.fr`,
`jacques.liandrat@centrale-marseille.fr`

Abstract. This work is devoted to the definition of stochastic subdivision schemes adapted to the reconstruction of non-regular data. These schemes are constructed in the framework of the Kriging theory. Thanks to the introduction of a zone-dependent error variance in the Kriging approach, they combine interpolatory and non interpolatory subdivision schemes according to a domain segmentation. Their originality relies on the introduction and coupling of three ingredients: a segmentation of the data, a local prediction according to the characteristics of the different zones and an adaption strategy near segmentation points. The convergence of the corresponding 4-point scheme is analyzed. Its behavior is compared with other subdivision schemes on various numerical experiments.

Keywords: subdivision, Kriging, interpolatory, non-interpolatory.

1 Introduction

Stochastic kriging-based approaches [3] for data modeling are classical methods in risk analysis since they integrate the spatial structure of the data in the prediction and allow to evaluate the precision of the estimation thanks to the underlying probabilistic model. However, they usually assume that the phenomenon to estimate is regular, which is not the case in practice.

Subdivision is a very powerful tool to construct smooth curves and surfaces starting from a given set of control points. Many different subdivision schemes such as linear (or not), position-dependent (or not) [1], interpolatory (or not) [4], stationary (or not) [7] schemes have been developed for specific situations.

This paper is devoted to the design and analysis of new stochastic subdivision schemes that aim to improve the accuracy of the reconstruction of non-regular data. These schemes are based on Kriging as in [2] but their originality rely on the introduction and the coupling of three main ingredients:

M. Floater et al. (Eds.): MMCS 2012, LNCS 8177, pp. 456–470, 2014.

- A segmentation of the data leading to a splitting in different zones containing or not non-regular data,
- The construction of a local predictor integrating the information coming from the segmentation and defining zone-dependent subdivision schemes that mix interpolatory and non-interpolatory predictions,
- The adaption of the schemes close to the segmentation points.

Our work is organized as follows: Section 2 provides a quick overview on subdivision scheme and ordinary Kriging theory. Section 3 is devoted to the construction of the new Kriging-based scheme and to the analysis of its convergence for the specific case of 4-point stencils. Finally in Section 4, several numerical experiments and comparisons with other classical subdivision schemes are performed.

2   Quick Overview on Subdivision Schemes and Kriging

2.1   Subdivision Schemes

In this paper only binary subdivision schemes are considered. We focus on subdivision on the real line and therefore consider grid points $x_k^j = k2^{-j}, j \geq 0, k \in \mathbb{Z}$. Starting from an initial set of control points $F^0 = \{f_n^0, n \in \mathbb{Z}\}$, subdivision iteratively reconstructs new set of points $F^j = \{f_n^j, n \in \mathbb{Z}\}, j \geq 1$ as

$$F^j = SF^{j-1}, \qquad j \geq 1,$$

with

$$f_k^j = \sum_{m \in \mathbb{Z}} a_{k-2m}^{j,k} f_m^{j-1}, \qquad k \in \mathbb{Z}.$$

The family $\{a_m^{j,k}\}_{m \in \mathbb{Z}}, j \geq 1, k \in \mathbb{Z}$ that satisfies $\forall j, \forall k, \sum_{m \in \mathbb{Z}} a_{k-2m}^{j,k} = 1$, is called the mask of the scheme. It controls the shape of the reconstructed points and the regularity of the limit function if the scheme is convergent.[1] A subdivision scheme is said to be stationary when the mask is independent of j and uniform when for any j it depends only on the parity of k.

Moreover since the mask coefficients $a_m^{j,k}, m \in \mathbb{Z}$ are supposed to be zero but a finite number of them, one can introduce $l_{j,k}$ and $r_{j,k}$ such that the values (f_{k+m}^{j-1}) involved in the evaluation of f_{2k}^j and f_{2k-1}^j are $S_{j,k} = \{f_{k-l_{j,k}}^j, \cdots, f_{k+r_{j,k}-1}^j\}$. This set is called the reconstruction stencil of index (j, k).

There exists many ways to construct the mask. Among classical schemes, translation-invariant and position-dependent strategies play an important role. The first one corresponds to $l_{j,k} = l_j, r_{j,k} = r_j$ and leads to uniform schemes.

---

[1] A subdivision scheme S is uniformly convergent if for any initial set of control point F^0, there exists a continuous function f such that $\|S^j F^0 - (f(k2^{-j}))_{k \in \mathbb{Z}}\|_\infty \longrightarrow_{j \to +\infty} 0$.

The second one corresponds to $(l_{j,k}, r_{j,k})$ depending on $k2^{-j}$ and leads to non-uniform schemes.

In [1], a first example of position-dependent Lagrange interpolatory subdivision scheme based on a segmentation of the real line has been constructed and fully analyzed. It is based on a local polynomial interpolation and a sequence of segmentation points $(y_i)_{i \in \mathbb{N}}$. More precisely, the sequences $\{a_{k-2m}^{j,k}\}_{m \in \{-l_{j,k}, \cdots, r_{j,k}-1\}}$ satisfy:

$$\begin{cases} a_{2k-2m}^{j,2k} = \delta_{m,k} \ , \\ a_{2k-1-2m}^{j,2k-1} = L_m^{l_{j,2k-1}, r_{j,2k-1}}(-\tfrac{1}{2}) \ , \end{cases}$$

where $L_m^{l_{j,2k-1}, r_{j,2k-1}}(x) = \Pi_{n=-l_{j,2k-1}, n \neq m}^{r_{j,2k-1}-1} \frac{x-n}{m-n}$. The position-dependent strategy is then fixed by the values of the couple $(l_{j,k}, r_{j,k})$: Fig.1 displays an example of such a strategy for a 4-point stencil and a single segmentation point y_0. In this case, it depends on the relative position of $k2^{-j}$ and y_0.

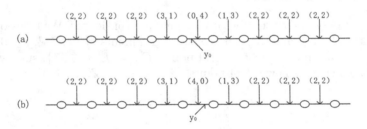

Fig. 1. Example of a 4-point stencil position-dependent strategy in the vicinity of a segmentation point y_0

In this paper, the construction of the scheme is performed in the Kriging framework which is recalled in the next section.

2.2   Ordinary Kriging

For sake of clarity, a quick overview on Ordinary Kriging is here provided independently of the subdivision framework. Starting from a set of n observations $\{f_i = f(x_i), i = 0, 1, \cdots, n-1\}$, Kriging theory ([3], [5]) assumes that each observation f_i is a realization of a random variable $\mathcal{F}(x_i)$ coming from a random process $\mathcal{F}(x)$. This random process satisfies $\mathcal{F}(x) = d + \delta(x)$, where d is the deterministic component of $\mathcal{F}(x)$ and $\delta(x)$ is a zero-mean random process.

Under the assumption of stationarity and of constant d, the spatial correlation of the data is identified by computing the semi-variogram:

$$\gamma(h) = \frac{1}{2} E((\mathcal{F}(x+h) - \mathcal{F}(x))^2) \ ,$$

where E is the mathematical expectation. In practice, since the number of observations is finite, it is approximated by the experimental semi-variogram:

$$\gamma_{exp}(h) = \frac{1}{2Card(N(h))} \sum_{(k,l)\in N(h)} (f(x_k) - f(x_l))^2 ,$$

where $N(h) = \{(x_k, x_l) \in \{0, 1, \cdots, n-1\}^2, |x_k - x_l| = h\}$ and $Card(N(h))$ is the number of elements of $N(h)$. The spatial structure is then identified by a least square fit of γ_{exp}. The candidates to this fit have to be chosen in a family of so-called valid semi-variogram models. In this paper, we focus on two classical models: the Gaussian type with $\gamma(h) = K \left(1 - e^{-(ah)^2}\right)$ (K and $a > 0$) and the Exponential type with $\gamma(h) = K \left(1 - e^{-ah}\right)$ (K and $a > 0$) (fig.2).

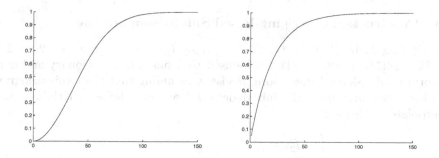

Fig. 2. Valid semi-variograms. Left, Gaussian type ($\gamma(h) = 1 - e^{-\frac{1}{50^2}h^2}$) and right, Exponential type ($\gamma(h) = 1 - e^{-\frac{1}{20}h}$).

The Kriging estimator of $\mathcal{F}$ at position x^* , $\mathcal{P}(\mathcal{F}, x^*)$, is then constructed as the unique linear unbiased predictor minimizing the estimation variance $\sigma^2_{OK} = var(\mathcal{F}(x^*) - \mathcal{P}(\mathcal{F}, x^*))$. It is written as

$$\mathcal{P}(\mathcal{F}, x^*)) = \sum_{i=0}^{n-1} \lambda_i \mathcal{F}(x_i) , \qquad (1)$$

where $\{\lambda_i\}_{i=0,1,\cdots,n-1}$ are called the Kriging weights and are the solutions of the following Ordinary Kriging system:

$$
\begin{bmatrix}
\gamma_{0,0} & \cdots & \gamma_{0,n-1} & 1 \\
\gamma_{1,0} & \cdots & \gamma_{1,n-1} & 1 \\
\cdots & \cdots & \cdots & 1 \\
\gamma_{n-1,0} & \cdots & \gamma_{n-1,n-1} & 1 \\
1 & 1 & 1 & 0
\end{bmatrix}
\begin{bmatrix}
\lambda_0 \\
\lambda_1 \\
\cdots \\
\lambda_{n-1} \\
\mu
\end{bmatrix}
=
\begin{bmatrix}
\gamma_{*,0} \\
\gamma_{*,1} \\
\cdots \\
\gamma_{*,n-1} \\
1
\end{bmatrix} , \qquad (2)
$$

where μ is the Lagrange multiplier enforcing the unbiasedness of the estimator, $\gamma_{i,j} = \gamma(|x_i - x_j|)$ and $\gamma_{*,i} = \gamma(|x^* - x_i|), i = 0, \cdots, n-1$.

The previous Kriging estimator is an exact interpolator. The constrain of exact interpolation can be relaxed by the introduction of an error variance on the diagonal of the left hand side matrix of (2) leading to the following more general Kriging system:

$$
\begin{bmatrix}
\gamma_{0,0} - c_0 \cdots & \gamma_{0,n-1} & 1 \\
\gamma_{1,0} & \cdots & \gamma_{1,n-1} & 1 \\
\cdots & \cdots & \cdots & 1 \\
\gamma_{n-1,0} & \cdots \gamma_{n-1,n-1} - c_{n-1} & 1 \\
1 & 1 & 1 & 0
\end{bmatrix}
\begin{bmatrix}
\lambda_0 \\
\lambda_1 \\
\cdots \\
\lambda_{n-1} \\
\mu
\end{bmatrix}
=
\begin{bmatrix}
\gamma_{*,0} \\
\gamma_{*,1} \\
\cdots \\
\gamma_{*,n-1} \\
1
\end{bmatrix},
\tag{3}
$$

where $\mathcal{C} = (c_0, ..., c_{n-1})$ denotes the vector of error variance [6]. In practice, it allows to take into account measurement errors at location $(x_0, \cdots, x_{n-1})$. As a result, the Kriging estimator based on (3) is no more interpolating.

2.3   Construction of Kriging-Based Subdivision Scheme

Substituting x_i by $(k + i - l_{j,k})2^{-(j-1)}$, n by $l_{j,k} + r_{j,k}$ and x^* by $2k \times 2^{-j}$ or $(2k - 1)2^{-j}$, expression (1) can be used to define a non-stationary and non-uniform subdivision scheme. More precisely, assuming that the semi-variogram has been identified from the initial data F^0 one can define a Kriging-based interpolatory scheme as:

$$
\begin{cases}
f^j_{2k} = f^{j-1}_k , \\
f^j_{2k-1} = \sum_{m=-l_{j,k}}^{r_{j,k}-1} \lambda^{j,2k-1}_m f^{j-1}_{k+m} ,
\end{cases}
\tag{4}
$$

where $\{\lambda^{j,2k-1}_m\}_{m=-l_{j,k},\cdots,r_{j,k}-1}$ are the Kriging weights solutions of system (2). Similarly, a non-interpolatory Kriging-based scheme can be constructed as:

$$
\begin{cases}
f^j_{2k} = \sum_{m=-l_{j,k}}^{r_{j,k}-1} \lambda^{j,2k}_m f^{j-1}_{k+m} , \\
f^j_{2k-1} = \sum_{m=-l_{j,k}}^{r_{j,k}-1} \lambda^{j,2k-1}_m f^{j-1}_{k+m} ,
\end{cases}
\tag{5}
$$

where $\{\lambda^{j,2k}_m\}_{m=-l_{j,k},\cdots,r_{j,k}-1}$ (resp. $\{\lambda^{j,2k-1}_m\}_{m=-l_{j,k},\cdots,r_{j,k}-1}$) are the Kriging weights solutions of (3) for $x^* = 2k \times 2^{-j}$ (resp. $x^* = (2k - 1)2^{-j}$).

The interpolatory subdivision scheme (4) has already been introduced in [2]. As in [7] and [1] (for Lagrange-type prediction), it can be translation-invariant and position-dependent. Moreover, the coupling with Kriging theory allows to define the mask according to the underlying spatial structure of the data. It provides a more accurate prediction when the data cannot be assumed to come from the discretization of a piecewise-polynomial function. The scheme proposed in this paper is constructed in the same framework. Its originality compared to the Kriging-based scheme of [2] relies on the possibility to relax the constrain of exact interpolator by adding an error variance leading to (5) and, more precisely, to relax this constraint according to the position of the prediction. The next sections are devoted to the definition and analysis of a zone dependant Kriging based subdivision scheme where the position dependant strategy is applied to the local error variance: the result is a stochastic scheme with position modulated interpolatory/non interpolatory property.

3   Zone-dependent Kriging-Based Subdivision Scheme

This section is devoted to the definition and the convergence analysis of a subdivision scheme aimed to be adapted to the reconstruction of non-regular data. It combines the interpolatory and non-interpolatory approaches previously introduced and is based on an a priori segmentation of the data in different zones characterized by different regularity. According to this regularity and therefore to the zone, error variance vectors are affected to the Kriging system, leading to a zone-dependent subdivision scheme. For sake of simplicity, we restrict our work in this paper to three zones (two zones of "regular data" and one zone $]y_0, y_1[$ containing "non-regular" data as displayed by fig.3, top, left). Moreover, we focus on predictions involving 4-point stencils i.e $\forall(j, k), l_{j,k} + r_{j,k} = 4$.

3.1   Construction of the Zone-dependent Scheme

The construction relies on three essential ingredients that are fully specified in the sequel:

- Segmentation of the data:
  The real line is supposed to be split in three zones separated by the segmentation points $y_0 < y_1$ belonging to the coarse grid $((y_0, y_1) \in I\!\!N^2)$. The zones $] - \infty, y_0]$ and $[y_1, +\infty[$ correspond to regular zones while $]y_0, y_1[$ correspond to a non-regular one.
- Local predictors (in the interior of the zones):
  Local Kriging-based subdivision predictors are constructed independently in each zone. The semi-variogram of the zone is first identified according to Section 2.2. Then an error variance function $C(x)$ is defined on the real line. This function provides for each point $x_k^j = k2^{-j}$ the corresponding error variance $C(x_k^j)$ involved in (3). In this paper, $C = \chi_{]y_0,y_1[}$ (fig.3, top, right)where χ_A denotes the characteristic function of the set A.[2]
  In the interior of each zone, a 4-point centered scheme is applied (i.e $l_{j,k} = r_{j,k} = 2$). For regular zones a subdivision scheme of type (4) is used whereas for $]y_0, y_1[$ the scheme corresponds to (5) with $c_i = 1, 0 \leq i \leq 3$.
- Adaption at the edges of the zones:
  These schemes can not be applied close to the edges of the zone and a specific adaption is required. For sake of clarity, we describe this adaption around the segmentation point y_0 written for any j as $y_0 = k_0^j 2^{-(j-1)}$. The adaption close to y_1 is performed similarly.
  Regular zone: the adaption is performed following the position-dependent strategy proposed in [1] and recalled on fig.1. As shown on fig.3 (bottom,

---

[2] $\chi_A(x) = 1$ if $x \in A$ and $\chi_A(x) = 0$ otherwise.

left), the stencil associated to the prediction at position $(2k_0^j - 1)2^{-j}$ is shifted to the left.

Non-regular zone: the interior scheme can not be used to predict the two first points of the zone at positions $(2k_0^j + 1)2^{-j}$ and $2(k_0^j + 1)2^{-j}$ (fig.3, bottom, right). We propose to estimate the first one by a 4-point Kriging extrapolation from the regular zone and the second one as an average between the extrapolation from the regular zone and the value at the previous level.

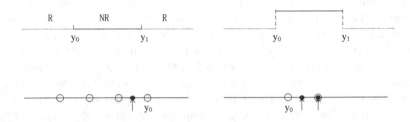

Fig. 3. From left to right and top to bottom: segmentation of the line (R:Regular; NR: Non-Regular); $C(x)$; adaption in the regular zone near the segmentation point y_0; adaption in the non-regular zone near the segmentation point y_0.

3.2 Convergence Analysis of the Zone-dependent Scheme

Clearly, Kriging based schemes are scale-dependent and therefore belong to the class of non-stationnary schemes. The convergence of a non-stationary scheme can be established as soon as its mask converges, when j goes to infinity, towards to the mask of a convergent stationary subdivision scheme called the asymptotical subdivision scheme. This important result is recalled by the following theorem ([2]):

Theorem 1. *Let S be a non-stationary subdivision scheme defined by its masks $\{a_m^{j,k}\}_{m \in \mathbb{Z}}$, $(j, k) \in \mathbb{Z}^2$. We suppose that there exists two constants $K < K'$, independent of j and k such that $a_m^{j,k} = 0$ for $m > K'$ or $m < K$. If there exists a convergent stationary subdivision scheme SS of masks $\{a_m^k\}_{m \in \mathbb{Z}}$, $k \in \mathbb{Z}$ with $a_m^k = 0$ for $m > K'$ or $m < K$ and such that*

$$\lim_{j \to +\infty} ||a^{j,k} - a^k||_\infty = 0 ,$$

then S is convergent.

The convergence analysis of our scheme can therefore be reached provided the limit of the coefficients of the mask exists and the corresponding stationary subdivision scheme is convergent. We recall the following theorem related to the convergence of stationary subdivision schemes [7]:

Theorem 2. *Let S be a stationary subdivision scheme such that there exists a subdivision scheme S_1 for the differences $(df^j)_k = f_{k+1}^j - f_k^j$,*

$$df^j = S_1 df^{j-1} ,$$

The scheme S is uniformly convergent if and only if S_1 converges uniformly to the zero function for all initial data F^0.

The following sections are first devoted to the evaluation of the (zone dependen) asymptotical subdivision scheme. Then, based on Theorem 2, its convergence is proved, leading, thanks to Theorem 1 to the convergence of our zone-dependent Kriging-based subdivision scheme.

Limit of the Kriging Weights. For the regular zone $]-\infty, y_0]$, the following proposition is borrowed from [2]:

Proposition 1. *The Kriging weights satisfy:*

- *Prediction in the interior zone:*
  - *for Gaussian type semi-variogram:*

$$\lim_{j \to +\infty} \left(\lambda_{-2}^{j,2k-1}, \lambda_{-1}^{j,2k-1}, \lambda_{0}^{j,2k-1}, \lambda_{1}^{j,2k-1} \right) = \left(-\frac{1}{16}, \frac{9}{16}, \frac{9}{16}, -\frac{1}{16} \right)$$

  - *for Exponential type semi-variogram:*

$$\lim_{j \to +\infty} \left(\lambda_{-2}^{j,2k-1}, \lambda_{-1}^{j,2k-1}, \lambda_{0}^{j,2k-1}, \lambda_{1}^{j,2k-1} \right) = \left(0, \frac{1}{2}, \frac{1}{2}, 0 \right)$$

- *Prediction near the edge:*
  - *for Gaussian type semi-variogram:*

$$\lim_{j \to +\infty} \left(\lambda_{-3}^{j,2k_0^j-1}, \lambda_{-2}^{j,2k_0^j-1}, \lambda_{-1}^{j,2k_0^j-1}, \lambda_{0}^{j,2k_0^j-1} \right) = \left(\frac{1}{16}, -\frac{5}{16}, \frac{15}{16}, \frac{5}{16} \right)$$

  - *for Exponential type semi-variogram:*

$$\lim_{j \to +\infty} \left(\lambda_{-3}^{j,2k_0^j-1}, \lambda_{-2}^{j,2k_0^j-1}, \lambda_{-1}^{j,2k_0^j-1}, \lambda_{0}^{j,2k_0^j-1} \right) = \left(0, 0, \frac{1}{2}, \frac{1}{2} \right)$$

As for the non-regular zone $]y_0, y_1[$, we have:

Proposition 2. *The Kriging weights satisfy:*

- *Prediction in the interior of the zone: for both Gaussian and Exponential type semi-variograms:*

$$\lim_{j \to +\infty} \left(\lambda_{-2}^{j,2k}, \lambda_{-1}^{j,2k}, \lambda_{0}^{j,2k}, \lambda_{1}^{j,2k} \right) = \left(\frac{1}{4}, \frac{1}{4}, \frac{1}{4}, \frac{1}{4} \right) ,$$

$$\lim_{j \to +\infty} \left(\lambda_{-2}^{j,2k-1}, \lambda_{-1}^{j,2k-1}, \lambda_{0}^{j,2k-1}, \lambda_{1}^{j,2k-1} \right) = \left(\frac{1}{4}, \frac{1}{4}, \frac{1}{4}, \frac{1}{4} \right)$$

- *Prediction near the edge:*

- *for Gaussian type semi-variogram:*

$$\lim_{j \to +\infty} \left(\lambda_{-4}^{j,2k_0^j+1}, \lambda_{-3}^{j,2k_0^j+1}, \lambda_{-2}^{j,2k_0^j+1}, \lambda_{-1}^{j,2k_0^j+1} \right) = \left(-\frac{5}{16}, \frac{21}{16}, -\frac{35}{16}, \frac{35}{16} \right) ,$$

$$\lim_{j \to +\infty} \left(\lambda_{-4}^{j,2k_0^j+2}, \lambda_{-3}^{j,2k_0^j+2}, \lambda_{-2}^{j,2k_0^j+2}, \lambda_{-1}^{j,2k_0^j+2} \right) = (-1, 4, -6, 4)$$

- *for Exponential type semi-variogram:*

$$\lim_{j \to +\infty} \left(\lambda_{-4}^{j,2k_0^j+1}, \lambda_{-3}^{j,2k_0^j+1}, \lambda_{-2}^{j,2k_0^j+1}, \lambda_{-1}^{j,2k_0^j+1} \right) = (0,0,0,1) ,$$

$$\lim_{j \to +\infty} \left(\lambda_{-4}^{j,2k_0^j+2}, \lambda_{-3}^{j,2k_0^j+2}, \lambda_{-2}^{j,2k_0^j+2}, \lambda_{-1}^{j,2k_0^j+2} \right) = (0,0,0,1)$$

We prove Proposition 2:

Proof. The computation of the asymptotical weights when predicting near the edge is just an extension of the result provided by Proposition 1 of [2]. Therefore, we focus in this proof on the behavior of the Kriging weights associated to the interior prediction. Moreover, we only study in the sequel the limit of $\{\lambda_m^{j,2k-1}\}_{m=-2,\cdots,1}$ since the proof for $\{\lambda_m^{j,2k}\}_{m=-2,\cdots,1}$ is similar.

According to the Kriging system (3), the left hand-side matrix and the right-hand side vector are written respectively as:

$$\Gamma_j = \begin{bmatrix} -1 & \gamma(2^{-j}) & \gamma(2 \times 2^{-j}) & \gamma(3 \times 2^{-j}) & 1 \\ \gamma(2^{-j}) & -1 & \gamma(2^{-j}) & \gamma(2 \times 2^{-j}) & 1 \\ \gamma(2 \times 2^{-j}) & \gamma(2^{-j}) & -1 & \gamma(2^{-j}) & 1 \\ \gamma(3 \times 2^{-j}) & \gamma(2 \times 2^{-j}) & \gamma(2^{-j}) & -1 & 1 \\ 1 & 1 & 1 & 1 & 0 \end{bmatrix} , \gamma_j = \begin{bmatrix} \gamma(\frac{3}{2} \times 2^{-j}) \\ \gamma(\frac{1}{2} \times 2^{-j}) \\ \gamma(\frac{1}{2} \times 2^{-j}) \\ \gamma(\frac{3}{2} \times 2^{-j}) \\ 1 \end{bmatrix} ,$$

where γ is either the Gaussian or the Exponential semi-variogram model.

Let us first notice that: $\lim_{j \to +\infty} \Gamma_j = G$, and $\lim_{j \to +\infty} \gamma_j = g$ with

$$G = \begin{bmatrix} -1 & 0 & 0 & 0 & 1 \\ 0 & -1 & 0 & 0 & 1 \\ 0 & 0 & -1 & 0 & 1 \\ 0 & 0 & 0 & -1 & 1 \\ 1 & 1 & 1 & 1 & 0 \end{bmatrix} \text{ and } g = (0,0,0,0,1)' .$$

Denoting Λ the solution of (3), since G is invertible, $\lim_{j \to +\infty} \Lambda = G^{-1}g$, and one can easily check that the unique solution of $GU = g$ is $U = \{\frac{1}{4}, \frac{1}{4}, \frac{1}{4}, \frac{1}{4}, \frac{1}{4}\}$. That concludes the proof.

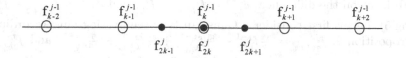

Fig. 4. Prediction in the interior of the non-regular zone: available values at level $j-1$ (o) and predicted values at the next level (•)

Convergence of the Asymptotical Scheme. The convergence of the asymptotical scheme in the regular zone has already been proved in [1]. Therefore, this section is devoted to the convergence analysis in the non-regular one.

– Convergence in the interior of the zone:
Since the scheme is uniform, it is enough to focus on the differences $d_1^{j,k} = f_{2k}^j - f_{2k-1}^j$ and $d_2^{j,k} = f_{2k+1}^j - f_{2k}^j$ (fig.4). According to Proposition 2, $f_{2k-1}^j = f_{2k}^j = \frac{1}{4}(f_{k-2}^{j-1} + f_{k-1}^{j-1} + f_k^{j-1} + f_{k+1}^{j-1})$ and $f_{2k+1}^j = \frac{1}{4}(f_{k-1}^{j-1} + f_k^{j-1} + f_{k+1}^{j-1} + f_{k+2}^{j-1})$.
Then,

$$\begin{cases} d_1^{j,k} = f_{2k}^j - f_{2k-1}^j = 0 \,, \\ d_2^{j,k} = f_{2k+1}^j - f_{2k}^j, \\ \quad = \frac{1}{4} \times [(f_{k+2}^{j-1} - f_{k+1}^{j-1}) + (f_{k+1}^{j-1} - f_k^{j-1}) + (f_k^{j-1} - f_{k-1}^{j-1}) + (f_{k-1}^{j-1} - f_{k-2}^{j-1})] \,. \end{cases}$$

Since $\forall j, k, d_1^{j,k} = 0$,
• if k is even $(k = 2m)$, $f_{k+2}^{j-1} - f_{k+1}^{j-1} = f_{2m+2}^{j-1} - f_{2m+1}^{j-1} = d_1^{j-1,m+1} = 0$, $f_k^{j-1} - f_{k-1}^{j-1} = f_{2m}^{j-1} - f_{2m-1}^{j-1} = d_1^{j-1,m} = 0$. Therefore,

$$d_2^{j,k} = \frac{1}{4} \times [(f_{k+1}^{j-1} - f_k^{j-1}) + (f_{k-1}^{j-1} - f_{k-2}^{j-1})] \,,$$

• if k is odd $(k = 2m + 1)$, $f_{k+1}^{j-1} - f_k^{j-1} = f_{2m+2}^{j-1} - f_{2m+1}^{j-1} = d_1^{j-1,m+1} = 0$, $f_{k-1}^{j-1} - f_{k-2}^{j-1} = f_{2m}^{j-1} - f_{2m-1}^{j-1} = d_1^{j-1,m} = 0$. Therefore,

$$d_2^{j,k} = \frac{1}{4} \times [(f_{k+2}^{j-1} - f_{k+1}^{j-1}) + (f_k^{j-1} - f_{k-1}^{j-1})] \,.$$

Since $|\frac{1}{4}| + |\frac{1}{4}| < 1$, $d_2^{j,k} \to_{j \to +\infty} 0$.

– Convergence near the edge:

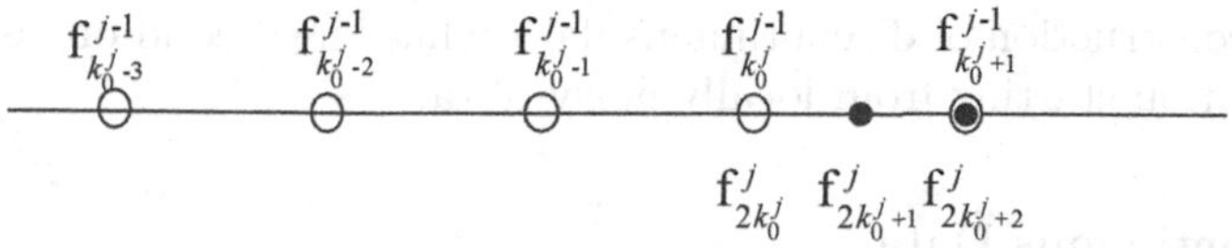

$$\begin{array}{ccccc} f_{k_0^j-3}^{j-1} & f_{k_0^j-2}^{j-1} & f_{k_0^j-1}^{j-1} & f_{k_0^j}^{j-1} & f_{k_0^j+1}^{j-1} \\ \circ & \circ & \circ & \circ & \bullet \\ & & & f_{2k_0^j}^j \; f_{2k_0^j+1}^j \; f_{2k_0^j+2}^j \end{array}$$

Fig. 5. Prediction near the edge in the non-regular zone: available values at level $j-1$ (o) and predicted values at the next level (•)

We focus on the differences $d_1^{j,k_0^j} = f_{2k_0^j+1}^j - f_{2k_0^j}^j$ and $d_2^{j,k_0^j} = f_{2k_0^j+2}^j - f_{2k_0^j+1}^j$ (fig.5). Let us first consider a Gaussian type semi-variogram. According to Proposition 2, $f_{2k_0^j+1}^j = \frac{35}{16} f_{k_0^j}^{j-1} - \frac{35}{16} f_{k_0^j-1}^{j-1} + \frac{21}{16} f_{k_0^j-2}^{j-1} - \frac{5}{16} f_{k_0^j-3}^{j-1}$ and $f_{2k_0^j+2}^j = \frac{1}{2} \times [(4f_{k_0^j}^{j-1} - 6f_{k_0^j-1}^{j-1} + 4f_{k_0^j-2}^{j-1} - f_{k_0^j-3}^{j-1}) + f_{k_0^j+1}^{j-1}]$. Therefore

$$
\begin{cases}
d_1^{j,k_0^j} = f_{2k_0^j+1}^j - f_{2k_0^j}^j = \frac{19}{16}(f_{k_0^j}^{j-1} - f_{k_0^j-1}^{j-1}) - (f_{k_0^j-1}^{j-1} - f_{k_0^j-2}^{j-1}) + \frac{5}{16}(f_{k_0^j-2}^{j-1} - f_{k_0^j-3}^{j-1}) , \\[2mm]
d_2^{j,k_0^j} = f_{2k_0^j+2}^j - f_{2k_0^j+1}^j \\[2mm]
\qquad = \frac{1}{2}(f_{k_0^j+1}^{j-1} - f_{k_0^j}^{j-1}) + \frac{5}{16}(f_{k_0^j}^{j-1} - f_{k_0^j-1}^{j-1}) - \frac{1}{2}(f_{k_0^j-1}^{j-1} - f_{k_0^j-2}^{j-1}) + \frac{3}{16}(f_{k_0^j-2}^{j-1} - f_{k_0^j-3}^{j-1}) .
\end{cases}
$$

Since the terms $(f_{k_0^j}^{j-1} - f_{k_0^j-1}^{j-1})$, $(f_{k_0^j-1}^{j-1} - f_{k_0^j-2}^{j-1})$ and $(f_{k_0^j-2}^{j-1} - f_{k_0^j-3}^{j-1})$ are differences between data of the regular zone, their limit is zero since the Kriging subdivision scheme is convergent in this zone [2]. Therefore $d_1^{j,k_0^j} \to_{j\to+\infty} 0$. Moreover, since $f_{k_0^j+1}^{j-1} - f_{k_0^j}^{j-1} = f_{2k_0^{j-1}+1}^{j-1} - f_{2k_0^{j-1}}^{j-1} = d_1^{j-1,k_0^{j-1}}$, we also get that $d_2^{j,k_0^j} \to_{j\to+\infty} 0$.

If we consider now an Exponential type semi-variogram, according to Proposition 2, $f_{2k_0^j+1}^j = f_{k_0^j}^{j-1}$ and $f_{2k_0^j+2}^j = \frac{1}{2} \times (f_{k_0^j}^{j-1} + f_{k_0^j+1}^{j-1})$. Therefore we have:

$$
\begin{cases}
d_1^{j,k_0^j} = f_{2k_0^j+1}^j - f_{2k_0^j}^j = 0 , \\[2mm]
d_2^{j,k_0^j} = f_{2k_0^j+2}^j - f_{2k_0^j+1}^j = \frac{1}{2}(f_{k_0^j+1}^{j-1} - f_{k_0^j}^{j-1}) .
\end{cases}
$$

Noticing that $f_{k_0^j+1}^{j-1} - f_{k_0^j}^{j-1} = f_{2k_0^{j-1}+1}^{j-1} - f_{2k_0^{j-1}}^{j-1} = d_1^{j-1,k_0^{j-1}} = 0$, we also get that $d_2^{j,k_0^j} = 0$.

From Theorem 2, the asymptotical subdivision scheme is then convergent. Applying Theorem 1 leads to the convergence of our zone-dependent Kriging-based scheme.

4   Numerical Results

This section is devoted to two applications of the zone-dependent Kriging-based subdivision scheme in order to evaluate its capability. The first application deals with the reconstruction of discontinuous data while the second one is related to curve generation starting from locally noisy data.

4.1   Discontinuous Data

There exists many subdivision schemes that are suitable to reconstruct non-regular data. Among them, the position-dependent Lagrange scheme introduced in [1] and recalled in Section 2.1 has provided promising results when its construction integrates precisely the information related to the discontinuity position. However as shown in [1], the quality of this scheme strongly deteriorates as soon as the segmentation of the line does not coincide with the discontinuity points. In this section we investigate how the zone-dependent Kriging-based scheme can circumvent this limitation. More precisely, we analyze the influence of the zone segmentation on the limit function, starting from a fixed initial data set exhibiting a single discontinuity point. We consider the following test function:

$$f = \begin{cases} 5 * sin(\pi * (x - 0.1)) + \frac{1}{6} & \text{if } x \in [0, 0.65] \\ 5 - 5 * sin(\pi * (x - 0.1)) + \frac{1}{6} & \text{if } x \in [0.65, 1] \end{cases}$$

with one discontinuity at $x_0 = 0.65$. The initial sequence is $F^{J_0} = (f(k2^{-J_0}), 0 \le k \le 2^{J_0} - 1$.

In the following tests, it is assumed that the segmentation procedure has led to an inaccurate estimation leading to the segmentation point $y_0' = 0.68$ and we compare the reconstruction for the three following schemes:

- Zone-dependent Kriging-based subdivision scheme(ZDK): the segmentation in different zones takes into account that a bad detection can occur. Therefore, 3 zones are defined: 2 regular ($[0, 0.6]$ and $[0.7, 1]$) and one non-regular ($]0.6, 0.7[$) corresponding to $y_0 = 0.6$, $y_1 = 0.7$ with $y_0 < x_0 < y_1$. The construction of Kriging-based scheme requires to identify the spatial structure of the data through the computation of the semi-variogram. For the rest of the section, this spatial structure is assumed to be the same in each zone and is estimated once and for all. The identified semi-variogram is of Gaussian type and is provided by fig.6, right.
- Position-dependent Lagrange-based subdivision scheme(PDL) assuming that the segmentation position is located at $y_0' = 0.68$, reminding that $y_0' \neq x_0$.
- Translation-invariant Lagrange-based subdivision scheme(TIL) since it is a classical example of schemes that are not adapted to singularities.

Fig.7 displays the limit functions reached by the different schemes starting from F^{J_0}, $J_0 = 4$. Table 1 provides the l_2-error $||f_k^{Jmax} - f(2^{-Jmax}k)||_{l_2}$, $k \in [0, 2^{Jmax}]$ for $Jmax = 12$ and different values of J_0.

From fig.7 it appears that the position-dependent Lagrange reconstruction (PDL) does not exhibit the oscillations generated by the translation-invariant one (TIL) thanks to its adaption to segmentation point. However it leads to a large l_2-error due to the mismatch between y_0' and x_0. On the contrary, the

Table 1. l_2-error for TIL, PDL, ZDK

l_2-error	$J_0 = 4$	$J_0 = 5$	$J_0 = 6$	$J_0 = 7$
PDL	0.0401	0.0293	0.0094	0.0127
TIL	0.0227	0.0227	0.0114	0.0057
ZDK	0.0203	0.0143	0.0098	0.0059

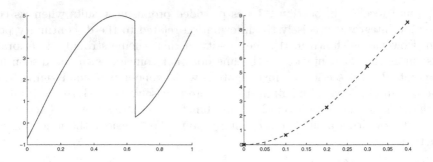

Fig. 6. Test function (left) and its experimental semi-variogram(right): the cross signs are the experimental values of the semi-variogram and the dashed line is the fitted γ

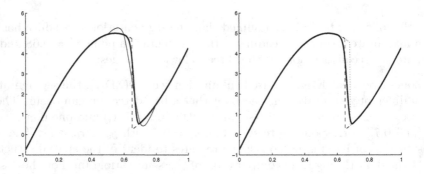

Fig. 7. Limit functions starting from $F^{J_0}, J_0 = 4$: From left to right, comparison between the test function (dashed line), TIL (thin line) and ZDK (thick line), comparison between the test function (dashed line), PDL (dotted line) and ZDK (thick line)

zone-dependent Kriging strategy (ZDK) is not penalized by the poor estimate of the discontinuity point (as soon as $y_0 < x_0 < y_1$). As a result, it keeps the interesting property of Gibbs phenomenon reduction of position-dependent approaches leading to an acceptable l^2 error while damping the strong dependence on the segmentation precision.

4.2   Curve Generation

Our zone-dependent Kriging-based subdivision scheme (ZDK) is here applied to curve generation. More precisely, we evaluate the capability of this approach in presence of noisy control points. Fig.8 displays an example of curve reconstruction, where some control points have been polluted by a white noise. It is assumed that a zone segmentation has been performed, including the noisy data in a single "non regular" zone $[y_0, y_1]$. As it is well known, translation-invariant interpolatory subdivision schemes such as Lagrange-based ones are not tailored

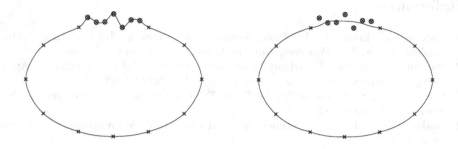

Fig. 8. Curve generation from a set of control points (cross signs), where several of them are affected by white noise (circled cross signs): left, reconstruction by a Lagrange interpolatory scheme ; right, reconstruction by a zone-dependent Kriging-based subdivision.

to adapt to noisy data and lead to undesirable oscillations (fig.8 left). On the contrary, since the zone-dependent Kriging strategy allows to combine interpolatory (in the zone without noise) and non-interpolatory (in the noisy zone) schemes, the oscillations are reduced and the reconstructed curve is more satisfactory (fig.8 right).

5   Conclusion

A new Kriging-based subdivision scheme has been constructed. It is based on a data segmentation in different zones that triggers the combination of interpolatory and non-interpolatory predictions thanks to the introduction of a zone-dependent error variance in the Kriging model. A full convergence analysis in the case of a 4-point stencil has been performed, establishing a connection between this non-stationary scheme and a convergent stationary one. The applications to the reconstruction of non-regular data and to curve generation have pointed out three main advantages of this type of approach:

- Adaption to data: it allows to integrate information related to discontinuities in the data which is necessary to avoid the so called Gibbs phenomenon.
- Weak dependance on the discontinuity detection precision: our zone-dependent scheme is not affected by inaccuracy of the segmentation that is the main weakness of previously introduced position-dependent schemes.
- Efficiency of the reconstruction in presence of noisy data: combining interpolatory (in zones without noise) and non-interpolatory (in noisy zones) predictions leads to smooth reconstructed curves reducing the oscillations associated to fully interpolatory schemes.

References

1. Baccou, J., Liandrat, J.: Position-dependent Lagrange Interpolating Multi-resolutions. Int. J. of Wavelets, Multiresol. and Inf. 5(4), 513–539 (2005)
2. Baccou, J., Liandrat, J.: Kriging-based Interpolatory Subdivision Schemes. Appl. Comput. Harmon. Anal (2012), doi:10.1016/j.acha. 2012.07.008
3. Cressie, N.A.: Statistics for Spatial Data. Wiley Series in Probability and Mathematical Statistics (1993)
4. Deslauriers, G., Dubuc, S.: Symmetric iterative interpolation processes. Const. Approx. 5, 49–68 (1989)
5. Wachernagel, H.: Multivariate geostatistics. Springer (1998)
6. Chiles, J.P., Delfiner, P.: Geostatistics: modelling spatial uncertainty. Wiley Series in Probability and Mathematical Statistics (1999)
7. Dyn, N.: Subdivision Schemes in Computer-aided Geometric Design. Advances in Numerical analysis II, Wavelets, Subdivision algorithms and Radial Basis Functions 20(4), 36–104 (1992)

Deriving Novel Formulas and Identities for the Bernstein Basis Functions and Their Generating Functions

Yilmaz Simsek*

Department of Mathematics, Faculty of Science,
Akdeniz University, Campus, 07058, Antalya-Turkey
ysimsek63@gmail.com

Abstract. By using the generating functions for the Bernstein basis functions, we derive various functional equations, differential equations and second order partial differential equations. By using these equations, we give new proofs of various identities, relations, integrals and derivatives of the Bernstein basis functions. Using second order partial differentia equation of the generating functions, we also obtain new derivative formulas for the Bernstein basis functions. By applying the Fourier transform and the Laplace transform to the generating functions, we derive series representations for the Bernstein basis functions. We also give the p-adic Volkenborn integral representations of the Bernstein basis functions.

Keywords: Bernstein polynomials, Generating function, Bezier curves, Fourier transform, Laplace transform, Functional equations, Partial differential equations, p-adic Volkenborn integral
AMS Subject Classification: 14F10, 12D10, 26C05, 26C10, 30B40, 30C15, 42A38, 44A10.

1 Introduction

Bernstein [4] first introduced and investigated the extended form of the (Bernstein) polynomials, which are now popularly known as the Bernstein polynomials. Recently, the Bernstein polynomials and the Bezier Curves have been studied by many scientists. There are various generalizations of the Bernstein polynomials (cf. [2], [7], [6], [9], [10], [11], [24], [25], [26], [21], [27], [39], [38], [35]), and other authors (see also the references cited in each of these earlier works).

The Bernstein polynomials are used to approximate a curve. In numerical analysis, the Bernstein polynomial is a polynomial in the Bernstein form, that is a linear combination of the Bernstein basis functions. There are many methods to evaluate polynomials in the Bernstein form. One of them is a numerically stable way to evaluate polynomials in the Bernstein form which is the de Casteljau's algorithm. It is also well known that polynomials in the Bernstein form

---

* Corresponding author.

M. Floater et al. (Eds.): MMCS 2012, LNCS 8177, pp. 471–490, 2014.
© Springer-Verlag Berlin Heidelberg 2014

were first used by Bernstein in a constructive proof for the Stone-Weierstrass approximation theorem. The Bernstein polynomials have many applications: in approximations of functions, in statistics, in numerical analysis, in p-adic analysis, in the solution of differential equations and in Computer Aided Geometric Design (CAGD). In CAGD, polynomials are often expressed in terms of the Bernstein basis functions. These polynomials are called the Bezier curves. The Bernstein polynomials, which are related to the Bezier curves, are also used to determine tight bounds on the range of a multivariate polynomial over a closed rectangle (cf. [2], [7], [6], [9], [10], [11], [13], [24], [25], [26], [21], [28], [27], [29], [38], [35], [39]).

Many of the known identities for the Bernstein basis functions are currently derived in an *ad hoc* fashion, using either the binomial theorem, the binomial distribution, tricky algebraic manipulations or blossoming. In [34]- [38], we use functional equations and differential equations of the generating functions for the Bernstein basis functions, we provide a new approach to derive both standard identities and new identities for the Bernstein basis functions.

We now give some well-known properties of the Bernstein basis functions and their generating functions. The Bernstein basis functions $B_k^n(x)$ are defined as follows:

Definition 1.1. *Let* $x \in [0,1]$. *Let* $n \in \mathbb{N}_0 := \{0,1,2,3,\cdots\}$. *The Bernstein basis functions* $B_k^n(x)$ *can be defined by*

$$B_k^n(x) = \binom{n}{k} x^k (1-x)^{n-k}, \tag{1}$$

where

$$k = 0,1,\ldots,n,$$

and

$$\binom{n}{k} = \frac{n!}{k!(n-k)!}.$$

(*cf.* [2], [8], [7], [6], [9], [10], [11], [12], [13], [14], [17], [23], [24], [25], [26], [21], [28], [27], [29], [34], [36], [40]).

Generating functions for the Bernstein basis functions can be defined as follows:

Definition 1.2. *Let* $t \in \mathbb{C}$ *and* $x \in [0,1]$. *The Bernstein basis functions are defined by means of the following generating functions*

$$f_{\mathbb{B},k}(x,t) := \sum_{n=0}^{\infty} B_k^n(x) \frac{t^n}{n!}, \tag{2}$$

where $k = 0,1,\ldots,n$.

Generating functions for the Bernstein basis functions are given by the following theorem:

Theorem 1.3. *Let $t \in \mathbb{C}$ and $x \in [0,1]$. Then we have*

$$f_{\mathbb{B},k}(x,t) = \frac{t^k x^k e^{(1-x)t}}{k!},$$ (3)

where $k = 0, 1, \ldots, n$. (cf. [38], [39], [3], [34], [35]).

Remark 1.4. *The formulas and identities we derive for the interval $[0,1]$ can easily be extended to arbitrary intervals $[a,b]$. That is, if we replace x by $\frac{x-a}{b-a}$, then Definition 1.1 yields the corresponding well known results concerning the Bernstein basis functions $B_k^n(x,a,b)$:*

$$B_k^n(x;a,b) = \binom{n}{k} \left(\frac{x-a}{b-a}\right)^k \left(\frac{b-x}{b-a}\right)^{n-k},$$ (4)

where $k = 0, 1, \cdots, n$ and $x \in [a,b]$ (cf. [4], [8], [7], [9, p. 384, Eq.(24.6)]). In [36], we modified Equation (3) as follows:

$$\frac{t^k \left(\frac{x-a}{b-a}\right)^k e^{\left(\frac{b-x}{b-a}\right)t}}{k!} = \sum_{n=0}^{\infty} B_k^n(x;a,b)\frac{t^n}{n!}.$$ (5)

The Bernstein polynomial $\mathcal{P}(x)$ is a polynomial represented by the Bernstein basis functions:

$$\mathcal{P}(x) = \sum_{k=0}^{n} c_k^n B_k^n(x),$$ (6)

(*cf.* [2], [8], [7], [6], [9], [10], [11], [12], [13], [14], [17], [23], [24], [25], [26], [21], [28], [27], [29]).

By using the Bernstein polynomials, one can easily find an explicit polynomial representation for the Bezier curves. The Bezier curves $B(x)$, with control points $P_0, \ldots, P_n$, are defined as follows:

$$B(x) = \sum_{k=0}^{n} P_k B_k^n(x)$$ (7)

(*cf.* [2], [8], [7], [6], [9], [10], [11], [28], [27], [29], [14], [24], [25], [26], [21]).

The organization of this paper is given as follows:

In Section 2; we define alternative forms of the generating functions for the Bernstein basis functions. We give functional equations of these generating functions. By using these equations, we derive some identities related to the Bernstein basis functions. In Section 3, by using the same method in [38], we derive various functional equations of the generating functions. By using these equations, we derive some novel and interesting identities for the Bernstein basis functions. In Section 4, we give a proof of Marsden identity using generating functions for the Bernstein basis functions. In Section 5, we give partial differential equations (PDEs) for the generating functions. By using these equations, we derive some new derivative formulas for the Bernstein basis functions. In Section 6, we give integral representations of the Bernstein basis functions. By using these representations, we derive some identities related to the Bernstein basis functions. In Section 7, by applying the Fourier transform and the Laplace transform to the generating functions, we derive some new series representations of the Bernstein basis functions. In Section 8, we give the p-adic Volkenborn integral representations of the Bernstein basis functions.

2   Alternative Forms of the Generating Functions

In this section, we give some alternative forms for the generating functions. By using these functions, we derive some functional equations. By applying these functional equations and generalized multinomial identity, we derive some identities for the Bernstein basis functions. Our new results are also generalized in Theorem 4.1 in [38].

We give some alternative forms of the generating functions in (3) as follows: Let $k_1, \cdots, k_v \in \mathbb{N}_0$ and $v \in \mathbb{N}$. We define

$$G_{\mathbb{B}, k_1 + \cdots + k_v}(t, x) = \frac{(tx)^{k_1 + \cdots + k_v}}{(k_1 + \cdots + k_v)!} e^{vt(1-x)} \tag{8}$$

$$= \sum_{n=0}^{\infty} v^{n - (k_1 + \cdots + k_v)} B_{k_1 + \cdots + k_v}^n (x) \frac{t^n}{n!}.$$

From (2) and (8), we get

$$\frac{k_1! \cdots k_v!}{(k_1 + \cdots + k_v)!} \prod_{l=1}^{v} f_{\mathbb{B}, k_l}(x, t) = \sum_{n=0}^{\infty} B_{k_1 + \cdots + k_v}^n (x) v^{n - k_1 - \cdots - k_v} \frac{t^n}{n!}.$$

Therefore

$$\frac{k_1! \cdots k_v!}{(k_1 + \cdots + k_v)!} \left(\sum_{n=0}^{\infty} B_{k_l}^n (x) \frac{t^n}{n!} \cdots \sum_{n=0}^{\infty} B_{k_v}^n (x) \frac{t^n}{n!} \right) = \sum_{n=0}^{\infty} B_{k_1 + \cdots + k_v}^n (x) v^{n - k_1 - \cdots - k_v} \frac{t^n}{n!}.$$

From the above equation, we obtain

$$\sum_{m_1 + \cdots + m_{v-1} = n} \frac{B_{k_v}^{m_v - 1}(x)}{(m_v - 1)!} \frac{B_{k_{v-1}}^{m_v - 2}(x)}{(m_v - 2)!} \cdots \frac{B_{k_1}^{m_1}(x)}{m_1!} \frac{B_{k_2}^{n - m_1 - m_2 - \cdots - m_{v-1}}(x)}{(n - m_1 - m_2 - \cdots - m_{v-1})!} t^n$$

$$= \frac{(k_1 + \cdots + k_v)!}{k_1! \cdots k_v!} \sum_{n=0}^{\infty} B_{k_1 + \cdots + k_v}^n (x) v^{n - k_1 - \cdots - k_v} \frac{t^n}{n!},$$

where

$$\sum_{m_1 + \cdots + m_{v-1} = n} = \sum_{m_{v-1} = 0}^{n} \sum_{m_{v-2} = 0}^{n - m_{v-1}} \cdots \sum_{m_1 = 0}^{n - m_2 - m_3 - \cdots - m_{v-1}}.$$

By comparing the coefficients of t^n on both sides of the above equation, we arrive at the following theorem.

Theorem 2.1. *Let $n \geq k_1 + \cdots + k_v$. Then we have*

$$B_{k_1+\cdots+k_v}^n(x) = \frac{v^{k_1+\cdots+k_v-n} k_1! \cdots k_v!}{(k_1+\cdots+k_v)!}$$

$$\times \sum_{m_1+\cdots+m_{v-1}=n} C_{m_1,\cdots,m_{v-1}}^n B_{k_v}^{m_v-1}(x) B_{k_{v-1}}^{m_v-2}(x) \cdots B_{k_1}^{m_1}(x) B_{k_2}^{n-m_1-\cdots-m_{v-1}}(x),$$

where

$$C_{m_1,\cdots,m_{v-1}}^n = \binom{n}{m_1, m_2, \cdots, n-m_1-\cdots-m_{v-1}}.$$

Remark 2.2. *Substituting $v = 2$ into Theorem 2.1, we have*

$$B_{k_1+k_2}^n(x) = \frac{2^{k_1+k_2-n} k_1! k_2!}{(k_1+k_2)!} \sum_{m_1=0}^{n} \binom{n}{m_1} B_{k_1}^{m_1}(x) B_{k_2}^{n-m_1}(x)$$

(cf. [38, Theorem 4.1]).

Here, we need the following generalized multinomial identity.

Lemma 2.3. *(Generalized multinomial identity [5, p. 41, Equation (12m)]) If $x_1, x_2, \ldots, x_m$ are commuting elements of a ring ($\Leftrightarrow x_i x_j = x_j x_i$, $1 \leq i < j \leq m$), then we have for all real or complex variables α:*

$$(x_1 + x_2 + \cdots + x_m)^\alpha = \sum_{v_1, v_2, \ldots, v_m \geq 0} C_{v_1, \cdots, v_{m1}}^\alpha x_1^{v_1} x_2^{v_2} \cdots x_m^{v_m}, \tag{9}$$

where the last summation takes place over all positive or zero integers $v_i \geq 0$, and

$$C_{v_1, \cdots, v_{m1}}^\alpha = \binom{\alpha}{v_1, v_2, \ldots, v_m} := \frac{(\alpha)_{v_1+v_2+\cdots+v_m}}{v_1! v_2! \cdots v_m!}$$

are called generalized multinomial coefficients, where

$$(n)_k = n(n-1)\cdots(n-k+1).$$

and $(x)_0 = 1$.

Remark 2.4. *The following multinomial identity is equivalent to (9):*

$$(x_1 + x_2 + \cdots + x_v)^k = \sum_{m_1+m_2+\ldots+m_{v-1}=k} \frac{x_1^{m_1}}{m_1!} \frac{x_2^{m_2}}{m_2!} \cdots \frac{x_v^{k-(m_1+m_2+\ldots+m_{v-1})}}{(k-(m_1+m_2+\ldots+m_{v-1}))!}$$

$$= \sum_{m_1+m_2+\ldots+m_{v-1}=k} C_{m_1,\cdots,m_{v-1}}^k x_1^{m_1} x_2^{m_2} \cdots x_v^{k-(m_1+m_2+\ldots+m_{v-1})},$$

where

$$\sum_{m_1+m_2+\ldots+m_{v-1}=k} = \sum_{m_1=0}^{k} \sum_{m_2=0}^{k-m_1} \cdots \sum_{m_v=0}^{k-m_1-m_2-\cdots-m_{v-1}}.$$

We set

$$F_{\mathbb{B},k}(t, x_1 + \cdots + x_v) = \frac{t^k(x_1 + \cdots + x_v)^k}{k!}e^{t(v-x_1-\cdots-x_v)} \qquad (10)$$

$$= \sum_{n=0}^{\infty} v^n B_k^n \left(\frac{x_1 + \cdots + x_v}{v}\right)\frac{t^n}{n!}.$$

From Lemma 2.3, we have the following functional equation for the generating functions of the Bernstein basis functions:

$$\sum_{m_1+m_2+\ldots+m_{v-1}=k}\;\prod_{l=1}^{v} f_{\mathbb{B},m_l}(x_l,t) = \sum_{n=0}^{\infty} v^n B_k^n \left(\frac{x_1 + \cdots + x_v}{v}\right)\frac{t^n}{n!}.$$

Thus we have

$$\sum_{m_1+m_2+\ldots+m_{v-1}=k} (f_{\mathbb{B},m_1}(x_1,t)f_{\mathbb{B},m_2}(x_2,t)\cdots f_{\mathbb{B},m_v}(x_v,t)) = \sum_{n=0}^{\infty} v^n B_k^n \left(\frac{x_1 + \cdots + x_v}{v}\right)\frac{t^n}{n!}.$$

By (2), we obtain

$$\sum_{m_1+m_2+\ldots+m_{v-1}=k} \left(\sum_{n=0}^{\infty} B_{m_1}^n(x_1)\frac{t^{m_1}}{m_1!}\cdots\sum_{n=0}^{\infty} B_{m_l}^n(x_l)\frac{t^{m_l}}{m_l!}\right)$$

$$= \sum_{n=0}^{\infty} v^n B_k^n \left(\frac{x_1 + \cdots + x_v}{v}\right)\frac{t^n}{n!}$$

where $0 \le \frac{x_1+\cdots+x_v}{v} \le 1$.

Hence

$$\sum_{n=0}^{\infty}\left\{ \sum_{m_1+m_2+\ldots+m_{v-1}=k}\sum_{j_1+\ldots+j_{v-1}=n} C_{j_1,j_2,\cdots,j_{v-1},n-j_1-j_2-\cdots-j_{v-1}}^n \right.$$

$$\left. \times B_{m_1}^{j_1}(x_1)B_{m_2}^{j_2}(x_2)\cdots B_{k-m_1-m_2-\ldots-m_{v-1}}^{n-j_1-j_2-\cdots-j_{v-1}}(x_v)\frac{t^n}{n!}\right\}$$

$$= \sum_{n=0}^{\infty} v^n B_k^n \left(\frac{x_1 + \cdots + x_v}{v}\right)\frac{t^n}{n!}.$$

Comparing the coefficients of t^n on both sides of the above equation, we arrive at the following theorem.

Theorem 2.5. *Let* $0 \le \frac{x_1+\cdots+x_v}{v} \le 1$. *Then we have*

$$B_k^n \left(\frac{x_1 + \cdots + x_v}{v}\right) = \frac{1}{v^n}\sum_{m_1+m_2+\ldots+m_{v-1}=k}\sum_{j_1+\ldots+j_{v-1}=n} C_{j_1,j_2,\cdots,j_{v-1},n-j_1-j_2-\cdots-j_{v-1}}^n$$

$$\times B_{m_1}^{j_1}(x_1)B_{m_2}^{j_2}(x_2)\cdots B_{k-m_1-m_2-\ldots-m_{v-1}}^{n-j_1-j_2-\cdots-j_{v-1}}(x_v).$$

We note that proofs of Theorem 2.1 and Theorem 2.5 can also be given by induction method on v.

Remark 2.6. *Substituting $v = 2$ into Theorem 2.5, we get the following identity:*

$$B_k^n\left(\frac{x_1 + x_2}{2}\right) = \frac{1}{2^n} \sum_{m_1=0}^{k} \sum_{j_1=0}^{n} \binom{n}{j_1} B_{m_1}^{j_1}(x_1) B_{k-m_1}^{n-j_1}(x_2).$$

3   New Identities for the Bernstein Basis Functions

In this section, by using the same method in [38], we give some functional equations for the generating functions. By using these equations, we derive some new identities for the Bernstein basis functions.

We set the following generating function:

$$g(x,y;t) = \sum_{n=0}^{\infty} (x - y)^n \frac{t^n}{n!}. \tag{11}$$

Using the series expansion for the exponential function, we can write (11) as follows

$$g(x,y;t) = e^{t(x-y)}.$$

By (11), we derive the following functional equation:

$$g(x,0;t)g(1,x;t) = g(1,0;t).$$

From the above equation, we get

$$\sum_{n=0}^{\infty} x^n \frac{t^n}{n!} \sum_{n=0}^{\infty} (1-x)^n \frac{t^n}{n!} = \sum_{n=0}^{\infty} \frac{t^n}{n!}.$$

Therefore

$$\sum_{n=0}^{\infty} \sum_{k=0}^{n} x^k (1-x)^{n-k} \frac{t^n}{k!(n-k)!} = \sum_{n=0}^{\infty} \frac{t^n}{n!}.$$

Equating the coefficients of t^n on both sides of the above equation, we obtain the following sum of the Bernstein basis functions:

Theorem 3.1

$$\sum_{k=0}^{n} B_k^n(x) = 1$$

Remark 3.2. *From (3), in [38], we found the following functional equation:*

$$\sum_{k=0}^{\infty} f_{\mathbb{B},k}(x,t) = e^t,$$

and by using this functional equation, we also proved the sum of the Bernstein basis functions [38].

By (11), we derive the following functional equation:

$$g(0, x; t)g(1, x; t) = g(1, 2x; t).$$

From the above equation, we get

$$\sum_{n=0}^{\infty} (-x)^n \frac{t^n}{n!} \sum_{n=0}^{\infty} (1-x)^n \frac{t^n}{n!} = \sum_{n=0}^{\infty} (1-2x)^n \frac{t^n}{n!}.$$

Therefore

$$\sum_{n=0}^{\infty} \sum_{k=0}^{n} (-1)^k x^k (1-x)^{n-k} \frac{t^n}{k!(n-k)!} = \sum_{n=0}^{\infty} (1-2x)^n \frac{t^n}{n!}.$$

Equating the coefficients of t^n on both sides of the resulting equation, we obtain the following alternating sum for the Bernstein basis functions:

Theorem 3.3

$$\sum_{k=0}^{n} (-1)^k B_k^n(x) = (1-2x)^n. \tag{12}$$

Remark 3.4. *In [38], we found the following functional equation:*

$$\sum_{k=0}^{\infty} (-1)^k f_{\mathbb{B},k}(x, t) = e^{(1-2x)t}, \tag{13}$$

and by using this equation, we also proved (12). The proof of (12) was also given by Goldman ([9], [10], [11]).

In [38], we derived the following functional equation:

$$f_{\mathbb{B},k}(xy, t) = f_{\mathbb{B},k}(x, yt)e^{(1-y)t} \tag{14}$$

By using (14), we proved the subdivision property for the Bernstein basis functions by the following theorem.

Theorem 3.5. *(Subdivision property)*

$$B_j^n(xy) = \sum_{k=j}^{n} B_j^k(x)B_k^n(y).$$

The proof of this theorem is also given by the following references (*cf.* [7], [9], [10], [11], [38]).

By using (14), we derive the following identity related to the Bernstein basis functions.

Theorem 3.6

$$\sum_{l=0}^{n} \binom{n}{l} y^{n-l} B_k^l(xy) = \sum_{j=0}^{n} \binom{n}{j} B_k^j(x)y^j. \tag{15}$$

Proof. By using (14), we obtain the following functional equation:

$$f_{\mathrm{B},k}(xy,t)e^{ty} = f_{\mathrm{B},k}(x,yt)e^{t}. \tag{16}$$

Combining (2) with this equation, we get

$$\sum_{n=0}^{\infty} B_k^n(xy)\frac{t^n}{n!} \sum_{n=0}^{\infty} y^n\frac{t^n}{n!} = \sum_{n=0}^{\infty} B_k^n(x)\frac{(ty)^n}{n!} \sum_{n=0}^{\infty} \frac{t^n}{n!}.$$

Therefore

$$\sum_{n=0}^{\infty} \left(\sum_{l=0}^{n} \binom{n}{l} B_k^l(x)y^{n-l} \right) \frac{t^n}{n!} = \sum_{n=0}^{\infty} \left(\sum_{j=0}^{n} \binom{n}{j} y^j B_k^j(x)y^l \right) \frac{t^n}{n!}.$$

Comparing the coefficients of $\frac{t^n}{n!}$ on both sides of the above equation, we arrive at the desired result. ∎

Substituting $x = 1$ into (15), we get the following corollary:

Corollary 3.7

$$\sum_{l=0}^{n} \binom{n}{l} y^{n-l} B_k^l(y) = \binom{n}{k} y^k. \tag{17}$$

From (2), we obtain the following functional equation:

$$e^{xt} \sum_{n=0}^{\infty} B_k^n(x)\frac{t^n}{n!} = \frac{t^k x^k}{k!}e^t. \tag{18}$$

By using (18), we can give another proof of (17) as follows:
Second proof of (17). By using (18), we get

$$\sum_{n=0}^{\infty} x^n \frac{t^n}{n!} \sum_{n=0}^{\infty} B_k^n(x)\frac{t^n}{n!} = \frac{t^k x^k}{k!} \sum_{n=0}^{\infty} \frac{t^n}{n!}.$$

Therefore

$$\sum_{n=0}^{\infty} \left(\sum_{j=0}^{n} \binom{n}{j} x^{k-j} B_k^j(x) \right) \frac{t^n}{n!} = \frac{x^k}{k!} \sum_{n=0}^{\infty} \frac{t^{n+k}}{n!}.$$

Comparing the coefficients of $\frac{t^n}{n!}$ on both sides of the above equation, we arrive at the the desired result. ∎

By using (14), we proved the following identity.

Theorem 3.8. ([36, Theorem 10])

$$\sum_{j=0}^{n}(-1)^{n-j} \binom{n}{j} B_k^j(xy) = y^n \sum_{l=0}^{n}(-1)^{n-l} \binom{n}{l} B_k^l(x). \tag{19}$$

By substituting $x = 1$ into (19), then we obtain the following identity:

Corollary 3.9

$$\sum_{j=0}^{n}(-1)^{n-j} \binom{n}{j} B_k^j(y) = (-1)^{n-k} \binom{n}{k} y^n.$$

4   Proof of the Marsden Identity

In this section, using the generating functions, we prove the Marsden identity. This identity was also proved in (*cf.* [2], [7], [6], [9], [10], [11], [12], [13], [14], [17], [23], [24], [25], [26], [21], [28], [27], [29]).

By (11), we derive the following functional equation:

$$g(y, x; t) = (-1)^j k!j!(xyt)^{-k-j} f_{\mathbb{B},k}(x, yt) f_{\mathbb{B},j}(y, -xt). \tag{20}$$

By using the above equation, we prove the Marsden identity by the following theorem:

Theorem 4.1

$$(y - x)^n = \sum_{k=0}^{n}(-1)^k \frac{1}{\binom{n}{k}} B_{n-k}^n(y)B_k^n(x). \tag{21}$$

Proof. By substituting (2) and (11) into (20), we get

$$\sum_{n=0}^{\infty}(y - x)^n \frac{t^n}{n!}$$

$$= \sum_{n=0}^{\infty} B_k^n(x) \frac{(ty)^{n-k-j} k!j!}{n!} \sum_{n=0}^{\infty}(-1)^{n+j} B_j^n(y) \frac{x^{n-k-j}t^n}{n!}.$$

Summing firs over $j + k = n$ and then over n, the right hand side of the above equation is written as follows:

$$\sum_{n=0}^{\infty}(x - y)^n \frac{t^n}{n!}$$

$$= \sum_{n=0}^{\infty} \left(\sum_{k=0}^{n}(-1)^k \frac{k!(n - k)!}{n!} B_{n-k}^n(y)B_k^n(x) \right) \frac{t^n}{n!}.$$

Comparing the coefficients of $\frac{t^n}{n!}$ on both sides of the above equation, we arrive at the desired result.   ∎

5 Partial Differential Equations (PDEs) for the Generating Functions

In this section, we give PDEs for the generating functions of the Bernstein basis functions. By using these equations, we derive some new derivative formulas for the Bernstein basis functions.

By differentiating the generating functions in (3) with respect to x, we get the following partial derivative equations:

$$\frac{\partial}{\partial x} f_{\mathbb{B},k}(x,t) = \left(\frac{k}{x} - t\right) f_{\mathbb{B},k}(x,t), \tag{22}$$

$$\frac{\partial}{\partial x} f_{\mathbb{B},k}(x,t) = \frac{k}{x} f_{\mathbb{B},k}(x,t) - \frac{k+1}{x} f_{\mathbb{B},k+1}(x,t), \tag{23}$$

and

$$\frac{\partial}{\partial x} f_{\mathbb{B},k}(x,t) = t\left(f_{\mathbb{B},k-1}(x,t) - f_{\mathbb{B},k}(x,t)\right). \tag{24}$$

To derive formulas for the derivatives of the Bernstein basis functions, we use the above PDEs in (22)-(24).

Theorem 5.1. *We have*

$$\frac{d}{dx} B_k^n(x) = \frac{k}{x} B_k^n(x) - B_k^{n-1}(x). \tag{25}$$

Proof. From (22) and (2), we obtain

$$\sum_{n=0}^{\infty} \frac{d}{dx} B_k^n(x) \frac{t^n}{n!} = \left(\frac{k}{x} - t\right) \sum_{n=0}^{\infty} B_k^n(x) \frac{t^n}{n!}.$$

Hence

$$\sum_{n=0}^{\infty} \frac{d}{dx} B_k^n(x) \frac{t^n}{n!} = \frac{k}{x} \sum_{n=0}^{\infty} B_k^n(x) \frac{t^n}{n!} - \sum_{n=0}^{\infty} B_k^n(x) \frac{t^{n+1}}{n!}.$$

Comparing the coefficients of t^n on both sides of the above equation, we arrive at the desired result. ∎

Theorem 5.2. *We have*

$$\frac{d}{dx} B_k^n(x) = \frac{k}{x} B_k^n(x) - \frac{k+1}{x} B_{k+1}^n(x). \tag{26}$$

Proof. From (23) and (2), we get

$$\sum_{n=0}^{\infty} \frac{d}{dx} B_k^n(x) \frac{t^n}{n!} = \frac{k}{x} \sum_{n=0}^{\infty} B_k^n(x) \frac{t^n}{n!} - \frac{k+1}{x} \sum_{n=0}^{\infty} B_{k+1}^n(x) \frac{t^n}{n!}.$$

Hence

$$\sum_{n=0}^{\infty} \frac{d}{dx} B_k^n(x) \frac{t^n}{n!} = \sum_{n=0}^{\infty} \left(\frac{k}{x} B_k^n(x) - \frac{k+1}{x} B_{k+1}^n(x)\right) \frac{t^n}{n!}.$$

Comparing the coefficients of t^n on both sides of the above equation, we arrive at the desired result.   ∎

By using (24), we arrive at the following theorem.

Theorem 5.3. *We have*

$$\frac{d}{dx}B_k^n(x) = n\left(B_{k-1}^{n-1}(x) - B_k^{n-1}(x)\right). \tag{27}$$

By using the generating functions, proof of formula (27) was also given by the author [38].

By differentiating the generating functions in (3) with respect to x and t, respectively, we get the following second order PDE:

$$\frac{\partial^2}{\partial x \partial t} f_{\mathbb{B},k}(x,t) = -(1+t-xt)f_{\mathbb{B},k}(x,t) + (1+t-2xt)f_{\mathbb{B},k-1}(x,t) \tag{28}$$
$$+txf_{\mathbb{B},k-2}(x,t).$$

By combining (2) with the above equation, we arrive at the following theorem, which gives us another new derivative formula for the Bernstein basis functions.

Theorem 5.4. *We have*

$$\frac{d}{dx}B_k^{n+1}(x) = B_{k-1}^n(x) + n\left(x-1\right)B_k^{n-1}(x) + nxB_{k-2}^{n-1}(x)$$
$$+ n\left(1-2x\right)B_{k-1}^{n-1}(x) - B_k^n(x).$$

Proof. From (28) and (2), we have

$$\sum_{n=1}^{\infty} \frac{d}{dx}B_k^n(x)\frac{t^{n-1}}{(n-1)!}$$

$$= -\sum_{n=0}^{\infty} B_k^n(x)\frac{t^n}{n!} + (x-1)\sum_{n=0}^{\infty} B_k^n(x)\frac{t^{n+1}}{n!} + \sum_{n=0}^{\infty} B_{k-1}^n(x)\frac{t^n}{n!}$$

$$+ (1-2x)\sum_{n=0}^{\infty} B_{k-1}^n(x)\frac{t^{n+1}}{n!} + x\sum_{n=0}^{\infty} B_{k-2}^n(x)\frac{t^{n+1}}{n!}.$$

Hence

$$\sum_{n=0}^{\infty} \frac{d}{dx}B_k^{n+1}(x)\frac{t^n}{n!}$$

$$= -\sum_{n=0}^{\infty} B_k^n(x)\frac{t^n}{n!} + (x-1)\sum_{n=0}^{\infty} nB_k^{n-1}(x)\frac{t^n}{n!} + \sum_{n=0}^{\infty} B_{k-1}^n(x)\frac{t^n}{n!}$$

$$+ (1-2x)\sum_{n=0}^{\infty} nB_{k-1}^{n-1}(x)\frac{t^n}{n!} + x\sum_{n=0}^{\infty} nB_{k-2}^{n-1}(x)\frac{t^n}{n!}.$$

Comparing the coefficients of t^n on both sides of the above equation, we arrive at the desired result.   ∎

6 Integral Representations of the Bernstein Basis Functions

In this section, we give integral representations of the Bernstein basis functions.

Theorem 6.1

$$\int_a^b B_k^n(x; a, b)dx = \binom{n}{k} (b-a) B(k+1, n-k+1),\tag{29}$$

where $B_k^n(x; a, b)$ is defined in (4) and $B(\alpha, \beta)$ denotes the Beta function which is defined by

$$B(n, m) = \frac{\Gamma(n)\Gamma(m)}{\Gamma(n+m)} = \frac{(n-1)!(m-1)!}{(n+m-1)!}\tag{30}$$

$n, m \in \mathbb{N}$ (cf. [41, p. 9, Eq-(62)]).

Proof. Setting $t = \frac{x-a}{b-a}$ in the following well known result:

$$B(l, v) = \int_0^1 t^{l-1}(1-t)^{v-1}dt,\tag{31}$$

where $l, v \in \mathbb{N} = \{1, 2, 3, \cdots\}$ (cf. [41, p. 9, Eq-(60)]), we obtain

$$B(l, v) = \frac{1}{b-a} \int_a^b \left(\frac{x-a}{b-a}\right)^{l-1} \left(\frac{b-x}{b-a}\right)^{v-1} dx\tag{32}$$

where

$$a < b.$$

Substituting $l - 1 = k$ and $v - 1 = n - k$ into (32), we have

$$\int_a^b \left(\frac{x-a}{b-a}\right)^k \left(\frac{b-x}{b-a}\right)^{n-k} dx = (b-a) B(k+1, n-k+1).$$

Multiplying both sides of the above equation by $\binom{n}{k}$, we arrive at the desired result. ∎

Remark 6.2. *By using (32), we arrive at the following definite integrals related to the Beta function:*

$$\int_a^b (x-a)^{l-1} (b-x)^{v-1} dx = (b-a)^{l+v-1} B(l, v)\tag{33}$$

(cf. see also [41, p. 10, Eq-(69)]).

Theorem 6.3

$$\int_a^b B_k^n(x; a, b)dx = \binom{n}{k} \sum_{j=0}^k \sum_{l=0}^{n-k} (-1)^{n-j-l} \binom{k}{j} \binom{n-k}{l}$$

$$\times \frac{a^{k-j}b^{n+j-k+1} - a^{n-l+1}b^l}{n+j-k-l+1}. \tag{34}$$

Proof. From (1),

$$\int_a^b B_k^n(x; a, b)dx = \binom{n}{k} \sum_{j=0}^k \sum_{l=0}^{n-k} (-1)^{-j-k-l} \binom{k}{j} \binom{n-k}{l}$$

$$\times a^{k-j}b^l \int_a^b x^{n+j-k-l} dx.$$

Therefore, we arrive at the desired result. ∎

7 Application of the Fourier Transform and the Laplace Transform to the Generating Functions

By applying the Fourier transform and the Laplace transform to the generating functions for the Bernstein basis functions, we obtain some interesting series representations for the Bernstein basis functions. From (10), we obtain the following functional equation:

$$\frac{t^k (x_1 + \cdots + x_v)^k}{k!} e^{-t(x_1 + \cdots + x_v)} = F_{\mathbb{B},k}(t, x_1 + \cdots + x_v)e^{-vt}. \tag{35}$$

Integrating this equation with respect to t from 0 to ∞, we get

$$\sum_{n=0}^{\infty} \frac{v^n B_k^n \left(\frac{x_1 + \cdots + x_v}{v}\right)}{n!} \int_0^{\infty} t^n e^{-vt} dt = \frac{(x_1 + \cdots + x_v)^k}{k!} \int_0^{\infty} t^k e^{-t(x_1 + \cdots + x_v)} dt.$$

Substituting the following well known formula for the Laplace transform in the above equation

$$\int_0^{\infty} t^n e^{-vt} dt = \frac{n!}{v^{n+1}},$$

we arrive at the following Theorem.

Theorem 7.1. *Let* $0 \le \frac{x_1 + \cdots + x_v}{v} \le 1$. *Then*

$$\sum_{n=0}^{\infty} B_k^n \left(\frac{x_1 + \cdots + x_v}{v}\right) = \frac{v}{x_1 + \cdots + x_v}.$$

Remark 7.2. *Substituting $v = 1$ into Theorem 7.1, we have*

$$\sum_{n=0}^{\infty} B_k^n(x_1) = \frac{1}{x_1} \tag{36}$$

(cf. [38, Theorem 5.1]).

From (14), we obtain functional equation

$$f_{\mathbb{B},k}(xy,t)e^{-t} = f_{\mathbb{B},k}(x,yt)e^{-ty} \tag{37}$$

(cf. [36]. By applying the Laplace transform to Equation (37), we arrive at the following Theorem (cf. [36]):

Theorem 7.3. *Let $x, y \in [0,1]$. Then*

$$\sum_{n=0}^{\infty} B_k^n(xy) = \sum_{n=0}^{\infty} \frac{1}{y} B_k^n(x). \tag{38}$$

Remark 7.4. *Substituting $x = 1$ into (38), we arrive at Equation (36).*

By applying the Fourier transform to Equation (35), we have

$$\frac{(x_1 + \cdots + x_v)^k}{k!} \int_0^{\infty} t^k e^{-t(x_1 + \cdots + x_v + is)} dt$$

$$= \sum_{n=0}^{\infty} v^n B_k^n \left(\frac{x_1 + \cdots + x_v}{v} \right) \frac{1}{n!} \int_0^{\infty} t^n e^{-(v+is)t} dt.$$

Substituting the following well known formula for the Fourier transform in the above equation

$$\int_0^{\infty} t^n e^{-(v+is)t} dt = \frac{n!}{(v+is)^{n+1}},$$

we arrive at the following Theorem.

Theorem 7.5. *Let $0 \le \frac{x_1 + \cdots + x_v}{v} \le 1$ and $s \in \mathbb{R}$. We have*

$$\sum_{n=0}^{\infty} \frac{v^n B_k^n \left(\frac{x_1 + \cdots + x_v}{v} \right)}{(v+is)^{n+1}} = \frac{(x_1 + \cdots + x_v)^k}{(x_1 + \cdots + x_v + is)^{k+1}}, \tag{39}$$

where $\left| \frac{v - (x_1 + \cdots + x_v)}{v(v+is)} \right| < 1$.

8 p-adic Volkenborn Integral Representations of the Bernstein Basis Functions

In this section, we give the p-adic integral representations of the Bernstein basis functions. Applying the p-adic Volkenborn integral on $\mathbb{Z}_p$ to the Bernstein basis functions, we give relations between the Bernstein basis functions, the Bernoulli numbers and the Euler numbers.

In order to prove our results, we recall work of Kim [15] that the p-adic q-Volkenborn integral is defined below. It is well known that

$$\mu_q(x + p^N \mathbb{Z}_p) = \frac{q^x}{[p^N]_q}$$

is a distribution on $\mathbb{Z}_p$ for $q \in \mathbb{C}_p$ with $\mid 1 - q \mid_p < 1$ (*cf.* [15]). Let $UD(\mathbb{Z}_p)$ be a set of uniformly differentiable functions on $\mathbb{Z}_p$. The p-adic q-integral of the function $f \in UD(\mathbb{Z}_p)$ is defined by Kim [15] as follows:

$$\int_{\mathbb{Z}_p} f(x)d\mu_q(x) = \lim_{N \to \infty} \frac{1}{[p^N]_q} \sum_{x=0}^{p^N - 1} f(x)q^x,$$

where

$$[x] = \frac{1 - q^x}{1 - q}.$$

From this equation, the *bosonic* p-adic integral (p-adic Volkenborn integral) was considered from a physical point of view to the bosonic limit $q \to 1$, as follows ([15]):

$$\int_{\mathbb{Z}_p} f(x) \, d\mu_1(x) = \lim_{N \to \infty} \frac{1}{p^N} \sum_{x=0}^{p^N - 1} f(x), \tag{40}$$

where

$$\mu_1(x + p^N \mathbb{Z}_p) = \frac{1}{p^N}.$$

The p-adic q-integral is used in many branch of mathematics, mathematical physics and other areas (*cf.* [1], [15], [18], [30], [31], [32], [33], [42], [43]).

By using (40), we have the Witt's formula for the Bernoulli numbers B_n as follows:

$$\int_{\mathbb{Z}_p} x^n d\mu_1(x) = B_n \tag{41}$$

(*cf.* [1], [15], [16], [19], [18], [30], [33], [32], [40], [42], [43]).

We consider the *fermionic* integral in contrast to the bosonic integral, which is called the fermionic p-adic Volkenborn integral on $\mathbb{Z}_p$ cf. [16]. That is

$$\int_{\mathbb{Z}_p} f(x) \, d\mu_{-1}(x) = \lim_{N \to \infty} \sum_{x=0}^{p^N - 1} (-1)^x f(x) \tag{42}$$

where

$$\mu_1\left(x+p^N\mathbb{Z}_p\right)=\frac{(-1)^x}{p^N}$$

(*cf.* [16]). By using (42), we have the Witt's formula for the Euler numbers E_n as follows:

$$\int_{\mathbb{Z}_p} x^n d\mu_{-1}(x) = E_n, \tag{43}$$

(*cf.* [16], [19], [33], [32], [40], [42]).

The multiplication of v basis functions $B_k^n(x)$ is given by the following formulas:

$$\prod_{j=1}^{v} B_{k_j}^{n_j}(x) = \frac{\prod\limits_{j=1}^{v}\binom{n_j}{k_j}}{\binom{n_1+\ldots+n_v}{k_1+\ldots+k_v}} B_{k_1+\ldots+k_v}^{n_1+\ldots+n_v}(x), \tag{44}$$

Hence, we get the following Lemma:

Lemma 8.1

$$\int_{\mathbb{Z}_p}\prod_{j=1}^{v} B_{k_j}^{n_j}(x)d\mu_1(x) = \frac{\prod\limits_{j=1}^{v}\binom{n_j}{k_j}}{\binom{n_1+\ldots+n_v}{k_1+\ldots+k_v}} \int_{\mathbb{Z}_p} B_{k_1+\ldots+k_v}^{n_1+\ldots+n_v}(x)d\mu_1(x). \tag{45}$$

By applying the bosonic p-adic Volkenborn integral and (45) to Theorem 2.1, we have

$$\int_{\mathbb{Z}_p} B_{k_1+\cdots+k_v}^{n}(x)d\mu_1(x)$$

$$= \frac{v^{k_1+\cdots+k_v-n}k_1!\cdots k_v!}{(k_1+\cdots+k_v)!}\sum_{m_1+\cdots+m_{v-1}=n} C_{m_1,\cdots,m_{v-1}}^{n}$$

$$\times \int_{\mathbb{Z}_p} B_{k_v}^{m_v-1}(x)B_{k_{v-1}}^{m_v-2}(x)\cdots B_{k_1}^{m_1}(x)B_{k_2}^{n-m_1-\cdots-m_{v-1}}(x)d\mu_1(x).$$

By applying (41) to the above equation, we get the following theorem.

Theorem 8.2

$$\int_{\mathbb{Z}_p} B_{k_1+\cdots+k_v}^{n}(x)d\mu_1(x)$$

$$= \frac{v^{k_1+\cdots+k_v-n}k_1!\cdots k_v!}{(k_1+\cdots+k_v)!}\sum_{m_1+\cdots+m_{v-1}=n} C_{m_1,\cdots,m_{v-1}}^{n}\binom{n_{v-1}}{k_v}\cdots\binom{m_1}{k_1}\binom{n-m_1-\cdots-m_{v-1}}{k_j}$$

$$\sum_{j=0}^{n-k_1-\cdots-k_v}(-1)^{n-k_1-\cdots-k_v}\binom{n-k_1-\cdots-k_v}{j}B_{n-j},$$

where B_{n-j} *denotes the Bernoulli numbers.*

By applying the fermionic p-adic Volkenborn integral and (45) to Theorem 2.1, we have

$$\int_{\mathbb{Z}_p} B_{k_1+\cdots+k_v}^n(x)d\mu_{-1}(x)$$

$$= \frac{v^{k_1+\cdots+k_v-n}k_1!\cdots k_v!}{(k_1+\cdots+k_v)!} \sum_{m_1+\cdots+m_{v-1}=n} C_{m_1,\cdots,m_{v-1}}^n$$

$$\times \int_{\mathbb{Z}_p} B_{k_v}^{m_{v-1}}(x)B_{k_{v-1}}^{m_{v-2}}(x)\cdots B_{k_1}^{m_1}(x)B_{k_2}^{n-m_1-\cdots-m_{v-1}}(x)d\mu_{-1}(x).$$

By applying (43) to the above equation, we get the following theorem:

Theorem 8.3

$$\int_{\mathbb{Z}_p} B_{k_1+\cdots+k_v}^n(x)d\mu_1(x)$$

$$= \frac{v^{k_1+\cdots+k_v-n}k_1!\cdots k_v!}{(k_1+\cdots+k_v)!} \sum_{m_1+\cdots+m_{v-1}=n} C_{m_1,\cdots,m_{v-1}}^n \binom{n_{v-1}}{k_v}\cdots\binom{m_1}{k_1}\binom{n-m_1-\cdots-m_{v-1}}{k_j}$$

$$\sum_{j=0}^{n-k_1-\cdots-k_v}(-1)^{n-k_1-\cdots-k_v}\binom{n-k_1-\cdots-k_v}{j}E_{n-j},$$

where E_{n-j} denotes the Euler numbers.

Remark 8.4. *By using (45), we also have the following results:*

$$\int_{\mathbb{Z}_p}\prod_{j=1}^v B_{k_j}^{n_j}(x)d\mu_1(x) = \prod_{j=1}^v\binom{n_j}{k_j}\sum_{j=0}^k\sum_{l=0}^{n-k}(-1)^{n-j-l}\binom{k}{j}\binom{n-k}{l}B_{n+j-k-l},$$

and

$$\int_{\mathbb{Z}_p}\prod_{j=1}^v B_{k_j}^{n_j}(x)d\mu_{-1}(x) = \prod_{j=1}^v\binom{n_j}{k_j}\sum_{j=0}^k\sum_{l=0}^{n-k}(-1)^{n-j-l}\binom{k}{j}\binom{n-k}{l}E_{n+j-k-l}$$

(cf. [12], [17], [23], [20]).

Acknowledgements. The present investigation was supported by the *Scientific Research Project Administration of Akdeniz University*. I would like to thank for referees for their valuable comments.

References

1. Amice, Y.: Integration p-adique, selon A. Volkenborn, Seminaire Delange-Pisot-Poitou. Theorie des Nombres 13(2), G1–G9 (1971-1972)
2. Alfeld, P., Neamtu, M., Schumaker, L.L.: Bernstein-Bézier polynomials on spheres and sphere-like surfaces, Comput. Aided Geom. Des 13(4), 333–349 (1996)

3. Acikgoz, M., Araci, S.: On generating functions of the Bernstein polynomials. In: Numerical Analysis and Applied Mathematics, Amer. Inst. Phys. Conf. Proc., vol. CP1281, pp. 1141–1143 (2010)
4. Bernstein, S.N.: Démonstration du théorème de Weierstrass fondée sur la calcul des probabilités. Commun. Kharkov Math. Soc. 13, 1–2 (1912)
5. Comtet, L.: Advanced Combinatorics: The Art of Finite and Infinite Expansions. Reidel, Dordrecht (1974)
6. Busé, L., Goldman, R.: Division algorithms for Bernstein polynomials. Comput. Aided Geom. Design. 25(9), 850–865 (2008)
7. Farouki, R.T., Goodman, T.N.T.: On the optimal stability of the Bernstein basis. Math. Comput. 65, 1553–1566 (1996)
8. Farouki, R.T.: The Bernstein polynomials basis: a centennial retrospective. Computer Aided Geometric Design 29, 379–419 (2012)
9. Goldman, R.: An Integrated Introduction to Computer Graphics and Geometric Modeling. CRC Press, Taylor and Francis (2009)
10. Goldman, R.: Pyramid Algorithms: A Dynamic Programming Approach to Curves and Surfaces for Geometric Modeling. R. Academic Press, San Diego (2002)
11. Goldman, R.: Identities for the Univariate and Bivariate Bernstein Basis Functions. In: Paeth, A. (ed.) Graphics Gems V, pp. 149–162. Academic Press (1995)
12. Jang, L.C., Kim, W.-J., Simsek, Y.: A study on the p-adic integral representation on $\mathbb{Z}_p$ associated with Bernstein and Bernoulli polynomials. Advances in Difference Equations, Article ID 163217, 6 pages (2010)
13. Jetter, K., Stöckler, J.: An identity for multivariate Bernstein poynomials, Comput. Aided Geom. Design. 20, 563–577 (2003)
14. Joy, K.I.: Bernsein polynomials, On-Line Geometric Modeling Notes, http://www.idav.ucdavis.edu/education/ CAGDNotes/Bernstein-Polynomials.pdf
15. Kim, T.: q-Volkenborn integration. Russian J. Math. Phys. 19, 288–299 (2002)
16. Kim, T.: q-Euler numbers and polynomials associated with p-adic q-integral and basic q-zeta function. Trend Math. Information Center Mathematical Sciences 9, 7–12 (2006)
17. Kim, T.: A note on q-Bernstein polynomials, arXiv:1009.0097v1 (math.NT)
18. Kim, T., Rim, S.-H., Simsek, Y., Kim, D.: On the analogs of Bernoulli and Euler numbers, related identities and zeta and L-functions. J. Korean Math. Soc. 45, 435–453 (2008)
19. Kim, T., Kim, M.S., Jang, L.C.: New q-Euler numbers and polynomials associated with p-adic q-integrals. Adv. Stud. Contemp. Math. 15, 140–153 (2007)
20. Kim, T.: q-Bernstein polynomials, q-Stirling numbers and q-Bernoulli polynomials, arXiv:1008.4547
21. Mazure, M.-L.: Chebyshev-Bernstein bases. Comput. Aided Geom. Des. 16(7), 649–669 (1999)
22. Kim, M.S., Kim, D., Kim, T.: On the q-Euler numbers related to modified q-Bernstein polynomials. Abstr. Appl. Anal., Art. ID 952384, 15 pages (2010)
23. Kim, T., Choi, J., Kim, Y.H., Ryoo, C. S.: On the fermionic p-adic integral representation of Bernstein polynomials associated with Euler numbers and polynomials, arXiv:1008.5207v1 (math.NT)
24. Lewanowicz, S., Woźny, P.: Generalized Bernstein polynomials. BIT Numer. Math. 44, 63–78 (2004)
25. Lorentz, G.G.: Bernstein Polynomials. Chelsea Pub. Comp., New York (1986)
26. Lyche, T., Scherer, K.: On the L_1-condition number of the univariate Bernstein basis. Constructive Approx 18, 503–528 (2002)

27. Phillips, G.M.: Interpolation and approximation by polynomials. CMS Books in Mathematics/ Ouvrages de Mathématiques de la SMC, vol. 14. Springer, New York (2003)
28. Phillips, G.M.: Bernstein polynomials based on the q-integers. Ann. Numer. Math. 4, 511–518 (1997)
29. Oruc, H., Phillips, G.M.: A generalization of the Bernstein polynomials. Proc. Edinb. Math. Soc. 42, 403–413 (1999)
30. Schikhof, W.H.: Ultrametric Calculus: An Introduction to p-Adic Analysis. Cambridge Studies in Advanced Mathematics, vol. 4. Cambridge University Press, Cambridge (1984)
31. Shiratani, K.: On Euler numbers. Mem. Fac. Kyushu Univ. Series A, Mathematics 27, 1–5 (1973)
32. Simsek, Y.: Twisted (h, q)-Bernoulli numbers and polynomials related to twisted (h, q)-zeta function and L-function. J. Math. Anal. Appl. 324, 790–804 (2006)
33. Simsek, Y.: Complete sum of products of (h, q) -extension of Euler polynomials and numbers. J. Differ. Equ. Appl. 16, 1331–1348 (2010)
34. Simsek, Y.: Construction a new generating function of Bernstein type polynomials. Appl. Math. Comput. 218, 1072–1076 (2011)
35. Simsek, Y.: Interpolation function of generalized q −Bernstein-type basis polynomials and applications. In: Boissonnat, J.-D., Chenin, P., Cohen, A., Gout, C., Lyche, T., Mazure, M.-L., Schumaker, L. (eds.) Curves and Surfaces 2011. LNCS, vol. 6920, pp. 647–662. Springer, Heidelberg (2012)
36. Simsek, Y.: Generating functions for the Bernstein type polynomials: a new approach to deriving identities and applications for these polynomials to appear in Hacettepe. J. Math. Stat.
37. Simsek, Y.: Unification of the Bernstein-type polynomials and their applications. Boundary Value Problems 2013, 56 (2013)
38. Simsek, Y.: Functional equations from generating functions: a novel approach to deriving identities for the Bernstein basis functions, Fixed Point Theory and Applications 2013, 80 (2013)
39. Simsek, Y., Acikgoz, M.: A new generating function of $(q-)$ Bernstein-type polynomials and their interpolation function. Abstr. Appl. Anal., 1–12 (2010)
40. Simsek, Y., Bayad, A., Lokesha, V.: q-Bernstein polynomials related to q-Frobenius-Euler polynomials, l-functions, and q-Stirling numbers. Math. Meth. Appl. Sci. 35, 877–884 (2012)
41. Srivastava, H.M., Choi, J.: Series Associated with the Zeta and Related Functions. Kluwer Academic Publishers, Dordrecht (2001)
42. Srivastava, H.M.: q-Bernoulli numbers and polynomials associated with multiple q-zeta functions and basic L-series. Russian J. Math. Phys. 12, 241–268 (2005)
43. Volkenborn, A.: On generalized p-adic integration. Memoires de la S. M. F. 39-40, 375–384 (1974)

Efficient Pixel-accurate Rendering
of Animated Curved Surfaces

Young In Yeo, Sagar Bhandare, and Jörg Peters

University of Florida, USA,
jorg@cise.ufl.edu

Abstract. To efficiently animate and render large models consisting of bi-cubic patches in real time, we split the rendering into pose-dependent, view-dependent (Compute-Shader supported) and pure rendering passes. This split avoids recomputation of curved patches from control structures and minimizes overhead due to data transfer – and it integrates nicely with a technique to determine a near-minimal tessellation of the patches while guaranteeing sub-pixel accuracy. Our DX11 implementation generates and accurately renders 141,000 animated bi-cubic patches of a scene in the movie 'Elephant's Dream' at more than 300 frames per second on a 1440×900 screen using one GTX 580 card.

1   Introduction

Curved, smooth, piecewise polynomial surfaces have become standard in high end, movie-quality animation. Subdivision surfaces [1,2], spline (NURBS) surfaces or Bézier patch-based surfaces are chosen over polygonal, polyhedral, or faceted-based representations both for aesthetic reasons and for their ability to represent models more compactly. In particular, curved surfaces yield more life-like transitions and silhouettes and, in principle, support arbitrary levels of resolution without exhibiting polyhedral artifacts (see Fig. 1). But while curved surfaces are commonly used in cinematic production and geometric design, they are not commonly used for interactive viewing. Animation artists and designers typically work off faceted models at a given resolution and have to call special off-line rendering routines to inspect the true outcome of their work. At the other end of the spectrum, game designers opt for coarsely-faceted models, made more acceptable by careful texturing, to achieve real-time rendering with limited resources under competing computational demands, e.g. computing game physics. In an attempt to narrow the gap, a number of mesh-to-surface conversion algorithms have been developed in the past years that run efficiently on the GPU (see Section 2). But so far their rendering has depended on screen projection heuristics without guarantees of accuracy.

The present paper explains how to render, at interactive rates, and on high-resolution screens, a substantial number of animated curved surfaces free of perceptible polyhedral artifacts, parametric distortion and pixel dropout. The paper leverages and extends the authors' approach [3] for efficiently determining the near-minimal tessellation density required for *pixel-accurate rendering* (see Section 2). Determining the near-minimal tessellation density requires, depending on the model, between 1% and 5% extra work. However, by avoiding overtessellation, pixel-accurate rendering is often faster than rendering based on heuristics (see Fig. 2, middle and right). Specifically, the paper shows

M. Floater et al. (Eds.): MMCS 2012, LNCS 8177, pp. 491–509, 2014.

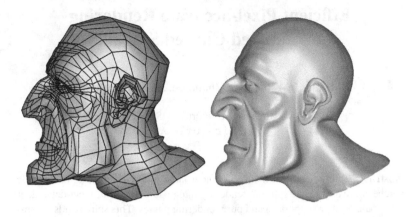

Fig. 1. Faceted versus smooth: Proog's head

how to integrate the approach into animation the animation pipeline to make it inter-active. Rather than repeatedly sending large control nets from the CPU to the GPU for rendering, we load the base mesh(es) once and we apply morph-target and skeletal-animation transformations to the characters' mesh model on the GPU and convert it into a curved surface. We then use the natural partition of animated sequences into pose-dependent, view-dependent and pure rendering frames to compute both the an-imation and the pixel-accurate patch-tessellation in a combination of two, one or no Compute Shaders preceding each standard rendering pass on the GPU.

As proof of concept, we animated and rendered 141K patches of a scene of the open-source movie Elephants Dream. In 2006, each frame of the movie required 10 minutes of CPU time at full-HD resolution [4]. We can now render the higher-order surfaces and textures (leaving out post effects) on the GPU at more than 300 frames per second Fig. 15 thanks to parallelism and new algorithms that take advantage of this parallelism. To wit, doubling processor speed every year since 2006 would reduce the time per frame only to ca 10 seconds per frame, three orders of magnitude slower.

Overview. In Section 2 we review the definition of pixel-accurate rendering of curved surfaces, animation basics and the conversion of faceted to smooth curved surfaces. Section 3 presents the idea and formulas for enforcing pixel-accurate rendering. Sec-tion 4 presents the algorithm and an efficient implementation, including pseudo-code, of pixel-accurate rendering of animated curved surfaces. In Section 5 we analyze the implementation's performance and discuss trade-offs and alternative choices. We also compare to a similar widely-available DX11 sample program.

2   Background

To efficiently pixel-accurately render the surfaces of Elephants Dream on the GPU, our proposed animation framework has to near-optimally set the tessellation factor for Bézier patches after replicating linear skeletal animation, relative shape-key animation (morph targets), and mesh-to-surface conversion on the GPU.

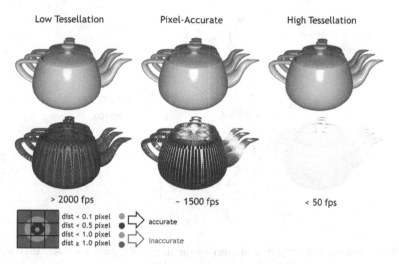

Fig. 2. Balancing pixel-accuracy and rendering speed. No red or green colors should be visible in the lower row if the tessellation is sufficiently fine for pixel-accuracy. The red and the green spots indicate a parametric distortion of more than 1/2 pixel (cf. the color coding *lower left*). Additional objects are analyzed in Fig. 12 of [3].

Tessellation and Pixel-accuracy. A key challenge when working with a curved surface is to set the density of evaluation so that the surface triangulation is a good proxy of the smooth surface. The density, or tessellation, has to be sufficiently high to prevent polyhedral artifacts and parametric distortions, and sufficiently low to support fast rendering. In modern graphics pipelines, the level of detail can be prescribed by setting the tessellation factor(s) τ of each patch $\mathbf{p} : (u, v) \in U \to \mathbb{R}^3$ of the curved surface. In 3D movie animation, it is common practice to over-tessellate and shade a very high number of fragments. Real-time animation cannot afford this since each fragment is evaluated, rasterized and shaded. Since the camera is free to zoom in or out of the scene, fixed level of tessellation results in faceted display or overtessellation. This disqualifies approaches that require setting τ a priori. Popular screen-based heuristics based on measuring edge-length or estimating flatness (see e.g. [5], [6, Sec 7]) do not come with guarantees or require an a priori undetermined number of passes to recursively split patches and verify that the measure falls below a desired tolerance.

Pixel-accurate rendering, Fig. 2, middle, determines the tessellation density (just) fine enough to guarantee correct visibility, prevent parametric distortion or pixel-dropout. Pixel-accuracy has two components: covering (depth) accuracy and parametric (distortion) accuracy [3, Section 3]. *Covering accuracy* requires that each pixel's output value be controlled by one or more unoccluded pieces of patches whose projection overlaps it sufficiently and *parametric accuracy* requires that for each pixel the following holds (cf.Fig. 3). Let $\left[\begin{smallmatrix} x \\ y \end{smallmatrix}\right]$ be the pixel's center, $\mathbf{p} : \mathbb{R}^2 \to \mathbb{R}^3$ a surface patch and (u, v) a parameter pair. Then the surface point $\mathbf{p}(u, v) \in \mathbb{R}^3$ must project into the pixel:

$$\|P(\mathbf{p}(u, v)) - [\begin{smallmatrix} x \\ y \end{smallmatrix}]\|_\infty < 0.5. \tag{1}$$

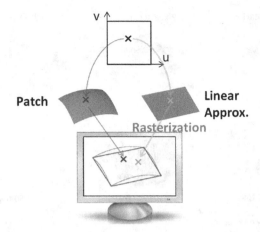

Fig. 3. Triangulation and projection distorting the image of a curved surface. Pixel-accurate rendering guarantees that the distortion is at below pixel level.

Inequality (1) prevents *parametric distortion*: if $P(\mathbf{p}(u, v))$ lies outside the pixel associated with parameters (u, v) then the wrong texture, normal or displacement is computed causing artifacts incompatible with accurate rendering. Parametric inaccuracy is color-encoded in Fig. 2: lack of accuracy is shown in red and green. Predictably, too coarse a tessellation yields a high frame rate and too fine a tessellation slows down rendering. The largely grey coloring of the teapot in Fig. 3, row two, under pixel-accurate rendering, indicating a distortion just below the pixel threshold, is therefore highly desirable. Work similar to [3], but based on the bounds in [7] includes [8] and most recently [9].

Skeletal animation. The most common technique for character animation, used by the artists of Elephants Dream, is linear blend skinning, also known as linear vertex blending or skeletal subspace deformation [10]. Here a character is defined by a template, a faceted model, called skin. The models animation or deformation is defined by a time-varying set of rigid transformations, called bones, that are organized into a tree structure, called skeleton. Any vertex position in a linear blend skin is expressed as a linear combination of the vertex transformed by each bone's coordinate system: at time t_i, a convex combination ω_k of bone transformations $\mathbf{R}_k$ is applied to each skin vertex initial position $\mathbf{v}(0)$:

$$\mathbf{v}(t) = \Big(\sum_k \omega_k \mathbf{R}_k(t)\Big)\mathbf{v}(0), \quad \sum_k \omega_k = 1. \tag{2}$$

The weights ω_k are assigned by the artist. Section 4 provides pseudo-code.

Since this direct linear combination of rotation matrices generically does not yield a valid rotation, a number of improvements have been suggested [11,12]. In particular dual quaternions [12] are sufficiently simple to have been implemented in Blender. Our framework is agnostic to the choice of animation since its implementation as a Compute Shader allows alternative animation techniques to be substituted such as deformation

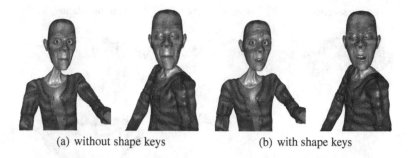

(a) without shape keys   (b) with shape keys

Fig. 4. Emo's mouth opened with **shape keys**

of the mesh points with respect to control cages (see e.g. [13,14,15,16]). However, since the artists of Elephants Dream used linear blend skinning, and compensated for its shortcomings, our real-time rendering applies linear, skeletal animation.

Shape Keys. For more nuanced, say facial expressions, Elephants Dream, and hence our implementation, additionally applies shape keys, also known as morph targets or blend shapes. Shape keys average between morph targets representing standard poses (see e.g. [17] for a detailed explanation.)

Mesh-to-Surface Conversion. In recent years, a number of algorithms have been developed to use polyhedral meshes as control nets of curved surfaces and efficiently evaluate these curved surfaces on the GPU. Such algorithms include conversions to piecewise polynomial and rational representation [18,19,20,21] as well as subdivision [22,23,6,24]. Our framework is agnostic to the choice of conversion algorithm. To be able to compare our GPU implementation to a widely accessible implementation, we chose Approximate Catmull-Clark (ACC) [20]: optimized shader code of ACC animation, SubD11, is distributed with MicroSoft DX11 [25]. The output of ACC is one bi-cubic patch patch for each face of the (refined) control mesh (plus a pair of tangent patches to improve the impression of smoothness as in [26]). Note that parametric accuracy is not concerned with whether ACC provides a good approximation to subdivision surfaces, an issue of independent interest (cf. [27,28]).

3   Computing Near-Minimal Accurate Tessellation Levels

The two main ingredients that make pixel-accurate rendering efficient are avoiding recursion and triangulating as coarsely as possible while guaranteeing pixel-accuracy (see Fig. 5). This section explains how to address both challenges by computing a near-minimal tessellation factor τ in a single step according to the approach in [3]. The tessellation factor is computed with the help of slefe-boxes [29]. Bilinear interpolants of these slefe-boxes, called slefe-tiles, sandwich the curved surface and the triangulation as illustrated in Fig. 6. Such slefe-boxes are not traditional bounding boxes enclosing a patch. Rather the maximal width of slefe-boxes gives an upper bound on the variance of the exact curved surface from triangulation. This reflects the goal: to partition the domain sufficiently finely so that the variance and hence the 'width' of the all projected

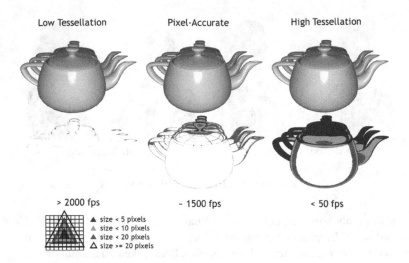

Fig. 5. Optimal tessellation of curved surfaces. Fewer, hence bigger triangles improve efficiency. (Note the different use of color-coding from Fig. 2).

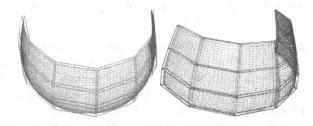

Fig. 6. The bi-linear interpolants to groups of four slefe-boxes define **slefe-tiles** that locally enclose the surface. Note that the tiles, while useful of collision, are **never explicitly computed** for the pixel-accurate rendering.

slefe-boxes and therefore of the slefe-tiles falls below a prescribed tolerance, e.g. half the size of a pixel.

Since knot insertion stably converts NURBS patches of degree (d_1, d_2) to tensor-product patches in Bézier-form (*glMap2* in OpenGL) with coefficients $c_{ij} \in \mathbb{R}^3$ and basis functions b_j^d,

$$p(u, v) := \sum_{i=0}^{d_1} \sum_{j=0}^{d_2} c_{ij} b_j^{d_2}(v) b_i^{d_1}(u), \qquad (u, v) \in [0..1]^2, \tag{3}$$

and since subdivision surfaces can be treated as nested rings of such patches, we focus on tensor-product Bézier patches. (Knot insertion can be a pre-processing step or done on the fly on the GPU. Rational patches are rarely used in animation; if needed, for strictly positive weights, bounds in homogeneous space plus standard estimates of interval arithmetic do the trick.) Moreover, slefe-boxes for patches in tensor-product form can be derived from bounds in one variable and the computations for building slefe-boxes are separate in each x, y and z coordinate. We can therefore simplify the discussion in the next subsection to one univariate polynomial piece p in Bézier-form with coefficients $c_j \in \mathbb{R}$ and parameter $u \in [0 \,..\, 1]$:

$$p : \mathbb{R} \to \mathbb{R}, u \mapsto p(u) := \sum_{j=0}^{d} c_j b_j^d(u), \quad b_j^d := \binom{d}{j}(1-u)^{d-j}u^j.$$

Subdividable Linear Efficient Function Enclosures, abbreviated as **slefes**, tightly sandwich non-linear functions p, such as polynomials, splines and subdivision surfaces, between simpler, piecewise linear, lower and upper functions, $\underline{p}$ and $\overline{p}$:

$$\underline{p} \leq p \leq \overline{p},$$

[30,31,32,33,29,34,35]. Specifically, in one variable, [30] shows that (cf. Fig. 7, *left*)

$$p(t) \leq \overline{p}(t) := \ell(t) + \sum_{j=1}^{d-1} \max\{0, \nabla_j^2 p\}\, \overline{a_j^d}^m(t) \tag{4}$$

$$+ \sum_{j=1}^{d-1} \min\{0, \nabla_j^2 p\}\, \underline{a_j^d}_m(t).$$

with the matching lower bound $\underline{p}$ obtained by exchanging min and max operators. Here

$$a_j^d, \quad j = 1, \ldots, d-1,$$

are polynomials that span the space of polynomials of degree d minus the linear functions $\ell(t)$; $\overline{a_j^d}^m$ is an m-piece upper and $\underline{a_j^d}_m$ an m-piece lower bound on a_j^d; and $\nabla_j^2 p := c_{j-1} - 2c_j + c_{j+1}$ is a second difference of the control points. If p is a linear function, upper and lower bounds agree. The tightness of the bounds is important since loose bounds result in over-tessellation. Fig. 7b shows an example from [3], where the min-max or AABB bound is looser by an order of magnitude than the slefe-width $w := \max_{t \in [0..1]} \overline{p}(t) - \underline{p}(t)$.

Being piecewise linear, the bounding functions $\overline{a_j^d}^m$ and $\underline{a_j^d}^m$ in (4) are defined by their values at the uniformly-spaced break points. These values can be pre-computed. Since, for $d = 3$, i.e. cubic functions, $a_2^3(1-t) = a_1^3(t)$, Table 1 lists all numbers needed to compute Fig. 7, e.g., for $t = 1/3$, the upper and lower breakpoint values $-.370370..$

Table 1. Values at **breakpoints of a** $m = 3$**-piece slefe.** This table and the tables for higher degree can be downloaded [36]. Similar slefe-tables exist for splines with uniform knots [30].

$t =$	0	1/3	2/3	1
$\overline{a_1^3}^3$	0	-.370370..	-.296296..	0
$\underline{a_1^3}_3$	-.069521..	-.439891..	-.315351..	-.008732..

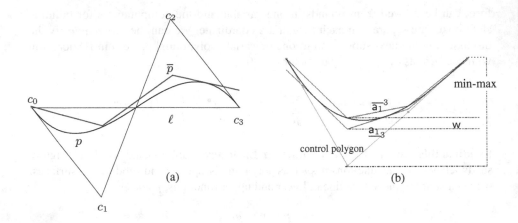

Fig. 7. The **slefe-construction** from [29]. (a) The function $p(t) := -b_1^3(t) + b_2^3(t)$ and its upper bound $\overline{p}$. (b) The lower bound $\underline{a}_{1_3}$ and the upper bound $\overline{a_1}^3$ tightly sandwiching the function $a_1 := -\frac{2}{3}b_1^3(t) - \frac{1}{3}b_2^3(t)$, using $m = 3$ segments. Table 1 shows $w = \max_{[0..1]} \overline{p} - \underline{p}$ to be < 0.07. The corresponding number for [7] (not illustrated) is $\frac{6}{8} = 0.75$ and for the min-max-bound $\frac{2}{3}$.

and $-.439891...$ Moreover, by tensoring, the 8 numbers suffice to compute all bounds required for ACC patches: the tensor-product patch (3) can be bounded by computing the upper values $\tilde{c}_{ij}$, $i = 0, \ldots, d_1$ (for each $j = 0, \ldots, m_2$) of the 1-variable slefe in the v direction and then treat the values as control points when computing the upper slefe in the u direction:

$$p(u,v) \leq \sum_{i=0}^{d_1} \sum_{j=0}^{m_2} \tilde{c}_{ij} b_j^1(v) b_i^{d_1}(u) \leq \sum_{j=0}^{m_2} \sum_{i=0}^{m_1} \bar{c}_{ij} b_i^1(u) b_j^1(v).$$

Ensuring Pixel-accuracy. The slefes just discussed are for functions, i.e. one coordinate of the image. Since we want to control the variance of the surface patches from their triangulation we now consider a patch $\mathbf{p} : \mathbb{R}^2 \to \mathbb{R}^3$ with three coordinates bounded by bilinear interpolants to upper and lower values at the grid points (u_i, v_j), $i, j \in \{0, 1, \ldots, m\}$. For each (u_i, v_j), abbreviating $\overline{\mathbf{p}_{ij}} := \mathbf{p}(u_i, v_j)$, $\underline{\mathbf{p}}_{ij} := \mathbf{p}(u_i, v_j)$, a *slefe-box* is defined as

$$\underline{\overline{\mathbf{p}}}(u_i, v_j) := \frac{\overline{\mathbf{p}_{ij}} + \underline{\mathbf{p}}_{ij}}{2} + [-\frac{1}{2}..\frac{1}{2}]^3 (\overline{\mathbf{p}_{ij}} - \underline{\mathbf{p}}_{ij}), \tag{5}$$

where $[-\frac{1}{2}..\frac{1}{2}]^3$ is the **0**-centered unit cube. That is, the slefe-box is an axis-aligned box in $\mathbb{R}^3$ (see red boxes in Fig. 8) centered at the average of upper and lower values.

To measure parametric accuracy, we define the minimal screen-coordinate-aligned rectangle that encloses the screen projection $[\begin{smallmatrix} \mathtt{x} \\ \mathtt{y} \end{smallmatrix}] := P(\underline{\overline{\mathbf{p}}}(u_i, v_j))$ of to the slefe-box with index i, j (see the blue dashed rectangles in Fig. 8):

$$q_{ij} := [\underline{\mathtt{x}}_{ij}..\overline{\mathtt{x}}_{ij}] \times [\underline{\mathtt{y}}_{ij}..\overline{\mathtt{y}}_{ij}] \supsetneq P(\underline{\overline{\mathbf{p}}}(u_i, v_j)). \tag{6}$$

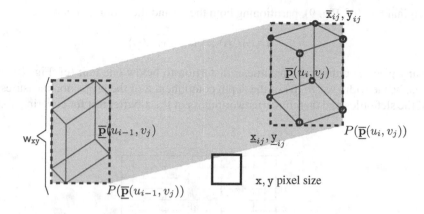

Fig. 8. Projected slefe-boxes. The projected slefe-boxes (red) are enclosed by axis-aligned rectangles (blue, dashed) whose linear interpolant (grey area) encloses the image (here of $\mathbf{p}([u_{i-1}..u_i], v_j)$). The (square-root of the) maximal edge-length of the dashed rectangles, in pixel size, determines the tessellation factor $\tau_{\mathbf{p}}$.

The maximal edge length over all q_{ij} is the parametric *width* $\mathsf{w_{xy}}$. This width is a close upper bound on the variance from linearity in the parameterization since the width of the projected boxes dominates the width of the slefe-tiles – that therefore need not be computed. The width shrinks to zero when the parameterization becomes linear.

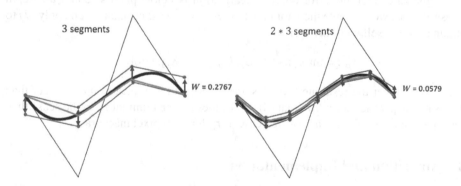

Fig. 9. Shrinkage of the width for a curve segment under subdivision. *black:* cubic curve, control polygon, *blue:* piecewise linear interpolant, *red:* slefe

We want to determine the tessellation factor $\tau_{\mathbf{xy}} \in \mathbb{R}$ so that $\mathsf{w_{xy}} < 1$. Let $\mathsf{w}_m(\mathbf{p})$ be the width of the projection of patch $\mathbf{p}$ measured for a slefe with m pieces and k_m a constant between 1.5 and 1, depending only on m. Since partitioning the u-domain into $1/h$ segments, and re-representing the function over the smaller interval before re-applying the bound, scales the maximal second difference down quadratically to h^2 its original size (cf. Fig. 9), partitioning both the u- and the v-domain into

$$\tau_{xy}(m, \mathbf{p}) := k_m \sqrt{\mathsf{w}_m(\mathbf{p})} \tag{7}$$

many pieces, confines the parameter distortion to below one unit (cf. Fig. 10) Analogously, the width $\mathsf{w}_z(m, \mathbf{p})$ of the depth component z of the projection measures depth of the slefe-tiles and therefore trustworthiness of the z-buffer test for covering accuracy.

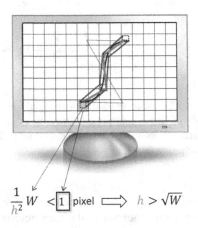

$$\frac{1}{h^2}W < \boxed{1}\ \text{pixel} \implies h > \sqrt{W}$$

Fig. 10. Shrinkage of slefe under h-fold subdivision

To guarantee that any error due to linearization is below pixel size and the depth threshold tol_z, we compute the width for low m, say $m = 2$ or 3, and then apply (7) to obtain a safe tessellation factor of

$$\tau_{\mathbf{p}} := \max\{\tau_{xy}(m, \mathbf{p}), k_m \sqrt{\mathsf{w}_z(m, \mathbf{p})}/\text{tol}_z\}. \tag{8}$$

Fig. 5 shows that the resulting triangles are, as hoped for, typically much larger than pixels and experiments confirm that (8) determines a near-minimal $\tau_{\mathbf{p}}$ in the sense that, for typical models, already a 10% decrease in $\tau_{\mathbf{p}}$ leads to pixel inaccuracy.

4   Algorithm and Implementation

The main costs, that our algorithm for rendering animated curved surfaces seeks to minimize, are the conversion of the mesh to the surface patch coefficients and rendering the patches with pixel accuracy. For details of the implementation of pixel accuracy as a Compute Shader pre-pass, we refer to [3]. The key to minimizing the conversion cost is to restrict conversion to pose changes of the animated character. The key to efficient pixel-accurate rendering is to integrate the control of the variance of the curved patch

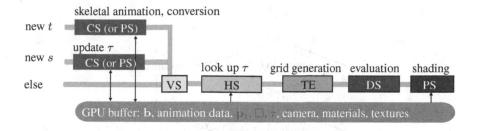

Fig. 11. Mapping of animation and conversion to curved surfaces to the DX11 graphics pipeline. CS=Compute Shader, VS=Vertex Shader, HS= Hull Shader, TE=Tessellation Engine, DS= Domain Shader, PS=Pixel Shader.

geometry from its triangulation, as just explained in Section 3, with the conversion to minimize overhead. Specifically, we split the work as follows.

- For every *pose* (geometry or mesh connectivity) change, re-compute the control mesh, all affected patches and slefe-boxes.
- For every *view* change, measure the width w_{xy} of the boxes' screen projections and their depth variance w_z.
- Determine the tessellation factor τ according to (8), i.e. a low as possible while still guaranteeing pixel-accuracy to make best use of the efficient rasterization stage on the GPU.

Pose and View Change. To minimize conversion and τ computation cost, our implementation calls either two, one or no Compute Shader passes followed by a standard DX11 rendering pass. This is illustrated in Fig. 12 and the details are as follows.

(a)   If the scene does not change in view or pose then the stored animated curved surface at time step t, $\mathbf{p}_t$, is rendered with the existing tessellation factors.

(b)   For each view change at time step s that is not an animation step, the modelview transformations are applied to the saved $\mathbf{p}_t$ and the tessellation factors τ are updated to guarantee pixel-accuracy for the new viewpoint. Then (a) is executed.

(c)   For each pose change (animation step t), the coefficients of the animated curved surface $\mathbf{p}_t$ are computed by executing the animation and conversion steps. The coefficients of $\mathbf{p}_t$ are stored in the GPU buffer. Then the slefe-boxes are re-computed and stored and the same computations are executed as in (b).

Throughout, only modified patches are updated.

Mapping to GPU Shader Code. In modern graphics APIs the triangulation density is set by up to six tessellation factors per surface patch. The two interior tessellation factors are set to τ, while the other four tessellation factors, corresponding to the boundaries, are set to the maximum of the interior factors of the patches sharing the boundary. This coordination in the Compute Shader pass guarantees a consistent triangulation by avoiding mismatch along boundaries between differently tessellated patches.

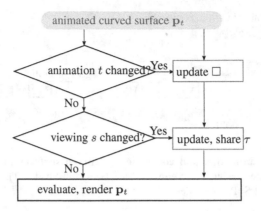

Fig. 12. Updating slefe-boxes □ and the tessellation factors τ is only required when the input mesh is animated or the view is changed

The pseudo-code of the Compute Shaders is given below. Detailed pseudo-code of pixel-accurate slefe-estimates is presented in Section 6 of [3]. The rendering pass is standard DX11 rendering.

The data flow outlined in Fig. 12 is made concrete by the following pseudocode. The mapping of the pseudocode to the DX11 graphics pipeline is shown in Fig. 11. Recall that each bi-cubic patch has $4 \times 4 = 16$ coefficients.

```
function MAIN(t, s)
    if new t then COMPUTE_SHADER_POSE_CHANGE(t)
    end if
    if new s then COMPUTE_SHADER_VIEW_CHANGE(s)
    end if
end function
```

```
[shared_mem cpts[16]]
[num_threads 16]
    function COMPUTE_SHADER_POSE_CHANGE(t)
        vtx_id ← thread_id + (patch_id * 16)
        SHAPE_KEY(vtx_id, t)
        SKELETAL_ANIMATION(vtx_id, t)
        CONVERT_TO_ACC(vtx_id)
    end function
```

```
[shared_mem width[16]]
[num_threads 16]
    function COMPUTE_SHADER_VIEW_CHANGE(s)
        vtx_id ← thread_id + (patch_id * 16)
        width[thread_id] ← project_slefe(vtx_id)
        synchronize_threads()
        if thread_id = 1 then
            TF ← pick_max_width(width)
```

```
    save_to_gpu(TF_buffer, patch_id, TF)
  end if
end function


function SHAPE_KEY(vtx_id, t)
  base_sk ← get_base_shape_key(vtx_id)
  shaped_vtx ← (0, 0, 0)
  for sk in shape_keys[vtx_id] do
    sk_wt ← get_shape_key_wt(vtx_id, sk, t)
    shaped_vtx += sk_wt * (sk.v[vtx_id] − base_sk.v[vtx_id])
  end for
  rest_vtx[vtx_id] ← shaped_vtx + base_sk.v[vtx_id]
end function


function SKELETAL_ANIMATION(vtx_id, t)
  tot_wt ← sum_influence_weights(vtx_id)
  final_mat ← zero_matrix(4, 4)
  for bone_i in influencing_bones[vtx_id] do
    posed_bone_mat ← pose_mat(bone_i, t)
    rest_bone_mat_inv ← rest_mat_inv(bone_i)
    bone_wt ← get_bone_wt(vtx_id, bone_i)/tot_wt
    final_mat += (rest_bone_mat_inv * posed_bone_mat * bone_wt)
  end for
  posed_vtx[vtx_id] ← rest_vtx[vtx_id] * final_mat
end function


function CONVERT_TO_ACC(vtx_id)
  cpts[thread_id] ← (0, 0, 0)
  for i ← 0 to stencil_size[vtx_id] do
    stencil_vtx ← posed_vtx[stencil_lookup[vtx_id, i]]
    cpts[thread_id] += stencil_wt[vtx_id, i] * stencil_vtx
  end for
  normalize(cpts[thread_id])
  save_to_gpu(cpt_buffer, vtx_id, cpts[thread_id])
  synchronize_threads()
  slefe ← update_slefe(thread_id)
  save_to_gpu(slefe_buffer, vtx_id, slefe)
end function
```

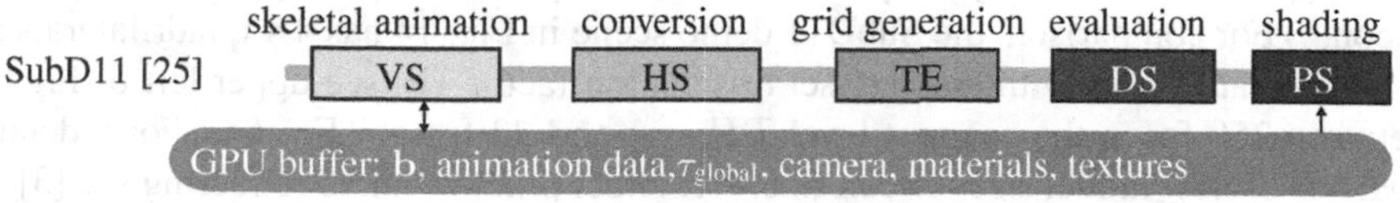

Fig. 13. DX11 SubD11 implementation [25]. CS=Compute Shader, VS=Vertex Shader, HS= Hull Shader, TE=Tessellation Engine, DS= Domain Shader, PS=Pixel Shader.

Table 2. Distribution of work per frame among pose, view and rendering. Pose change dominates

GPU processing	% of total
skeletal animation + conversion	58
slefe bounds	4
pose change total	62
view change	6
rendering pass	32

5   Discussion and Comparison

Performance. Table 2 shows the work distribution of a rendering cycle. The pose change consists of mesh animation and conversion plus recomputation of slefe-box vertices. The pose change dominates the work, but the recomputation of the slefe bounds accounts for less than 4%. The slefe bounds and their projection make up ca 10% of the overall work. According to measurements in Section 7 of [3], the bounds are within 12% of the optimal for widely-used, representative test examples in computer graphics (the tessellation factor in the implementation of [3] was inadvertently scaled by $\sqrt{2}$). Given that tight bounds reduce work when accurate rendering is required, it is not surprising that 10% computational overhead buys a considerable speedup compared to the overtessellation of conservatively-applied heuristics.

We used an NVidia GeForce GTX 580 graphics card with Intel Core 2 Quad CPU Q9450 at 2.66GHz with 4GB memory to render the geometry of the movie Elephants Dream. Elephants Dream is a 10-minute-long animated movie whose source is open. In 2006 it was reported to have taken 125 days to render, consuming up to 2.8GB of memory for each frame in Full-HD resolution (1920×1080) [4]. That is, each frame took on the order of 10 minutes to render. Since the Elephants Dream character meshes of Proog and Elmo contain triangles, but ACC requires a quadrilateral input mesh, we applied the standard cure of one step of Catmull-Clark subdivision yielding 140,964 curved surface patches for Proog and Emo together. In our implementation, we replicated Elephants Dream except that we did not apply post effects so as to isolate the effect of improved patch rendering. The 141K textured bi-cubic ACC patches render at over 300 frames per second (fps) with full pixel-accuracy. (We also used a variant of ACC that avoids the increase in patches and rendered 32K quads and 350 triangles at 380 fps when animating every frame and 1100 fps when animating at 33 frames per second.) For comparison, the SubD11 demo scene in Fig. 14 has 4K quadrilaterals and its frame rate varies with the user-set tessellation factor TF (see upper left of Fig. 14) between 250 fps at the coarsest level $TF = 1$ and 23 fps at TF= 64. For a detailed analysis of how model size, screen size, etc. affect pixel-accurate rendering see [3].

Memory Usage and Data Transfer. By placing the animation and the conversion from the quad mesh to the Bézier patches onto the GPU, the approach is memory efficient and minimizes data transfer cost. For example, one frame in the Proog and Emo scene has up to 0.25 million bi-cubic Bézier patches requiring 206.5 MB of GPU memory.

Traditional CPU-based animation would transfer this amount of data to the graphics card at every frame. In our approach, for the same scene, just once at startup, the static mesh of 4MB plus 9MB of shape key data are transferred; also the skeletal animation data per frame (45kB for 684 'bones') and the 289 shape keys (1kB) are packed into GPU buffers at startup. Moreover, the near-optimal *ephemeral triangulation* via the tessellation engine saves space and transfer cost compared to massive, 'pre-baked' triangulations.

Relation to Micro-polygonization. An established alternative for high-quality rendering, used in 3D movie animation, is micro-polygonization. *Micro-polygonization* owes its prominence to the Reyes rendering framework [37]. Since canonical implementations of micro-polygonization are recursive (cf. [5]), micro-polygonization is harder to integrate with current graphics pipelines [38] and leads to multiple passes as refinement and testing are interleaved. Even on multiple GPUs, there is a trade-off between real-time performance and rendering quality [39] (RenderAnts). Micro-polygonization aims to tessellate the domain U of a patch into (u, v) triangles so that the *size* of the screen projection of their image triangles is less than half a pixel. By contrast, pixel-accurate rendering aims at minimally partitioning the patches, just enough so that the difference, under projection, between the triangulated surface and the true non-linear surface is less than half a pixel: pixel-accurate rendering forces the *variance*, between the displayed triangulated surface and the exact screen image, to below the visible pixel threshold.

Comparison with the DX11 ACC SubD11 distribution. Our implementation is similar to that of SubD11 [25]: both implement skeletal animation and apply mesh conversion by accessing a 1-ring neighborhood of each quadrilateral. However, our implementation uses a sequence of Compute Shaders to animate and convert while SubD11 uses the Vertex Shader and the Hull Shader. See Fig. 11 for the execution pipeline of our algorithm and compare to that of SubD11, Fig. 13.

Since SubD11 executes in a single pass it appears to be more efficient. However, the Vertex Shader (VS) animation and Hull Shader (HS) conversion that perform the bulk of the work in SubD11 need to be synchronized by the index buffer mechanism

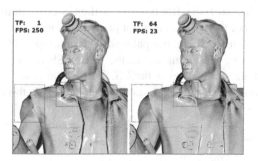

Fig. 14. DX11 SubD11 model from [25] consisting of 3,749 ACC patches (plus 150,108 flat triangles). The screen is captured at 1440x900 resolution. Setting $TF = 1$ results in polyhedral artifacts, at the shoulder and neck, while setting it high to remove these artifacts, decreases the frames per second by an order of magnitude.

Table 3. Performance in frames per second when placing animation and computation of τ onto the **CS or PS** or* just the animation onto the VS

Anim Updates/Sec	CS	PS	VS*
33	311	184	253
every frame	130	53	75

to prevent conversion before every vertex of a surface patch is animated; and SubD11 does not support interactive adaptive tessellation (without cracks) and must re-execute animation and conversion steps even when no view or pose change occur.

In our approach the main work, apart from rendering, is executed in the Compute Shader (CS). This automatically provides the necessary synchronization and allows co-ordination for interactive GPU-based *adaptive* tessellation without cracks. Using the Compute Shader also allows saving partial work in the GPU buffer (the animated surface p_t and the tessellation factors τ) and thereby reduces data transfer and communicates edge tessellation factors for adaptive rendering without mismatch. Executing only the appropriate type of the CS avoids re-computation, and guarantees sub-pixel accuracy. The end of the next section compares timings. A further advantage of using the Compute Shader is that it allows an indexed list rather than a fixed-size array when accessing neighbors. The Hull Shader limitation on primitives in SubD11 constrains the vertex valence, i.e. the number of points that can be accessed to construct the ACC patches. This matters for Proog and Emo models which contain 256 vertices of valence 32.

Compute Shader vs. Pixel Shader. We explored executing animation and τ-computation in a Pixel Shader (PS) pass. For large data sets, our CS implementation was clearly more efficient (see Table 3; Note that the CS has less overhead than a extra pass.). This can partly be attributed to higher parallelism: we can use 16 threads per patch in the CS as opposed to one per patch on the PS. (We could use 16 pixels in the PS, but would then have to synchronize to be able to compute τ). We also tried to use the Hull Shader (HS). But not only is the HS computationally less efficient on current hardware, but the HS also can not provide the necessary communication of adaptive tessellation factors to neighbor patches. The rightmost column VS* of Table 3 shows that just executing the animation in the Vertex Shader is already slower than executing animation and conversion in the CS. This explains why our code is considerably faster than SubD11, even though our code guarantees sub-pixel accuracy while SubD11does not.

6   Conclusion

To optimally leverage the approach to pixel-accurate rendering of [3] to skeleton-based animation, we partitioned the work for pixel-accurate rendering into stages that match animation-dependent transformations and view-time dependent camera motions. This allocation is as natural as it is practically powerful: it allows us to combine interactive

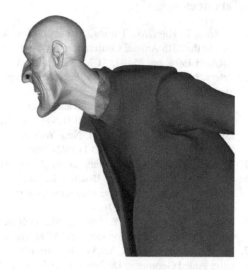

Fig. 15. Proog and Emo scene rendered in 7 seconds by Blender on a Intel Core 2 Duo CPU at 2.1GHz with 3GB memory; and in 3×10^{-3} seconds by our GPU algorithm

animation with high-quality rendering of curved surfaces. For gaming and animation it is crucial to spend minimal effort in redrawing static images since many other operations, say physics simulations, compete for compute resources. Also, in the game setting, the user often pauses to react to new information – so there is not continuous animation. The result is accurate for the given bi-cubic patches – distortion is below half a pixel, i.e. the error is not visible; it is efficient – there is no recursion and triangles are of maximal size; the adaptation is automatic – there is no need for manually setting the level of detail; and our implementation is fast, rendering 141k patches at more than 300 frames per second.

We tested the framework by rendering scenes of the movie Elephants Dream at $10\times$ real-time, leaving enough slack for larger data sets, complex pixel shaders and the artists' other work. Since the final pass is a generic DX11 rendering pass, it is fully compatible with displacement mapping (not used in Elephants Dream) and post effects. (We are not claiming pixel-accurate displacement, since this notion is not well-defined: displacement maps prescribe discrete height textures that require interpretation.) The rendering speed can provide high visual quality under interactive response. This may be useful for interactive CAD/CAM design in that the user no longer has to guess a suitable level of triangulation.

Acknowledgements. This work was supported in part by NSF Grant CCF-1117695. We thank the contributors to Elephants Dream for creating this wonderful resource and the creators of SubD11 to provide source code and model. Georg Umlauf's insightful question after the conference presentation prompted the inclusion of the constant k_m in the paragraph following (6).

References

1. DeRose, T., Kass, M., Truong, T.: Subdivision surfaces in character animation. In: Proceedings of the 25th Annual Conference on Computer Graphics and Interactive Techniques, SIGGRAPH 1998, pp. 85–94. ACM, New York (1998)
2. Peters, J., Reif, U.: Subdivision Surfaces. Geometry and Computing, vol. 3. Springer, New York (2008)
3. Yeo, Y.I., Bin, L., Peters, J.: Efficient pixel-accurate rendering of curved surfaces. In: Proceedings of the ACM SIGGRAPH Symposium on Interactive 3D Graphics and Games, I3D 2012, pp. 165–174. ACM, New York (2012), http://doi.acm.org/10.1145/2159616.2159644, doi:10.1145/2159616.2159644
4. Blender, Foundation, Elephants dream (2006), http://orange.blender.org
5. Fisher, M., Fatahalian, K., Boulos, S., Akeley, K., Mark, W.R., Hanrahan, P.: DiagSplit: parallel, crack-free, adaptive tessellation for micropolygon rendering. ACM Transactions on Graphics 28(5), 1–8 (2009)
6. Nießner, M., Loop, C.T., Meyer, M., DeRose, T.: Feature-adaptive GPU rendering of Catmull-Clark subdivision surfaces. ACM Trans. Graph. 31(1), 6 (2012)
7. Filip, D., Magedson, R., Markot, R.: Surface algorithms using bounds on derivatives. Computer Aided Geometric Design 3(4), 295–311 (1986)
8. Guthe, M., Balázs, A., Klein, R.: GPU-based trimming and tessellation of NURBS and T-Spline surfaces. ACM Transactions on Graphics 24(3), 1016–1023 (2005)
9. Hjelmervik, J.: Hardware based visualization of b-spline surfaces, presentation. In: Eighth International Conference on Mathematical Methods for Curves and Surfaces Oslo, June 28-July 3 (2012)
10. Magnenat-Thalmann, N., Laperrière, R., Thalmann, D.: Joint–dependent local deformations for hand animation and object grasping. In: Graphics Interface 1988, pp. 26–33 (1988)
11. Cordier, F., Magnenat-Thalmann, N.: A data-driven approach for real-time clothes simulation. Computer Graphics Forum 24(2), 173–183 (2005)
12. Kavan, L., Collins, S., Zára, J., O'Sullivan, C.: Geometric skinning with approximate dual quaternion blending. ACM Trans. Graph. 27, 105:1–105:23 (2008)
13. Ju, T., Schaefer, S., Warren, J.D.: Mean value coordinates for closed triangular meshes. ACM Trans. Graph. 24(3), 561–566 (2005)
14. Zhou, K., Huang, X., Xu, W., Guo, B., Shum, H.-Y.: Direct manipulation of subdivision surfaces on GPUs. ACM Trans. Graph. 26(3)
15. Joshi, P., Meyer, M., DeRose, T., Green, B., Sanocki, T.: Harmonic coordinates for character articulation. ACM Trans. Graph. 26(3), 71 (2007)
16. Lipman, Y., Levin, D., Cohen-Or, D.: Green Coordinates, ACM Transactions on Graphics 27 (3), 78:1–78:10 (2008)
17. Blender, Foundation, Shape keys, http://wiki.blender.org/index.php/Doc:2.4/Manual/Animation/Techs/Shape/Shape_Keys
18. Myles, A., Ni, T., Peters, J.: Fast parallel construction of smooth surfaces from meshes with tri/quad/pent facets. Computer Graphics Forum 27(5), 1365–1372 (2008)
19. Yeo, Y.I., Ni, T., Myles, A., Goel, V., Peters, J.: Parallel smoothing of quad meshes. The Visual Computer 25(8), 757–769 (2009)
20. Loop, C.T., Schaefer, S.: Approximating Catmull-Clark subdivision surfaces with bicubic patches. ACM Trans. Graph. 27(1)
21. Loop, C., Schaefer, S., Ni, T., Castano, I.: Approximating subdivision surfaces with Gregory patches for hardware tessellation. ACM Trans. Graph. 28, 151:1–151:9 (2009)
22. Bolz, J., Schröder, P.: Rapid evaluation of Catmull-Clark subdivision surfaces. In: Web3D 2002: Proceeding of the Seventh International Conference on 3D Web Technology, pp. 11–17. ACM Press, New York (2002)

23. Bunnell, M.: GPU Gems 2: Programming Techniques for High-Performance Graphics and General-Purpose Computation, ch. 7. Adaptive Tessellation of Subdivision Surfaces with Displacement Mapping. Addison-Wesley, Reading (2005)
24. Nießner, M., Loop, C.T., Greiner, G.: Efficient evaluation of semi-smooth creases in catmull-clark subdivision surfaces, p. 4 (2012)
25. MicroSoft, Subd11 sample (direct3d11) (November 2008), `http://preview.library.microsoft.com/en-us/library/ee416576`
26. Vlachos, A., Peters, J., Boyd, C., Mitchell, J.L.: Curved PN triangles. In: Symposium on Interactive 3D Graphics. Bi-Annual Conference Series, pp. 159–166. ACM Press (2001)
27. Boier-Martin, I., Zorin, D.: Differentiable parameterization of Catmull-Clark subdivision surfaces. In: Scopigno, R., Zorin, D. (eds.) Symp. on Geom. Proc., Eurographics Assoc., Nice, France, pp. 159–168 (2004)
28. He, L., Loop, C., Schaefer, S.: Improving the parameterization of approximate subdivision surfaces. In: Bregler, C., Sander, P., Wimmer, M. (eds.) Pacific Graphics, pp. xx–xx (2012)
29. Peters, J.: Mid-structures of subdividable linear efficient function enclosures linking curved and linear geometry. In: Lucian, M., Neamtu, M. (eds.) Proceedings of SIAM Conference, Seattle (November 2003); Nashboro (2004)
30. Lutterkort, D.: Envelopes of nonlinear geometry. Ph.D. thesis, Purdue University (August 2000)
31. Lutterkort, D., Peters, J.: Tight linear bounds on the distance between a spline and its B-spline control polygon. Numerische Mathematik 89, 735–748 (2001)
32. Lutterkort, D., Peters, J.: Optimized refinable enclosures of multivariate polynomial pieces. Computer Aided Geometric Design 18(9), 851–863 (2002)
33. Peters, J., Wu, X.: On the optimality of piecewise linear max-norm enclosures based on slefes. In: Schumaker, L.L. (ed.) Proc. Curves and Surfaces, St Malo (2002); Vanderbilt Press (2003)
34. Wu, X., Peters, J.: Interference detection for subdivision surfaces. Computer Graphics Forum, Eurographics 2004 23(3), 577–585 (2004)
35. Wu, X., Peters, J.: An accurate error measure for adaptive subdivision surfaces. In: Proceedings of the International Conference on Shape Modeling and Applications, pp. 51–57 (2005)
36. Wu, X., Peters, J.: Sublime (subdividable linear maximum-norm enclosure) package (2002), `http://surflab.cise.ufl.edu/SubLiME.tar.gz` (accessed January 2011)
37. Cook, R.L., Carpenter, L., Catmull, E.: The Reyes image rendering architecture. In: Stone, M.C. (ed.) Computer Graphics (SIGGRAPH 1987 Proceedings), pp. 95–102 (1987)
38. Fatahalian, K., Boulos, S., Hegarty, J., Akeley, K., Mark, W.R., Moreton, H., Hanrahan, P.: Reducing shading on GPUs using quad-fragment merging. ACM Trans. Graphics 29(3) (2010); (Proc. ACM SIGGRAPH 2010) 29(4), 67, 1–8 (2010)
39. Zhou, K., Hou, Q., Ren, Z., Gong, M., Sun, X., Guo, B.: Renderants: interactive Reyes rendering on GPUs. ACM Trans. Graph 28(5)

Author Index